P9-DHB-818

Scientific Style and Format

The Council of Biology Editors was established in 1957 by joint action of the National Science Foundation and the American Institute of Biological Sciences. It enjoys close relations with other organizations in scientific publishing, both national and international, but it functions autonomously.

The Council aims to improve communication in the life sciences by educating authors, editors, and publishers; by providing efficient means of cooperation among persons interested in publishing in the life sciences; by promoting effective communication practices in primary and secondary publishing in any form; and by supporting, devising, and disseminating standards for scientific style and format.

Style Manual Committee: Edward J Huth, Chairman; Bruce P Dancik; Thor Kommedahl; David E Nadziejka; Peggy Robinson; Winfield Swanson

Manuscript Editor: Frances H Porcher

Other Publications of the Council of Biology Editors
Scientific Writing for Graduate Students
Illustrating Science: Standards for Publication
Ethics and Policy in Scientific Publication
Peer Review in Scientific Publishing
Financial Management of Scientific Journals
Editorial Forms: A Guide to Journal Management

Inquiries should be directed to Council of Biology Editors, Inc.
11 South LaSalle Street, Suite 1400, Chicago IL 60603, USA
Telephone 312 201-0101; fax 312 201-0214

Scientific Style and Format

The CBE Manual for Authors, Editors, and Publishers

6th Edition

Style Manual Committee
Council of Biology Editors

Published for the Council of Biology Editors
by the Press Syndicate of the University of Cambridge
The Pitt Building, Trumpington Street, Cambridge CB2 1RP, UK
40 West 20th Street, New York NY 10011-4211, USA
10 Stamford Road, Oakleigh, Melbourne 3166, Australia

6th edition 1st published 1994
Reprinted 1995, 1996, 1997

Printed in the United States of America

Library of Congress Cataloging-in-Publication Data

Scientific style and format : the CBE manual for authors, editors, and publishers / Style Manual Committee, Council of Biology Editors. — 6th ed.
p. cm.
Rev. ed. of: CBE style manual / CBE Style Manual Committee. 5th ed., rev. and expanded. c1983.
Includes index.
ISBN 0-521-47154-0
1. Technical writing—Handbooks, manuals, etc. I. CBE Style Manual Committee. II. CBE Style Manual Committee. CBE style manual.
T11.S386 1994 94-22001
808'.0666-dc20 CIP

A catalog record for this book is available from the British Library

ISBN 0-521-47154-0 Hardback

Contents

PART 5 THE PUBLISHING PROCESS

Dedication

We wish to have this new edition of the CBE style manual stand as a memorial to Marianne Brogan, a late member of the Style Manual Committee. Her sudden death before the manual was completed brought to a sad end her valuable part in our work. She applied her keen and encyclopedic mind not only to her chapters on the complexities of chemical nomenclature and notation but also just as critically to all that the rest of us wrote. Despite her months-long struggle with a crippling malignancy, she worked relentlessly for the benefit of all who will use this manual. She lives on as an exemplar for any person who wishes to be a dedicated, thorough, well-informed, and critical editor.

—The Style Manual Committee

Preface

Virtually all of this edition was written by the members of the CBE Style Manual Committee, but their work was greatly helped by many persons, including other members of the Council and others engaged in scientific publishing. Their help was various: some provided useful materials, others advised on the scope and content of the manual, others critically reviewed draft manuscripts. The listing of their names here does not, however, imply that any of them endorses any of the content of the manual.

Arly Allen; Lawrence, Kansas
Michael S Altus; Baltimore, Maryland
Ernest E Banttari; University of Minnesota, St Paul, Minnesota
Mike Boroczki; National Research Council of Canada, Ottawa, Ontario, Canada
Joseph J Bosik; BIOSIS, Philadelphia, Pennsylvania
Charles Buck; American Type Culture Collection, Rockville, Maryland
Murrie W Burgan; Applied Physics Laboratory, The Johns Hopkins University, Laurel, Maryland
A Judith Butcher; Cambridge, UK
Patricia A Butler; World Health Organization, Geneva, Switzerland
David J Chitwood; US Department of Agriculture, Beltsville Agricultural Research Center, Beltsville, Maryland
Edward H Coe; ARS-USDA and University of Missouri, Columbia, Missouri
Lois Ann Colaianni; National Library of Medicine, Bethesda, Maryland
Robert R Compton; Stanford University, Stanford, California
John O Corliss; Albuquerque, New Mexico
Perry B Cregan; US Department of Agriculture, Beltsville Agricultural Research Center, Beltsville, Maryland
A Jamie Cuticchia; The Johns Hopkins University School of Medicine, Baltimore, Maryland

Ted Dachtera; National Geographic Society, Washington, DC
Kevin Davies; *Nature Genetics,* Washington, DC
Muriel T Davisson; The Jackson Laboratory, Bar Harbor, Maine
Wallace D Dawson; University of South Carolina, Columbia, South Carolina
Marie-Louise Desbarats-Schönbaum; Venhorst, Netherlands
Judith H Dickson; Science Editing, Rockville, Maryland
Dong Geng; Zhejiang Academy of Agricultural Sciences, Hangzhou, Zhejiang, China
J D Dowell; University of Birmingham, Birmingham, UK
Mary Jane Edwards; American Type Culture Collection, Rockville, Maryland
John Emsley; Imperial College of Science, Technology and Medicine, London, UK
Claude Fauquet; Scripps Research Institute, La Jolla, California
Larry W Finger; Carnegie Institution of Washington, Washington, DC
Bernard T French; Chemical Abstracts Service, Columbus, Ohio
Robert L Gherna; American Type Culture Collection, Rockville, Maryland
John W Glen; University of Birmingham, Birmingham, UK
Donald E Groom; Lawrence Berkeley Laboratory, Berkeley, California
John Grossman; University of Chicago Press, Chicago, Illinois
J P Gustafson; University of Missouri, Columbia, Missouri
Harold D Hatt; American Type Culture Collection, Rockville, Maryland
Robert K Herman; University of Minnesota, Minneapolis, Minnesota
Virginia Hitchcock; World Bank, Washington, DC
John E Huth; Harvard University, Cambridge, Massachusetts
Barbara Jasny; *Science,* American Association for the Advancement of Science, Washington, DC
Penelope Kaiserlian; University of Chicago Press, Chicago, Illinois
Gurdev S Khush; International Rice Research Institute, Manila, Philippines
Lammegien Kok-Noorman; Academisch Medisch Centrum, Amsterdam, Netherlands
Sheldon Kotzin; National Library of Medicine, Bethesda, Maryland
Jane M Krauhs; KRUG Life Sciences, Houston, Texas
Vicki Lawrence; *American Mineralogist,* Ann Arbor, Michigan
María L Lebrón; American Physical Society, College Park, Maryland

William D Lipe; Crow Canyon Archeological Center, Cortez, Colorado
William R Luellen; American Society of Agronomy, Madison, Wisconsin
Beth Magura; Radiation Effects Research Foundation, Hiroshima, Japan
Don Marlette; American Institute of Physics, College Park, Maryland
Robert H Masterson; Radiation Effects Research Foundation, Hiroshima, Japan
Phyllis J McAlpine; University of Manitoba, Winnipeg, Manitoba, Canada
R A McIntosh; University of Sydney, Cobbity, Australia
Derek McNally; International Astronomical Union, Paris, France
David W Meinke; Oklahoma State University, Stillwater, Oklahoma
Joy E Merritt; Chemical Abstracts Service, Columbus, Ohio
Wladyslaw V Metanomski; Chemical Abstracts Service, Columbus, Ohio
Francis Molina; American Type Culture Collection, Rockville, Maryland
Frederick A Murphy; University of California at Davis, Davis, California
Bertram G Murray Jr; Rutgers University, New Brunswick, New Jersey
Thomas A Nerad; American Type Culture Collection, Rockville, Maryland
Stephen J O'Brien; National Cancer Institute, Frederick, Maryland
Maeve O'Connor; European Association of Science Editors, London, UK
Henrik Olesen, Rigshospitalet, Copenhagen, Denmark
Jennifer Owens; American Society for Microbiology, Washington, DC
Mary E Palm; US Department of Agriculture, Beltsville Agricultural Research Center, Beltsville, Maryland
Karen Patrias; National Library of Medicine, Bethesda, Maryland
Shirley M Peterson; Wayne, Pennsylvania
Mary Polacco; University of Missouri, Columbia, Missouri
Christopher J Porter; The Johns Hopkins University School of Medicine, Baltimore, Maryland
Carl A Price; Commission on Plant Gene Nomenclature, Waksman Institute, Piscataway, New Jersey
Marta Pulido; Institut Municipal d'Investigació Mèdica, Barcelona, Spain

J Margaret Rabbinowitz; CERN, Geneva, Switzerland
Samuel M Rankin III; American Mathematical Society, Providence, Rhode Island
Johanna M Reinhart; Allen Press Incorporated, Lawrence, Kansas
Charles M Rick; University of California at Davis, Davis, California
Amy Y Rossman; US Department of Agriculture, Beltsville Agricultural Research Center, Beltsville, Maryland
Rosemary Russell; National Research Council of Canada, Ottawa, Ontario, Canada
Lois E Schmitt; *Science,* American Association for the Advancement of Science, Washington, DC
Eugene H Schmitz; University of Arkansas, Fayetteville, Arkansas
John T Scott; American Institute of Physics, Woodbury, New York
James B Shaklee; Washington Department of Fish and Wildlife, Olympia, Washington
Karen Shashok; Granada, Spain
Ethan Shevach; National Institutes of Health, Bethesda, Maryland
Albert Simmonds; R R Bowker, New York, New York
Sally K Sinn; National Library of Medicine, Bethesda, Maryland
Elmer F Smith III; US Geological Survey, Reston, Virginia
Bridget Snyder; National Geographic Society, Washington, DC
David M Stelly; Texas A & M University, College Station, Texas
Elwin L Stewart; Pennsylvania State University, University Park, Pennsylvania
Jane E C Sykes Bär; Utrecht, Netherlands
Thomas G Trippe; Lawrence Berkeley Laboratory, Berkeley, California
Frances A Uecker; US Department of Agriculture, Beltsville Agricultural Station, Beltsville, Maryland
Penny von Wettstein-Knowles; Carlsberg Laboratory and University of Copenhagen, Copenhagen, Denmark
John H Wiersema; US Department of Agriculture, Beltsville Agricultural Research Center, Beltsville, Maryland
George A Wilkins; University of Exeter, Exeter, Devon, England
Robyn Woollcott; Australian Government Publishing Service, Canberra, Australia
Georgia Yerk-Davis; University of Missouri, Columbia, Missouri
Christine Felter Yglesias; National Geographic Society, Washington, DC
Ellis L Yochelson; National Museum of Natural History, Washington, DC
Frances R Zwanzig; National Academy of Sciences, Washington, DC

Some of the most helpful recommendations came from anonymous reviewers of virtually the entire manuscript at a late stage of its preparation. The Committee and all users of this manual owe these listed persons and the anonymous reviewers unending thanks for their help. If we have omitted the names of any persons who helped us in preparing this manual, we offer them our apology for that error.

—Style Manual Committee, Council of Biology Editors
Edward J Huth, Chairman; University of Pennsylvania
Marianne Brogan [deceased], American Chemical Society
Bruce P Dancik, University of Alberta
Thor Kommedahl, University of Minnesota
David E Nadziejka, Argonne National Laboratory
Peggy Robinson, Canadian Medical Association
Winfield Swanson, National Geographic Society

PART 1 Introduction

1 About This Manual

> Style is interpreted broadly to mean forms of expression in scholarly writing, and the general technical requirements of journals, such as details for typing manuscripts, standard abbreviations, and citation of references.
>
> —Preface, *Style Manual for Biological Journals,* 1960

This manual is the 6th edition of the style manual issued by the Council of Biology Editors through 3 decades; it differs greatly from the 1st manual, *Style Manual for Biological Journals* (CBE 1960), and the subsequent 4 editions (CBE 1964, 1972, 1978, 1983). They included content on writing and submitting papers for journal publication as well as on publication style and format. Recommendations on scientific style were limited mainly to the plant sciences, zoology, microbiology, and the medical sciences.

SCOPE AND AIMS

GENERAL SCOPE

1·1 This new edition of the CBE style manual covers all scientific disciplines; technologic fields not closely related to experimental and observational science are generally not represented. The contents of the preceding editions on how to write and submit papers for journal publication have been dropped to make room for the coverage of all sciences. This edition is limited to recommendations on how scientific papers, journals, and books should be styled and formatted for publication. In most parts of this manual, "style" means "publication style": the conventions such as punctuation, abbreviation, capitalization, symbolization, and reference citation, format, and content. Where "style" is meant in its literary sense, the term "prose style" is used.

SCIENTIFIC SCOPE

Note that Part 3, *Special Scientific Conventions,* is not organized by sciences but by their subjects because the older boundaries of scientific disciplines are breaking down. Genetics and biochemistry, for example, were for many years quite separate in the subjects they covered. No longer—the molecular structure of genes and how they chemically produce their effects are subjects complexly intertwined with the older subjects of genetic inquiry. In deciding what sequence to apply in organizing Part 3, the Committee settled on a rising scale of the dimensions of the subjects. Therefore, Part 3 starts with the fundamental units of the universe and proceeds up the scale of dimensions, through chemical and cellular components, the cell, microorganisms, more complex organisms, and human society, to Earth and the rest of the universe.

The recommendations on scientific style cover mostly proper style for nomenclature and symbols. The principles governing scientific nomenclature and the use of symbols are explained only sufficiently to make the basis and rationale for proper style clear; this is not a manual on how to form scientifically acceptable words and phrases, nor is it a catalogue of scientific names and symbols. Authoritative documents are available in many disciplines to guide authors responsible for coining new terms and symbols; references to such documents are given at the end of many of the chapters.

AIMS

1·2 The Style Manual Committee planned this new edition with 5 aims in mind.

Supporting Convergence in Style

Moving from differing styles to a uniform style for particular needs would benefit authors, editors, publishers, readers, and librarians; therefore, the Committee has sought to reduce differences in style. Some characteristically British preferences are recommended because of the more convincing logic for their basis, for example, the closing period following a closing quotation mark. Likewise, American styles with a better logical basis than corresponding British styles are given priority. Some users of the manual may dislike giving up some of their prejudices with regard to style, but we ask them to consider the benefits of having a truly common style, useful in the international world of science, not just in 1 nation.

Simplifying Formats for Citations and References

Science uses myriad formats for bibliographic references. We hope that by specifying the needed elements of bibliographic references and illustrating their virtually identical use in both the citation-sequence ("Vancouver") and name-year ("Harvard") systems, the difficulties posed by the many variations in formats for references may be reduced.

Simplifying Style Rules

The Committee's efforts toward simplifying style were based in part on its wish to bring about as much convergence in style as possible for the advantages of having a more nearly uniform style throughout science. But a further reason is the value derived from reducing the number of decisions on style that have to be made. Examples are eliminating the possessive form for eponymic terms, using numerals for all numbers modifying countable units—not just for measures and numbers above 9 or 10, and eliminating Latin abbreviations.

Offering Options

Despite the Committee's commitment to convergence and simplification of styles, it was aware that some differences could not be eliminated all at once. Therefore, alternatives are presented at many points. The most conspicuous example is the recommendations for both the name-year and citation-sequence systems of citations and references mentioned above. Some disciplines have deeply rooted conventions at odds with those in other fields; these differences are often pointed out and the preferences in the manual explained.

Reducing Work at the Keyboard

Some recommendations are aimed at reducing work at the typewriter or computer keyboard. Examples are the preferences for unpunctuated abbreviations where possible and eliminating the ampersand (&) in favor of "and".

THE BASIS FOR RECOMMENDATIONS

1·3 Recommendations are based as much as possible on authoritative documents on nomenclature and symbols developed for scientific publication in English. Examples of such documents are the standards published by the National Information Standards Organization and the International Organization for Standardization and the style manuals of the US Geological Survey (Hansen 1990) and the American Mathematical Society (AMS 1990). Where such documents are not available, well-established and widely used styles have been selected, as much as possible from general style manuals such as *The Chicago Manual of Style* (UCP 1993). At some points, however, the Committee chose to depart from a few recommendations in these sources.

FORMAT AND STYLE OF THIS MANUAL

1·4 The text is divided into chapters grouped in 5 parts. The 1st part has the 2 introductory chapters. Part 2 covers conventions widely applied in both general and scientific scholarly publishing. Part 3 covers details of style mainly applicable in the sciences. Part 4 recommends formats applicable in scientific journals and books, including formats for bibliographic citations and references. Part 5 presents procedural aspects of publishing.

Recommendations are illustrated at many points with examples. These are indented and set in a type size 1 point smaller than the main text.

Sections are numbered in the left margin to simplify finding text relevant to specific topics. The part of the section number preceding the raised period (raised point) represents the chapter number; the part following represents the section within the chapter. Note that the raised period was chosen rather than a period so as to preserve the use of the period in numbers as much as possible for its function as the decimal point.

References and their citations are in the name-year style as set forth in Chapter 30. This choice was based on the advantages of that system for a book of this kind and is not an endorsement of the name-year system as being preferable to the citation-sequence system, also described in Chapter 30.

References relevant mainly to particular chapters are listed at their close. For some chapters there are 2 groups of references: "Cited References" and "Additional References". "Cited References" (titled in some disciplines as "Literature Cited") are those indicated in the text as supporting documentation. "Additional References" are other sources relevant to the chapter's topic.

Words and phrases presented as explanatory or illustrative terms are enclosed within quotation marks rather than being italicized. The italic convention is reserved for book titles and for specific scientific meanings.

Care has been taken not only to avoid terms with undesirable biases, but also to avoid stylistic convolutions and loss of efficient statement.

RECOMMENDATIONS FOR FUTURE EDITIONS

1·5 Monitoring the scientific literature for documents establishing or recommending new nomenclature, notations, or formats is a huge task. The Council of Biology Editors Style Manual Committee urges scientific societies, committees, work groups, individual scientists, and other possible sources to send to the committee both formal and informal documents with recommendations on nomenclature, symbols, and style, whether they are new or simply not represented in this edition. Decisions on what recommendations will be represented in future editions will be made by the committee after reviewing them with consultants in relevant fields. Documents with recommendations should be sent to Style Manual Committee, Council of Biology Editors, 11 South LaSalle Street, Suite 1400, Chicago IL 60603, USA. The telephone number is 312 201-0101.

REFERENCES

Cited References

[AMS] American Mathematical Society. 1990. A manual for authors of mathematical papers. Providence (RI): AMS.

[CBE] Conference of Biological Editors, Committee on Form and Style. 1960. Style manual for biological journals. Washington: American Inst Biological Sciences.

[CBE] Conference of Biological Editors, Committee on Form and Style. 1964. Style manual for biological journals. 2nd ed. Washington: American Inst Biological Sciences.

[CBE] Council of Biology Editors, Committee on Form and Style. 1972. CBE style manual. 3rd ed. Washington: American Inst Biological Sciences.

[CBE] Council of Biology Editors, CBE Style Manual Committee. 1978. Council of Biology Editors style manual: a guide for authors, editors, and publishers in the biological sciences. 4th ed. Bethesda (MD): CBE.

[CBE] Council of Biology Editors, CBE Style Manual Committee. 1983. CBE style manual: a guide for authors, editors, and publishers in the biological sciences. 5th ed. Bethesda (MD): CBE.

Hansen WR, revisor and editor. 1990. Suggestions to authors of the reports of the United States Geological Survey. 7th ed. Washington: US Government Printing Office.

[UCP] University of Chicago Press. 1993. The Chicago manual of style. 14th ed. Chicago: UCP.

2 Scientific Notation: A Brief History

> By relieving the brain of all unnecessary work, a good notation sets it free to concentrate on more advanced problems, and in effect increases the mental power of the race. . . . Civilization advances by extending the number of important operations which we can perform without thinking about them. —A N Whitehead, *An Introduction to Mathematics,* 1911

> The first thing the intellect does with an object is to class it along with something else.
> —William James, *The Varieties of Religious Experience,* Lecture 1, 1902

The history of scientific conventions for precisely naming, symbolizing, and numbering the objects and actions, both visible and invisible, of our world and the rest of the universe is a long one. There is not space here for more than the briefest account of that history; this chapter sketches only its main lines and milestones. Tables 2·1 and 2·2 give a chronologic picture of some steps forward in scientific style, and their dates suggest the accelerating pace of innovations in style that began at the end of the Renaissance. The content of these tables is necessarily selective.

TIME, PLACE, AND NUMBER: THE ANCIENT AND MEDIEVAL WORLD

2·1 Men and women needed to communicate with each other even in the obscure origins of human society. Prey had to be distinguished from predators, and quickly. Nouns were not enough. The number of objects had to be conveyed; 10 wolves had to be distinguished from 1. Therefore,

Table 2·1 Scientific nomenclature, notation, and symbols: ancient times through the 18th century

Period or year	Development
5 millennia BCE[a]	Sumerians divide the year into 12 months of 30 days each and the day into 12 units, with each unit divided into 30 subunits; Egyptian year of 365 days and day of 24 hours.
Late centuries BCE	Roman numerals evolve, probably from Etruscan sources.
2nd century BCE	Hipparchus compiles the 1st known (through Ptolemy) astronomical catalogue: 1025 stars with their celestial coordinates, distribution in constellations, and estimates of relative brightnesses.
47 BCE	Julius Caesar decrees the Julian calendar.
2nd century CE[a]	Claudius Ptolemy's *Geography* establishes for geographic positions a grid of coordinates expressed in degrees: latitudes measured from the Equator and longitudes from the western extremity of his world, the "Fortunate Isles" (Hierro, formerly Ferro, the westernmost of the Canary Islands).
8th century CE	A circle as a symbol for the concept "zero" is introduced in China.
Premedieval and medieval	Symbols for the sun, moon, planets, and planetary positions.
10th–14th centuries	Arabic numerals enter use in Europe: 1st appearance in earliest European form, the late 10th century; well-developed form, 15th century.
15th century	The + (plus) and − (minus) symbols develop in Germany.
16th century	The parentheses, ( ), to show aggregation, as in 2(3 + 6), and the root symbol, $\sqrt{}$, originate in Germany. Use of ° ("degree" symbol), ′ ("minute" symbol), and ″ ("second" symbol) with values for angles.
1557	Recorde introduces the equals symbol, = .
1576	Tycho Brahe begins to compile his great star catalogue.
1582	Pope Gregory XIII establishes the Gregorian calendar.
17th century	Symbols in mathematics and symbolic logic begin to replace verbal statement; "log" for logarithm. Leibnitz introduces symbols for differential and integral calculus. Development of modern English punctuation.
1603	Bayer publishes atlas of constellations (in his *Uranometria*): stars designated by greek letters still used.

Table 2·1 *(cont.)*

Period or year	Development
1604	Robert Cawdrey's *A Table Alphabeticall,* "the 1st English dictionary" (Starnes and Noyes 1991).
1617	The comma (,) and the period (.) as decimal indicators (Napier).
1637	*z, y,* and *x* as unknown numbers (Descartes). Raised arabic numerals as indicators of powers (Descartes).
1651	Riccioli establishes in his *Almagestum novum* a nomenclature for features of the moon that continues in use.
1659	÷ as division symbol (Rahn).
1698	· (raised period) as multiplication symbol (Leibnitz).
Mid 18th century	Leonhard Euler's use of roman, greek, italic, and Fraktur letters in mathematical notation, for example, π (pi) for 3.1416.
Late 18th century	$ as the dollar symbol.
1781	Charles Messier's catalogue of star clusters, nebulae, and galaxies.
1786	Werner's *Kurze Klassification und Beschreibung der verschiedenen Gebirgsarten,* a pioneering document in rock classification that moved geology toward standardized nomenclature.
1787	With *Méthode de nomenclature chimique,* Guyton de Morveau, Lavoisier, Berthollet, and Fourcroy begin the development of modern chemical nomenclaure.

[a]BC "Before Common Era"; CE, "Common Era"—a dating system used by historians that replaces the traditional BC with BCE and AD with CE, reflecting cultural sensitivity to the feelings of the majority of the people of the world, who are not Christians. The divisions of time are the same.

the start of scientific notation, both nomenclature and numbering, probably lies well back in prehistory. The conventions that have survived to our time have much later origins: the calendars and other time divisions of the Sumerians, the Egyptians, the Jews; the numerals of the Romans. Calendars and numbers served practical needs in agriculture and commerce. But practical needs were not enough to satisfy the itch of some restless minds to look farther in the world around them and into the sky. Astronomical observations may have originated in searches for omens and prognostications, but eventually the observations pushed back the boundaries of the unknown; an example is the star catalogue of Hipparchus, compiled at the end of the 2nd century BCE (BC). The same compulsion to describe in detail the surrounding environment led to Ptole-

Table 2·2 Scientific nomenclature, notation, and symbols: 19th and 20th centuries

Period or year	Development
Early 19th century	$n!$ for factorial n.
1800	The metre introduced in France as the measure of length.
1814	Berzelius proposes symbols for chemical elements still used today, such as C (carbon), Co (cobalt), Cu (copper), Sn (tin), S (sulfur).
1816	William Smith's classification of rock strata identified by the fossils contained therein.
1820	First *United States Pharmacopoeia* published; establishes standard English and Latin names for drugs.
1822	Rock units named by characteristics: "Carboniferous" for coal units in England; "Cretaceous" for chalk formations.
1834	Terms for electrochemistry, such as "anion", "cation", "electrolyte", devised by Faraday and Whewell. Subscript numerals in chemical formulas, as in "H_2O", introduced by Liebig and Poggendorff.
1836	Thomsen publishes his 3-age system for periods of prehistory: Stone Age, Bronze Age, Iron Age.
1849–1952	d'Orbigny designates 5 major geologic terms still used: Paleozoic, Triassic, Jurassic, Cretaceous, Tertiary.
1853	Thaer classifies soils by textural and agricultural properties.
1862	Fallou classifies soils by geologic origin and lithologic composition.
1863	Argelander's *Bonner Durchmusterung,* positions and approximate brightnesses (down to the 11th magnitude) of more than 324 000 stars.
1865	Ring structure for benzene is proposed and illustrated by Kekulé; the concept is extended by Erlenmeyer and Graebe to napththalene (2 fused rings) and by Graebe and Lieberman to anthracene (3 fused rings).
1869	Mendeleev's periodic table of elements.
1875	International standardization of weights and measures is launched with the signing of the "Metre Convention" by 17 nations and the establishment of the International Bureau of Weights and Measures (Bureau International des Poids et Mesures, BIPM).
1881	Adoption of the centimetre-gram-second system at the International Congress of Electricians. Powell develops an innovative comprehensive scheme of colors and symbols for maps of the US Geological Survey.
1882	*Nomenclature and Notation,* The Chemical Society's (London) pioneering guidelines on uniform and systematic nomenclature.
1883	The 4 standard time zones for the United States are established, based on integral numbers of hours west of the Greenwich meridian.

Table 2·2 *(cont.)*

Period or year	Development
1884	The International Meridian Conference, held in Washington, establishes the Greenwich meridian as the prime meridian of the world and midnight there as the start of the Universal Day expressed in Greenwich Mean Time.
1886	The American Chemical Society's Committee on Nomenclature and Notation endorses, with minor changes, the British Chemical Society's 1882 guidelines.
	Dokuchaev's classification of soils based on their properties and on soil-forming factors.
1887	International Astrographic Congress, Paris, initiates an effort to prepare an *International Star Chart and Catalogue* as a consequence of Paul and Prosper Henry's *Carte du ciel.*
1888	US Geological Survey's 1st guide for authors.
1892	Basic nomenclatural principles for organic chemistry established in Geneva at the International Congress of Chemistry.
1895	*Basel Nomina Anatomica* (BNA), the 1st unified anatomical terminology published.
1897	The beginning of coordination compound nomenclature with Werner's publication in *Zeitschrift für Anorganische Chemie* 1897;14:23.
1905	*Règles internationales de la Nomenclature zoologique,* adopted at the Sixth International Congress of Zoology, Berne, 1904.
1906	*Règles internationales de la Nomenclature botanique,* adopted at the International Congress of Botany, Vienna, 1905.
1909	Whitney's American soil classification system.
	First edition of the US Geological Survey's *Suggestions to Authors.*
1914	Glinka's *The Types of Soil Formation, Their Classification and Geographical Distribution.*
1923	*Nomina Anatomica Veterinaria* (American Veterinary Medical Association) is published but fails to gain international acceptance.
1930	*Definitive Report of the Commission for the Reform of Nomenclature in Organic Chemistry of the International Union of Chemistry,* a revision and extension of the 1892 Geneva rules.
	Boundaries of the constellations defined by the International Astronomical Union.
1936	The Jena revision of *Basel Nomina Anatomica* widely known as the JNA.
1938	Classification of all known US soils (Baldwin, Kellogg, and Thorp).

Table 2·2 (*cont.*)

Period or year	Development
1940	Recommendations for nomenclature of inorganic compounds issued by the Committee on the Reform of Inorganic Nomenclature of the International Union of Pure and Applied Chemistry. Recommendations for symbolic representation of genes of the laboratory mouse issued by the Committee on Mouse Genetics Nomenclature.
1948	*International Code of Nomenclature of Bacteria.* *Symbols, Units and Nomenclature in Physics* 1st issued by the Commission for Symbols, Units, Nomenclature, Atomic Masses and Fundamental Constants of the International Union of Pure and Applied Physics.
1951	The American Institute of Physics publishes its 1st *Style Manual.*
1954	The International System of Units (Système international d'unités, or SI) established by the 10th General Conference of Weights and Measures (Conférence Générale des Poids et Mesures, or CGPM).
1957	*Nomenclature of Inorganic Chemistry, Definitive Rules for Nomenclature of Inorganic Chemistry,* an expansion of the 1940 recommendations. Standards for symbolic representation of genes established by a committee of the Union Internationale des Sciences Biologiques.
1960	The Conference of Biology Editors (later the Council of Biology Editors) publishes *Style Manual for Biological Journals,* the 1st predecessor of this manual. The Chemical Society's (London) 1st *Handbook for Chemical Society Authors.* The Denver Conference establishes "the basis for all subsequent nomenclature reports" for symbolization of normal and abnormal chromosomes.
1961	The International Astronomical Union publishes its 1st *IAU Style Book.*
1963	The London Conference classifies 7 groups of chromosomes as A to G.
1967	The American Chemical Society's 1st *Handbook for Authors.*
1968	*Nomina Anatomica Veterinaria* (International Committee on Veterinary Anatomical Nomenclature).
1971	*The Classification and Nomenclature of Viruses* published. First style manual of the American Mathematical Society.
1975	The Metric Conversion Act passed by the US Congress endorses the SI. *Soil Taxonomy* published by the US Department of Agriculture.
1976	*System of Astronomical Constants* published by the International Astronomical Union.

my's *Geography;* description was not enough, and his desire for precise statement of location led him to develop a system of geographic coordinates that is the ancestor of today's latitude and longitude.

Then for close to a thousand years most of science slept. Arabic (Islamic) science was awake and working, but for a long time it was little known outside its own culture. Science has never remained the property of the culture in which it arises, and Arabic astronomy, chemistry, optics, other physical sciences, and medicine began to seep into Europe. Carried in this slow current was the Arabic system of numerals, which gradually replaced the cumbersome Roman numerals in arithmetic operations. Other mathematical innovations began to develop. The 15th and 16th centuries CE (AD) saw the start in Europe of conventions in mathematical symbols still used in our time. Other conventions survived in their countries of origin for years but eventually found no place in the international world of science.

SYSTEMATIC NOMENCLATURE AND SYMBOLISM: THE RENAISSANCE AND MODERN SCIENCE

2·2 In the 15th, 16th, and 17th centuries European science began to stir and then to move. The stimuli were many. Greek and Arabic texts became available and offered a picture of science unknown or long forgotten. The invention of the printing press in the 15th century led to more rapid communication and reduced the costs of recording, transmitting, and accessing information. Geographic exploration brought the practical problems of navigation and mapping but also opened up unstudied parts of the world. New technical means such as the telescope and the microscope brought ways to see what had not been seen before in the skies and in the world of tiny things. Political stability in principalities and nation states made it safer for observational scientists to travel to long-unexamined regions. Descriptive science began to flourish, notably in astronomy, botany, and zoology. Astronomers compiled new catalogues of stars and of the moon's features. Catalogues of plants and animals were compiled. The biology of humans and animals advanced rapidly through the intense study of the microscopic world. But as Singer (1950) has pointed out, the very torrent of observations and description brought the need for control of the new information.

> The general tone of the biological writing that followed [the classical microscopists of the 17th century] is very different from that which precedes them. Variety and complexity now begin to overawe the naturalist.

> Amidst the multiplicity of phenomena, order must be sought if knowledge is not to lose itself in detail. So it is that in the age that follows, the importance of *classification* becomes greatly emphasized.

Classification itself does not necessarily produce a systematic and coherent nomenclature, but is a step toward it. Such a major step was the publication of Gaspard Bauhin's great botanical catalogue in 1623. He began to delineate the concepts of "genus" and "species", and his catalogue represented a system of binary classification, which, as Singer points out, "led ultimately to the 'binomial nomenclature' of Linnaeus". But before Linnaeus came Joachim Jung (1587–1657), who developed a morphologic terminology in botany that included terms such as "pinnate" and "digitate", "opposite" and "alternate", "perianth", and "stamen", which are still used. Indeed, "his manner of naming plants approaches extraordinarily near to a formal binomial system" in that he gave each plant he catalogued 2 names; the 1st, a noun serving in effect as a generic name, the 2nd, an adjective serving as a specific name. Our modern binomial nomenclature in biology finally took form with the publication by Linnaeus (1707–1778) of his *Species Plantarum,* published in 1753, and *Systema Naturae,* drafted in 1735, modified to its definitive 10th edition and published in 1758. These works established for both plants and animals the notion of the taxa of genera and species. The view of Linnaeus that genera and species are constant and immutable, having all been created at once, clearly has not survived the much later concepts of evolution and the mutable gene. The next major step toward our present binomial nomenclature in biology was Lamarck's *Philosophie zoologique,* published in 1809. Its 14 classes for the animal kingdom were arranged in a descending scale from mammals to infusoria, the top 4 classes designated vertebrate and the bottom 10, invertebrate (his coinage as "invertébrés").

Linnaeus's great contribution in the 18th century to systematic nomenclature in biology was paralleled in chemistry by Guyton de Morveau, Lavoisier, Berthollet, and Fourcroy's publication in 1787 of *Méthode de nomenclature chimique.* But chemistry had the further problem of systematically and coherently representing in some economical, and hence necessarily symbolic, way the chemical structures that could not be seen by microscopy. Progress with this problem began to be made with the atomic symbols (such as "Cu" for "copper") proposed by Berzelius in 1814, the subscript figures in chemical formulas (such as the subscript 2 in "H_2O") introduced by Liebig and Poggendorff in 1834, and graphic ring structures (Kekulé's for benzene) in 1865.

2·3 Accurate and reproducible measurement is as important in science as systematic, universally applied nomenclatures and symbolization. In the 18th century, units of measurement were a chaos of traditional, incoherent, and unstandardized national and local nonsystems. The 1st step toward a rational, coherent, and universal system of measurement came in 1790 as a fortunate byproduct of an upheaval that otherwise did great damage to French intellectual life, the French Revolution. Among the reforms that emerged in the Revolution but failed eventually were a new era with a new calendar having a year that began on the autumnal equinox, a 10-hour clock, and a 400-degree circle. The reform that in time did succeed was launched in 1790 by an act of the National Assembly that called on the Académie des Sciences to develop a new system of weights and measures (*Loi relative aux mesures à prendre pour arriver à la fixation de l'unité naturelle des poids et mesures*). France introduced the metre in 1799, but the metric system did not become a truly international system until 1875 with the convening of the Diplomatic Conference of the Metre in Paris and the signing of the "Metre Convention" by the 17 nations attending. The metric system reached its present degree of refinement with establishment of the Système internationale d'unités (SI) in 1954.

2·4 By the last quarter of the 19th century, scientists around the world had begun to see the need for standard and uniform conventions for nomenclature and symbols. Chemists and biologists led in efforts to develop national and international conventions, probably because of the numbers and complexities of the entities their literature had to represent. These efforts were concentrated first in international congresses that dealt with issues in measurement and nomenclature in chemistry. The great acceleration of such efforts came after World War II, running in parallel with the explosive expansion of science in all disciplines and the consequent expansion of new knowledge. The clearest example of this great growth in scientific style conventions is the very large number of recommendations developed by the International Union of Pure and Applied Chemistry since the mid-1960s; the references for Chapter 16 of this manual are dominated by the IUPAC documents. Hardly less comprehensive have been the conventions for symbolically representing chromosomes and genes in animal and plant genetics. Such efforts to standardize style conventions continue at a great pace. Whoever wants to keep fully informed on scientific style must periodically check for new relevant documents in the major indexing services of the United States such as those of BIOSIS, the Chemical Abstracts Service, the National Library of Medicine, and the Institute for Scientific Information.

SCIENTIFIC STYLE MANUALS

2·5 Despite the great steps forward in the 2nd half of the 19th century and early in the 20th in standardizing scientific style, many authors were unaware of, or ignored, established conventions. Journal editors were often presented with manuscripts containing antique or inadequate nomenclature, scientifically archaic units of measure, nonstandard abbreviations, and other hindrances to clear and accurate communication. By mid-century these problems had led many scientific societies to draft and publish style manuals defining the elements of desirable scientific style for publication. General style manuals had been widely available since early in the 20th century, such as that published in successive editions since 1906 by the University of Chicago Press (*The Chicago Manual of Style*) and that of the Oxford University Press (*Hart's Rules for Compositors and Readers at the University Press Oxford*) 1st published in 1904. But these general manuals, as good as they have been, have never represented scientific style in the detail needed by authors in science. A scientific style manual was published early in the 20th century by the US Geological Survey, but most of the scientific style manuals now widely used in the United States have been published since 1950 (see Table 2·2).

THE FUTURE OF SCIENTIFIC STYLE

2·6 Science continues to become more and more complex. Such fields as molecular biology, genetics, immunology, and virology have generated wealths of new information in the past 2 decades that could not have been imagined at mid-century. These advances have steadily called for new style conventions to represent their findings. And who can tell what the new biological science may do to present taxonomic and nomenclatural systems? Full knowledge of the genome for plants and animals could render today's taxonomic binomial nomenclature largely obsolete. The dissolution of boundaries among many scientific disciplines complicates agreement on style needs. Immunologic matters, for example, may be as much the concern of a biochemist as of a microbiologist, and gene structure as much the business of an academic physician as a geneticist. So those in science with responsibilities for developing new conventions of scientific style must seek to make them useful for new complex needs while striving to minimize divergence and promote convergence in styles among disciplines.

In the closing pages of his great history of mathematical notations, Cajori (1928, 1929) pointed to the self-centeredness of individuals, organizations, and national groups of mathematicians in developing their own notations and ignoring those of others. He described how this chaos had long hindered the development of uniform and internationally used and understood systems of signs and symbols for mathematics. That was in 1929. Fortunately, some disciplines, notably chemistry and some parts of biology and physics, have since improved their systems of nomenclature and symbols. But even today, many scientific groups develop systems within their own circles, ignoring the value of a coherent system that would better serve all disciplines. As conspicuous a failure is the lack of a central agency for conventions in scientific style that might guide scientists in all fields to develop new conventions that would be consistent in concept and presentation across all disciplinary groups. The authors of this manual have tried to reduce divergence in scientific style and promote convergence. Many minor differences in style among scientific disciplines seem pointless. The scientific world needs to take more steps toward a uniform and complete international system of scientific nomenclature and symbols.

A NOTE ON SOURCES

2·7 The content of the text of this chapter and its tables was drawn from many sources: standard reference works, prefaces to the style manuals of scientific organizations, handbooks, and histories of scientific disciplines. Especially valuable for their detail and documentation were the history of mathematical notation by Cajori (1928, 1929) and the histories of chemistry by Partington (1961), Ihde (1964), and Brock (1992). Most of the sources for this chapter are given in the reference list below; the list is definitely not a comprehensive bibliography of documents relevant to the history of scientific nomenclature and symbols.

REFERENCES

Cited References

Brock WH. 1992. The Norton history of chemistry. New York: WW Norton.

Cajori F. 1928. A history of mathematical notations. Volume 1. Chicago: Open Court.

Cajori F. 1929. A history of mathematical notations. Volume 2. Chicago: Open Court.

Ihde AJ. 1964. The development of modern chemistry. New York: Harper & Row.

Partington JR. 1961. A history of chemistry. London: Macmillan.

Singer C. 1950. A history of biology to about 1900: a general introduction to the study of living things. Republished from Abelard-Schuman edition. 1st ed 1931; rev ed 1950. Ames (IA): Iowa State Univ Pr.

Starnes DT, Noyes GE. 1991. The English dictionary from Cawdrey to Johnson 1604–1755. New edition with an introduction and a select bibliography by Gabriele Stein. Amsterdam: John Benjamins.

Additional References

Block BP, Powell WH, Fernelius RJ. 1990. Inorganic nomenclature: principles and practice. Washington: American Chemical Soc.

Buol SW, Hole FP, McCracken RJ. 1980. Soil genesis and classification. 2nd ed. Ames (IO): Iowa State Univ Pr.

Corliss JO. 1982. The history and role of nomenclature in the taxonomy and classification of organisms. In: Parker SP. 1982. Synopsis and classification of living organisms. New York: McGraw-Hill.

Crosland MP. 1978. Historical studies in the language of chemistry. New York: Dover.

de Vaucouleurs G. 1957. Discovery of the universe: an outline of the history of astronomy from the origins to 1956. New York: Macmillan.

Diringer D. 1968. The alphabet: a key to the history of mankind. 3rd ed. New York: Funk & Wagnalls.

Gillespie C, editor. 1970–1980. Dictionary of scientific biography. New York: Charles Scribner's Sons.

Haverton JL. 1978. Historical introduction. In: An international system for human cytogenetic nomenclature (1978). ISCN(1978): report of the Standing Committee on Human Cytogenetic Nomenclature. Cytogenet Cell Genet 21:312–403.

Hellemans A, Bunch B. 1988. The timetables of science: a chronology of the most important people and events in the history of science. New York: Simon & Schuster.

Howse D. 1980. Greenwich time and the discovery of the longitude. Oxford: Oxford Univ Pr.

[ICVGAN] International Committee on Veterinary Gross Anatomical Nomenclature. 1983. Nomina anatomica veterinaria. Ithaca (NY): ICVGAN.

Maltman A. 1990. Geological maps: an introduction. New York: Van Nostrand Reinhold.

Mayr E, Ashlock PD. Principles of systematic zoology. 2nd ed. New York: McGraw-Hill. The history of taxonomy; p 8–13.

Morton AG. 1981. History of botanical science: an account of the development of botany from ancient times to the present day. New York: Academic Pr.

Tanaka Y, chairman. 1957. Report of the International Committee on Genetic Symbols and Nomenclature. Union of International Sci Biol Ser B, Colloquium No. 30.

PART

2 General Style Conventions

3 Alphabets, Symbols, and Signs

As the development of the mind proceeds, symbols, instead of being employed to convey images, are substituted for them.

—T B Macaulay, *On John Dryden,* 1828

Compressions [of experience] seek to store lots of pieces of information in a single mark or representation. Symbols are the result of such a desire.

—J D Barrow, *Pi in the Sky,* 1992

Meaning is conveyed in scientific writing by words, numbers, and symbols. Words are built from letters of an alphabet. Symbols usually represent the opposite process, the representation of words by alphabet-derived or graphic symbols.

ALPHABETS

3·1 Most European languages are written and printed with letters of the roman alphabet. English uses, in general, unmodified roman letters. For special needs, scientific English also uses some characters formed from roman letters, greek letters, and some special typefaces for the roman alphabet.

In this chapter and elsewhere in this manual, the adjectives "greek", "italic", "roman", and "hebraic" are not capitalized when they modify the nouns "alphabet", "font", "letter", "numeral", and "typeface" used in typographic contexts that do not relate directly to the civilization, culture, history, people, or society of Greece, Italy, Rome, and ancient and modern Israel. The term "Latin alphabet" is preferred to "roman alphabet" by some historians (Diringer 1968). But the widespread use of this alphabet for languages other than Latin, notably the Romance languages, justifies the preference for "roman", a position also taken in *The Oxford Companion to the English Language* (McArthur 1992). See section 8·18 for further discussion of capitalization of "arabic", "greek", and "roman".

ROMAN

3·2 The classical Roman alphabet had 23 letters. The letters J, U, and W were developed in the Middle Ages to differentiate the consonantal J from the vowel I, the vowel U from the consonantal V, and the consonantal W from the consonantal V; therefore, the present roman alphabet has 26 letters. Each letter can be represented as a capital letter ("uppercase letter" in printer's terms) or small letter ("lowercase letter"); see Table 3·1. In some non-English European languages these letters are modified by diacritical marks or combined in ligatures to form additional characters; see Table 3·2 and section 4·39. These additional characters are not used in texts of scientific literature in English, with 2 exceptions.

1 Proper names and titles that must be presented in their original, non-English language, such as in bibliographic references not providing translations.
2 Non-English words, such as in quotations from a non-English language.

Until World War II much German literature was published with the roman alphabet in the Fraktur script version, also known as "black letter script" or "Gothic script". The use of Fraktur script for official publications was abolished in 1941, and roman letters are now generally used in German publications. Fraktur letters are still used for some conventions in mathematics and physics.

In general, bibliographic references with titles of papers and books originally published in a non-English language give these titles in translation within square brackets; see section 30·23. Titles in languages written in nonroman characters such as Russian and other Slavic languages (the Cyrillic alphabet and its variants), Chinese, and Japanese are usually represented in scientific literature in English by transliterated versions using roman letters or by translation.

For detailed presentations of special alphabetic characters in European languages, see *The Chicago Manual of Style* (UCP 1993) and Chapter 24, "Foreign Languages", in the *United States Government Printing Office Style Manual 1984* (USGPO 1984).

Table 3·1 Roman and greek letters

The roman alphabet		The greek alphabet[a]		
Capital (uppercase) letters	Small (lowercase) letters	Name of letter	Capital (uppercase) letters	Small (lowercase) letters
A	a	alpha	Α	α
B	b	beta	Β	β
C	c			
D	d	delta	Δ	δ
E	e	epsilon	Ε	ε
F	f			
G	g	gamma	Γ	γ
H	h	eta	Η	η
		theta	Θ	θ
I	i	iota	Ι	ι
J	j			
K	k	kappa	Κ	κ
L	l	lambda	Λ	λ
M	m	mu	Μ	μ
N	n	nu	Ν	ν
O	o	omicron	Ο	ο
P	p	pi	Π	π
Q	q			
R	r	rho	Ρ	ρ
S	s	sigma	Σ	σ
T	t	tau	Τ	τ
U	u	upsilon	Υ	υ
V	v			
W	w			
X	x	xi	Ξ	ξ
Y	y			
Z	z	zeta	Ζ	ζ
		phi	Φ	φ
		chi	Χ	χ
		psi	Ψ	ψ
		omega	Ω	ω

[a]Note that in the normal order of the greek alphabet the letter *gamma* follows *beta*, *zeta* follows *epsilon*, and *xi* follows *nu*. The order here was selected to show the relation of the greek alphabet to the roman.

Table 3·2 Common diacritical marks and special formations for the roman alphabet used primarily in languages other than English

Name	Mark	Example
acute accent	´	é
double acute accent	˝	ő
grave accent	`	è
breve	˘	Ğ
caron, wedge (haček [Czech, "little hook"])	ˇ	ž
circumflex (circonflex, French)	^	ô
cedilla	¸	ç
ligature		æ
dot	˙	İ
macron	¯	n̄
over-ring (krou¢ek [Czech, "little circle"])	˚	å
ogonek (Polish, "hook")	˛	ę
slash (stod, Danish)	/	o/
stroke	⁄	Ł
tilde	~	ñ
umlaut (diaeresis mark)	¨	ü

GREEK

3·3 Some capital letters of the classical Greek alphabet lent themselves to the forms of roman capital letters, but some did not; see Table 3·1. Greek letters, capital and lowercase, are used in many nomenclatures and notations for scientific disciplines, notably astronomy, biochemistry, chemistry, mathematics, and physics. These uses are presented in detail in other sections of this manual.

Among typefaces for greek letters there is no equivalent analogous to the italic typefaces developed for "sloping" versions of roman letters; normal greek lowercase letters are cursive, with a "flowing, often sloping" character (Bringhurst 1992).

HEBRAIC

Hebraic letters are rarely used in English-language scientific literature except for the aleph (א), which is occasionally needed in mathematics as a full-size symbol or a superscript character.

SYMBOLS

3·4 Notations representing quantities, objects, and actions have been in use through all of recorded history, for example, numerals for numbers of objects. As the pace of scientific discovery and description accelerated in the Renaissance and succeeding centuries, new notations were needed to represent efficiently the much more complex knowledge developed (see Chapter 2). These needs led to the present extensive systems of notation represented in detail elsewhere in this manual by symbols for chemicals, genes, animal functions, and other subjects of scientific inquiry, and for mathematical notation.

Many symbolic notations in science have been developed in a logical and coherent scheme for specific functional needs. They may serve either to represent what cannot be as economically expressed by a term or to represent the subject symbolized in a functional relation to other symbolized subjects, as in mathematical equations. In contrast, abbreviations have been developed mainly to eliminate the effort that would go into writing out what they represent or to save space; they usually represent simply the shortening of a term.

TYPES OF SYMBOLS

Symbols are characters or other graphic units representing a quantity, unit, element, unit structure, relation, or function. They are of 2 general kinds.

1 Single alphabetic or numeric characters or 2 or more alphabetic, numeric, or alphanumeric characters unspaced. Note that a symbol may be an abbreviation (such as "*V*" for "volume" and "cos" for "cosine"), but it need not be (such as "φ" for "electric flux"); note, too, that not all abbreviations need serve as symbols. A few symbols are graphic units derived from an alphabetic origin, for example, "ℙ" for "Pennsylvanian" in geology.
2 Graphic units designated as specific representations, for example, the equals symbol (=) that represents the relation "equality".

A symbol may stand alone or be incorporated into a larger symbol. For examples of varieties of scientific symbols see Table 3·3.

An abbreviation can function as a symbol when it is part of a group of symbols that together are used for the same conceptual purpose. For example, abbreviations of geologic-period names (such as "Pz" for "Paleozoic" and J for "Jurassic") serve as symbols when they are used with other alphabet-derived symbols (such as Ꞓ, for "Cambrian" and ℙ for "Pennsylvanian") in representing geologic systems in relations such as those depicted on geologic maps. Chemical symbols may be abbreviations (such as "C" for "carbon" and "O" for "oxygen"), but they serve as symbols in indicating relations as in a unit structure (as in "CO_3^-", the symbolic representation of "the carbonate anion"). When abbreviations

Table 3·3 Examples of scientific and common symbols

Symbol	Name of representation	Type of subject
1	one	unit
x	unknown quantity	quantity
n	number	quantity
m	mass	quantity
P	probability	quantity
V	volume	quantity
CO	cardiac output	rate
3	1 + 1 + 1	unit structure
♂	male in a family tree	unit structure
Al	aluminum, aluminium	unit, chemical
$AlBr_3$	aluminum bromide	unit structure, chemical
sis	oncogene related to simian sarcoma	unit structure, genetic
Pz	Paleozoic Era	unit, geologic time
m	meter, metre	unit, measurement
$	dollar	unit, monetary
£	pound sterling	unit, monetary
cos	cosine	mathematical function
+	plus (for addition)	mathematical function

do not serve such purposes but are used solely for shortening a term, they remain designated as abbreviations.

Symbols for specific scientific fields are presented in this manual in the chapters on style in those fields; see Table 3·4.

TYPE STYLES FOR SYMBOLS

3·5 The appropriate uses of roman (upright) and italic (sloping) type for symbols and the appropriate styling are specified in the international standard *Quantities and Units—Part 0: General Principles* (ISO 1992) (available in the 1981 version in *ISO Standards Handbook 2: Units of Measurement* [ISO 1982]); see Table 3·5. Note that boldface type is used for symbols for vectors and tensors; see section 11·15. These conventions are also summarized in Chapter 9.

Another detailed consideration of symbols is *A Guide to International Recommendations on Names and Symbols for Quantities and on Units of Measurement* (Lowe 1975).

PHONETIC SYMBOLS

3·6 Linguists and phonologists use a complex group of special symbols to represent sounds in transcriptions of spoken language. Some of these are standard roman or greek letters, but many are special formations derived

Table 3·4 Scientific and common symbols presented in this manual

Subject	Section	Table
Algebra	11·14, 11·20	11·12
Alicyclic compounds	16·7	
Allergens	24·22	
Amino acids	16·16–16·20	16·1
Animal strains	24·5	
Archeological inscriptions	25·8	25·2
Astronomy	27·7	27·5
Atomic and molecular states	15·8, 15·9	
Biothermodynamics	17·2, 17·3	
Blood groups	24·9	
Bone histomorphometry	24·12	
Calculus	11·16	11·9
Carbohydrates: configuration and conformation	16.21	
Chemical elements	15·5	15·1
Chemical formulas	15·7, 16·4	
Chemical kinetics	17·2–17·6	
Chromosomes	20·2–20·22	20·2, 20·3
Chromosomes: components and units	20·19–20·22	
Complement	24·19	
Crystallography	26·10–26·12	26·2, 26·3
Cyclic compounds	16·8–16·10	
Drug receptors	19·15	
Electrocardiographic recordings	24·14	
Elementary particles	15·1–15·3	
Enzymes	16·28	
Enzyme kinetics	17·10	17·4
Folic acids	16·29	
Genes and phenotypes	20·23–20·51	20·4–20·14
Geochronology	26·1	26·1
Hemodynamic functions	24·13	24·3
Hemoglobins	24·10	
Immunoglobulins	24·17, 24·18	
Immunologic systems	24·16–24·21	
Ion movement	17·8	17·3
Isotopic modifications	16·31–16·33	
Mathematical functions	11·14	11·8
Mathematical operators	11·8	11·10
Mathematical symbols	11·13–11·23	11·12
Matrices	11·17	11·11
Microbial processes	17·7	17·2
Monetary units	11·7	11·2
Neoplasia	24·31	
Nuclear particles	15·4	
Nucleic acids and nucleosides	16·38–16·40	16·2
Nuclides	15·6	
Number theory		11·12

Table 3·4 (*cont.*)

Subject	Section	Table
Pedigrees	24·6	24·2
Pharmacokinetics	19·17	19·2
Plant physiology	23·23	23·4
Prenols	16·43–16·45	
Purines, pyrimidines	16·38	16·3
Renal function	24·26	
Renin–angiotensin system	24·25	
Respiratory functions	24·27	24·6
Roman numerals	11·1	11·1
Set theory notations	11·16	11·10
SI base units	11·8	11·3
SI-derived units	11·8	11·4
SI-derived units with special names	11·8	11.4
SI-unit prefixes	11·9	11·5
Non-SI units	11·11, 11·12	11·6, 11.7
Soil horizons	26·20	26·5, 26·6
Solutions	17·9	
Sounds in phonetics	3·6	3·6
Statistics	11·24	11·13
Thermal regulation	24·28	24·7
Thermodynamics	17·1	17·1

from roman letters by structural change or addition of diacritical marks. These symbols appear in the literatures of anthropology, audiology, linguistics, phonology, and sociology; some of them are used in pronunciation guides in dictionary entries. The most widely used symbols are those developed by the International Phonetic Association. See Table 3·6 for examples of phonetic symbols with their names. A detailed source on phonetic symbols is *Phonetic Symbol Guide* (Pullum and Ladusaw 1986).

SIGNS

3·7 Graphic notations that stand for relations and operations, especially in mathematics, are often called "signs". Examples are +, the "plus sign", and −, the "minus sign". Because such notations represent functions and relations they are presented in this manual by the term "symbol". The ISO handbook on units of measurement (ISO 1982) maintains a terminologic distinction between the terms "sign" and "symbol" but does not define the distinction. Note that the decision for this manual to include mathematical "signs" within the category "symbols" is consistent with usage in the style manuals of the American Institute of Physics (AIP

Table 3·5 Style conventions for alphanumeric symbols

Symbol category	Typeface	Examples and notes
Quantity	Italic[a] (sloping)	Examples: *y*, *V*. In italic type[a] regardless of the type used for the rest of the text; usually a single roman or greek letter; no following punctuation mark except as needed for normal punctuation of or within a sentence; may be modified by a subscript. For a frequent exception, see Footnote b. Note that if text must be in italic type, words or phrases that normally should be in italics are set in roman; see section 9·3.
Subscript modifier for a quantity symbol		
Quantity symbol	Italic (sloping)	Example: P_x for probability, *P*, of an unknown quantity, *x*.
Other symbols	Roman (upright)	Example: V_{L} for "lung" (L) "volume" (*V*)
Unit of measurement	Roman (upright)	Examples: m (for "meter"), kg (for "kilogram"). In roman type regardless of the type used for the rest of the text; plurals keep the singular form; no following punctuation mark except as needed for normal punctuation of or within a sentence; follows the numeric value for a quantity by 1 space (by a hyphen when used as a compound modifier). Usually lowercase letters except for the initial capital for unit name derived from a proper name.
Number	Roman (upright)	Examples: 11, 64; XVIII, xviii. Arabic and roman numerals.
Mathematical operator	Roman (upright)	Examples: +, ×. Rare exceptions to upright convention.
Chemical element	Roman (upright)	Examples: K, Rb, Sn, $CaCl_2$. No following punctuation mark except as needed for normal punctuation of or within a sentence; modifying subscripts and superscripts (as in symbols for compounds and nuclides; see section 15·6) are also in roman type.
Chromosome	Roman (upright)	Example: 46,Yt(Xq+;16p−). See sections 20·2–20·19.
Gene	Italic (sloping)	Examples: *sis*, *HRCT*. See sections 20·20–20·41.
Phenotype	Roman (upright)	Example: HRCT. See sections 20·20–20·41.

[a]There is no italic type for greek letters; see section 3·3.

[b]Multiletter symbols that are derived from terms as contraction abbreviations are typeset in roman, rather than italic, type even when they represent quantities. An example is "CO", representing "cardiac output", a variable quantity defined with the unit "liters · minute^{-1}".

Table 3·6 Examples of phonetic symbols, with their names[a]

Symbol	Name
æ	ash
ɓ	hooktop b
ç	curly-tail c
ð	eth
ɠ	hooktop g
ħ	crossed h
ɮ	l-yogh ligature
ɳ	n with right tail
ɹ	turned r
ʃ	esh
ʐ	curly-tail z

[a]Based on Pullum and Ladusaw (1986).

1990) and the American Mathematical Society (AMS 1990; Swanson 1979), and in the NISO standard (NISO 1988) for electronic manuscript markup.

The term "sign" is retained in this manual for notations that convey a direction for the reader, for example, signs such as * and ¶ often used to direct the reader to footnotes.

REFERENCES

Cited References

[AIP] American Institute of Physics. 1990. AIP style manual. 4th ed. New York: AIP.

[AMS] American Mathematical Society. 1990. A manual for authors of mathematical papers. Providence (RI): AMS.

Bringhurst R. 1992. The elements of typographic style. Vancouver (BC): Hartley & Marks.

Diringer D. 1968. The alphabet: a key to the history of mankind. New York: Funk & Wagnalls.

[ISO] International Standards Organization. 1982. ISO standards handbook 2: units of measurement. 2nd ed. Geneva: ISO.

[ISO] International Standards Organization. 1992. Quantities and units—Part 0: general principles. Geneva: ISO. International standard ISO 31-0(E).

Lowe D. 1975. A guide to international recommendations on names and symbols for quantities and on units of measurement. Geneva: World Health Organization.

McArthur T. 1992. The Oxford companion to the English language. Oxford: Oxford Univ Pr.

[NISO] National Information Standards Organization. 1988. Electronic manuscript preparation and markup. New Brunswick (NJ): Transaction Publishers. American national standard Z39.59-1988. Available from NISO Press; see "Standards for Editing and Publishing" in Appendix 3.

Pullum GK, Ladusaw WA. 1986. Phonetic symbol guide. Chicago: Univ Chicago Pr.

Swanson E. 1979. Mathematics into type: copy editing and proofreading of mathematics for editorial assistants and authors. Revised edition. Providence (RI): American Mathematical Soc.

[UCP] University of Chicago Press. 1993. The Chicago manual of style. 14th ed. Chicago: UCP.

[USGPO] US Government Printing Office. 1984. United States Government Printing Office style manual 1984. Washington: USGPO.

4 Punctuation and Related Marks

> . . . the old stopping was frankly to guide the voice in reading aloud, while the modern is mainly in seeing through the grammatical construction.
>
> —Fowler and Fowler, *The King's English,* 1931

4·1 For many centuries, punctuation in English was applied mainly as a guide for readers to stops or pauses, its rhetorical use. Beginning in the late 17th century, its logical, or grammatical, use began to replace the rhetorical use. Today punctuation serves mainly to clarify the relation of prose elements to each other, as with periods demarcating sentences, and to support the meaning of statements, as with the question mark and the exclamation mark. With the growth of scientific literature, punctuation increasingly was needed in conventions for numeric and mathematical expression and science's specialized nomenclature and symbolization. Especially useful and clear discussions of the logic and development of punctuation can be found in Chapter 1, "Introductory", of *You Have a Point There* (Partridge 1953) and Chapter IV, "Punctuation", of *The King's English* (Fowler and Fowler 1931). *The Oxford Companion to the English Language* (McArthur 1992) has concise explanations of the uses of punctuation marks. A thorough history and analysis of punctuation from antiquity into the 19th century is *Pause and Effect: Punctuation in the West* (Parkes 1993). A complete catalogue of analphabetic characters (punctuation marks and related characters) and a discussion of them from a typographer's point of view are found in Appendix A, "Sorts & Characters", in *The Elements of Typographic Style* (Bringhurst 1992).

First considered here (Table 4·1) are the marks designated "stops" and applied mainly to indicate the close of a sentence or its equivalent: the period, question mark, and exclamation mark. Next come the marks within sentences or words: the colon, semicolon, comma, hyphen, dash, parentheses, square brackets, angle brackets, slash, apostrophe, and diacritical marks. Braces and ditto marks indicate relations between 2 or more lines of text. These marks are discussed in this chapter.

The main uses of quotation marks and the use of ellipsis points are discussed in Chapter 9 (see sections 9·10–9·13 and 9·14). The hyphen and apostrophe are additionally considered in Chapter 5 (see sections 5·17, 5·19, 5·21, and 5·29, 5·30, 5·33–5·35, 5·37, 5·39–5·42).

Each section of this chapter presents general uses and then, as appropriate, specialized uses in science. Note that many of the conventional punctuation marks serve together in applications of the Standard Generalized Markup Language (SGML) for marking electronic manuscripts (NISO 1991). Only the SGML angle brackets and the vertical line (bar) used in manuscript markup are described in this chapter, see sections 4·29 and 4·36; Chapter 32 includes more detail.

Table 4·1 Punctuation marks grouped by their general functions

<table>
<tr><th>Mark</th><th>Term and synonyms</th><th>Sections in this chapter</th></tr>
<tr><td colspan="3">Stops</td></tr>
<tr><td>.</td><td>period (full stop, full point)</td><td>4·3–4·6</td></tr>
<tr><td>?</td><td>question mark (interrogation mark, interrogation point)</td><td>4·7, 4·8</td></tr>
<tr><td>¿</td><td>interrogation mark in Spanish</td><td>4·48</td></tr>
<tr><td>!</td><td>exclamation mark (exclamation point)</td><td>4·9</td></tr>
<tr><td colspan="3">Intrasentence marks</td></tr>
<tr><td>:</td><td>colon</td><td>4·10, 4·11</td></tr>
<tr><td>;</td><td>semicolon</td><td>4·12, 4·13</td></tr>
<tr><td>,</td><td>comma</td><td>4·14–4·17</td></tr>
<tr><td>“ ” ‘ ’</td><td>quotation marks: double, single</td><td>4·18</td></tr>
<tr><td>—</td><td>em-dash</td><td>4·19, 4·20</td></tr>
<tr><td>——</td><td>2-em dash</td><td>4·21</td></tr>
<tr><td>–</td><td>en-dash</td><td>4·22</td></tr>
<tr><td>( )</td><td>parentheses (parenthesis marks, round brackets)</td><td>4·23, 4·24</td></tr>
<tr><td>)</td><td>single parenthesis (right or closing parenthesis)</td><td>4·25</td></tr>
<tr><td>[ ]</td><td>square brackets (brackets, square parentheses)</td><td>4·26, 4·27</td></tr>
<tr><td>{ }</td><td>braces (curly brackets)</td><td>4·28</td></tr>
<tr><td>< ></td><td>angle brackets</td><td>4·29</td></tr>
<tr><td>‾‾‾</td><td>vinculum</td><td>4·30</td></tr>
<tr><td>« »</td><td>guillemets, chevrons (quotation marks in French)</td><td>4·48</td></tr>
<tr><td colspan="3">Term and word marks</td></tr>
<tr><td>-</td><td>hyphen</td><td>4·31–4·33</td></tr>
<tr><td>/</td><td>slash (oblique bar, oblique mark, oblique stroke, shilling mark, slant line, virgule)</td><td>4·34, 4·35</td></tr>
<tr><td>|</td><td>vertical bar (vertical line)</td><td>4·36</td></tr>
<tr><td>’</td><td>apostrophe</td><td>4·37</td></tr>
<tr><td>′</td><td>prime sign</td><td>4·38</td></tr>
<tr><td></td><td>diacritical marks</td><td>4·39</td></tr>
<tr><td>*</td><td>asterisk</td><td>4·40</td></tr>
<tr><td>&</td><td>ampersand</td><td>4·41</td></tr>
<tr><td>@</td><td>“at” symbol</td><td>4·42</td></tr>
<tr><td>#</td><td>octothorp (numeral sign, pound sign, space sign)</td><td>4·43</td></tr>
<tr><td colspan="3">Marks for line relations</td></tr>
<tr><td>{</td><td>brace</td><td>4·44</td></tr>
<tr><td>″</td><td>ditto marks</td><td>4·45</td></tr>
<tr><td>.....</td><td>dots</td><td>4·46</td></tr>
<tr><td>¶</td><td>paragraph mark (pilcrow)</td><td>4·47</td></tr>
<tr><td>§</td><td>section mark</td><td>4·47</td></tr>
</table>

Note that the relations of punctuation marks in quotations are summarized in Chapter 9 (see sections 9·12 and 9·13).

The terms used here for punctuation marks are those found in *Electronic Manuscript Preparation and Markup* (NISO 1991).

4·2 Three general principles are helpful in reaching decisions on punctuation.

1 Punctuation should be used when called for by well-established and internationally accepted conventions of style. But note that some "well-established" uses change as time goes by.
2 When the need for a mark (or for omitting it) is not clear, the mark should be used if it will reduce or eliminate possible ambiguity.
3 If a format or text structure logically arranges text elements and thereby makes the meaning clear, punctuation that might otherwise be used should be omitted, for example, in vertical lists or after introductory phrases.

STOPS

PERIOD ("FULL STOP" IN BRITISH USAGE) (.)

General Uses

4·3 1 To close a declarative sentence or a short equivalent of an implied complete declarative sentence.

> Four of the soluble vitamins of the B complex have precise roles in the citric acid cycle.
>
> What is the worst ethical sin in science? Fraud. What next? Plagiarism.

2 To close an imperative sentence.

> Filter the solution while it is hot.

Note that in scientific texts, imperative sentences are usually not closed with an exclamation mark; see section 4·9.

3 To close a complete sentence within parentheses.

> The first rat died. (It refused to eat.)

4 To close a sentence structured in a list format but punctuated as a complete sentence.

> The survey was carried out with 3 purposes in mind:
>
> to estimate the size of the rodent populations,
> to estimate environmental influences on population sizes, and
> to identify previously unrecognized species.

Such formats are sometimes used to emphasize the listed elements when they would lack emphasis in a running sentence.

5 To close a sentence heading a list when the listed items are not closed with continued punctuation (as in the example above).

Surveys of rodent populations were carried out in 4 cities.

Birmingham
Chicago
Manchester
New York

6 To separate elements of some kinds of abbreviations; to denote an abbreviation. Abbreviations of Latin phrases (but see Table 10·2 on the English equivalents).

e.g. i.e. op. cit. et seq.

Nontechnical abbreviations in style systems preferring traditional conventions.

Mr. Ph.D. F.R.S. No. Thomas A. Smith, Jr.

See section 10·2 on recommendations for omitting periods not needed in abbreviations (for example, "Mr Smith", "PhD", "FRS").

7 To separate from text an introductory element (usually a number or letter) in an ordered list.

1. Biochemistry a. New Jersey
2. Chemistry b. New York
3. Physics c. Pennsylvania

Also note the use of a closing parenthesis mark for this need; see section 4·25. Other functions for the period include ellipsis marking (see section 9·14), dot-line connectors (see section 4·46), and the raised period (see section 4·6).

Unnecessary and Undesirable Uses

4·4 1 To close a complete sentence that is within a sentence and enclosed by parentheses.

The first rat died (it had refused to eat).

2 To mark symbols and abbreviations with periods unless a period is needed to remove ambiguity or for conformity with an established scientific convention.

National Institutes of Health NIH [not "N.I.H."] DNA
before Christ BC [not "B. C."]
Russell Tomar PhD Ag

[but]

species spp. [plural]; not "spp"

See section 10·2 for more on punctuating abbreviations.

3 After a title, text heading or subheading, equation, or formula standing separate from text. See section 31·4.

4 After an item in a table unless it is needed because of an established scientific style convention (such as "spp." for "species").

5 After an item in a list unless it is recommended by a convention specified above.

6 In numbers except where it serves as a decimal point.

"17.30" for "17 hours, 30 minutes"; prefer "17:30" or "1730" (as in railway schedules)

"4.2" for "4th chapter, 2nd section"; prefer "4·2" [see section 4·6, item 6]

Specialized Uses: On-the-Line Period

4·5 1 The decimal point in American, British, and Canadian usage (also see section 11·3 on European use of the comma).

The infant weighed 10.7 kg at birth.

The asteroid was calculated to have passed at a distance of 0.00072 AU from Earth.

2 Indication of hierarchical divisions represented numerically.

4.2 Period

4.3 Question Mark

CENTRAL NERVOUS SYSTEM INFECTIONS	C10.228.228
AIDS DEMENTIA COMPLEX	C10.228.228.57
BRAIN ABSCESS	C10.228.228.114

Note that this use as illustrated in the 1st example can imply that the period is serving as a decimal point; this manual recommends that the raised period be used instead for this function; see sections 1·4 and 4·6.

3 End of a data field in a bibliographic reference; see section 30·18.

Smith A. The interview: a new method. Clin Med. 1998;76:11–5.

4 Indication of ring size in chemical nomenclature.

spiro[2.3]hexane bicyclo[2.2.2]octane

Specialized Uses: Raised Period (Also Sometimes Called Centered Dot)

4·6 1 A multiplication symbol in equations and other mathematical expressions and in compound units; see Table 11·12 and section 11·11.

$k \cdot g(a + 2)$ $1\ \mathrm{C} = 1\ \mathrm{A{\cdot}s}$

2 In an ellipsis symbol in a mathematical expression; see section 11·20.

$x_1 + x_2 + \cdots x_n$

3 Indicating associated base pairs of nucleotides.

G · C [for "guanine" and "cytosine"]

4 Connection with adducts (for example, water of hydration) in a chemical formula.

$Na_2B_4O_7 \cdot 10H_2O$

5 Three centered dots indicate a chemical association of unspecified type (Dodd 1986, p 96).

$F \cdots H\text{–}NH_3$ $C \cdots Pt$

6 Indication of hierarchical divisions represented numerically.

4·3 Period
4·7 Question Mark

QUESTION MARK (?)

General Uses

4·7 1 To close a freestanding question whether it is a complete sentence or not.

What is the government's policy on pollution of river headwaters?

The policy was overturned. Why? In view of the number of species lost, this change appears . . .

2 To close a declarative sentence and indicate it is a rhetorical question in its context.

The policy was overturned. The loss of 10 species was not too much? In the next session . . .

3 To close a declarative sentence ending with a complete question set off by a comma even if it is not a direct quotation.

The committee asked, why were so many species killed?
[rhetorical emphasis on the question]
[but "The committee asked why so many species were killed."]

4 To indicate uncertainty about an element within a sentence.

> Girolamo Fracastoro (1483?–1553) was in effect the father of the concept of infectious disease.

The question mark should be placed so that it appears to apply only to the queried element, not possibly to all elements in the group.

> 1172?–1221 [not "?1172–1221"]
>
> 1546–?1627 [not "1546–1627?"] ?1546–?1627

Specialized Uses

4·8 1 To indicate a questionable identification of a chromosome or chromosomal structure; see Tables 10·2 and 10·3.

> 45,XXY,–?8 [45 chromosomes, xx sex chromosomes, missing chromosome which is probably chromosome 8]

EXCLAMATION MARK (!)

General and Specialized Uses

4·9 1 To close a declarative, imperative, or interrogative sentence so as to give it rhetorical emphasis. Such uses are usually not justified in reporting scientific findings but can be useful in commentaries on findings or in documents about science.

> Freud's "science" was pure metaphysics!
>
> Wash your hands every time you handle possibly infected articles!
>
> Wasn't the withdrawal of funding for the SSC a major disaster for physics?!

The close of the 3rd example represents what Partridge calls "compound points" (1953).

2 A factorial symbol in mathematics; see Table 11·12.

> $n!$ 4! [represents $1 \times 2 \times 3 \times 4$]

3 Indication in botanical writing of specimens examined by the author; see section 23·7.

> Lectotype here designated: P!; isolectotypes: K!, NY!.

4 A phonetic symbol in some literature on phonetics, linguistics, or anthropology.

> His papers dealt almost exclusively with the !Kung.

5 In the document type declaration tag in the Standard Generalized Markup Language (SGML; see sections 32·8 and 32·10).

INTRASENTENCE MARKS

COLON (:)

4·10 The colon is a mark indicating by its general uses a stronger pause than that indicated by a semicolon or a comma. It implies that the reader should now pay close attention to the formal structure thus marked. A student caught this use in a musical metaphor: "A colon works like a drumroll: it prepares you for the cymbals' clash; a semicolon is more like a modulation" (Allen 1992).

General Uses

1 To separate an independent clause from a following independent clause.

> We had 2nd thoughts about the 1st run: the data lacked the needed precision.
>
> Slow-stop mutants complete the round of replication: they cannot start another.

This use is best reserved for sentences in which the 2nd clause amplifies or clarifies the 1st, a use akin to that of the colon preceding an explanatory list. The 1st word following the colon is usually capitalized if the 2nd clause is a formal statement or long quotation.

> This is the rule: On closing a file, back it up on a separate disk.
>
> Lewin captured the essence of the history of genetics thus: "The gene is . . ."

In general, 2 independent clauses that are simply related in meaning, the 2nd clause not being explanatory, are satisfactorily related with a semicolon; see section 4·12. The 2nd of the 1st pair of examples immediately above would preferably have a semicolon instead of a colon.

> Slow-stop mutants complete the round of replication; they cannot start another.

2 To introduce a list (series).

> The lectures covered 3 topics: carbohydrates, lipids, and proteins.
>
> The sequence of the 3 lectures was as follows: carbohydrates, lipids, proteins.

Note that the part of the sentence preceding the colon must be a complete independent clause; a colon must not come between a verb or preposition and its object.

> The lectures covered carbohydrates, lipids, and proteins.

[not "The lectures covered: carbohydrates, lipids, and proteins."]

His tastes ran to acorn squash, carrots, and spinach.

[not "His tastes ran to: acorn squash, carrots, and spinach."]

3 To introduce a long quotation, especially when the quotation clarifies or amplifies the preceding independent clause.

Lewin captured the essence of the history of genetics in a few sentences: "The gene is the unit of genetic information. The crucial feature of Mendel's work, a century ago, was the realization that the gene is a distinct entity. The era of the molecular biology of the gene began in . . ."

4 To close a formal salutation in a letter or the introductory remark in an address.

Dear Professor Carpenter:

Officers of the Association and Distinguished Guests:

A colon should not be used after a title, text heading or subheading, equation, or formula standing separate from text.

Specialized Uses

4·11 1 To couple elements of titles such as book titles, chapter titles, table titles.

Scientific Style and Format: The CBE Manual for Authors, Editors, and Publishers

Chapter 3: Pulmonary Functions

Note that the elements in the 2nd example could be distinguished simply through adequate spacing. A period is sometimes used instead of a colon in chapter, table, and figure designations.

2 To demarcate certain elements in bibliographic references; see section 30·18.

Smith A. The interview: a new method. Clin Med 1998;76:11–5.

3 To separate chapter and verse, or chapter and line, designators in biblical and other classical references.

Major surgery was 1st described in Genesis 2:21.

Has any insult ever exceeded that of Agamemnon by Achilles (*The Iliad* 1:264)?

4 To separate parts of a ratio or proportion.

The ratio of females to males was 3:1.

We recommended a 5:10:5 fertilizer.

5 To couple components of a time datum when the units are hours, minutes, or seconds. When the coupled element is a decimal fraction of the initial unit, a decimal point is used rather than a colon.

12:38 AM

at precisely 2345:15 [or] 23:45:15 [for "23 hours, 45 minutes, 15 seconds"]

[but "12.25" in decimal notation for "12 hours 15 minutes"]

Do not use a period to separate hour and minute elements; this is an undesirable British convention with the ambiguity as to whether the period represents a decimal point or not (also see section 12·2).

6 To symbolize a chromosomal break in a detailed description; see Table 20·2. A double colon symbolizes a break and reunion.

46,XX,del(1)(pter→q21::q31→qter) [breakage and reunion of bands 1q21 and 1q31 in the long arm of chromosome 1]

7 To group the locants in chemical names.

2,3:4,5-bis-*O*-(phenylmethylene)-*altro*-hexodialdose

(3,4:3,9:5,6:6,7:7,8-penta-μH)-(3-*endo*-*H*)-*nido*-nonaborane

pyrido[1′,2′:1,2]imidazo[4,5-*b*]quinoxaline

8 To separate indications of originating and sanctioning authors to protect names of fungi designated as "sanctioned"; see section 23·26.

Valsa coronata (Hoffm.:Fr.) Fr.   *Agaricus adustus* Pers.:Fr.

9 To separate, in archeological usage, the name of a ceramic type from its variety by a colon and 1 space if both are given; see section 25·7.

Mimbres Bold Face: Black-on-white

Urita Gouged-incised: Urita Variety

SEMICOLON (;)

4·12 As a mark of linkage, the semicolon is weaker than the colon but stronger than the comma. It is a mark of coordination.

General Uses

1 To separate 2 or more coordinate clauses whether or not joined by a conjunctive adverb such as "however", "besides", "therefore"), but generally not clauses joined by a simple conjunction such as "and" (which can be separated by a comma). Note that short clauses not needing separation with rhetorical force can be separated with commas.

I came; I saw; I conquered.
[but "I came to the table, I felt the mass, I cut."]

We proved his hypothesis; we failed to find support for ours.

The patient was 1st treated in the emergency room; shortly, however, she had to be transferred to the intensive care unit.

The nurse removed the sutures, and the patient was discharged.

[not "The nurse removed the sutures; and the patient was discharged."]

2 To separate 2 or more coordinate clauses when at least 1 of them has internal punctuation (comma, dash), even if connected by a simple conjunction.

The nurse, acting on an order from the surgeon, removed the sutures; and the patient, disregarding the surgeon's advice, insisted on leaving the hospital.

The uppermost formation, 1st identified by Smith, is sandstone; the next lower, identified by Jones, is a shale.

3 To separate 2 coordinate clauses when the 2nd is structurally parallel, even if grammatically elliptical.

In men the most important etiologic factor is a high-fat diet; in women, an estrogen deficiency.

4 To separate the elements of a complex series, specifically a series with elements carrying internal punctuation.

His ethnographic studies concentrated on 3 groups: Chinese, Japanese, and Taiwanese; French, Germans, and Austrians; and Inuit, Mexicans, and Peruvians.

Specialized Uses

4·13 1 To demarcate certain elements of bibliographic references; see section 30·18.

Jones R. Shigellosis. Rev Hosp Infect 1956;2:1–3.

2 To symbolize chromosomal rearrangements; see sections 20·3 and 20·4 and Tables 20·2 and 20·3.

46,XX,t(12;?)(q15;?) [rearranged chromosome 12; the segment of the long arm (q) distal to band 12q15 could not be identified]

3 To separate (with semicolon and space) symbols for mutant genes on nonhomologous chromosomes of the fruit fly; see section 20·28.

bw; e; ey

4 To separate subscripts in selectively labeled compounds.

$[1\text{-}^{2}H_{1;2}]SiH_{3}OSiH_{2}OSiH_{3}$

5 To separate sets of locants in which colons have already been used.

benzo[1˜,2˜:3,4;5˜,4˜:3′,4′]dicyclobuta[1,2-*a*:1′,2′-*a*′]diindene

COMMA (,)

General Uses

4·14 The comma should serve mainly to clarify grammatical structures.

GENERAL USES: INTRODUCTORY COMMA

1 To set off an introductory clause beginning with a subordinating conjunction (if, although, because, since, when, where, while).

When the physician has 2 equally efficacious and safe treatments, great weight should be put on the patient's preference.

2 To set off a transitional or parenthetic word or phrase.

To be sure, another geological survey should be conducted.

Of course, another geological survey should be done.

After all, 8 samples may not be enough.

Finally, the cost is excessive.

3 To set off a short word of address or emphasis.

Chemists, you cannot be held responsible for all environmental pollution.

No, state insurance is not the answer.

COMMA OF SEPARATION AND BRACKETING

4·15 1 To separate 2 independent clauses joined by a coordinating conjunction (and, but, neither, nor, or) unless the clauses are short and the absence of a comma would not cause ambiguity.

After a delay the survey was completed in June 1983, and the map was published in July 1984.

We finished in June and the map appeared in December.

2 To separate elements so as to clarify meaning.

In all, 8 experiments were performed. [not "In all 8 experiments were performed."]

Keys provide, except in the most specialized works, a useful means of identification.

3 To separate a nonrestrictive clause from the rest of the sentence.

The cells, which had been sent to us by the institute's depository, were infected.

4 To separate a nonrestrictive appositive from the rest of the sentence.

Raymond Turner, a mammalogist, described the 2 species.

William Applecart, President of the University, gave the commencement address.

Note that in the preceding 2 uses, paired commas are necessary.

5 To separate a parenthetic statement, term of address, or interjection from the rest of the sentence.

Ravdin, and I shall never forget his short and pudgy figure, was always the 1st man in the operating suite each morning.

Remember, students, you have come here to learn and will go forth to work.

Penicillin, alas, was in short supply in the region.

6 To separate the elements (words, phrases, clauses) of a simple series of more than 2 elements. A comma should precede a closing "and" or "or". This rule applies to adjectives each modifying the following noun.

The tomatoes, beans, and peppers were planted in April.

The choice of antibiotic can be penicillin, ampicillin, or erythromycin.

The tumor was bloody, necrotic, and malodorous.

He used a cheap, effective, and readily obtained remedy.

An adjective forming a noun phrase with the noun it modifies is not separated from the immediately preceding adjectives.

He was noted for waving a large, torn, faded American flag [noun phrase, "American flag"] at commencement.
[A test of whether commas are needed is to substitute "and" for the comma and check the sense.]

Some writers prefer to omit the comma before the closing conjunction (as in "the eye, ear, medulla and spinal cord"), but routine use of this comma saves having to take time to consider possible ambiguities. An apparent series may in fact represent a rhetorical coupling.

The postdoctoral fellows brought the beers, barrels and bottles.
[not "brought the beers, barrels, and bottles."; the closing pair "barrels and bottles" was intended by the author to expand alliteratively and rhetorically on "beers".]

7 To separate contrasting expressions and interdependent clauses.

It is orange, not red.

The greater the risks, the greater will be the probable gain from treatment.

8 To indicate an elliptical construction.

The fetuses exposed to the drug in the 3rd month have spinal defects; the others, no defects.

9 To set off adjacent and unrelated numbers.

By the end of 1935, 1000 experiments had been completed.

10 To separate different elements of an address or geographic designation.

The laboratory was moved to 133 Glenside Avenue, Philadelphia, Pennsylvania.

The specimens of species newly identified were deposited in the museum in Cairo, Egypt.

11 To set off the year in a date expressed as month day year. The commas should be omitted when only the month and year are stated.

The first celebration was set for October 12, 1991, before the facts of his birth were known.
[but "set for October 1991 before the facts . . ."]

Unnecessary and Incorrect Uses

4·16 1 To separate 2 relatively simple and short independent clauses connected by a coordinating conjunction if the lack of a comma would not produce ambiguity.

The survey was completed and we went home.

[not "The survey was completed, and we went home."]

2 To set off a short introductory phrase or clause if the comma would not contribute to clarity or ease of reading.

Despite the pause the run was successful.

[not "Despite the pause, the run was successful."]

3 To set off a restrictive appositive (a defining word or phrase needed for the desired meaning).

The species *Bombyx mori* is distinguished from other species by . . .

[not "The species, *Bombyx mori,* is distinguished from . . ."]

4 To separate digits in a page number, an address number, a year number, a patent number.

page 6984 10779 Glenwood Avenue AD 1066

US Patent 5 973 257

[not "page 6,984" "10,779 Glenwood Avenue" "AD 1,066"

"US Patent 5,973, 257"]

5 To separate digits of numbers with 5 or more digits into groups of 3. Use instead a thin space; see section 11·3.

12 578 896 [not "12,578,896"]

6 To separate a noun clause from the predicate of the sentence.

He said that the bird was a flicker.
[not "He said, that the bird was a flicker."]

Where the plant grew was fertile ground.
[not "Where the plant grew, was fertile ground."]

7 To separate name modifiers from the stem name.

Franklin D Roosevelt Jr [not "Franklin D Roosevelt, Jr"]
Homer Smith III [not "Homer Smith, III"]

8 To set off the year in a date expressed as day month year or as year month day.

The 1st celebration was set for 12 October 1991 before the facts of his birth were known.

Specialized Uses

4·17 1 To separate subelements of bibliographic groups in bibliographic references; see section 30·18.

Smith TH, Robbit Q, Tannen M. The molecular biology of . . .

2 To separate, unspaced, elements in symbolic representation of human chromosome aberrations; see section 20·6 and Table 20·30.

46,XX,t(4;13)(p21;q32)

3 To separate, unspaced, amino acid symbols in amino acid residues of unknown sequence; see section 16·20.

Asp-His-Pro-(Gly,Phe,Tyr)-Lys

4 To separate, unspaced, locants in chemical formulas (letters and numbers identifying the location of an atom or group in a molecule); see section 16·13.

4,5-difluoro-2-nitroaniline *N,N,N',N'*-tetramethyldiaaminomethane

5 In European conventions for numbers, as the decimal point. This convention is recommended by the standard ISO 31/0-1981 (E) (ISO 1982) but is generally not followed in the United States, Canada, and the United Kingdom (UCP 1993; DSSC 1985; USGPO 1984; McCoubrey 1991) for both their general and scientific literature.

[In Europe] The value for the constant was finally set at 5,064 531 on the basis of Razecki's studies.

[In Canada, the United Kingdom, and the US: "was finally set at 5.064 531 on . . ."]

QUOTATION MARKS, DOUBLE AND SINGLE (" " ' ')

Double Quotation Marks

4·18 Quotation marks are used mainly to delineate quoted words, terms, or longer elements of text. These uses are discussed in detail in sections 9·10–9·13. A few more specialized uses should be noted here.

1 To indicate the title of a journal article, a book chapter, or a series title, rather than by the use of italics (see section 9·2), which is generally confined to journal and book titles.

> *Astronomy* won the prize for its article "The Next Asteroid to Hit Earth".

This use does not apply to the same kinds of titles when they are part of a bibliographic reference; see section 30·17.

2 To suggest a nonstandard or ironic use of a word or term.

> The Chief Executive told us he "cared" about big science.

3 To indicate a word or phrase used in text as such (that is, as an example or explanation) when italics for such elements might be confused with proper scientific italicization; this use is applied in this manual.

> Avoid the sloppy use of "impact", "case", and "individual".

Single Quotation Marks

Single quotation marks are used to enclose the names of plant cultivars; see section 23.14. See section 9·10 for differences in American and British use of single quotation marks.

DASHES

4·19 Four lengths of dashes are available for printed text: the en-dash, the em-dash, the 2-em dash, and the 3-em dash. The hyphen serves as a connector rather than as an indicator of interruption or omission and thus does not belong to the family of functions generally served by dashes. Note that the en-dash can serve as a connector and thus is akin to the hyphen. The 3-em dash has been used in bibliographic references to represent names of authors when the authors are the same as those in the preceding reference; this use is not recommended.

These different dashes were generally not available in typewriter typefaces but are usually available in word-processing programs.

Em-Dash (—)

4·20 The length of the em-dash is equal to the size in points of the typeface; thus the em-dash in 10-point type is 10 points long (wide). The name "em-dash" indicates that this length approximates the width of the capital letter M. In typescripts the em-dash should be represented by 2 adjacent hyphens with no space on either side; a copy editor will mark it appropriately for the printer.

The em-dash has 4 main types of uses.

1 To set off elements within a sentence that express a parenthetic break in the sentence's line of meaning. The set-off statement usually defines, elaborates, emphasizes, explains, or summarizes and typically is a sharp break, tangential and not vital to the sentence's central message.

> Osteoporosis—and the diagnosis is hard to make early without expensive equipment—may be the most common disorder of postmenopausal women in the United States.
>
> `Smith spent 10 years at Johns Hopkins--they may have been the best years of his professional life--before he took up his post in San Francisco.` [em-dashes in typescript]

This function of the em-dash is akin to uses of the comma and parentheses, but the 3 marks usually represent different types and strengths of interruption. The em-dash, Partridge noted (1953), "resembles parentheses, in that, in one important function, it expresses rather more strongly, rather more abruptly, what parentheses express less strongly and much more smoothly". At the other end of rhetorical force, enclosing commas are less appropriate for a parenthetic or "aside" sense and more appropriate for an explanatory sense.

2 To set off introductory elements in a sentence that explains their significance.

> The Japanese beetle, the starling, the gypsy moth—these pests all came from abroad.

3 To indicate the source of a quotation or editorial statement.

> What is a doctor? A licensed executioner. —Mazarinade

> *Publication of this letter does not indicate that it represents a policy of the American Chemical Society. —The Editor*

4 To set off a parenthetic statement in text within parentheses and square brackets (see section 4·27) as a 3rd level of interpolation.

> and of the 17 species of birds (the house sparrow [common in bushes—especially those affording good perches—as well as in trees] is the most common), we were able to . . .

This type of construction should be avoided if possible.

Two-Em Dash (——)

4·21 The 2-em dash (4 hyphens in typed text) can be used to represent unknown or missing letters in words.

1 In printed versions of written documents with illegible letters in the original text.

> My dear Mr Darwin, Your st—— about our descent is . . .

```
My dear Mr Darwin, Your st---- about our descent
from . . .
```

> [the 2-em dash in typescript]

2 In quotations of incomplete statements.

> His chronic harping on ecology was a steady pain in my ——.

3 In statements preserving anonymity.

> Patient M—— was the first in this series to develop the complication of . . .

In most scientific documents the initial capital letter will suffice.

En-Dash (–)

4·22 The length of the en-dash approximates the width of the capital letter "N". The most common uses of the en-dash in scientific writing are as the minus symbol (see Table 11·12) and to link numbers representing a range of values. In typing, the en-dash is represented by a hyphen, but note that the en-dash is now readily produced by word-processing programs. In general the connective "to" is preferred to avoid possible ambiguity with the minus symbol.

> with temperatures of –5 to 25 °C the . . .
> [not " with temperatures of –5–27 °C . . ."]

with a confidence interval of 0.03 to 0.57 . . .
[not " a confidence interval of 0.03–0.57" . . .]

with weight changes of −3.5 to +4.7 kg

Do not use a minus symbol or the word "from" with an en-dash.

–4 to –6 °C [not "–4– –6 °C]
from page 6 to 10 [not "from page 6–10"]

The en-dash should be used in bibliographic references to connect numbers representing 1st and last pages, but hyphens are widely used instead; see section 30·27. It can be used in text to express "through" as in "pages 6–10".

Roberts T, Smalley R. The Cretaceous–Triassic boundary in Somalia. Science 1928;45:358–69.

The en-dash can be used to link 2 words or phrases representing items of equal rank and 2-word concepts, with the hyphen used to link the names of units of different ranks.

the main north–south avenues of the city were. . .
Michaelis–Menten kinetics
the Río San Juan–Lake Nicaragua route
gas–liquid chromatography

IUPAC–IUBMB
[en-dash linking the abbreviated names of 2 organizations of equal rank]

NC-IUB
[hyphen linking the abbreviations for the names "Nomenclature Committee" and "International Union of Biochemistry"; the committee is a unit of the union]

The en-dash can be used as a coordinate connector within a term including hyphenated elements.

the Winston-Salem–Raleigh group of scientists
["Winston-Salem" and "Raleigh", 2 cities]

Columbia-Presbyterian–Brigham cases
["Columbia-Presbyterian" and "Brigham", 2 medical institutions]

Woody–Mia–Soon-Yi episode
[3 persons: "Woody", "Mia", and "Soon-Yi"]

the lake–river-mouth location [but "river-mouth" location"]

sugar-maple–dominated forest [but "maple-dominated forest"]

This kind of construction should be avoided if possible because of the frequent difficulty for readers in distinguishing between the hyphen and the en-dash.

Specifically scientific uses include linking the names of components of mixed solvents and representing chemical bonds.

hexane–benzene solvent $C_6H_5CO–O–COCH_3$

PARENTHESES, DOUBLE AND SINGLE (PARENTHESIS MARKS, ROUND BRACKETS) (( ))

4·23 As a term in rhetoric, "parenthesis" means "a word, phrase, or sentence inserted as an aside in a sentence complete in itself" (Lanham 1991), but its most frequent meaning now is that of "parenthesis mark". The plural form "parentheses" refers to paired parenthesis marks used to enclose a parenthetic element. The text enclosed by parentheses is generally referred to as a "parenthetic statement" or by some similar phrase.

General Uses

1 To enclose a parenthetic word, phrase, or sentence.

The most common use of parenthesis marks (parentheses) is . . .

Urinary incontinence (and disorders discussed in other chapters) is a frequent problem . . .

In County Armagh (we lacked time for studies in the other counties) we found that . . .

This function is sometimes better carried out with commas or em-dashes when different strengths of interruption are needed; see sections 4·15 and 4·20. Note that the parenthetic text is grammatically independent of the sentence carrying it; in the 2nd example above the subject "urinary incontinence" and the verb "is" are singular; a plural verb "are" would be incorrect.

2 To indicate the meaning of a symbol or abbreviation to be used later in the text.

and ethylenediaminetetracetate (EDTA) was used as . . .

Persons found to be carrying the human immunodeficiency virus (HIV) should . . .

3 To enclose directive text.

As noted recently (Smith 1992), the . . .
[a citation directing readers to the reference]

As discussed elsewhere (see section 17·3), we . . .
[cross reference to other text]

Specialized Uses

4·24 1 To enclose certain data in bibliographic references; see section 30·19.

Mastri AR. Neuropathy of diabetic neurogenic bladder. Ann Intern Med 1980;92(2 Pt 2):316–8.

2 To enclose mathematical elements that must be grouped for certain functions; see section 11·19.

$z = k(a + b + c)$ $y = 3.47(x + z)^3$

3 To group molecular components to indicate the application of a subscript number; see sections 15·7 and 16·4.

$K_4Fe(CN)_6{\cdot}3H_2$ $(CH_3)_2CHCH_2CH(NH_2)COOH$

4 To enclose complex substituent prefixes; see section 16·17.

N-(methoxycarbonyl)alanine

5 To enclose oxidation numbers in text. In formulas they are placed superscript; see sections 15·6 and 16·4.

Pb(IV)

6 To enclose unknown sequences of amino acids in polypeptides; see sections 4.17 and 16·20.

(Asp,His,Pro)

7 To enclose the stereochemical descriptors *R*, *S*, *E*, and *Z*; see section 16·13.

(2*S*)-alanine (*E*)-2-butene

8 For immunoglobulin notations; see section 24·17.

$F(ab')_2$ IgG(Pr)

9 For notation of specificities of histocompatibility antigens; see section 24·17.

HLA-Bw56(w22)

10 To enclose symbols indicating structural alteration of a chromosome; see sections 20·3, 20·4, 20·7, and 20·14; also Tables 20·2 and 20·3.

46,XX,t(4;13)(p21;q32)

11 To enclose herbarium abbreviations in plant specimen citations; see section 23·7.

Mrs. Thompson s. n. (Holotype: GH!)

12 To enclose the name of the author of the original taxonomic description when a species is transferred to another genus; see sections 23·5 and 24·3.

Haplopappus radiatus (Nutt.) Cronq. from *Pyrrocoma radiata* Nutt.
Blatella germanica (Linnaeus 1767)

Single (Right) Parenthesis ( ))

4·25 The 2nd mark of parentheses can be used alone after numerals or letters introducing elements of a list or series, usually in a sentence.

Three projects were funded.
1) Philadelphia: Archeology of the Late 17th Century
2) New York: Sites of Black Cemeteries
3) San Francisco: Pre-Gold-Rush Buildings

Three diseases almost wiped out the South Sea populations in the early 19th century: 1) syphilis, 2) tuberculosis, and 3) gonorrhea.

The series in the 2nd example could be demarcated with commas or semicolons; the demarcated numbers represent emphasis by a device of format and may be preferable in sentences with 3 or more elements in the list or confusingly complex elements.

SQUARE BRACKETS (BRACKETS) ([ ])

4·26 "Brackets" is the term widely used in the United States for this pair of marks. But "bracket" can also be applied to other kinds of punctuation marks: parentheses ("round brackets"), the bracket pair {} (often called "braces" or "curly brackets"), and angle brackets (<>). Hence the British term "square brackets" is preferable to "brackets".

General Uses

1 To demarcate text or letters added to quoted text to amplify or clarify the original text.

Cushing commented, "When Osler moved [to Baltimore], he was not risking his future".

The last entry in his journal was "I took lunch with D[arwin] and spotted P[otter] with his mistress E[laine Smythe]".

The results of an analysis by Neyman, Scott, and Smith [Science 163:1445–9] of a carefully conducted experiment were released to the press.

2 To demarcate an editorial comment.

> His diary included this note: "When I was in London I briefly met Darwynne [*sic*], that horrible chap who claims we are descended from monkeys".

This use of *sic* ("thus" in Latin; see Table 10·2) presumes that the reader knows the correct form. A more helpful notation is a correction of the error.

> "I briefly met Darwynne [Charles Darwin], that horrible chap who . . ."

3 To enclose a parenthetic statement within a parenthetic statement enclosed by parentheses.

> and immunization was carried out with DTP (the CDC recommendation [17]) before . . .

Specialized Uses

4·27 1 Multiple bracketing in mathematical expressions and chemical names; see sections 11·19 and 16·4.

> $z = k[(a + b) - y(c + d)]$ bis(bicyclo[2.2.2]octadiene)platinum

The general sequence for multiple bracketing in mathematics is {[( )]}, which differs from the usual sequence in text represented in the 3rd general use described in the section immediately above. In mathematical usage brackets may be referred to as "fences", "closures", or "signs of aggregation"; see section 11·19 for more detail.

In chemistry the Chemical Abstracts Service uses such sequences as [[[()]]], that is, no braces except for special notation; the International Union of Pure and Applied Chemistry uses whatever sequence is needed for the circumstances, for example, [{()}]. In some chemical usages, internal brackets must be used and parentheses may surround the brackets, as ([ ]); see the example above.

2 Enclosing chemical concentrations.

> $[Na^+]$ $[HCO_3^-]$

3 Enclosing isotopic prefixes; see section 16·32.

> $[^{14}C_2]$glycolic acid $[^{32}P]$AMP

4 Enclosing complex substituent prefixes containing internal parentheses.

> *N,N*-bis[(phenylmethoxy)carbonyl]alanine

5 Enclosing connecting points in fusion nomenclature; see section 16·9.

benz[*a*]anthracene

6 Enclosing numbers used as ring-size indicators in spiro and bicyclo names; see section 16·8.

spiro[2.3]hexane   bicyclo[2.2.2]octane

BRACES ({})

4·28 Paired left and right braces are used as a symbol for aggregation in mathematics for fencing elements already enclosed by parentheses and square brackets; see sections 4·27 and 11·19. A single multiline brace can be used to aggregate elements in adjacent lines, but this use is now uncommon outside of mathematics; see section 4·42.

ANGLE BRACKETS (<>)

4·29 These brackets are known singly as the "less-than symbol" and the "greater-than symbol", but as a pair they have specialized uses.

1 Opening and closing delimiters in the Standard Generalized Markup Language (NISO 1988) applied to electronic manuscript marking; see section 32·8 and Table 32·30.

<fn><no>5<bb>Entropy, A New World View<au>Jeremy Rifkin<obi>(New York: Bantam Books, 1981), pp.6–10</fn> [marking for a footnote]

2 Enclosure of a statement of a key to be pressed in computer-program instructions.

Press <Enter> to start searching.

VINCULUM ($\overline{}$)

4·30 The vinculum ("a bond", from *vincire,* Latin "to bind") can be used in mathematics (Partridge 1953; Swanson 1979) to span algebraic symbols for the same function as parentheses or to span letter groups, but parentheses are preferred.

$a - \overline{b - c}$ $a - (b - c)$

TERM AND WORD MARKS

HYPHEN (-)

General Uses

4·31 Partridge (1953) sums up the 2 main uses of the hyphen as "dividing and compounding". The "compounding" use of the hyphen in word and term formation is discussed in detail in sections 5·17 and 5·19. The hyphen connects some prefixes and suffixes to stem words, forms compound terms from 2 or more stem words, and forms compound modifiers containing a numeral or spelled-out number. There is a strong tendency in scientific English either to form new single-word terms by combining 2 stem words or to not use the hyphen between 2 or more modifiers. The hyphen is also used to mark word divisions for line breaks and to indicate syllabication.

Some types of terms usually take the hyphen.

1 Terms formed by connecting certain prefixes and suffixes to stem words. Hyphens are generally used to connect prefixes to stem words that must be capitalized. Terms with pairs of letters that might be visually confusing with the joining of a prefix or suffix often retain the hyphen in British usage but not in American. The prefixes "ex-" and "self-" generally take the hyphen. The suffixes "-elect", "-type", and "-designate" take the hyphen.

a pre-Columbian civilization   a shell-like carapace   anticodon
chairman-elect   meta-analysis [or] metaanalysis   co-ordination
coordinates   cooperate   radioisotope   ex-husband
self-inflicted wounds

Standard dictionaries should be consulted to resolve uncertainties about such hyphenation.

2 Some compound terms, especially those in which absence of the hyphen would leave 1 of the stem words with its original meaning.

light-year
[the hyphenated "light" converts "year" from being a unit of time to "light-year", a unit of length (the distance traveled by light in 1 year), a distinctly different meaning from "light year", a "year" that is not "heavy"]

cure-all   has-been
[the verb "cure" is converted to a noun; the progressive verb "has been" is converted to a noun]

3 Compound coordinate modifiers.

cost-benefit analysis

a black-white-Hispanic coalition of patients or the Stevens-Johnson syndrome

The en-dash is preferred by many authors for connecting compound coordinate modifiers; see section 4·22. The hyphen is often preferred when the coordinate characteristic does not need emphasis and the keyboard of the author does not have a key for the en-dash. Note that nonhyphenated double proper names are not hyphenated when they function as modifiers.

Bence Jones protein [from "Henry Bence Jones"]

Proper names in compound modifiers are not hyphenated.

Fisher exact test   Wilcoxon matched-pairs test

4 Compound modifiers in which the 2nd element is a past or present participle.

well-established rules of prediction   seizure-inducing drugs

the well-known physicist   all-encompassing   ill-advised

better-represented

The hyphen is omitted when the modifier is a predicate modifier and the 1st element is an adverb.

The rules are well established.
He was well known for years in his field.

This rule follows from the lack of ambiguity in such constructions as to whether an adverb such as "well" might modify a noun it precedes rather than the adjective. Adverbial elements in compound modifiers are not hyphenated when they end in "ly".

The widely applauded plan for conservation

clearly described new species

This exception to linking an adverbial element to the adjective it modifies appears to be based on the identifiability of the adverb by its "ly" ending. The adverb "well" may not be readily identified as an adverb when it is not in a predicate construction.

5 Compound modifiers when a nonhyphenated form could have ambiguous meaning.

low-frequency amplitudes   a large-bowel obstruction

Two or more modifiers of the same adjectival noun retain the hyphen.

Low- and high-frequency amplitudes were the same.

sodium- and potassium-conserving drugs

6 Modifiers with numeric values and units.

a 5-g dose   the 50-km circumference   a 10-woman team

But age terms take a double hyphen.

a 3-year-old child   a 50-year-old patient

7 A spelled-out fraction (a hyphen between the numerator and denominator) unless either component is already hyphenated.

one-third of the population   thirty-two hundredths of an inch

8 Compound cardinal and ordinal numbers from 21 through 99 when spelled out.

He repeated the experiment ninety-nine times.

He pointed to the Sixty-Sixth Congress.

9 Verbs needing hyphens for correct meaning.

He re-covered the explored well.
[but "He recovered quickly from the operation."]

Such patients are usually re-treated.
[but "Washington retreated to Valley Forge."]

Some authorities recommend rewriting sentences with such formations.

Unnecessary Uses

4·32 1 Well-established compound terms without a hyphen. Standard scientific dictionaries are reliable guides.

freezing point determination   amino acid residues

potassium chloride absorption

2 Latin phrases used adjectivally.

a post hoc hypothesis   in vitro testing
a quid pro quo arrangement

3 Letters used as modifiers in scientific terms; hyphens are used when the terms are adjectival.

LE cells [but "LE-cell rosettes"]

T lymphocytes [but "T-cell lymphocyte functions"]

Specialized Uses

4·33 1 Representing single bonds in chemical or molecular formulas or names; for double bonds represented by a double line see section 16·7.

$(CH_3)_2$-CH-CH_2-CH(NH_2)-COOH

2 Between element symbols (for example, C for "carbon") and the numeral designating a particular atom; also between the element name and the mass number of an isotope in spelled-out form.

C-3 iodine-131

3 Between amino acid symbols in amino acid residues of known sequence; each hyphen represents a peptide bond; see section 16·18.

Gly-Lys-Ala-His

4 Between a prefix that specifies molecular configuration, or that serves as a locant, and the name of the chemical compound; see section 16·12.

S-benzyl-*N*-phthaloylcysteine

9-(1,3-dihydroxy-2-propoxymethyl)guanine

5 Linkage of nucleotides in polynucleotides; see section 16·39.

pG-A-C-C-T-T-A-G-C-A-A-T-Gp

SLASH (OBLIQUE BAR, OBLIQUE MARK, OBLIQUE STROKE, SLANT LINE, VIRGULE) (/)

4·34 The main use of the slash is as a symbol for the mathematical operation of division; the symbol equals the phrase "divided by". Unfortunately, the slash has come into general use from legal and business sources as a lazy substitute for the comma or hyphen or full expression. A series should generally be punctuated with commas (see section 4·15); coordinate modifiers, with hyphens or en-dashes (see sections 4·22 and 4·31).

The route of the geologic tour was New York, Pittsburgh, Chicago, and Salt Lake City.
[not "The route of the geologic tour was New York/Pittsburgh/Chicago/Salt Lake City."]

The medical school finally opened the new Hematology–Oncology Unit.
[not "The medical school finally opened the new Hematology/Oncology Unit"]

should be accepted by the patient, but if he or she objects to . . .
[not "should be accepted by the patient, but if he/she objects to . . ."]

In scientific writing the slash should be reserved for mathematical expressions and its few other specialized uses.

Specialists in typography (Bringhurst 1992) distinguish between the solidus (shilling mark) and the slash (virgule). The solidus (⁄) has a greater slant than the virgule (/) and has been conventionally used to separate units of traditional British currency (pounds, shillings, and pence) and the numerator and denominator of fractions. The virgule has tended to displace the solidus in representation of fractions on the line.

Specialized Uses

4·35 1 A mathematical symbol for division as an operation or implied by a unit; see sections 11·9, 11·11, and 11·21 and Table 11·4.

$1/4$ $y = 3.5x/(a + b)$
$kg{\cdot}m/s^2$ [*for* kilogram-meter per second squared]

Care must be taken to use no more than a single slash in an expression because of the resulting mathematical ambiguity.

$1.5\ pCi{\cdot}km^{-2}{\cdot}yr^{-1}$ [or] $1.5\ pCi/(km^2{\cdot}yr)$ [not "$1.5\ pCi/km^2/yr$"]

2 In expressions of rate or concentration.

5 m/s   20 mol/L

3 To separate symbols for mutant genes on homologous chromosomes of the fruit fly; see section 20·28.

y w f/B

4 To separate cell lines in describing mosaics or chimeras; see Tables 20·2 and 20·3.

mos45,X/46,XY

5 To separate genotype designations; see Table 20·7.

*ADA*1/*2*   *G6PD*A/G 6PD*B*   *CHE1*A/CHE1*Q0*

6 In end tags of the Standard Generalized Markup Language (SGML); see section 32·9.

VERTICAL BAR (VERTICAL LINE) (|)

4·36 The vertical bar has specialized uses in chemical and electrochemical notation and in mathematics. For example, it is used to indicate loosely associated ion pairs.

$X^+||Y^-$

APOSTROPHE (')

4·37 The apostrophe is used to form possessives and, occasionally, to form plurals. For information on possessives, see sections 5·33–5·35, 5·37, and 5·39–5·41; on plurals, 5·29 and 5·30. The apostrophe indicates contractions in informal writing ("I'd prefer" for "I would prefer").

PRIME SIGN (′)

4·38 The prime sign must not be confused with the apostrophe. If it must be represented in typescript by the apostrophe, the author should indicate that use with a marginal note.

Scientific Uses

1 With locants in chemical names.

N,N′-dimethylurea   *ω,ω′*-dibromopolybutadiene

2 To indicate "minute" and, doubled, "second" in geographic coordinates; see section 13·13.

The meteorite was found at latitude 52°33′05″ N, longitude 13°21′10″ E.

DIACRITICAL MARKS

4·39 Some words of non-English origin retain diacritical marks in English usage, others have dropped diacritical marks but retain the original pronunciation, while others have shifted in spelling and dropped diacritical marks to reflect their original pronunciation.

garçon   résumé   facade [from "façade"]   canyon [from "cañon"]

Standard English-language dictionaries indicate current standard usage. See Table 3·2 for the most widely used diacritical marks in European languages. Diacritical marks should be retained in personal names and place names if they have not been anglicized.

Authors using diacritical marks not available on their typewriters or word-processing programs must produce the marks by hand and should indicate in the typescript's margin the name or names of marks used by writing the mark's name and encircling it.

ASTERISK (*)

4·40 The asterisk has been used to indicate ellipsis (Partridge 1953; Swanson 1979), but that use is discouraged (UCP 1993); ellipsis points are now the standard device for this function. The asterisk does still serve as a footnote sign (see section 3·7), but this use is discouraged (see section 31·13).

A scientific use of the asterisk is in symbolization of genes; see Tables 20·5 and 20·7.

AMPERSAND (&)

4·41 As noted by Partridge (1953), "The ampersand (a slurring of 'and per se and')[makes] a convenient symbol for 'and'". The symbol was formed by a combination of the letters "e" and "t" of the Latin *et,* meaning "and". It is still occasionally used in commercial names, but it should not be used in titles and text of scientific writing. It should be retained when it is part of a proper name, such as that of a publisher or law firm.

A specialized use is designating entity references for special characters in Standard Generalized Markup Language (SGML); see section 32·8 and Table 32·3.

"AT" SYMBOL (@)

4·42 The symbol "@" for "at" has been widely used in commercial documents in such phrases as "100 copies @ $47.50".

Some scientific uses for "@" have recently been developed. It has been introduced to link the symbol for a chemical element with that of a fullerene to indicate that the fullerene molecule contains an atom of the indicated element (for example, $He@C_{60}$); see section 16·8. A 2nd use is placing it after a gene symbol to indicate reference to a family of genes; see Table 20·7.

OCTOTHORP (NUMERAL SIGN, POUND SIGN) (#)

4·43 The octothorp ("8 fields") has been used in cartography as a symbol for "village" and in the avoirdupois system of weights as a symbol for "pound". These uses are archaic for scientific publishing, but this sign is still used by proofreaders and copy editors to indicate "space" and by printers to specify weight per 1000 sheets of paper stock.

MARKS FOR LINE RELATIONS

BRACE

4·44 A multiline brace can be used as a fence for aggregation in mathematics.

$$\begin{cases} x = a + b \\ y = c + d \\ a = e + f \end{cases}$$

A 2nd use is to cluster text elements of a like kind, but such relations are usually more readily and efficiently shown by a tabular list or a text statement.

Typical Differences in Spelling

$$\text{American}\begin{cases} \text{color} \\ \text{leukemia} \\ \text{theater} \end{cases} \qquad \text{British}\begin{cases} \text{colour} \\ \text{leukaemia} \\ \text{theatre} \end{cases}$$

Typical differences in spelling are represented by the American spellings *color, leukemia, theater* and the British *colour, leukaemia,* and *theatre.*

Typical Differences in Spelling

American	color, leukemia, theater
British	colour, leukaemia, theatre

DITTO MARKS (")

4·45 Ditto marks have been widely used in lists and tables to indicate that at the mark's location the word or term directly above in the preceding line should be understood to also apply to the location of the mark. This device may be seen by typists and other keyboarders as saving time and effort in not having to repeat the dittoed text, but to avoid esthetically unattractive and ragged-looking tables and to support exact statement it should not be used in scientific writing and publishing.

DOTS (LEADERS) (.....)

4·46 A line of closely spaced period marks is sometimes used as a formatting device to lead the reader's eye from a text element at or near the lefthand margin to a text element at or near the righthand margin, as in a table of contents.

Punctuation Marks 12
  Period 13
  Colon 14
Spelling 19
  American 20
  British 23

A decision on whether to use this device is generally best left to the designer of the book or journal.

Short dotted lines should not be used in tables to indicate the absence of an entry for a cell because of the resulting ambiguity between no observation for that cell and a datum of no importance; see section 31·12.

See section 9·14 for the use of spaced dots (. . .) to indicate an omission (ellipsis) in a quotation.

PARAGRAPH MARK (PILCROW)(¶)

4·47 This mark is still used as a symbol in legal literature (HLRA 1992) to mark the opening of a paragraph or to refer to a particular paragraph. It also serves as a footnote sign, but this use is discouraged (see section 31·13). It is used as a proofreader's mark to indicate the beginning of a paragraph.

SECTION MARK (§)

This mark, a scribal form of double "s", is still used in legal literature (HLRA 1992) as an abbreviation or symbol for "section". The plural form, for "sections", is §§.

PUNCTUATION MARKS IN NON-ENGLISH LANGUAGES

4·48 Some punctuation marks not used in English may have to be preserved within parts of scientific texts in English such as block quotations from non-English texts. Examples are the guillemets (chevrons) used in French to mark quotations and the inverted question mark that opens an interrogative statement in Spanish.

> N'oubliez pas ces mots: «Guérir quelquefois, soulager souvent, consoler toujours».

¿Quién o qué estaba «interpretando»?

Details on punctuation signs in non-English languages and their uses can be found in *The Chicago Manual of Style* (UCP 1993).

REFERENCES

Cited References

Allen JS. 1992. Educating performers. The Key Reporter; Spring:5–9.

Bringhurst R. 1992. The elements of typographic style. Vancouver (BC): Hartley & Marks.

Dodd JS. 1986. The ACS style guide: a manual for authors and editors. Washington: American Chemical Soc.

[DSSC] The Department of the Secretary of State. 1985. The Canadian style: a guide to writing and editing. Toronto: Dundurn Pr.

Fowler HW, Fowler FG. 1931. The King's English. 3rd ed. Oxford: Oxford Univ Pr.

[HLRA] Harvard Law Review Association. 1992. A uniform system of citation. 15th ed. Cambridge (MA): HLRA.

[ISO] International Organization for Standardization. 1982. General principles concerning quantities, units, and symbols : ISO 31/0-1981 (E). In: Units of measurement: ISO standards handbook 2. 2nd ed. Geneva: ISO.

Lanham RA. 1991. A handlist of rhetorical terms: a guide for students of English literature. 2nd ed. Berkeley (CA): Univ California Pr.

McArthur T, editor. 1992. The Oxford companion to the English language. Oxford: Oxford Univ Pr.

McCoubrey AO. 1991. Guide for the use of the International System of Units: the modernized metric system. National Institute of Standards and Technology Special Publication 811. Washington: US Government Printing Office.

[NISO] National Information Standards Organization. 1991. Electronic manuscript preparation and markup: American national standard for electronic manuscript preparation and markup ANSI/NISO Z39.59-1988. New Brunswick (NJ): Transaction Publishers. Now available from NISO Press, Bethesda, MD; see the section "Standards for Editing and Publishing" in Appendix 3.

Parkes MB. 1993. Pause and effect: punctuation in the West. Berkeley (CA): Univ California Pr.

Partridge E. 1953. You have a point there: a guide to punctuation and its allies. London: Routledge & Kegan Paul.

Swanson E. 1979. Mathematics into type: copy editing and proofreading of mathematics for editorial assistants and authors. Rev ed. Providence (RI): American Mathematical Soc.

[UCP] University of Chicago Press. 1993. The Chicago manual of style. 14th ed. Chicago: UCP.

[USGPO] US Government Printing Office. 1984. United States Government Printing Office style manual 1984. Washington: USGPO.

Additional References

[AGPS] Australian Government Publishing Service. 1988. Style manual for authors, editors and printers. 4th ed. Canberra: AGPS Pr.

Gower E. 1988. The complete plain words. Revised by S Greenbaum and J Whitcut. Boston: DR Godine.

Lauther H. 1991. Lauther's complete punctuation thesaurus of the English language. Boston: Branden.

Lefevre PG. 1977. Using hyphens. Nature 269:556.

Morse JM. 1985. Webster's standard American style manual. Springfield (MA): Merriam-Webster.

[OUP] Oxford University Press. 1983. Hart's rules for compositors and readers at The University Press Oxford. 39th ed. Oxford: OUP.

[OUP] Oxford University Press. 1991. The Oxford dictionary for scientific writers and editors. Oxford: OUP.

Tichy HJ. 1988. Effective writing for engineers, managers, scientists. 2nd ed. New York: Wiley Interscience.

van Loon AJ. 1990. Correct English is not enough: a plea for more hyphenation. European Science Editing (39):5–6.

Wilkinson AM. 1991. The scientist's handbook for writing papers and dissertations. Englewood Cliffs (NJ): Prentice Hall.

5 Spelling, Word Formation and Division, Plurals, and Possessives

---

Take care that you never spell a word wrong. Always before you write a word consider how it is spelt, and if you do not remember it, turn to a dictionary. —Thomas Jefferson, *To Martha Jefferson,* 1783

. . . the change [in the United States] from *-our* to *-or* in words of the *honor* class was a mere echo of an earlier English uncertainty. In the first three folios of Shakespeare, 1623, 1632 and 1663–6, *honor* and *honour* were used indiscriminately and in almost equal proportions; English spelling . . . was then still fluid, and the *-our* form was not used consistently until the Fourth Folio of 1685. —H L Mencken, *The American Language,* 1957

In English the structure of words and their combinations in terms has changed through time, and the governing rules tend simply to codify current practice. Although scientific English is almost uniform throughout the world, there are differences in some of its details. The 2 main varieties of English can be loosely designated as American English and British English, but even within these 2 the current practices are not uniform.

SPELLING: AMERICAN AND BRITISH DIFFERENCES

5·1 The spelling of words in English is not governed by any national or international authority, as spelling in French is specified by the Académie française. Hence many words in English are spelled in variant forms that can be characterized as American or British preferences. British forms tend to be preferred in the Commonwealth countries, notably Australia (AGPS 1988), Canada (DSSC 1985), India, and New Zealand. Some of the preferences can be rationalized; others cannot and simply represent national preferences stemming from historical influences and cultural trends. In general the British forms tend to reflect the continuing influence of French (stemming from the Norman Conquest) and of education stressing Latin and Greek, while American forms tend to reflect trends in clearer representation of pronunciations of individual words and simplified spelling. Both groups of variant forms are usually represented in the major dictionaries of English published in the United States and the Commonwealth countries. An extensive summary of differences can be found in the entry "American English and British English" of *The Oxford Companion to the English Language* (McArthur 1992).

Choices for American or British spellings should be governed by what forms an expected majority of readers of a publication would prefer, and authors can be reasonably expected to try to follow the preferences selected by the publication when that policy is made clear to them. Journals can specify whether an American, Australian, British, or Canadian dictionary should be the standard for spelling in papers to be submitted. They can help authors adhere to the expected spellings by recommending that papers be prepared with a word-processing program that includes a spelling checker adhering to a national preference.

The generally preferred variant forms are described below with examples.

NOUNS ENDING IN "CTION" OR "XION"

5·2 Some nouns derived from verbs ending in "ect" may, in British usage, have "x" rather than the "ct" in American usage.

[American]	[British]
connection	connexion
deflection	deflexion

NOUNS ENDING IN "ER" OR "RE"

5·3 The American "er" ending reflects a preference for a phonetically based form; the British forms reflect French derivations. Note that the "er" form is used in both American and British English for verbs and for some terms for personal agents.

[American]	[British]
center	centre
fiber	fibre
liter	litre
maneuver	manoeuvre
meter	metre [unit of measure]
	meter [instrument]
theater	theatre

["manager", "mediocre", "timbre" (for tone) in both]

NOUNS ENDING IN "OR" OR "OUR"

5·4 The American preference for abstract nouns in this group is the "or" ending; the British "our" reflects French origins (for example, "honour" from *honneur*). The "or" ending is preferred in both systems for some terms representing personal agents and for scientific and medical nouns.

[American]	[British]
Abstract Nouns	
behavior	behaviour
color	colour
humor	humour

[but note "horror", "stupor", and "terror" in both]

Nouns for Personal Agents

director	director
monitor	monitor

[but note nouns such as "interpreter", "manager" in both]

Scientific and Medical Nouns

pallor	pallor
tremor	tremor

For some derivatives from such nouns, British and American preference is for the "or" form.

coloration honorary laborious valorous

NOUNS ENDING IN "ENSE" OR "ENCE"

5·5 The British preference is for the "ence" form; both forms are used in the United States, but there is, in general, a preference for the "ense" form.

[American]	[British]
defense, defence	defence [but "defensible" in both]
license, licence	licence [but "license" as a verb]
offense	offence [but "offensive" in both]
practice	practice [but "practise" as a verb]

[but note the adjectives "immense" and "intense" in both]

NOUNS ENDING IN "LOG" OR "LOGUE"

5·6 The American preference for the "log" form, especially in scientific and technical usage, stems from the trend to simpler and more phonetic spelling. The British "logue" reflects French origins.

[American]	[British]
analogue [general use]	analogue
analog [scientific and technical use]	
catalog, catalogue	catalogue
dialogue, dialog	dialogue

[but note the forms "ideologue" and "secretagogue" in both]

NOUNS WITH DIGRAPHS "AE" AND "OE"

5·7 British usage retains the digraphs in nouns (and derivatives) of Latin or Greek origin.

[American]	[British]
anesthesia	anaesthesia
cesium	caesium
diarrhea	diarrhoea
edema	oedema
encyclopedia	encyclopaedia
esophagus	oesophagus
estrogen	oestrogen
etiology	aetiology
fetus	foetus
hematology	haematology
leukemia	leukaemia

"Digraph" should not be confused with "diphthong". A digraph is a pair of letters that represent 1 sound; a diphthong (in Greek, "a double sound") is a vowel that shifts its quality as pronounced; an example is "oy" in "toy" (McArthur 1992).

NOUNS REFLECTING AMERICAN PREFERENCE FOR SIMPLER FORMS

5·8 Many nouns in American usage reflect preferences for simpler and phonetic spellings.

[American]	[British]
aluminum	aluminium [reflects the chemist Humphrey Davy's original "alumium"]
artifact [and "artificial"]	artefact
check	cheque
draft	draught
mold	mould
program	programme
sulfur	sulphur
through, thru	through

WORDS ENDING IN A SILENT "E"

5·9 American derivatives formed from verbs ending in "e" generally drop the silent "e" when a suffix is added.

[Root verb]	[American]	[British]
acknowledge	acknowledgment	acknowledgement
age	aging	ageing
judge	judgment	judgement
like	likable	likeable

VERBS ENDING IN "IZE" AND "YZE" OR "ISE" AND "YSE" AND THEIR NOUN DERIVATIVES

5·10 The reasons for the national preferences are not clear for some choices. British usage prefers "ise" and "yse" endings for verbs derived from "lysis" such as "analyse" ("analysis") and "catalyse" ("catalysis"). British usage tends to accept the "ize" ending for more recently formed verbs like "transistorize". Both usages prefer "ise" in some other formations, even in words pronounced with a "z" (British "zed"), rather than an "s" sound. Consult a standard dictionary to verify a national preference for a verb and its noun derivative. Note that some of the British forms below are given as 2nd choices in some British dictionaries.

[American]	[British]
Different Forms	
analyze (analysis)	analyse (analysis)
catalyze (catalysis)	catalyse (catalysis)
civilize (civilization)	civilise (civilisation)
organize (organization)	organise (organisation)
rationalize (rationalization)	rationalise (rationalisation)
Same Forms	
advertise (advertisement)	advertise (advertisement)
advise	advise
compromise	compromise
transistorize	transistorize

Because of the uncertainties on preferred usage likely to arise with unfamiliar verbs and the frequent absence of clear preferences, a publishing house or journal office should develop its own list of preferences to reduce the need for consulting dictionaries too frequently.

VERBS ENDING IN "L" OR "LL"

5·11 The terminal letter "l" may be single or double in verbs ending with a stressed syllable in both American and British usage, although the British tendency is toward the single "l". The "l" is doubled in both usages when a suffix ("able", "ant", "ed", "ing") is added.

[American]	[British]
compel, compelling	compel
distill	distil
enroll	enrol
fulfill	fulfil
propel, propellant	propel

Derivatives formed by adding "ment" retain the double "l" in American usage, but not in British usage.

forestall, forestallment	forestall, forestalment
install, installment	install, instalment

In British usage when endings such as "ed" and "ing" are added to form participial forms, the "l" is doubled.

fulfil → fulfilled   propel → propelled

WORDS DERIVED FROM NON-ENGLISH WORDS AND ROOTS

5·12 Words transliterated or derived from other languages can have variant spellings. In general, the variant that is phonetically unambiguous should be preferred.

[American]	[British]
leukemia	leukaemia

[preferred to "leucemia" and "leucaemia"]

When the spelling of a word in English is identical with that in the language of origin except for 1 or more diacritical marks and the word is in standard English dictionaries, the mark or marks should generally be omitted. With anglicization the pronunciation is likely to change.

[English]	[French]
brassiere	brassière
emotion	émotion

In a foreign-language phrase used by anglophones but pronounced as in the language of origin and not widely used as an English phrase, it is customary to retain the original spelling, including diacritical marks.

aide-mémoire   chargé d'affaires   pièce de résistance

VERBS BEGINNING WITH "IN" OR "EN"

5·13 Some verbs and their noun derivatives beginning with "en" or "in" are near-homophones; unless they are pronounced carefully, they may be misspelled. But some of these words are spelled with either opening syllable, and the resulting words may have different meanings.

enclose, inclose   endorse, indorse   endue, indue
ensure, insure [differing meanings in American English]

DOUBLED CONSONANTS BEFORE SUFFIXES

5·14 American usage has tended toward not doubling terminal consonants in the formation of derivative words ending in "ed", "er", "est", and "ing", but current practices are generally similar to those in British usage. A good concise description of the generally accepted British rules can be found in *Hart's Rules for Compositors and Readers at the University Press Oxford* (OUP 1983).

Doubled Consonants

A single terminal consonant is doubled when it follows a single vowel of a monosyllable and the suffix begins with a vowel.

sag, sagging   rot, rotted   dig, digging

A single terminal consonant is doubled when it follows a final accented syllable, or unaccented syllable with a short vowel sound, of a multisyllable word and the suffix begins with a vowel.

control, controlling   forget, forgetting   occur, occurring
format, formatting   input, inputting

Not-Doubled Consonants

If the terminal accent of the stem moves forward in the derived form, the consonant is not doubled.

infer, inference   refer, reference

When the final syllable is not accented in the stem, the terminal consonant is not doubled.

travel, traveled   cancel, canceled

A terminal consonant "h", "w", "x", or "z" is not doubled.

watch, watched   stew, stewed   flaw, flawed   sex, sexed

WORD FORMATION

5·15 The rapid growth of knowledge in the sciences has necessarily generated many new terms and words: some single words coined from Latin or Greek roots, but more often compound terms created from existing stem words or stem words combined into single words. Some terms originally compound have shifted to a combined form, but many compound terms retain their original compound form. Editorial offices should select a preferred standard dictionary, specify it to authors, and adhere to its recommendations.

bloodstream [but] blood vessel   database [but] data flow
headphone [but] head shield

Some scientific fields have specific rules for vernacular terms; for recommendations in the plant sciences see section 23·18, in the animal sciences see section 24·4. Concise summaries of general usage can be found in *Webster's Standard American Style Manual* (Morse 1985) and *United States Government Printing Office Style Manual 1984* (USGPO 1984). Good Canadian sources are *Editing Canadian English* (FEAC 1987) and *The Canadian Style: A Guide to Writing and Editing* (DSSC 1985).

The formation of new words in English by adding prefixes or suffixes to stem words is not governed by firm and unequivocal rules, and usage tends to shift in time.

FORMATIONS WITH PREFIXES

Nonhyphenated Prefixes

5·16 Many scientific terms are formed by the nonhyphenated addition of prefixes to stem words. In general, these are prefixes that do not stand alone as words; some are derivatives in an adjectival form. In their formations they indicate action, character, location, number, state, or time.

[Prefix]	[Example]	[Prefix]	[Example]
aero	aerostatics	inter	interface
after	aftershock	iso	isohexane
ante	antepartum	macro	macroflora
anti	anticodon	meta	metaanalysis
astro	astrophysics	micro	microfossil
auto	autoimmunity	mid	midbrain
bi	bivalve	milli	millisecond
bio	biomechanics	mini	minicomputer
chemo	chemotherapy	multi	multiprocessor
co	coenzyme	non	nonconductor
counter	counterimmunoelectrophoresis	over	overtone
de	denitrification	para	paramyxovirus
di	diketone	photo	photochemistry
electro	electrosurgery	physio	physiotherapy
exo	exopathogen	phyto	phytopathology
extra	extrasystole	poly	polyarthritis
geo	geochemistry	post	postprecipitation
hemi	hemianesthesia	pre	prediabetes
hemo	hemodialysis	pro	proinsulin
hyper	hyperventilation	pseudo	pseudogout
hypo	hypomenorrhea	re	recombination
in	incoordination	semi	semischist
infra	infrared	stereo	stereochemistry

sub	subspecies	tri	tribromoethanol
super	supersaturation	ultra	ultracentrifuge
supra	suprascapula	un	unconformity
trans	transpolarizer	under	undernutrition

When there is uncertainty about whether to hyphenate new terms formed with this pattern, the bias should be for the closed-up form, not the hyphenated form. Some computer programs for searching text may not recognize the hyphenated form as a term and thus fail to find it. Note, however, that readers may object to the doubled vowels that can result from the general rule to not hyphenate, as in "semiindependent" and "metaanalysis". Such formations often become, in time, acceptable to the eye, as in "coordination", now generally not hyphenated in American usage.

Authors can be confused about homonymic or near-homonymic prefixes; an editorial office may find it useful to maintain a list of such errors.

ante, anti	for, fore
antediluvian [not "antidiluvian"]	forward [not "foreward"]
antifreeze [not "antefreeze"]	foreword [not "forword"]

Hyphenated Prefixes

5·17 Prefixes must be joined to the stem word with a hyphen in some formations.

1 When the stem is capitalized.

pre-Columbian civilization   post-Copernican astronomy
sub-Saharan Africa

2 When omitting the hyphen changes the meaning of the formation.

Re-cover the flask after adding the reagent.
You may not recover enough precipitate.

For recommendations on hyphenation in compound nouns and adjectives, see section 4·31.

FORMATIONS WITH SUFFIXES

Suffixes That Are Not Words

5·18 Many nouns and adjectives are formed by adding suffixes that are in general not complete words; these formations are not hyphenated. Note that some of these suffixes are similar, and the words they form may readily be misspelled.

able	ible		eous	ous
ance	ence		erous	orous
ant	ent		ful	full
ative	ive		ified	yfied
cede	ceed	sede	efy	ify

Most misspellings that arise from confusion about correct suffixes will be caught by spelling checkers of word-processing programs, but an editorial office may find it valuable to have its own compilation of frequently confused suffixes in words heavily used in its field.

Words as Suffixes

5·19 Many nouns have been formed by adding to a stem verb or noun a suffix that is itself a complete word. Which are hyphenated is mainly a usage of habit, but in such formations, as with prefix formations, the hyphen is now often omitted; some exceptions must be noted.

[Suffix]	[Formation]
away	runaway reactor
down	breakdown of the varieties of species
in	cave-in at the excavation site [the hyphen is generally desirable]
like	wormlike lower vertebrate [but the hyphen is preferred when the stem word ends in a single or double "l" or another ascender, as in "a mold-like excrescence", "a shell-like carapace"]
off	heavy runoff causing a flood
out	turnout greater than expected
over	turnover in the faculty
up	breakup of the USSR

Note that the form of the word may depend on its use as a noun, verb, or adjective.

They shut down the reactor. ["shut down", a phrasal verb]

The shut-down reactor needed extensive repairs.
["shut-down", an adjective]

The announcement of the shutdown was delayed several days.
["shutdown", a noun]

Variant Suffixes

5·20 Some suffixes have variant forms that convey the same meaning. In general, the shorter form should be preferred for American usage.

histologic [not "histological"]
microscopic [not "microscopical"]

But note that some potential variant forms are not idiomatic, for example, "chemic" is not a proper short form of "chemical". Other variant forms carry different meanings.

astronomic ["an astronomic budget deficit"]
astronomical ["an astronomical observatory"]

economic ["his economic theory"]
economical ["an economical administration"]

ethic ["His was a demanding ethic."]
ethical ["a violation of ethical principles"]

historic ["10 historic sites in the state"]
historical ["an outmoded historical theory"]

statistic ["that statistic was false"]
statistical ["a better statistical analysis"]

Adjectives formed from stem words ending with "ology" should, in general, be used only to refer to the discipline or function and not to the subjects of the discipline.

pathologic anatomy [anatomy applied to the study of the subjects of the discipline pathology]

abnormal lungs [not "pathologic lungs"]

WORD AND TERM DIVISION

5·21 For consistency in dividing words at the ends of lines to maintain satisfactory spacing within lines, apply the divisions indicated in a standard dictionary. American divisions are generally based on syllable structure; British, on etymologic elements.

Hyphenated words and terms should be broken after the hyphen so that the need for another hyphen is avoided.

cost-benefit analysis   cost-   benefit analysis

Long chemical names can be divided by syllables, by etymologic units, or by the hyphen rule, but at least 4 or more characters should appear on each line. The break should not be at a hyphen connecting a locant or descriptor prefix.

2-acetylaminofluorene 2-acetyl- aminofluorene
[not "2- acetylaminofluorene"]

The part carried over to the next line should not look like a separate word.

path- ologic [not "patho- logic"] into [not "in- to"]

Numbers should not be separated from their unit designations.

a concentration of 250 g/L [not "a concentration of 250 g/L"]

One-syllable and very short 2-syllable words should not be divided.

as also but by if to when

Word-processing programs and computer-driven composition have reduced the need for word divisions, especially in manuscripts with left-margin justification (ragged right). Note that manuscripts to be marked for electronic methods of publication should not have end-of-line word divisions indicated by hyphens, which should be used only in compound words needing hyphenation.

PLURALS

GENERAL PRINCIPLES

5·22 The plural of most common nouns is formed by adding "s" or "es" to the singular form; "es" is added to nouns that end in sibilants (soft "ch", "s", "sh", "x", or "z").

chemicals	axes	foxes
expressions	beaches	grasses
monarchs	branches	losses
organisms	buses	washes

The plural of most nouns ending in "i" is usually formed by adding "s".

alibis rabbis skis alkalis [or] alkalies

The plural of nouns ending in "o" preceded by another vowel is formed by adding "s" to the singular.

cameos duos ratios scenarios tattoos taboos

For nouns ending in "o" preceded by a consonant, there is no strict rule; some words take only "s" and others take "es"; for yet others, either form is acceptable.

echoes halos, haloes infernos mosquitos, mosquitoes
mottos, mottoes potatoes tomatoes zeros, zeroes

The plural of nouns ending in "y" preceded by a consonant or a consonant sound is formed by changing the "y" to "i" and adding "es".

berry, berries city, cities colloquy, colloquies
duty, duties equity, equities fly, flies

PROPER NOUNS

5·23 The plurals of proper nouns are formed in generally the same manner as those of common nouns.

the Charnyshes the Joneses the Churchills the Roosevelts

It should be noted, however, that the spelling of the name must not change; for example, most proper nouns ending in "y" retain the "y" in the plural form.

before their unification, both Germanys [not "Germanies"]
the Perrys [not "the Perries"]
3 Marys [not "3 Maries"]

Exceptions can be found in shortened forms of geographic terms.

the Alleghenies the Rockies the Sicilies

If adding "s" or "es" to a proper noun would result in a false pronunciation, the plural is the same as the singular; this situation arises mainly with French names that end in an unpronounced "s", "z", or "x"; in the plural form the "s" is pronounced:

the 16 King Louis of France [not "Louises"]
the 3 Gervais [not "Gervaises"]

Furthermore, if adding "s" or "es" would result in an awkward formation, in particular in polysyllabic Spanish names ending in sibilants, it is probably easier to recast the sentence to avoid the plural form.

2 Velasquezes [prefer "2 paintings by Velasquez"]

IRREGULAR FORMS

5·24 Many nouns have irregular plural forms, notably English nouns of pre-Norman Conquest origin. There are no formal rules for these plurals.

child, children   ox, oxen   die, dice   mouse, mice

For some single-syllable nouns the plural is formed by changing vowels within the word.

foot, feet   goose, geese   man, men   tooth, teeth

[Note that "mongoose" ends in "goose" but is not a compound word; its plural is "mongooses".]

The plurals of compound words that end in these words are formed in the same way.

clubfoot, clubfeet   dormouse, dormice   workman, workmen

The plurals of words ending in "man" that are not compounds are formed simply by adding "s".

human, humans   German, Germans   Norman, Normans

Although the plurals of most nouns that end in "f", "ff", or "fe" are formed in the regular way, a few change the "f", "ff", or "fe" to "ves" in the plural.

calf, calves   dwarf, dwarves[a]   half, halves

hoof, hooves[a]   leaf, leaves   life, lives   loaf, loaves

self, selves   sheaf, sheaves   staff, staffs, staves[a]

thief, thieves   wife, wives   wolf, wolves

[[a] See section 5·28 on nouns with more than 1 plural form; note that some plural forms can have different meanings.]

NOUNS WITH A SINGLE FORM

5·25 Some nouns have only 1 form, which can indicate singular or plural depending on the context.

aircraft   corps   forceps   goods [meaning "material things"]

pains [meaning "effort"]   progeny   remains

series   spacecraft   species   sperm

The names of many nationalities have only 1 form, which is used to indicate both singular and plural.

Chinese   Japanese   Portuguese

Some nouns cannot represent a plural concept and therefore have no plural form.

equipment horsepower information manpower

For some nouns the sole form appears to be a plural but is construed as a singular.

heaves measles mumps rickets
shingles [the skin eruption, herpes zoster]
whereabouts works ["a factory"] news
premises [meaning "property"]

NOUNS BOTH SINGULAR AND PLURAL

5·26 For some nouns, mainly the names of animals, the singular form is used to denote both 1 and more than 1 individual.

deer moose sheep swine
fish [and individual species: trout, cod, haddock, herring]

For some of these nouns, as well as others that in normal scientific usage have a singular form (such as "oat", "wheat"), the regularly formed plural is used to indicate more than 1 species, strain, or variety.

6 experimental wheats 10 mutant oats
3 fishes of interest

For the names of many large mammals and some other organisms either the singular or the regularly formed plural may be used to indicate the plural.

antelope, antelopes bighorn, bighorns buffalo, buffalos, buffaloes
caribou, caribous crab, crabs elk, elks gazelle, gazelles
giraffe, giraffes lobster, lobsters walrus, walruses

When in doubt as to whether the singular form is acceptable as the plural, authors and editors should consult a dictionary.

PLURAL ENDINGS FROM NON-ENGLISH LANGUAGES

5·27 English scientific and medical vocabularies have many terms that are taken directly from other languages. The plurals of such words are formed by the rules of the original language. In some cases, an anglicized form that follows the general rules for plural formation is also acceptable (see also section 5·28 on nouns with more than 1 plural form).

The acceptability of the English plural forms often differs with the type of publication and its audience. Editorial offices should create their own lists of preferred plural forms.

1 The Latin ending “a” changes to “ae”.

abscissa, abscissae [also “abscissas”]
alga, algae [also “algas”] alumna, alumnae
amoeba, ameba; amoebae, amebae [also “amoebas”, “amebas”]
formula, formulae [also “formulas”] larva, larvae [also “larvas”]
medulla, medullae [also “medullas”]
mucosa, mucosae [also “mucosas”]
papilla, papillae [also “papillas”]
sequela, sequelae [also “sequelas”]
theca, thecae [also “thecas”]

2 The French endings “eu” and “eau” change to “eux” and “eaux”, respectively.

milieu, milieux [also “milieus”] rouleau, rouleaux [also “rouleaus”]

3 The Latin ending “en” changes to “ina”.

foramen, foramina [also “foramens”]
lumen, lumina [also “lumens”]
nomen, nomina rumen, rumina [also “rumens”]

4 The Latin endings “ex” and “ix” change to “ices”.

appendix, appendices [also “appendixes”, especially for closing sections of books]
calix, calices [also “calixes”]
calyx, calyces [also “calyxes”]
fornix, fornices [also “fornixes”]
index, indices [in economics and mathematics, “indices”; “indexes” for closing sections of books]
matrix, matrices [also “matrixes”] vortex, vortices [also “vortexes”]

5 The Greek or Latin ending “is” changes to “es”.

analysis, analyses basis, bases crisis, crises
dialysis, dialyses hydrolysis, hydrolyses hypothesis, hypotheses
metastasis, metastases mitosis, mitoses
parenthesis, parentheses

6 The Greek or Latin ending “itis” changes to “itides”.

arthritis, arthritides encephalitis, encephalitides
enteritis, enteritides meningitis, meningitides
vasculitis, vasculitides

7 The Italian ending "o" changes to "i".

virtuoso, virtuosi [also "virtuosos"]

8 The Greek ending "on" or "oan" changes to "a".

criterion, criteria [also "criterions"]   enteron, entera
mitochondrion, mitochondria
phenomenon, phenomena [also "phenomenons"]
protozoon, protozoa
protozoan, protozoa [also "protozoans"]

9 The Latin ending "um" changes to "a".

addendum, addenda   antiserum, antisera
bacterium, bacteria   datum, data
erratum, errata   flagellum, flagella [also "flagellums"]
inoculum, inocula   maximum, maxima [also "maximums"]
medium, media [also "mediums"]
memorandum, memoranda [also "memorandums"]
minimum, minima [also "minimums"]
ovum, ova   mycelium, mycelia   phylum, phyla
pudendum, pudenda   septum, septa   serum, sera [also "serums"]
symposium, symposia [also "symposiums"]

Note that the English form of the plural may be required for some uses. The US Geological Survey (Hansen 1990, p 163) specifies "datums" (rather than "data") for benchmarks and time markers.

The Latin ending "us" usually changes to "i".

bacillus, bacilli   bronchus, bronchi   focus, foci [also "focuses"]
fungus, fungi [also "funguses"]   locus, loci   rhonchus, rhonchi

For some words with the Latin ending "us" the plural idiomatic in scientific English is formed by adding "es".

apparatus, apparatuses   consensus, consensuses   hiatus, hiatuses
prospectus, prospectuses   sinus, sinuses   [but "genus", "genera"]

OTHER NOUNS WITH 2 PLURAL FORMS

5·28 There are some nouns that have more than 1 plural form, and sometimes the 2 forms have different meanings. The author or the editor should be sure to use the form correct for the context.

brother: brothers [males with the same parents],
brethren [members of a society]

die: dies [devices for stamping], dice [gambling pieces]

genius: geniuses [persons of extraordinary intelligence], genii [supernatural spirits]

index: indexes [alphabetic lists of topics or names; also numeric expressions], indices [numeric expressions]

os: ossa [bones], ora [mouths]

staff: staffs [groups of employees or assistants], staves [poles]

For some other nouns, the 2 forms of the plural are identical in meaning. Regular and specialized dictionaries may suggest a preferred form, may indicate that either form is acceptable, or may list only 1 of the variants. In the list that follows, the 1st plural form is the variant recommended by this manual. Authors, editors, and publishing houses should create their own lists of preferred forms. See also section 5·26.

[Singular]	[Plural]	[Singular]	[Plural]
biceps	biceps, bicepses	hoof	hooves, hoofs
femur	femora, femurs	scarf	scarves, scarfs
gladiolus	gladioli, gladioluses	thorax	thoraxes, thoraces

NUMBERS, LETTERS, AND ABBREVIATIONS

5·29 The plural of a single-digit numeral is formed by adding an apostrophe and "s".

1's [not "1s"]   2's [not "2s"]   His 3's usually looked like 8's.

The plural of a numeral higher than 9 is formed by adding only "s".

expressed in 100s [not "expressed in 100's"]

expressed in 1000s [not "1000's"]

the 1990s   the 1800s   persons in their 50s

The plural of a number to be expressed as a word is formed by following the general rules given at the beginning of the section on plurals (5·22).

ones   at sixes and sevens   counting by tens   in her teens

twenties   hundreds of specimens   thousands of species

The plural of all single letters, both capital and lowercase, is formed by adding an apostrophe and "s" for consistency and to avoid possible confusion with widely accepted abbreviations (for example, "Ps" for "photosynthesis") or true words ("as", "is").

He forgot to plot some of the x's and y's.

The plurals of the numeral 1 (1's) and the lowercase letter "l" (l's) could be confused. Authors and editors should consider recasting a sentence if the context does not make the meaning clear. The plurals of abbreviations and initialisms that do not contain periods are formed by adding only "s".

DNAs ELISAs MDs PCBs PhDs

But if the abbreviated term itself is a plural, do not add the "s".

NIH [for "National Institutes of Health", as in "The NIH were allotted far more in the 1993 budget."] But "CDC [Centers for Disease Control and Prevention] is located in Atlanta."

Although the use of periods in abbreviations should be limited as much as possible (see section 10·2), periods are sometimes needed to prevent misreading (for example, "c.o.d." for "collect on delivery" and not "cod"). In such cases or when the abbreviation without periods forms a word that has another meaning (for example, "SIN" for "social insurance number") an apostrophe and "s" should be used for the plural. The apostrophe can be omitted if a capitalized form of abbreviation is used or the context makes clear the meaning of the abbreviation

c.o.d.'s SIN's CODs SINs

The plurals of several abbreviations frequently used in references are usually formed irregularly.

line, lines: l., ll. page, pages: p., pp. chapter, chapters: c., ch.
section, sections: s., ss. manuscript, manuscripts: ms, MS; mss, MSS
"and the one following", "and those following": f., ff.

Note, however, that this manual recommends that in references the abbreviation for "page" or "pages" be simply "p", with its meaning as singular or plural determined by its position and whether the associated numeral(s) is a single numeral or a span of numbers; see section 30·37.

SI AND OTHER UNITS OF MEASUREMENT

See section 11·10 for recommendations on plurals of symbols for SI units.

WORDS AND SYMBOLS REFERRED TO AS SUCH

5·30 The plurals of words and symbols referred to as such are formed by adding an apostrophe and "s".

There are too many however's in this paragraph.

The printer has difficulty setting ±'s.

In a graph, for example, ●'s and ○'s might be used to designate points representing data from control and experimental groups, respectively.

In non-textual contexts, the plural is formed by adding "s" only.

♂s, nr [in a table heading or axis label]
♀s, nr [in a table heading or axis label]

The plurals of the symbols AND and OR used in the context of a computer language should be formed by adding "s" only.

ANDs and ORs

PLURALS OF COMPOUND WORDS AND TERMS

5·31 For compounds that are spelled as a single word, the plural is formed regularly, by adding the appropriate ending to the end of the word.

workman, workmen jumpsuit, jumpsuits spoonful, spoonfuls

Hyphenated and open compounds take the plural form of the noun that is the basis of the term.

aide-de-camp, aides-de-camp
attorney general, attorneys general
brother-in-law, brothers-in-law
chargé d'affaires, chargés d'affaires
chief of staff, chiefs of staff
coup d'état, coups d'état
court-martial, courts-martial
deputy chief of staff, deputy chiefs of staff
director general, directors general
hanger-on, hangers-on
major general, major generals
man-of-war, men-of-war
right-of-way, rights-of-way
surgeon general, surgeons general
table d'hôte, tables d'hôte

If the compound contains no nouns or if none of the nouns is significant in the context, "s" is added to the last component.

forget-me-not, forget-me-nots go-between, go-betweens
jack-in-the-pulpit, jack-in-the-pulpits

If the components of a compound term are more or less equivalent, the plurals of both are used.

woman scientist, women scientists [or] woman scientists

COMMON NAMES TAKEN FROM SPECIES NAMES

5·32 For some plants, the vernacular English name is the genus name set in roman type without an initial capital letter (also see section 23·19). For such nouns the plural is usually formed by the general rules given at the beginning of the section on plurals (section 5·22).

camellia, camellias crocus, crocuses iris, irises
rhododendron, rhododendrons
[Although the plural "gladioluses" for "gladiolus" is correct, "gladioli" is preferred.]

Some microorganisms also have a common plural designation based on the name of the genus. If a form of this kind is not in general use, add "organisms" to the italicized genus name (for example, "*Escherichia* organisms").

bacilli [for organisms in the genus *Bacillus*]
chlamydiae [for organisms in the genus *Chlamydia*]
mycobacteria [for organisms in the genus *Mycobacterium*]
pseudomonads [for organisms in the genus *Pseudomonas*]
salmonellae [for organisms in the genus *Salmonella*]
staphylococci [for organisms in the genus *Staphylococcus*]
streptococci [for organisms in the genus *Streptococcus*]
treponemes [for organisms in the genus *Treponema*]

POSSESSIVES

GENERAL PRINCIPLES

5·33 The possessive of most singular common and proper nouns and of some indefinite pronouns is formed by adding an apostrophe and "s".

the patient's condition	Canada's agreement with the United States
the wolf's territory	Pettigrew's study
the ax's blade	one's own view
the fox's den	everyone's attendance
the beach's location	someone's responsibility

Some indefinite pronouns, including "any", "few", "many", "none", and "such", do not have a possessive form.

Authors and editors who take the view that inanimate objects cannot "possess" something can avoid the possessive by recasting the sentence or phrase to an "of" phrase.

the leaf's color → the color of the leaf
the mineral's characteristics → the characteristics of the mineral
the solution's boiling point → the boiling point of the solution

SINGULAR NOUNS THAT END IN "S"

5·34 The general principle of adding an apostrophe and "s" holds for most nouns, including proper nouns, that end in "s". Pronunciation can serve as a guide: if one would pronounce the possessive "s", it should appear in the written form.

the bus's upholstery	Charles's suggestion
the grass's texture	Paris's allure
the moss's means of reproduction	Dickens's works of fiction
Williams's work on this topic	Professor Harris's viewpoint
the lens's properties	Dr Jones's interpretation

If in a particular case the double sibilant sounds awkward, the sentence should be recast to avoid the possessive form altogether.

the texture of the grass   the sisters' singing → the singing of the sisters

The possessive forms of Greek and hellenized names of more than 1 syllable ending in "s" (which often have an unaccented ending pronounced "eez"), as well as those of "Jesus" and "Moses", are formed by adding an apostrophe only.

Archimedes' screw	Ulysses' adventures
Euripides' plays	Hercules' labors
Hippocrates' teachings	Ramses' tomb
Jesus' time	Moses' followers

Achilles' heel [an expression meaning "weakness", not an anatomic part]

If the final "s" (or "x" or "z") is silent, an apostrophe and "s" must be used to yield the correct pronunciation.

Agassiz's theories of glaciation	Arkansas's geography
Descartes's essays	Lemieux's reasons for proceeding

PROPER NAMES SET IN ITALIC TYPE

5·35 In the possessive form of names of books and journals, the name is set in italic type and the apostrophe and the "s" in roman type.

The Lancet's reputation   *Surgery*'s 20th volume

POSSESSIVE PRONOUNS

5·36 Possessive pronouns are formed by adding only "s" to the pronoun, without an apostrophe.

His [a modification of "he" plus "s"] is the most visible display.

Here is mine; have you found hers yet?

This laboratory is ours, theirs is across the hall, and yours is in the other building.

I collected the specimen; its measurements have already been recorded. [The possessive pronoun "its" must not be confused with the contraction "it's" for "it is".]

PLURAL NOUNS

5·37 The possessive form of common and proper plural nouns that end in "s" is formed by adding only an apostrophe.

the patients' histories	the lenses' characteristics
the animals' behavior	the Harrises' family tree
the doctors' privileges	the workers' contract

the United States' relationship with its allies [a plural construction for the possessive is used, although "United States" is usually construed as singular]

The possessive of plural nouns that do not end in "s" is formed by adding an apostrophe and "s".

the bacteria's growth patterns   the data's complexity
the results of the men's pulmonary capacity tests
the children's eating habits

When the noun ends in a sibilant sound, the resulting phrase may sound awkward; in such cases, the sentence can be recast to avoid the possessive form.

the mice's nesting material → the nesting material used by the mice

the geese's migratory formation → the migratory formation of the geese

EPONYMIC TERMS

5·38 The scientific and medical vocabularies contain many compound terms that incorporate a proper name referring to a theoretician, a researcher, a physician, a patient, or a place. Such terms refer to a wide variety of laws, theories, methods, anatomic parts, conditions, diseases, reagents, syndromes, and tests, among other entities. For many of the terms that incorporate the name of a researcher the practice has long been to use the possessive form (Avogadro's number, Down's syndrome, Hodgkin's disease, Wilms's tumour, Bareggi's reaction); for names that represent a place or a patient the possessive form has not generally been used (Minamata disease, Chicago disease, Hageman factor). It is recommended that the possessive form be eliminated altogether from eponymic terms so that they can be clearly differentiated from true possessives. Some comprehensive reference works recognize the trend toward eliminating the possessive form but have found it unfeasible to eliminate all possessive forms at once. Therefore, possessive forms may continue to appear even though the nonpossessive form is preferred, and variation in form does not necessarily indicate a variation in preferred usage. For hyphenated eponymous terms that incorporate the names of more than 1 person, the possessive is never used.

McCune-Albright syndrome
Miller-Abbott tube
Stanford-Binet test

For further discussion of eponyms for diseases and syndromes, see sections 8·18 and 24·30.

COMPOUND EXPRESSIONS

5·39 The possessive of a compound expression is formed by adding an apostrophe and "s" to the final element.

someone else's proposal
anyone else's questions
everybody else's preferences
the editor-in-chief's guidelines
the attorney general's assistant
the mother-in-law's request
the surgeon general's warning
the go-between's instructions
the workmen's safety procedures
the deputy chief of staff's response

If the expression is more than a few words long, there is a risk of making understanding difficult, as in "the speaker who gave the workshop's hotel arrangements", where it might appear that the hotel arrangements were arranged by the workshop and not by the speaker. In such cases, rewording is highly recommended for clarity.

For nouns in a series, the form of the possessive is determined by whether joint or individual ownership is intended; for joint ownership an apostrophe and "s" are added only to the second name; for individual ownership an apostrophe and "s" are added to both names.

Watson and Crick's landmark paper   Watson's and Crick's memoirs
the student and her tutor's appointment
the student's and her tutor's phone numbers

When 1 of the nouns is a possessive pronoun, the others take the possessive form too.

Dr Denmore's and your grant proposal
my student's and my assessment of the data

ORGANIZATION NAMES

5·40 The possessive of an organization's name is formed by adding an apostrophe and "s" to the last element of the name (or an apostrophe only if the last element is a plural that ends in "s").

the American Psychological Association's author guidelines
the Council of Biology Editors' annual meeting

The names of many organizations and institutions incorporate a possessive or a plural; for example, the possessive form may be used to indicate "for", or the plural may be used as an adjective. Correct usage is determined by the official name of the organization itself, so these names should be checked carefully.

American Medical Writers Association [plural as adjective]
Freelance Editors' Association of Canada [possessive]

The apostrophe must be retained in names that include the possessive of a plural noun that does not end in "s"; otherwise a nonsense word is formed.

Women's Christian Temperance League
the Children's Hospital of Eastern Ontario

TRADITIONAL PHRASES

5·41 For reasons of tradition and euphony, several phrases that incorporate a possessive form and the word "sake" take only an apostrophe.

for appearance' sake   for goodness' sake   for conscience' sake
for righteousness' sake   for convenience' sake

EXPRESSIONS OF DURATION

5·42 Expressions of duration based on the genitive case are analogous to possessives and are formed in the same way; in many instances another form that does not require the possessive is preferable.

a week's vacation a week of vacation [or] a 1-week vacation
an hour's delay a delay of an hour a few hours' drive
3 months' therapy 3 months of therapy
after many years' experience after many years of experience
an hour's time an hour in 3 days' time in 3 days

If the word "of" cannot be inserted into the expression, the possessive is incorrect.

She was 6 months pregnant. [not "6 months' pregnant"]
The report was 8 days late. [not "8 days' late"]

REFERENCES

Cited References

[AGPS] Australian Government Publishing Service. 1988. Style manual for authors, editors and printers. 4th ed. Canberra: AGPS Pr.

[DSSC] Department of the Secretary of State of Canada. 1985. The Canadian style: a guide to writing and editing. Toronto: Dundurn Pr.

[FEAC] Freelance Editors' Association of Canada; Burton L, Cragg C, Czarnecki B, Paine SK, Pedwell S, Phillips IH, Vanderlinden K. 1987. Editing Canadian English. Vancouver (BC): Douglas & McIntyre.

Hansen WR. 1990. Suggestions to authors of the reports of the United States Geological Survey. 7th ed. Washington: US Government Printing Office.

McArthur T. 1992. The Oxford companion to the English language. Oxford: Oxford Univ Pr.

Morse JM, editor. 1985. Webster's standard American style manual. Springfield (MA): Merriam-Webster.

[OUP] Oxford University Press. 1983. Hart's rules for compositors and readers at The University Press Oxford. 39th ed. Oxford: OUP.

[USGPO] United States Government Printing Office. 1984. United States Government Printing Office style manual 1984. Washington: USGPO.

Additional References

American Medical Association. 1989. Manual of style. 8th ed. Baltimore: Williams & Wilkins.

Dodd JS. 1986. The ACS style guide: a manual for authors and editors. Washington: American Chemical Soc.

Huth EJ. 1987. Medical style & format: an international manual for authors, editors, and publishers. Philadelphia: ISI Pr. Available through Williams & Wilkins, Baltimore.

Todd L, Hancock I. 1987. International English usage. New York: New York Univ Pr.

[UCP] University of Chicago Press. 1993. The Chicago manual of style. 14th ed. Chicago: UCP.

6 Prose Style and Word Choice

Good prose is like a window pane. — George Orwell, *Why I Write,* 1946

Prose is bad when people stop to look at it.
— T E Lawrence, *Men in Print,* 1940

Effective scientific prose is accurate, clear, economical, fluent, and graceful. These qualities depend on myriad details: how paragraphs are linked, how sentences relate to each other, the length and flow of individual sentences, the choice of words. Only those details that can be economically improved during redaction are considered here; few editorial offices can afford to rewrite extensively. Authors seeking detailed guidance in writing clear, efficient scientific prose will find help in books noted in the section "Guides to Usage and Prose Style" of Appendix 3,

"Annotated Bibliography", and in the reference list at the end of this chapter.

GRAMMATICAL ERRORS

AGREEMENT IN NUMBER BETWEEN SUBJECT AND PREDICATE

Latin Nouns

6·1 Lack of knowledge of Latin frequently leads writers to fail to distinguish between singular and plural forms of Latin-derived nouns and hence to mismatch subjects and predicates and make other inappropriate choices. This fault may not hurt the meaning in popular, nonscientific texts, but scientific prose should maintain the distinction.

> The best culture medium for *Legionella* . . . is . . .
>
> The most widely used culture media are . . .
>
> The data support the view that . . . [not "data supports . . ."]
>
> A faulty datum in the survey unfortunately . . . [not "A faulty data . . ."]

For plural forms of Latin-based nouns widely used in science and their anglicized equivalents, see section 5·27. Note that the English form of the plural may be required for some uses. The US Geological Survey (Hansen 1990; p 163) specifies "datums" (rather than "data") for benchmarks and time markers.

Collective Nouns

Some collective nouns such as "committee" can take a singular or plural verb in the related predicate. The choice should be made according to the noun's meaning defined by the action represented by the verb.

> The committee announces decisions every Friday.
>
> [A single announcement comes from the committee; its members do not each issue an announcement.]
>
> The committee argue all points with care.
>
> [The members of the committee argue among themselves.]

The plurality of a subject may be concealed by abbreviation, but such a plural may be reasonably treated as a collective noun.

The NIH were appropriated their budgetary requests.

["NIH" stands for "National Institutes of Health"; each institute received its own appropriation.]

The NIH was headed by Dr Healy.

[The director is head of the group of institutes in the NIH.]

DANGLING PARTICIPLES

6·2 Present and past participles frequently open an adjectival phrase that is expected to modify the noun or noun phrase serving as the subject of the sentence or clause. The noun modified must be present to indicate what agent is responsible for the action represented by the participle. If no agent is apparent, the participle is said to be "dangling".

Reviewing the available data, the cause of the accident was mechanical, not chemical.

[Who did the reviewing? Putting in the agent changes the sentence to "Reviewing the available data, the committee concluded that the cause of the accident was mechanical, not chemical." This is still not a felicitous form; putting the agent first produces a stronger sentence: "The committee concluded from its review of the data that . . ."]

Such sentences can often be revised by using the present participle as a gerund (verb form used as a noun).

Reviewing the data led to the judgment that the cause of the accident was mechanical, not chemical.

["Reviewing the data" is now a noun phrase serving as the subject of the sentence; "the committee" as agent may be clear from the context.]

A dangling participle frequently seen in scientific writing is "based".

Based on the evidence, the accident was caused by a mechanical, not chemical, failure. [What is based "on the evidence", "the accident"? No, a conclusion or judgment or decision.]

[2 potential revisions]
The decision, based on the evidence, was that the accident was caused by a mechanical, not chemical, failure.

Basing its decision on the evidence, the committee decided the accident was caused by a mechanical, not chemical, failure.

One way to test for dangling is to move the opening phrase to the interior of the sentence so that it follows the noun it should modify, according to the original structure of the sentence.

> The accident, based on the evidence, was that . . .
> [The participle is dangling because "the accident" was not "based on the evidence".]

That a participle is dangling may not be apparent when it is not at the beginning of the sentence.

> The county was surveyed using a Wehrtopf pocket altimeter.
> [Who used the altimeter? Possible revision: "The county surveyor used a Wehrtopf pocket altimeter."]

CONFUSED AND MISUSED PAIRS

HOMOPHONES AND NEAR-HOMOPHONES

6·3 Homophones, pairs of words with the same sound but different meanings, can be misused by authors uncertain of the correct spelling of the intended word.

discrete, discreet	principle, principal	sheer, shear	here, hear
council, counsel	albumen, albumin	complement, compliment	

Such misuses may not be detected by spelling checkers in word-processing programs, and editorial offices should consider compiling lists of homophones frequently misused in their authors' texts.

A related kind of error is the misspelling of words ending in "able" or "ible" and "ance" ("ant") or "ence" ("ent").

IMPRECISELY APPLIED WORDS

6·4 Careful writers and editors in science strive to select and use the word that accurately, precisely, and correctly conveys the intended meaning. Examples of such words follow. The emphasis of the definitions here is on usage in science and scientific contexts and on distinctions that can clarify and sharpen meaning. Some of these words may have a more specific meaning in some scientific disciplines; these meanings can be found in scientific dictionaries such as the *McGraw-Hill Dictionary of Scientific and Technical Terms* (Parker 1989) and the *Academic Press Dictionary of Science and Technology* (Morris 1992). Nuances of, and preferences for, many common words are well discussed in the synonym and usage notes in *The American Heritage Dictionary of the English Language* (Soukhanov 1992).

a, an, the: "a" or "an" (preceding a vowel sound) is the correct article when the subject of the noun it precedes can exist in more than 1 form or as more than 1 case, for example, "A new species of *Escherichia* was identified by . . ."; other species exist or other new species may be identified in the future. When no more than 1 instance exists [as in "The new species of *Escherichia* announced recently in *Microbiology News* was identified by . . ."] or is likely to exist in the future, **the** is appropriate, as in "The organism identified as responsible for this outbreak was *Escherichia coli.*"

absorbance, absorptance, absorptivity: absorbance is the logarithm of the ratio of the intensity of light entering a solution to the intensity it transmits; **absorptance** and **absorptivity** refer to the ratio of energy absorbed by a body to the energy striking it.

absorption, adsorption: absorption is the process, ongoing or completed, of taking up by capillary, osmotic, chemical, or solvent action; **adsorption** is the holding of something by the surface of a solid or liquid through physical or chemical forces.

accuracy, precision: accuracy is the degree of correctness of a measurement or a statement; **precision** is the degree of refinement with which a measurement is made or stated ["The number 3.43 is more precise than 3.4, but it is not necessarily more accurate."]; when applied to a statement, **precision** implies the qualities of definiteness, terseness, and specificity.

adduce, deduce, induce: adduce is to bring forward as an example or as evidence for proof in argument; **deduce** is to reason to a conclusion or infer it from a principle; **induce** is to reach a conclusion through inductive reasoning (going from particular facts to a principle); **induce** also means to bring about an effect, as in "to induce labor".

aetiology: see **etiology.**

affect: see **effect.**

after, following: after can mean simply later than a particular time or event [as in "He died of anaphylactic shock soon after you saw him."]; **after** can also imply cause and effect [as in "He died of anaphylactic shock after swallowing a capsule of the wrong antibiotic."]; **following** is a preposition that can be used synonymously with "after" but is best reserved in science as an indicator of position not related to time, for example, "The following authors are frequently cited in genetics textbooks: Mendel, Morgan, Wright, . . ."

aliquot, sample: aliquot is the part of a total amount of a gas, liquid, or solid that has been completely divided into equal parts; for example, 10 mL is an aliquot of 100 mL of a liquid that could be, or has been, divided into 10 equal parts; **sample** is a part taken as representative of its source, for analysis or study.

alternate, alternative: alternate is to pass back and forth successively from 1 state, action, or place to another ["The alternate states are ice and water."]; **alternative** is a choice between 1 state or place and another, the 2 being mutually exclusive ["The alternative to the present method is to use a catalyst."].

although: see **while.**

among, between: among is a preposition indicating a relationship involving more than 2 units of the same kind ["Among the antibiotics of this class, antrotomycin is the best choice."]; **between** is a preposition indicating a relationship involving 2 units of the same kind ["The choice was between penicillin and ampicillin."].

analogous: see **homoeologous.**

anatomy, morphology, structure: anatomy is a scientific discipline that studies and describes structure, especially of living things; the term "anatomy" should not be used as a synonym for "structure"; **morphology** is a discipline that studies and describes the shape, structure, or formation of living things, especially external shape and form; "morphology" should not be used for "structure" or "structural characteristics"; **structure** relates to the parts of living or nonliving things as they relate to each other.

ante, anti: ante is a prefix indicating "before" in a sequence or location ["antedate", "antebrachial artery"]; **anti** is a prefix indicating "against", "opposed to", "in contrast to" ["antibody", "antiparticle"].

article: see **manuscript.**

as, because, since: as is preferably used only in the temporal sense ["As we were completing the paper, new evidence came to light."] and not in the causal sense, to avoid ambiguity [not "As we were away in Africa, he thought he could pilfer our data."]; **because** is used to show cause; **since** is used to show a temporal relation, although it is technically not incorrect to use "since" in the causal sense.

assess: see **determine.**

average, typical: in scientific usage **average** should be reserved, if possible, as a synonym for "statistical mean"; **typical** or "characteristic" is often a suitable adjective in place of a nonstatistical "average".

axenic, gnotobiotic: axenic is an adjective indicating organisms kept in isolation from other living things or indicating such an environment; **gnotobiotic** is an adjective indicating laboratory animals reared to be free of infectious agents except for any agent deliberately introduced.

because: see **as.**

before: see **prior to.**

believe, feel, think: the 3 verbs can connote an author's convictions or persuasions with different strengths of basis; **feel** implies an intuitive, or a not fully reasoned, conviction; **think**, a view based on evidence or logic; **believe**, a definite conviction on the view regardless of the strength of evidence. Because of the subjectivity of opinion their use implies, they should not be used to describe the convictions of others, who preferably can be objectively said to "state", "say", "claim", or "describe" something.

benchmark: see **criterion.**

between: see **among.**

bi, semi: the prefix **bi** can indicate "2" [as in "bicorn", the adjective for an animal with 2 horns] or at intervals of 2 units [as in "bimonthly publication" (every 2 months) and "bicentennial celebration"]; **semi** indicates "half" ["semicircle"] or "partial" ["semiretirement"], but "semimonthly publication" is less ambiguously phrased as "twice-a-month publication" or "biweekly publication"; also see **quasi.**

case, patient: case is an instance, example, or episode [as of disease, "She had a case of measles"; or "The worst case would be 2 earthquakes within 2 days"]; **patient** is a person ["We saw 12 patients in the clinic", not "We saw 12 cases . . ."].

cause: see **etiology.**

circadian, diurnal: circadian is an adjective for occurrence approximately every 24 hours; **diurnal** is an adjective that means in meteorology a 24-hour cycle occurring every 24 hours; in biology, occurring daily or every day in the daytime; in botany, open in the daytime and closed at night.

comparable, similar: comparable should be reserved as an adjective indicating an item lending itself to comparison with a similar item, [as in "Because the methods of ascertainment differed, the mortality statistics of Sweden and Chile are not comparable" ("cannot be compared")]; **similar** is better as the adjective indicating likeness ["The mortality rates in Sweden and Chile are similar", not "are comparable"].

compared to, compared with: compared to may only imply finding resemblances or similarities in dissimilar objects but tends to connote a stronger, subjective desire in making the comparison, looking for contrast, or judging against a standard ["Compared to the Japanese and Germans, Americans are poor savers."]; **compared with** implies looking for similarities or differences in compared objects ["In the clinical trial, the low-fat diet was compared with a high-fat diet."]; also see **versus.**

compose, comprise, constitute: compose as an active verb following a plural subject means to form, to make up a single object, to go together ["Forty-eight states compose the contiguous United States of America."], but as a passive verb is synonymous with "comprise" ["The contiguous United States of America is composed of 48 states."]; **comprise** is a verb meaning to include, to contain, to be made up of ["The United States comprises 50 states."]; **constitute** is in many contexts synonymous with "compose" ["Forty-eight states constitute the contiguous United States of America."], but also can mean "to amount to", "to equal", "to set up", "to establish".

congenital, genetic: congenital means "born with", "present at birth"; **genetic** means having to do with genes, chromosomes, or their effects in producing phenotypes or determining heritable characteristics. A disease or abnormality caused by genetic effects is not necessarily congenital (apparent at birth).

conjecture: see **law.**

connote, denote: connote is used to imply a meaning beyond the usual specific, exact meaning; **denote**, to indicate the presence or existence of.

constitute: see **compose.**

contagious: see **infectious.**

continual, continuous: continual means a prolonged succession or recurrence going on in time with no, or only brief, interruption; it functions as an adjective ["The continual clatter outside the hospital was nerve-wracking for the patients."]; **continuous** implies without interruption, never ceasing even briefly.

conventional: see **customary.**

criterion, standard, benchmark: criterion may be synonymous with "standard" but is often restricted in meaning to be synonymous with "measure", a specific type of basis for a judgment but with possibly different specifications for favorable or unfavorable judgment; **standard** is often reserved for a criterion assigned a specific value against which a judgment is made; **benchmark** is used in different contexts as a synonym for "criterion" (a benchmark without an assigned value) or for "standard" (a benchmark assigned a value).

customary, conventional, traditional, normal, norm: customary means long used, accepted for a long period, used as a habit; **conventional** means an established and generally agreed-upon practice or characteristic; **traditional** means long used or applied, and tending to connote a long-standing and general acceptance in a social or professional group or community; **normal** is an adjective indicating that the noun it modifies has the characteristic(s) of a satisfactory or desirable majority. Note that **norm** implies not simply a normal but a desired characteristic. To imply "usual", either "customary" or "conventional" should be preferred to "normal".

database, data set, data: database is a formal structure (computer file, printed document) carrying data organized for retrieval and analysis and representing a conceptually coherent subject; **data set** is a particular coherent body of data maintained in a database; **data** is a group of individual items of information.

deduce: see **adduce.**

definite, definitive: definite means clearly limited, firmly established; **definitive** means conclusive, defining.

demonstrate, exhibit, reveal, show: the 2 longer terms are often used as inflated versions of **show; demonstrate** should be reserved for a deliberate action intended to illustrate an action or procedure [as in "The technician demonstrated how to operate the pH meter."] and **exhibit** for a deliberate action to make visible [as in "He exhibited the mineral specimens at the last congress."]; these 2 terms should not be used for passively carrying something apparent [as in "The patient demonstrated (or exhibited) a rash." for "The patient had a rash."]. The verb **reveal** represents an action to make visible what can then be considered to have been hidden or inapparent; it should not be used as a synonym for "report". Note that an inanimate agent cannot "demonstrate" anything [not "The data demonstrated an increase in the blood pressure when the dose was lowered."].

denote: see **connote.**

determine, evaluate, assess, examine, measure: these terms are frequently used synonymously or almost so, but distinct and differing meanings for them can make for clearer expression; **determine** is best reserved to mean "set a limit on" or "establish conclusively" (despite its common use in chemical jargon as a synonym for "measure"); **evaluate** means to ascertain or fix a value on the object of the action; **assess** is best reserved to mean "to estimate a value, as for taxation, or set the amount of a payment", as for a fine; to **examine** is to look at, or inquire into, or test closely; **measure** is "to proceed to examine an object for its quantitative characteristic" [as in "measuring blood glucose"].

different, diverse, disparate: different means having at least some dissimilar characteristics ["The fossil evidence establishes that this outcropping is a different formation."]; **diverse,** having a notable range of differences ["His department of chemistry has a staff with diverse interests and skills."]; **disparate,** distinctly different ["They reached disparate conclusions from the same evidence."]. Also see **varying.**

different from, different than: the "from" phrase is usually the better choice because of its parallel with "differs from" ["The flora of the Quarternary is distinctly different from that of the Cretaceous." "The flora of the Quarternary differs distinctly from that of the Cretaceous."].

differing: see **varying.**

digit: see **number.**

disparate: see **different.**

diverse: see **different.**

dosage, dose: dosage implies not only an amount but also frequency of administration; it is not synonymous with a single dose; **dose** is the amount of a drug administered at 1 time.

effect, affect: effect as a verb means to bring about a change ["He effected a budgetary change."], but as a noun means the result of some action ["The effect was a cut in the budget."]; **affect** as a verb means to influence ["His budget cut affected all members of the staff."] but as a noun means the impression of feeling or emotion conveyed by a person's demeanor, action, or speech [The diagnostic feature in this case was the patient's flat affect."]. Also see **impact.**

employ: see **utilize.**

enable, permit: enable means to give someone or something an ability; **permit** is to allow an action previously restricted or forbidden; "allow" and "let" are synonyms for **permit** but with less of the connotation of preceding restraint.

etiology, cause: etiology should be reserved as a term for the study and description of **cause**; it should not be used as a pompous equivalent of "cause".

evaluate: see **determine.**

examine: see **determine.**

execute: see **perform.**

exhibit: see **demonstrate.**

farther, further: farther as an adverb means "to a more distant or advanced point", either in physical dimension or nonphysical dimension [as in "concept"], and as an adjective, "at a more distant or advanced point"; **further** is best reserved for use as a transitive verb, to move along, to develop [as in "His theory did little to further our knowledge of the oldest galaxies."].

feel: see **believe.**

few, fewer, less: few and **fewer** are adjectives indicating small and smaller in number, or infrequent; they are used with counted or countable items; **less,** a cognate of "little", is used with uncounted or uncountable quantities applied to mean a smaller total amount ["We have fewer astrologers per capita now than in 1900." "We have less information on the genome of the horse than on that of man."].

following: see **after**.

frequent: see **regular.**

fungus, fungal, fungoid: fungus, the noun, is an organism with nucleated cells having rigid walls but no chlorophyll, a subject of the discipline mycology; **fungal** is the adjective form of fungus ["He has a fungal infection."]; **fungoid** means having the character of a fungus.

further: see **farther.**

general, generally, generic, generically, usual, usually: general and **generally,** adjective and adverb, indicate as modifiers a broad or group-typical application, relevance, or characteristic; **generic** and **generically** are 1st cousins of **general** and **generally** but imply "of the same category" as in "generic drugs", drugs of a group representing a particular type in effect, in contrast to specifically trademarked individual drugs; **usual** and **usually** are preferably applied with the connotation of likely, expected, more frequent.

genetic: see **congenital.**

gnotobiotic: see **axenic.**

heterogenous, heterogeneous: see **homogenous.**

homoeologous, homologous, analogous: homoeologous is an adjective used to characterize partially homologous chromosomes; **homologous,** in biology, means corresponding in structure, position, origin, or other characteristics but not in function, and is similarly applied in other sciences; **analogous** means similar in function but dissimilar in structure and origin.

homogenous, homogeneous: having closely similar or identical characteristics, such as components, structure, origin, and so on; **homogeneous** is generally preferred in science (because of its derivation from "homo", meaning "same", and "genesis", meaning "origin"). **Homogenous**, as an adjective derived from "homogeny", has been used in biology specifically to mean having similar structures derived from common origins. Similar distinctions can be drawn between **heterogenous** and **heterogeneous** ("hetero", meaning "different").

hypothesis: see **law.**

hypothesize, hypothecate: hypothesize means to form a hypothesis; **hypothecate** is a verb meaning to pledge property as security without transfer of rights.

ic, ical: see **ologic.**

identical to, identical with: the 2 forms are regarded as equally acceptable in some usage guides, but the "with" form should be preferred because of the parallel form in the use of the noun "identity" ["A German botanist clearly showed the plant's identity with the plant described 10 years before in the Canadian survey."].

impact: impact should be reserved for the meaning of the striking of 1 body against another and not be used in its clichéd, hyperbolic, and jargon sense to mean simply **effect.** Also see **effect.**

imply, infer: imply is to indicate or suggest that the subject can be interpreted to a following but not necessarily correct conclusion; **infer** is to draw a conclusion from evidence ["These data imply that the sensors are defective." "I infer from the data that the sensors are defective."].

incidence, prevalence, point prevalence, period prevalence: incidence is the number of new cases occurring in a population of stated size during a stated period of time; it is not synonymous with **prevalence**, the number of cases existing in a population of stated size at a particular time; **point prevalence** indicates cases known to exist on a particular date, and **period prevalence**, all cases known to exist during a stated period. For a detailed discussion see the document "Prevalence and Incidence" (WHOEC 1966) issued by a World Health Organization expert committee; for longer definitions see the relevant entries in *A Dictionary of Epidemiology* (Last 1988).

individual, person: individual, a noun or adjective, indicates a unit considered specifically and distinguished from a group; for an individual woman or man, **person**, with its connotation of a particular human being having his or her own personality, is to be preferred to the simpler and dehumanized connotation of "individual". Also see **people, persons.**

induce: see **adduce.**

infectious, infective, contagious: a useful distinction is to apply "infectious" to mean harboring a potentially infecting agent or having been caused by an infecting agent ["an infectious disease"] and "infective" to indicate an agent that can cause infection ["Not all the bacteria likely to be found in this environment will be infective."]; **contagious**

is an adjective meaning that the infecting agent in an infectious disease has a high probability of being transmitted.

infer: see **imply.**

infested, infected with: infested means harboring or carrying visible lower organisms, notably worms or insects, not causing inflammation or other immunologic consequences of their presence as viruses or bacteria would, for which the parallel term would be **infected with.**

inherent, intrinsic: synonymous adjectives used to mean a characteristic necessarily in and of the noun modified, but note the use of **intrinsic** in anatomy to mean entirely within a structure or organism.

kind: see **type.**

law, theory, hypothesis, conjecture: terms for concepts with decreasing degrees of certitude. A **law**, in science, is a concept with a high degree of certitude, sufficient for confidence in its use for predicting phenomena; a **theory** is a broad concept based on extensive observation, experimentation, or reasoning and expected to account for a wide range of phenomena covered by its scope; a **hypothesis** is a narrow concept, generally postulated as a potential explanation for phenomena and to be tested by experiment or observation; a **conjecture** is a speculation, a concept akin to a hypothesis but not proposed for testing.

less: see **few.**

localize, locate: localize means to confine, restrict, or attribute to a particular place or to have the characteristic resulting from such action ["The infection localized in the antecubital space."]; **locate** means to specify, place, or find in a particular place ["We finally located the infection in the right pleural cavity."]; **localize** should not be used to mean "find" [not "We localized the primary site of the disseminated cancer in the pancreas."].

majority, most: majority, a number of items greater than half the total items of a particular class [general and scientific use] or, in politics, the difference between the votes the winning candidate received and the total votes for the other candidate or candidates; frequently used as a synonym for **most**, but "most" is the preferred term when a quantitative expression is not needed and a preponderance needs to be implied. ["Most physicians are licensed in only 1 state." rather than "A majority of physicians are licensed in only 1 state."].

manuscript, paper, article: manuscript should refer only to the physical representation (paper sheets typed or written on; word-processing file on a diskette) of a scientific **paper** (the intellectual document itself, its words and numbers); a peer reviewer reviews a paper (the intellectual material), not a manuscript; the reviewer is sent the paper in the form of its manuscript; a journal publishes a paper [often preferably called an **article**], not a manuscript. Also see **report.**

measure: see **determine.**

meiosis, mitosis, miosis: meiosis is cellular division resulting in production of cells with a haploid number of chromosomes; **mitosis** is cellular division producing the diploid number of chromosomes; **miosis** is excessive smallness of the ocular pupil.

method: see **technique.**

morphology: see **anatomy.**

most: see **majority.**

mucus, mucous, mucoid: mucus is a thick, slimy secretion produced by body membranes and glands; **mucous** is the adjective form indicating that something has the character of mucus or produces mucus [as in "mucous membranes"]; **mucoid** means mucus-like.

mutant, mutation: mutant as a noun refers to an organism carrying or expressing 1 or more genetic mutations, but it is also applied as an adjective as in a "mutant gene"; **mutation** is a stable and heritable change in a nucleotide sequence in DNA or RNA.

need: see **require.**

norm, normal: see **customary.**

number, numeral, digit: number is the count of some class of objects; a **numeral** is a single character in the group of numbers from zero (0) to nine (9). The number 345 is represented by the arabic numerals 3, 4, and 5 in that sequence. **Digit** can be used synonymously with **numeral** but is sometimes used in science to refer to the number of numerals and their representation of magnitudes in the decimal system ["He reported his data with 3 digits."].

nutrition, nutritional, nutritious: nutrition is the discipline concerned with desirable foodstuffs and feeding, also desirable feeding itself; **nutritional** means having to do with nutrition; **nutritious** means having the character associated with desirable nutrition.

ologic, ological: both forms are used as suffixes in adjectives formed from nouns ending in "**ology**"; this manual recommends general use of the **ologic** form, but note such firm idiomatic distinctions with **ical** and **ic** as those between "historic" and "historical" and "economical" and "economic". See section 5·20.

outbreak: an imprecise term often applied to mean a sudden appearance of a disease, especially of an infectious disease, or of some other kind of social phenomenon; alternative terms (nouns) could be "episode", "sudden occurrence", "epidemic", "epizootic".

paper: see **manuscript, paper, article.**

parameter: parameter means a potential variable to which a particular value can be assigned to determine the value of other variables; it should not be used loosely and pompously as a jargon term synonymous with "variable", "index", "indicator".

part: see **portion.**

pathology: pathology should be reserved to mean the discipline that studies diseases, disorders, and other abnormalities in plants, animals, or humans; **pathology** is not a synonym for "abnormality", "disease", "disorder", or "lesion".

patient: see **case.**

people, persons: people means not-numbered groups of persons having in common some characteristic such as nationality ["the French people"] or location ["the various peoples living east of the Volga"]; **persons** is the term to use when individualities should be emphasized ["Persons with impaired vision can use Easy Access on the Macintosh."] or when referring to a numbered group ["An ambulance took 3 injured persons to the same hospital."].

percent, percentage: percent is a term meaning "units per 100 units", often represented by the symbol % ["45 percent" or "45%" standing for "45 units per 100 units"]; **percentage** is a statement of a quantity or rate expressed as the unit percent ["45%" is a percentage]. Note that the difference between 2 percentages should be stated as a difference in percentage units or points, not as a percent; the difference between 25% and 50% is 25 percentage points, not a "25% difference".

perform, carry out, execute: such verbs can often be replaced by a more specific verb such as **analyze, operate, do**, or other possible choices [not "He performed an appendectomy despite the patient's unstable condition" but "He did an appendectomy despite . . ." or "He removed the appendix despite . . ."].

permit: see **enable.**

person: see **individual, person.**

portion, part: portion should be reserved for the more specific meanings of a part separated from an entirety or a part allotted to an entity [person, organization]; **part** is an adequate term for what is too often designated by **portion** serving as an inflated substitute ["The Triassic forms a major part of central Connecticut" instead of "The Triassic forms a major portion of . . ."; "Tornadoes are frequent in the coastal parts of the Gulf states", not "in the coastal portions of . . ."].

precision: see **accuracy.**

presently, at present: presently, used by some as a synonym for "currently", is reserved by careful writers to mean "soon", "shortly", "in the near future"; **at present** means "now".

prevalence: see **incidence.**

prior to, before: prior to should be reserved for an event preceding in time or sequence in which being 1st has an importance or value above the event that comes next ["He recalculated all of his data prior to making new observations."]; **before** is adequate to introduce a simple sequence ["He had had no illnesses before his 1st stroke", not "He had had no illnesses prior to his . . ."].

proven, proved: proven as a verb [irregular past participle] should generally be replaced by **proved** as the past participle with regular form ["It has been proved that a retrovirus is responsible for the recently described syndrome", not "It has been proven that a . . ."].

quasi, semi: quasi is a prefix modifying the stem term to indicate "to a degree", "to some extent"; **semi** is a prefix that modifies a stem term with the same meanings as **quasi** but also sometimes carries the more specific meaning of "half" as in "semicircle" or "semimonthly" ["every half month"]; also see **bi.**

regime, regimen: these terms can be used synonymously to mean a regular pattern of occurrence, but in medical contexts **regimen** implies a stipulated or controlled program or scheme for treatment or activity [as in "a dietary regimen aiming to reduce the risk of atherosclerosis"]; **regime** is also used to mean "a government in power".

regular, frequent: regular as an adjective indicates "ordered", "consistent", "at fixed times or points"; **frequent** means occurring at relative short intervals or often.

relationship, relation: relationship should be reserved to mean relations between 2 or more persons; **relation** is adequate to describe a connection between inanimate objects ["the cause-and-effect relation of the human immunodeficiency virus to the acquired immunodeficiency syndrome"].

report, paper, study, trial: a **report** (or **paper**) describing a study or clinical trial should be distinguished from the research itself ["Jones published a **report** of his clinical **trial** of licorice in *The New England Journal of Medicine* last year", not "Jones published his trial in *The* . . ."]. Also see **manuscript, paper, article.**

require, need: require should be reserved for use as a transitive verb with the stronger meaning of an active agent's setting obligatory or compelling expectations ["The journal requires submission of 2 copies of a manuscript."]; **need** is appropriate for a passive agent ["Green-leaved plants need sunlight."].

reveal: see **demonstrate.**

sacrifice: a euphemism for "kill", "exsanguinate", and other terms that do not imply a religious rite.

sample: see **aliquot.**

semi: see **bi**; also **quasi.**

sensitivity, specificity: when applied to describe the use of diagnostic tests, **sensitivity** is the capacity to detect desired or predefined findings even if undesired findings are also found in relatively large numbers; **specificity** is the capacity to detect only desired findings and exclude undesired findings.

show: see **demonstrate.**

significant: significant should be used to mean serving as a sign of or pointing to; this precise use is especially needed in contexts with proper use of **significant** in the statistical sense of reaching a predefined numeric threshold and hence pointing to a specific statistical conclusion ["The mean blood pressure was significantly lowered, with a *P* value of 0.05."]. Other adjectives such as "great", "important", "influential", "major", "valuable", "useful", "desirable" are preferred choices when "pointing to" or "indicating" is not intended.

similar: see **comparable.**

since: see **as.**

specificity: see **sensitivity.**

standard: see **criterion.**

structure: see **anatomy.**

study: see **report.**

technique, method: both words are widely used to mean an analytic, quantitating, observational, or another similar kind of procedure, but a valuable distinction is made by using **method** for "procedure" and reserving **technique** for the skill, good or bad, applied in carrying out a procedure ["This bioassay is a reliable method when the analyst applies careful technique." "Horowitz always showed impeccable technique at the keyboard."]. The form **technic** is a variant spelling, but has been used as a synonym for "technology".

that: see **which.**

the: see **a.**

theory: see **law.**

think: see **believe.**

traditional: see **customary.**

trial: see **report.**

trophic, tropic: trophic is a suffix indicating a stimulating, nourishing, or supporting function in the growth or development of an agent's target [as in "adrenocorticotrophic hormone", a hormone stimulating increased activity of the adrenal cortex] or indicating growth or development itself; **tropic** is a suffix indicating capacity to respond to a modifying or changing agent ["Algae in that genus are all chemotropic", that is, they respond to chemical stimuli]. The **trophic** form is tending to disappear from endocrinologic usage.

type, kind: type is often used synonymously with **kind** but should be reserved in science to mean an inanimate object or a specific animal, plant, or microorganism characteristic of or standing for a larger group of closely related items; a "type species" is one that establishes a genus or subgenus name.

typical: see **average.**

upon, on: in most contexts **on** is preferred by writers striving to write economical and unpretentious prose.

usual, usually: see **general.**

utilize, use, employ: these words are often used synonymously, but in most uses of **utilize** the less stuffy and more straightforward **use** is adequate for the meaning of applying or drawing on for a purpose; when "consumption" through a use is implied, either the phrasal verb "use up" or the verb "consume" is more specific. Reserve **employ** for its meaning of putting a person to work.

vaccinate, immunize: these 2 terms are often used as synonyms, but to vaccinate means to purposely expose a person or animal to an antigen in hopes of eliciting protective antibody; to **immunize** implies that exposure to an antigen through infection or vaccination successfully elicits protective antibody.

varying, differing, different: varying should be used to mean changing, and **differing** or **different** to mean having unlike characteristics; "Philadelphia and New York have varying mean annual temperatures in a long-term cycle" means each city has a mean temperature that changes in the cycle; "Toronto and Santa Fe have differing [or "different"] mean annual temperatures" means that their mean annual temperatures are not the same; **various** can be a near-synonym for **differing** ["Various ethnic groups make up the population of Los Angeles."].

versus, compared with: versus means "against"; for comparisons or contrasts as in a clinical trial of 2 drugs, **compared with** should be used ["Penicillin Compared with Ampicillin for the Treatment of

Pnenumococcal Pneumonia", not "Penicillin versus Ampicillin for the . . . "]. **Versus** is used correctly in legal titles ["Roe *vs* Wade"; "Roe v. Wade" (legal style)] and some terms in which it does represent the meaning "against" ["graft-versus-host disease"].

volume: volume should be used to mean a separately bound part of a longer work ["Volume 3 of the *Encyclopedia Britannica*"] and not as a synonym, pompous and needlessly long, for "book" [as in a book review beginning "This volume is a welcome addition to the literature of . . ."].

which, that: that should be used as the relative pronoun introducing a restrictive clause ["This is the house that Jack built."] and **which** to introduce a nonrestrictive clause ["This house, which Jack built, was of shoddy construction."]. Also see section 6·6.

while, although: while should be used to indicate a period of time under consideration ["While he was waiting for surgery, his angina pectoris became steadily more frequent."] and **although** for a conditional state ["Although he was being treated for the staphyloccal infection while waiting for surgery, he succumbed to rupture of his aortic valve."].

EXCESSIVELY LONG COMPOUND TERMS ("FREIGHT-TRAIN PHRASES", "GOODS-TRAIN GRAMMAR")

6·5 Readers of scientific texts are frequently called on to read a string of adjectival nouns modifying a noun that is the subject or object of a sentence. Often it is not clear whether some of the nouns are compound modifiers or single modifiers and hence whether particular nouns modify an adjacent modifier or the main noun.

> a new type motor skills college performance test
>
> a percentage transmission recording ultraviolet light absorption meter

The remedies can include rewriting to clarify meaning by phrasing some of the relationships and hyphenating directly related modifiers.

> a new kind of motor-skills test used in colleges
>
> an ultraviolet light-absorption meter for recording percentage transmission

Such constructions are especially burdensome for readers when they are repeated subsequently in the text where the main noun could be readily understood by itself, or with a single modifier, through its context.

They designed a new type of test to measure the motor skills of college students . . .

Users of the new test will find . . .

ACCURATE STATEMENT

"THAT" AND "WHICH"

6·6 The use of "that" to introduce a restrictive clause and "which" to introduce a nonrestrictive clause remains a device valuable for accurate statement.

> Tumors developed in the fish that survived the treatment.
>
> [The restrictive clause, "that survived the treatment", is needed to define precisely which fish had the tumors being discussed; in other words, the meaning of "fish" is restricted to the fish surviving the treatment.]
>
> Tumors, which were found only several weeks later, developed in the 3rd fish.
>
> [The nonrestrictive clause, "which were found . . .", supplies optionally needed additional information; in its context the sentence could adequately read "Tumors developed in the 3rd fish."]

A classic example comes from a nursery recitation.

> This is the house that Jack built.
>
> [Without "that Jack built", "This is the house" would make no sense. Contrast the restrictive, defining clause "that Jack built" with the "which" nonrestrictive clause in "The house, which was the last one Jack built, sold for a high price."]

This distinction is often ignored by even careful writers, but it can be useful in sharpening precise statement in science.

BIAS-FREE USAGE

6·7 Careful attention to avoiding terms that reflect stereotypic biases or habitual vocabulary is often justified mainly by the motive of not offending or insulting the persons who are the subjects of discourse. But along with this motive, which helps in promoting social harmony and the sense of community, is the value of scientifically accurate statement. Stereotypic or habitual statement is not scientifically acceptable because it ignores, or even simply obscures, the complexity of scientific questions. Entire books have been written on biased usage with recommendations for more accurate modes of statement (Maggio 1991, 1992; Miller and Swift

1988); this section only illustrates a few examples. Relevant advice is also available in some style manuals (APA 1983; FEAC 1987). Editorial offices may find it useful to compile guidelines on bias-free usage that are especially relevant to their readerships.

Gender and Sex

6·8 When both men and women are the subject of the text, this must be made clear by reference to both and not obscured by an assumption that a male referent is adequate.

> His analysis ignored the economic problems of the man in the street.
>
> [The stereotypic and trite phrase "man in the street" can be replaced by a more exact statement.]
>
> His analysis ignored the economic problems of ordinary men and women.

Some terms with apparent gender reference may be acceptable because of a long-established non-gender-specific definition, and the choices of terms may depend on the views of the intended audience.

> Archaeology is the science that strives to build new histories for humanity.
>
> [If the "man" in "humanity" is regarded as giving "humanity" an incomplete meaning, a substitute is in order.]
>
> Archaeology is the science that strives to build new histories for men and women of the past.

Such constructions as "s/he" and "she/he" have appeared as substitutes for the phrase "she or he", but the slash should be reserved for specific scientific uses; see section 4·34.

Note that "gender" was long applied mainly in reference to the grammatical categories of masculine, feminine, and neuter. In recent years its use has been extended to refer to the social, economic, and historical categories "man" and "woman", which are based mainly, though not entirely, on the sex of individuals, with "sex" referring to the biological categories.

Ethnicity and Race, Nationality and Citizenship, Religion

6·9 The term "race" does not have a precise definition in biological terms, and its use depends mainly on judgments about physical characteristics that can differ quite widely among members of a so-called "race". Whenever possible, descriptions of populations or large social groups should draw on more sharply definable criteria such as country of birth or habitation or self-description. "Nationality" (defined by citizenship or by place of birth) should be distinguished from "citizenship", the more narrow category, and ethnic terms should not be used with pseudoprecision: "Caucasian" is no more scientifically precise than "white"; also see section 25·3.

Religious designations may be inaccurate if they ignore criteria for proper categorizing.

> The British are a Christian people.
>
> [an erroneous statement; not all British persons profess belief in Christian doctrines even though a Christian church is the established national church.]
>
> The United Kingdom is a Christian nation.
>
> [might be acceptable in a context in which "nation" is defined in part by the official position of the government on its state religion.]

Disabilities and Health-Determined Categorization

6·10 Note that "disability" refers to a condition that limits the ability of a person to carry out satisfactorily some usual activity or function. "Handicap" is a judgmental term referring to an environmental or attitudinal barrier to usual functioning.

References to individuals should hold to maintaining their identity as persons and avoid such depersonalizing terms as "an alcoholic", "a diabetic", "an epileptic", "a schizophrenic"; prefer "a diabetic patient", "an epilectic child", "a boy with hemophilia".

EXCESSIVE ABBREVIATION

6·11 Much of the contemporary and growing difficulty in reading and understanding scientific literature (Hayes 1992) arises from the ad hoc coining of abbreviations for noun phrases and heavy use of them in the following text. Their use where they are not needed in the context slows the reader not closely familiar with the subject of the text and may even force the reader to return to the head of the text again and again for the key to the abbreviation.

> Over a course of weeks to months, Amadori products undergo further rearrangement reactions to form fluorescent, cross-linking moieties called advanced glycosylated end products or AGEs. These products remain irreversibly bound to long-lived proteins such as collagen and accumulate as a function of age. Hyperglycemia accelerates the formation of protein-

bound AGEs. Consequently, tissue AGE amounts increase rapidly in patients with diabetes mellitus.

[Consider this revised version.]

Through weeks and months, Amadori products undergo rearrangement reactions to form fluorescent, cross-linking moieties called advanced glycosylated end products (AGEs). These new moieties accumulate and remain irreversibly bound to long-lived proteins such as collagen. Hyperglycemia accelerates their formation, and consequently their amounts in tissues increase rapidly in diabetic patients.

[The coinage "AGEs" may be justified by some subsequent needs for textual or tabular economy, but many of the uses are not needed in the context of the entire article. Note in the upper example the ambiguous "age" and "AGE" of "AGEs".]

Such coinages clog current scientific literature. Unfortunately, many of them become implicitly sanctioned usages. Editors concerned with whether their texts can be understood by nonspecialists should require substantial cuts in such practices, with removal of abbreviations where the context makes them unnecessary.

UNNEEDED WORDS AND PHRASES

6·12 A text with unneeded words and phrases slows the reader, and they should be eliminated. A phrase such as "it is interesting to note that" adds no information and only delays getting to the point of the sentence. Expressions such as "It is reported by Smith that . . ." can be shortened, for example, to "Smith reported that . . .". Many such widely used wordy phrases can be shortened to simpler forms.

[Wordy]	[Concise]
a majority of	most
a number of	few, many, several, some
accounted for the fact that	because
along the lines of	like
an innumerable number of	innumerable, countless, many
an order of magnitude	10 times
are of the same opinion	agree
as a consequence of	because of
as far as our own observations are concerned, they show	we observed
ascertain the location of	find
at the present moment, at this point in time	now

[Wordy]	[Concise]
bright green in color	bright green
by means of	by, with
caused injuries to	injured
completely filled	filled
[We] conducted inoculation experiments on	inoculated
definitely proved	proved
despite the fact that	although
due to the fact that	because, due to
during the course of	during, while
during the time that	while, when
fewer in number	fewer
for the purpose of examining	to examine
for the reason that	because
future plans	plans
give rise to	cause
goes under the name of	is called
has the capability of	can, is able
if conditions are such that	if, when
in a satisfactory manner, in an adequate manner	satisfactorily, adequately
in all cases	always, invariably
in case	if
in close proximity to	near
in connection with	about, concerning
in [my, our] opinion it is not an unjustifiable assumption that	[I, We] think
in order to	to
in the course of	during, while
in the event that	if
in the near future	soon
in the vicinity of	near
in view of the fact that	because
is in a position to	can, may
it has been reported by Jones	Jones reported
it is believed that	[omit]
it is often the case that	often
it is possible that the cause is	the cause may be
it is this that	this
it is worth pointing out that	note that
it would thus appear that	apparently
lacked the ability to	could not

[Wordy]	[Concise]
large amounts of	much
large in size	large
large numbers of	many
lenticular in character	lenticular
located in, located near	in, near
masses are of large size	masses are large, large masses
necessitates the inclusion of	needs, requires
of such hardness that	so hard that
on account of	because
on behalf of	for
on the basis of	from, by, because
on the grounds that	because
original source	source
oval in shape, oval-shaped	oval
owing to the fact that	because, due to
past history	history
plants exhibited good growth	plants grew well
prior to [in time]	before
referred to as	called
results so far achieved	results so far, results to date
round in shape	round
serves the function of being	is
smaller in size	smaller
subsequent to	after
take into consideration	consider
the fish in question	this fish, these fish
the question as to whether	whether
the tests have not as yet	the tests have not
the treatment having been performed	after treatment
there can be little doubt that this is	this probably is
through the use of	by, with [not "via"]
throughout the entire area	throughout the area
throughout the whole of the experiment	throughout the experiment
two equal halves	halves
was of the opinion that	believed
with a view to getting	to get
with reference to	about [or omit]
with regard to	about, concerning [or omit]
with the result that	so that

Similarly, many modifiers can be omitted. The general rule is that of Herbert Read (1952): "omit all epithets that may be assumed and . . . admit only those which definitely further action, interest, or meaning."

> Careful hemodynamic monitoring is needed if one is to prevent . . .
>
> [Omit "careful"; would the author suggest careless monitoring?]

Unneeded modifiers are frequent in popular speech where they serve to emphasize though they are not needed for clear meaning.

> "Be sure to take your personal belongings when you leave the plane." ["Belongings" are items one owns; they cannot be impersonal; omit the "personal".]
>
> "your final destination . . ." [A destination is the final point of a trip.]
>
> "It's an actual fact." [A fact is actual.]

ABSTRACT NOUNS

6·13 The frequent use of nouns formed from verbs and ending in "ion" produces unnecessarily long sentences and dull, static prose. Examples of such nouns are "production" from "produce" and "interpretation" from "interpret". The sentence becomes long because of the length of "ion" nouns and the need to use unnecessary prepositions and verbs with them. The dullness comes from the lack of action that would be stated by the "ion" noun's verb equivalent and the presence of the passive verbs that are needed. This fault is remedied and the sentence often shortened by reworking it to introduce the verb equivalent of the abstract noun.

> If we interpret the deposition of chemical signals as initiation of courtship, then initiation of courtship by females is probably the usual case in mammals.
>
> [can be revised to]
>
> If we interpret the depositing of chemical signals as initiating courtship, mammalian courtship is probably initiated by females.

Such changes may also yield a clearer sequence.

> A direct correlation between serum vitamin B12 concentration and mean nerve conduction velocity was seen.
>
> The mean velocity of nerve conduction correlated directly with the vitamin B12 concentration in serum.

Replacing abstract nouns with their equivalent verbs may bring the agent into the sentence and make it more specific and vivid.

> Following termination of exposure to pigeons and resolution of the pulmonary infiltrates, there was a substantial increase in lung volume, some improvement in diffusing capacity, and partial resolution of the hypoxemia.
>
> After the patient stopped keeping pigeons, her pulmonary infiltrates partly resolved, lung volume greatly increased, diffusing capacity improved, and hypoxemia lessened.

JARGON

6·14 The technical vocabulary or informal idiom in a scientific field is a jargon. If it is based on good etymologic bases, it may be acceptable in formal reports. But when a word or phrase represents slang or obscures meaning for readers not familiar with the jargon of the field, it should be revised to yield clearer meaning. English-language jargon can be quite unintelligible to nonanglophone readers.

Jargon can be of several types.

1 Shortened forms of words arise in conversation and may enter formal communication.

> The lab data confirmed the diagnosis. [for "The laboratory data confirmed the diagnosis."]

2 Verb-object relations are ignored.

> We stocked trout in the stream. [for "We stocked the stream with trout."]

3 Nouns are not used with their proper formal meaning.

> No pathology was found in the lung. [for "No abnormalities were found in the lung."]

4 Euphemisms are used to soften harsh realities.

> Many health-care providers lack skill in communicating potential adverse effects of the planned procedure. [for "Many surgeons cannot tell patients about possible bad outcomes of surgery."]

There are no simple rules for allowing or excluding jargon in formal texts, and editorial offices can profitably develop and maintain their own lists of what is acceptable and what is not. These judgments should be based on the standard vocabularies of the readership: "robust test" has a clear meaning for a statistician but would probably not be understood by readers of a journal of sports medicine.

DIFFICULTIES FOR AUTHORS FOR WHOM ENGLISH IS A 2ND LANGUAGE

6·15 All languages have their own characteristic grammar and syntax, and their cultures have rhetorical conventions. An author who is not thoroughly familiar with English and whose culture has a different rhetorical style may not readily write scientific papers in idiomatically acceptable English.

GRAMMATICAL AND SYNTACTICAL PROBLEMS

Prepositions

6·16 The meanings of generally corresponding prepositions in English and other languages do not always coincide. The use of a preposition in a non-English language may not be idiomatic in English.

> Spanish *de,* English *of*: "soy de Barcelona", "I am from Barcelona"; not "I am of Barcelona"

Some nonanglophone authors skilled in writing clear and generally highly accurate English are prone to not-quite-accurate use of prepositions. These minor defects, fortunately, are usually readily detected by copy editors and corrected with little risk of introducing errors in the change.

Position of Verb

6·17 In non-English languages the predicate (verb indicating action or state) may be properly at the beginning or end of a sentence, whereas in English it is idiomatically in a different position.

> "It was found by Smith (1974) important activity against some dimorphic fungi."
>
> [for "Important activity against some dimorphic fungi was found by Smith (1974)."]

Progressive Tenses of Verbs

6·18 Some languages rarely use a progressive tense, while some variants of English, notably those in the Indian subcontinent and neighboring countries, may use them excessively.

> "We are finding fossil evidence that this formation belongs in the Cretaceous Period."

[for "We have found fossil evidence that this formation belongs in the Cretaceous Period." or "We have fossil evidence that this formation belongs in the Cretaceous Period."]

Phrasal Verbs

6·19 Many nonanglophone authors are not familiar with English phrasal verbs, verbs usually of action or movement followed by an adverbial or prepositional particle. Phrasal verbs are sometimes used in scientific writing for a figurative or metaphorical meaning. Omitting the particle usually changes the meaning of the verb.

Ethnic tensions made it difficult to patch up community spirit. ["patch up" as a metaphorical equivalent of "repair"]

[not "Ethnic tensions made it difficult to patch community spirit."; the missing "up" distorts the metaphorical meaning of "patch up" in this context]

Many phrasal verbs cannot be found in standard English dictionaries but are found in the *Oxford Dictionary of Current Idiomatic English: Volume 1: Verbs with Prepositions and Particles* (Cowie and Mackin 1975). Phrasal verbs tend to be used more in informal prose than in formal scientific writing. In general, nonanglophone authors should avoid their use and strive for more direct and less metaphorical statement.

Incorrect Use or Omission of Articles

6·20 Articles are sometimes omitted when needed in English.

Physician should be able to communicate readily with patient.

[for "The physician should be able to communicate readily with the patient." or "A physician should be able to communicate readily with a patient."]

The choice of "the" or "a" may depend on the rhetorical strength needed. This is a kind of nuance in style that depends more for a correct choice on long experience in English idiom than on knowledge of syntactical rules. Note that when a plural form implies that the noun represents a category rather than individuals, the article is not needed.

Physicians should be able to communicate with patients.

Inappropriate Use of Gender in English

6·21 English nouns are generally assigned feminine or masculine gender only when they represent persons or animals of female or male sex, the type of linguistic gender known as natural gender. Many other languages assign grammatical gender to nouns without sex reference. Authors native in these languages may erroneously carry over grammatical gender into English.

> This new product was naftifine and his potency was similar to clotrimazole against strains of *Candida albicans.*
> [for "This new product was naftifine, and its potency was similar to that of clotrimazole against strains of *Candida albicans.*"]

Gender is assigned in popular speech and its print equivalent to some English nouns lacking sex reference (such as ships and countries), but this idiom is rare in scientific writing.

> The Queen Mary sailed from Southampton on Friday; by Monday it was clear she would break the transatlantic speed record.

VOCABULARY

Many terms cannot be carried into English by apparently legitimate translation.

"False Cousins"

6·22 Many nouns in non-English languages resemble, or are even cognates of, English nouns but carry different meanings. French "actuelle", for example, means "current" in the title of a review of existing knowledge or practice; in English "actual" can mean "current" but more frequently means "existing" or "real", in contrast to "potential", and would be unidiomatic in the title of an article or book.

Transfer Coinages

6·23 Nonanglophone authors, especially those native in a Romance language, may attempt to "translate" a noun into an apparently English equivalent that, in fact, is not an English idiom.

> His work for many years was on the causalism of infectious diseases.
> ["Causalisme" is a French term meaning "theory of causation"; "etiology" is an idiomatic English term in medicine for the study of causation.]

When such coinages have an obscure meaning they can sometimes be deciphered by finding the apparent noun of origin in a dictionary of the author's native language and deducing from its definition there what the English equivalent should be.

SOLUTIONS FOR THESE PROBLEMS

6·24 Most of these errors are easily detected by English-language editors and readily corrected. Errors in vocabulary should, however, be recorded in the editorial office's in-house style manual for future reference and correction.

Nonanglophone authors not highly experienced in writing in English should seek a review of their papers by 1 or more readers with a strong knowledge of English idiom before they submit them to an English-language journal; such review should take place late in the writing process, preferably after the scientific content of the paper seems satisfactory and in the right sequence.

Multilanguage scientific dictionaries, too numerous to cite here, are available for many fields. Especially useful are special bilingual glossaries, such as those developed by the Medical Research Council of Canada; examples are its *Vocabulaire du génie génetique—Vocabulary of genetic engineering, Vocabulaire de la statistique et des enquêtes—Statistics and surveys vocabulary, Vocabulaire canadien du Quaternaire—Canadian Quaternary vocabulary, Vocabulaire du réchauffement climatique—Vocabulary of global warming*. These are available in Canada from Canada Communication Group–Publishing, Ottawa, ON K1A 0S9; in the United States from International Specialized Book Services, 5602 NE Hassalo Street, Portland OR 97213; and in Europe from Canadian Books Express, The Abbey Bookshop, 29, rue de la Parcheminerie, 75005 Paris, France.

REFERENCES

Cited References

[APA] American Psychological Association. 1983. Publication manual of the American Psychological Association. 3rd ed. Washington: APA.

Cowie AP, Mackin R. 1975. Oxford dictionary of current idiomatic English: Volume 1: Verbs with prepositions and particles. Oxford: Oxford Univ Pr.

[FEAC] Freelance Editors' Association of Canada: Burton L, Cragg C, Czarnecki B, Paine SK, Pedwell S, Phillips IH, Vanderlinden K. 1987. Editing Canadian English. Vancouver: Douglas & McIntyre.

Hansen WR. 1990. Suggestions to authors of the reports of the United States Geological Survey. 7th ed. Washington: US Government Printing Office.

Hayes DP. 1992. The growing inaccessibility of science. Nature 356:739–40.

Last JM, editor. 1988. A dictionary of epidemiology. 2nd ed. New York: Oxford Univ Pr.

Maggio R. 1991. The dictionary of bias-free usage: a guide to nondiscriminatory language. Phoenix (AZ): Oryx Pr.

Maggio R. 1992. The bias-free word finder: a dictionary of nondiscriminatory language. Boston: Beacon Pr.

Miller C, Swift K. 1988. The handbook of nonsexist writing: for writers, editors, and speakers. 2nd ed. New York: Harper & Row.

Morris C, editor. 1992. Academic Press dictionary of science and technology. San Diego (CA): Academic Pr.

Parker SP, editor. 1989. McGraw-Hill dictionary of scientific and technical terms. 4th ed. New York: McGraw-Hill.

Read H. 1952. English prose style. New York: Pantheon Books.

Soukhanov AH, editor. 1992. The American heritage dictionary of the English language. 3rd ed. Boston: Houghton Mifflin.

[WHOEC] WHO Expert Committee. 1966. Prevalence and incidence. Bull World Health Organ 35:783–4.

Additional References

Allbutt TC. 1984. Notes on the composition of scientific papers. London: British Med Assoc, Keynes Pr. Reprint, with introduction by A Paton.

Chartier A, editor. 1991. Glossaire de génétique moléculaire et génie génétique [Glossary of molecular genetics and genetic engineering]. Paris: Institut national de la recherche agronomique (INRA).

Gower SE. 1990. The complete plain words. Greenbaum S, Whitcut J, revisers. Boston: DR Godine.

Graves R, Hodge A. 1990. The use and abuse of the English language. New York: Paragon House. Formerly: The reader over your shoulder.

Greenbaum S, Whitcut J. 1988. Longmans guide to English usage. Burnt Mill, Harlow, Essex (UK): Longman.

Maddox J. 1992. Language for a polyglot readership. Nature 359:475.

McArthur T. 1992. The Oxford companion to the English language. Oxford: Oxford Univ Pr.

[MW] Merriam-Webster. 1989. Webster's dictionary of English usage. Springfield (MA): MW.

Perttunen JM. 1986. The words between: a handbook for scientists needing English, with examples mainly from biology and medicine. 2nd ed. Helsinki: Kustannus oy Duodecim.

[RDA] Reader's Digest Association. 1983. Success with words: a guide to the American language. Pleasantville (NY): RDA.

Todd L, Hancock I. 1987. International English usage. New York: New York Univ Pr.

Wilkinson AM. 1991. The scientist's handbook for writing papers and dissertations. Englewood Cliffs (NJ): Prentice Hall.

Williams JM. 1990. Style: toward clarity and grace. 3rd ed. Chicago: Univ Chicago Pr.

Wilson KG. 1993. The Columbia guide to standard American English. New York: Columbia Univ Pr.

7 Names, Terms of Address, Degrees, and Honors

> Names are the marks of things.
> —A legal maxim, cited by H L Mencken, *A New Dictionary of Quotations,* 1966

Using the correct form of personal names in their proper arrangement and of titles and honors and their abbreviations not only is the courteous thing to do but also serves accuracy and comprehension, 2 qualities even more vital in scientific names.

PERSONAL NAMES

7·1 For the formatting of bibliographic references, indexing, and other alphabetic arrangements, it may be important to determine the name or names that constitute a person's formal name, especially the surname ("family name"). For personal references in text, one may need to know the name, or shortest sequence of names, that can be used after a title such as Dr or Professor. The convention in the United States, Canada, the United Kingdom, Italy, the former USSR, the Scandinavian countries, and others is that given names precede the family name (patronymic name) and the family name is the formal name; therefore, "Alena Rosemary Bird" is properly referred to as "Professor Bird", and her name would be indexed under the letter B. For surnames that incorporate prefixes the correct choice may not be so obvious. Furthermore, in other cultures family names are positioned differently or the formal surname may be determined in some other way. Ideally, the preference of the named person should be followed; it is often impossible to ascertain this preference directly, but knowing how the name has appeared in previous publications may help.

Comprehensive guidelines for selecting the surname to be used as the entry element in bibliographic systems or indexes are provided in the International Development Research Centre's (IDRC) *Manual for Preparing Records in Microcomputer-based Bibliographic Information Systems* (Di Lauro and Brandon 1990) and *The Chicago Manual of Style* (UCP 1993). These and other sources disagree on some of the finer points. In addition, the treatment of parts of names, such as prefixes, differ according to the country of origin; for example, an English-speaking person with the name "Robert De La Salle" should be indexed under D ("De La Salle, Robert"), but for a French-speaking person the same name should appear under L ("La Salle, Robert de"). The recommendations in Table 7·1 are drawn from the IDRC manual and the 5th edition of the *CBE Style Manual* (CBE 1983). For the use of capital and small-capital letters to denote surnames, see section 9·7.

In bibliographic references, the initials to follow the spelled-out surname should be capitalized letters for the rest of the name.

Bimal C Sen Gupta → Sen Gupta BC

Maria Anna da Fonseca → Fonseca MA da

ACADEMIC DEGREES AND TITLES, HONORS, AND MILITARY TITLES

7·2 A wide range of degrees, honors, and other designations can be appended to, and serve as modifiers of, personal names. These modifiers indicate academic achievement, the recognition of peers in certain organizations, and other distinctions. Table 7·2 has examples of degrees and honors; more comprehensive listings can be found in the *Acronyms, Initialisms & Abbreviations Dictionary* (Mossman 1994). For medical honors, more comprehensive listings can be found in the American Medical Association *Manual of Style* (AMA 1989), the *Oxford Companion to Medicine* (Walton and others 1986), and the *Canadian Medical Directory* (CMA [annual]). Many standard dictionaries (see Appendix 3) identify academic degrees represented by abbreviations and initialisms.

Table 7·1 Guidelines for determining formal names for author indexing

European name, except Portuguese and Spanish
  Single family name (surname)
    Use the last element of the name. Examples: Bird, Rosemary [for "Rosemary Bird"]; Davidson, Jeff [for "Jeff Davidson"].
  Simple compound name
    Use both parts of the compound, whether they are hyphenated or not. Examples: Carson-Peters, Henriette; Bonham Carter, Mark.
  Compound containing prefixes (if the country of origin is not known)
    Include the following prefixes as part of the name: am, de, del, della, delle, des, di, du, l', la, las, le, les, li, los, ver, vom, zum, zur. Examples: Di Giacomo, Roberto; Ver Boven, Aja.
    Do not include the following prefixes: af, den, op de, ten, ter, van, van den, van der, von, von der. Examples: Beethoven, Ludwig van; Beek, Leo op de.
  Compound containing prefixes (if the country of origin is known)
    English-speaking countries, Italy: include the prefix.
    France: include prefixes, except "de". Examples: Beauvoir, Simone de; La Salle, Marie.
    Germany, Austria, Netherlands: same as when the country of origin is not known.
    Denmark, Norway, Sweden: do not include the prefix if it is of Germanic origin: von, der, af; otherwise, include the prefix.
  Hungarian name
    The family name precedes the given name; the names need not be transposed. Example: Bartok Bela becomes Bartok, Bela in an English-language index.
  Portuguese name
    Use the last element of the family name. Example: Silva, Ovidio Saraiva de Carvalho e.
    If the last element is a qualifier indicating family relationship (Filho, Neto, Sobrinho), include the next-to-last element as well: Example: Vidal Neto, Victor. Note that in former Portuguese colonies, the qualifier may constitute the family name.
    If a simple surname begins with a prefix, do not include it in the surname. Example: Fonseca, Maria Anna da.
  Spanish name
    For a compound Spanish family name consisting of the father's name followed by the mother's maiden name (and, if the person is a married woman, perhaps the preposition "de" and her husband's name), use all elements of the family name. Example: Perez y Fernandez, Juan [for "Juan Perez y Fernandez"].
    If the surname consists of 1 name with a prefix that is an article (la, el, las, los), include the prefix and capitalize it. Example: Las Heras, Manuel.
    If the prefix is a preposition or a preposition and an article, do not include the prefix. Example: Vega, José de la.
  Asian-Indian name
    In modern usage the family name is the last element. If "Sen" or "Das" precedes an Indian name, include it with the family name. Example: Sen Gupta, Bimal C

Table 7·1 (*cont.*)

Indonesian name
Many Indonesian names have only 1 element. Examples: Sukarno, Suharto.
Burmese name
Most Burmese names have only 1 element. If the term of respect "U" precedes the name, retain it. Example: Thant, U.
Chinese name
Use the family name, usually the 1st element of the person's name; if it can be determined that the person has adopted a Western form by putting the family name last, treat it as an English name. Note that given names usually have 2 syllables, which may be hyphenated or closed up. Examples: Lee, Hon-Ling [for "Lee Hon-Ling"]; Hu, JD [for "J Donald Hu"].
Japanese, Korean name
Use the family name, which is the last element of the name. Example: Yakamoto, Hiroko [for "Hiroko Yakamoto"].
Thai name
Use the family name, which is the last element of the name. Example: Tiep, Nguyen Lam [for "Nguyen Lam Tiep"]; Duangjai, Somskdi [for "Somskdi Duangjai"].
Arabian, Egyptian name
Use the Egyptian or other Arabic family name, which is the last element of the name. Example: Khalil, Hassan Fahmy [for "Hassan Fahmy Khalil"].
When a prefix or its variant (el, ibn, abdel, abdul, abdoul, abu, abou, aboul) precedes the family name, hyphenate it when the name is transposed. Example: Ibn-Saud, Aziz [for "Aziz Ibn Saud"].

When degrees and honors are listed after a personal name, the abbreviated form without periods should be used.

Cassandra Levinson MD DPH   Terrence Rolf ChB FRCCPS

If the degree, title, or honor is referred to in a context in which it must represent the correct full and formal title, it should be written out in full and capitalized.

Randolph Macon, Associate Professor of Phonology, Harvard University, has been appointed . . .

As a Fellow of the American College of Otolaryngology, James Bagninon was eligible for . . .

The prerequisites for the Master of Science program are outlined in . . .

If the term is a generic term or not a specific formal term, it need not be capitalized.

Table 7·2 Abbreviations of selected academic degrees, honors, and other professional designations

Abbreviation	Term
AB	Artium Baccalaureus (Bachelor of Arts)
AgrM	Agrégé de médecine
AgrSc	Agrégé de science
AM	Artium Magister (Master of Arts)
BA	Bachelor of Arts
BC, BCh, BChir	Baccalaureus Chirurgia (Bachelor of Surgery)
BM	Bachelor of Medicine
BPhar	Baccalauréat en pharmacie
BPharm	Bachelor of pharmacy
BS, BSc	Bachelor of Science
BVSc	Bachelor of Veterinary Science
CB, ChB	Chirurgia Baccalaureus (Bachelor of Surgery)
ChD	Chirurgia Doctor (Doctor of Surgery)
CM, ChM	Chirurgia Magister (Master of Surgery)
DCh	Doctor of Surgery
DDM	Doctor of Dental Medicine
D en M	Docteur en médecine
DHyg	Doctor of Hygiene
DM	Doctor of Medicine
DMSc	Doctor of Medical Science
DP, DPharm	Doctor of Pharmacy
DPaed	Doctor of Paediatrics
DPH	Doctor of Public Health
DPhil	Doctor of Philosophy
DS, DSc	Doctor of Science
DSW	Doctor of Social Work
DU(P)	Docteur de l'Université (de Paris)
DVM	Doctor of Veterinary Medicine
DVSc	Doctor of Veterinary Science
FAAAS	Fellow of the American Association for the Advancement of Science
FAAP	Fellow of the American Academy of Pediatrics
FACC	Fellow of the American College of Cardiology
FACOG	Fellow of the American College of Obstetricians and Gynecologists
FACP	Fellow of the American College of Physicians
FACR	Fellow of the American College of Radiology
FACS	Fellow of the American College of Surgeons
FAFPHM	Fellow of the Australian Faculty of Public Health Medicine
FANZCA	Fellow of the Australian and New Zealand College of Anaesthetists [from February 1992]
FAPHA	Fellow of the American Public Health Association
FCOG(SA)	Fellow of the South African College of Obstetricians and Gynaecologists
FCP(SoAf)	Fellow of the College of Physicians, South Africa
FCSSA	Fellow of the College of Surgeons, South Africa

Table 7·2 (*cont.*)

Abbreviation	Term
FFARACS	Fellow of the Faculty of Anaesthetists, Royal Australasian College of Surgeons [until February 1992; see "FANZCA"]
FInstP	Fellow of the Institute of Physics
FRACOG	Fellow of the Royal Australian College of Obstetricians and Gynaecologists
FRACP	Fellow of the Royal Australasian College of Physicians
FRACS	Fellow of the Royal Australasian College of Surgeons
FRCOG	Fellow of the Royal College of Obstetricians and Gynaecologists
FRCP	Fellow of the Royal College of Physicians of London
FRCPA	Fellow of the Royal College of Pathologists of Australasia
FRCPath	Fellow of the Royal College of Pathologists
FRCPC	Fellow of the Royal College of Physicians of Canada
FRCP(Edin)	Fellow of the Royal College of Physicians of Edinburgh
FRCPI	Fellow of the Royal College of Physicians of Ireland
FRCS	Fellow of the Royal College of Surgeons of England
FRCSC	Fellow of the Royal College of Surgeons of Canada
FRCS(Irel)	Fellow of the Royal College of Surgeons of Ireland
FRCVS	Fellow of the Royal College of Veterinary Science
FRIC	Fellow of the Royal Institute of Chemistry
FRS	Fellow of the Royal Society
FRSC	Fellow of the Royal Society of Canada
JD, JuD	Juris Doctor (Doctor of Jurisprudence [or] Doctor of Law)
LLB	Legum Baccalaureus (Bachelor of Laws)
LLD	Legum Doctor (Doctor of Laws)
MA	Master of Arts
MB	Medicinae Baccalaureus (Bachelor of Medicine)
MC, MCh, MChir	Magister Chirurgia (Master of Surgery)
MD	Medicinae Doctor (Doctor of Medicine)
MPH	Master of Public Health
MS, MSc	Magister Scientia (Master of Science)
PhD	Philosophiae Doctor (Doctor of Philosophy)
ScD	Scientia Doctor (Doctor of Science)
SM	Scientia Magister (Master of Science), Science Maître
VMD	Veterinary Medical Doctor

Students in the master's programs will present . . .

John Smith, a professor of chemistry at Stanford, has . . . [his formal title is "Associate Professor of Chemistry"]

Because designations such as Junior (Jr), Senior (Sr), the Second (II or 2nd), and the Third (III or 3rd) are part of the person's name, they precede any academic degrees or honors and are not separated from the personal name by a comma.

Patrick Elliott II MD PhD   David Garrison Sr MB FRCPC

Use such designations only with the full personal name.

Dr James Kelly Jr [not "Dr Kelly Jr"]

Academic titles placed before a personal name (for example, "Professor", "Doctor") and academic degrees appearing after are not included for indexing.

ACADEMIC DEGREES

7·3 A term of address that represents an academic degree and appears before a personal name should not be used with an abbreviation representing the degree after the name.

Dr J B Wingler [or] J B Wingler MD [not "Dr J B Wingler MD"]

Dr P R Cole [or] P R Cole PhD [not "Dr P R Cole PhD"]

The abbreviation of academic degrees should be capitalized according to Table 7·2. Abbreviations are used after full personal names in bylines and in text, and the complete forms are used when referring to the degrees in more general terms.

Lyell Carrington MD represented the association . . .

students in the Doctor of Public Health program . . .

the Bachelor of Medicine degree may be a prerequisite for the Doctor of Medicine program at some institutions . . .

Degrees that are not readily verifiable should be confirmed with the author or another appropriate source.

HONORS

7·4 Professional honors such as fellowships and presidencies indicate levels of membership and particular achievements of members of academic and professional societies and disciplines. These honors are appropriately included in some contexts, for example, in society news articles, biograph-

Table 7·3 Abbreviations of selected honors[a] for meritorious service awarded by the British and Canadian governments

Abbreviation	Honor
BEM	British Empire Medal
CB	Commander of the Order of the Bath
CBE	Commander of the Order of the British Empire
CC	Companion of the Order of Canada
CH	Companion of Honour
CM	Member of the Order of Canada
DBE	Dame Commander of the Order of the British Empire
DCB	Dame Commander of the Order of the Bath
GBE	Knight or Dame Grand Cross of the Order of the British Empire
GCB	Knight Grand Cross of the Order of the Bath
KBE	Knight Commander of the Order of the British Empire
KCB	Knight Commander of the Order of the Bath
KG	Knight of the Order of the Garter
Kt	Knight
OBE	Officer of the Order of the British Empire
OC	Officer of the Order of Canada
OM	Order of Merit
VC	Victoria Cross

[a]Spelled "honours" in British and Commonwealth publications.

ical essays, and obituaries; however, they are usually not appropriate for author bylines in research articles (see section 28·33).

Titles of distinction awarded by governments to honor national service or especially meritorious accomplishments (Table 7·3) may also be appropriate for news articles, biographical material, historical articles, and obituaries.

Abbreviations of degrees come before those representing other honors. Academic degrees should be sequenced in order from the lowest academic level to the highest. In the anglophone world the following sequence is usual: bachelor degrees, master degrees, professional degrees (such as DDS, MD), other doctoral degrees, and honorary degrees. Note that advanced medical degrees awarded after the MD should follow it, MD then MPH. If more than 1 degree at 1 level must be represented, they should come in chronologic order if it is known, and if it is not known, in alphabetic order. Similarly, abbreviations of honors should appear in order of increasing distinction, that is, the least important should appear 1st. Certifications and degrees that are prerequisite for higher degrees should not be listed unless they show the person's specialty.

Academic degrees and honors need not be separated from the personal name with a comma. When more than 1 abbreviation follows a name, they can be separated by commas, but when 2 or more names

Table 7·4 Military ranks with official abbreviations to be used in terms of address for commissioned officers in the Canadian, British, and US military services[a,b]

Rank	Canada	United Kingdom	United States
ARMY			
General	Gen	Gen	GEN
Lieutenant-General[b]	LGen	Lt Gen	LTG
Major-General	MGen	Maj Gen	MG
Brigadier-General	BGen	Brig	BG
Colonel	Col	Col	COL
Lieutenant-Colonel	LCol	Lt Col	LTC
Major	Maj	Maj	MAJ
Captain	Capt	Capt	CPT
Lieutenant or 1st Lieutenant	Lt	Lt	1LT
2nd Lieutenant	2Lt	2nd Lt	2LT
NAVY[c]			
Admiral	Adm	Adm	ADM
Vice-Admiral	VAdm	Vice Adm	VADM
Rear-Admiral	RAdm	Rear Adm	RADM
Commodore	Cmdre	Cdre	COMO[d]
Captain	Capt(N)	Capt	CAPT
Commander	Cdr	Cdr	CDR
Lieutenant-Commander	LCdr	Lt Cdr	LCDR
Lieutenant	Lt(N)	Lt	LT
Sub-Lieutenant or Lieutenant, Junior Grade	S/Lt	Sub Lt	LTJG
Acting Sub-Lieutenant or Ensign	A/S/Lt	[No equivalent]	ENS
AIR FORCE[e]			
General	Gen	ACM (Air Chief Marshal)	Gen
Lieutenant-General	LGen	AM (Air Marshal)	Lt Gen
Major-General	MGen	AVM (Air Vice-Marshal)	Maj Gen
Brigadier-General	BGen	Air Cdre (Air Commodore)	Brig Gen [Air Force] BGen [Marine Corps]
Colonel	Col	Gp Capt (Group Captain)	Col
Lieutenant-Colonel	LCol	Wg Cdr (Wing Commander)	Lt Col
Major	Maj	Sqn Ldr (Squadron Leader)	Maj
Captain	Capt	Flt Lt (Flight Lieutenant)	Capt
Lieutenant	Lt	Fg Off (Flying Officer)	1st Lt
2nd Lieutenant	2Lt	Plt Off (Pilot Officer)	2nd Lt

[a] Sources: Canadian Forces Administration Order 3-5, Annex A. [No date]. Ottawa (ON): Canadian Department of National Defence. Defense Intelligence Agency Manual, DIAM 10-1. 1979. Washing-

appear in a sentence, the commas separating abbreviations could lead to a confusing sequence of commas.

> The appointee to the chairmanship of the department is Ian Maclaughlan BM PhD DSc(hon), whose elegant work on . . .
>
> Eleanor Jones MD FAAP, Robert Baron PhD, and Stephanie Russell DDS FACD will receive awards from Temple University for their teaching skills.

If an author is on active military service, only the service designation should be used; for example, MC for Medical Corps.

MILITARY TITLES

7·5 In text, it is preferable to spell out titles denoting rank or position. The unabbreviated forms of military ranks that precede a personal name are capitalized.

> The award went to Captain Desmond Grover . . .
>
> The researchers were greeted by Lieutenant Colonel Jennifer Ford . . .

The abbreviations of military titles or ranks (Table 7·4) are appropriate in tables, addresses, or military text or when many such titles must appear and space is limited. The abbreviation of a title or rank precedes, and that of a military service (if appropriate) (Table 7·5) follows, the personal name. A comma separates the name and the abbreviation representing military service.

> COL Adrian Locke, MC, USA CPT Susanna Fort, USA
>
> CDR Roberta Simpson, MC, USN LCDR Jeffrey Malone, USCG
>
> Maj David Eldridge, USMC Brig Gen Colleen Drysdale, USAF

Because the official abbreviations of ranks used by the military services may be difficult for a nonmilitary reader to understand, it may be appropriate to use a longer but more easily comprehended form. Suggested

Notes to Table 7·4 *(cont.)*

ton (DC): US Defense Intelligence Agency. p 151. Personal communication, Sqn Ldr Jeff Clyde, British High Commission, Ottawa (ON), January 1992.

[b] The Canadian and UK services use a hyphen for most of the compound ranks (Brigadier-General, Lieutenant-General); the US services do not.

[c] Includes the Coast Guard for US forces.

[d] In the United States this rank is used only in time of war or national emergency.

[e] Includes the Marine Corps for US forces. The full names of ranks in the Royal Air Force are given within parentheses.

Table 7·5 Abbreviations to be used for US military services

Abbreviation	Military service
MC, ANG	Air National Guard Medical Corps
MC, ARNG	Army National Guard Medical Corps
MC, USAF	US Air Force Medical Corps
MC, USA	US Army Medical Corps
MC, USN	US Navy Medical Corps
PHS	Public Health Service
USAF	US Air Force
USA	US Army
USCG	US Coast Guard
USMC	US Marine Corps
USN	US Navy
USAFR	US Air Force Reserve
USAR	US Army Reserve
USCGR	US Coast Guard Reserve
USMCR	US Marine Corps Reserve
USNR	US Naval Reserve

unofficial forms can be found in *The Chicago Manual of Style* (UCP 1993).

Lt Gen [instead of "LTG" for "Lieutenant General"]

Vice Adm [instead of "VADM" for "Vice Admiral"]

Lt Col [instead of "LTC" or "LCol" for "Lieutenant Colonel"]

The unabbreviated names of the military services are capitalized (Table 7·5).

The US Army Medical Corps is proud of its record . . .

NAMES OF ORGANIZATIONS

7·6 The complete form of the name of an organization should be given the 1st time it is mentioned in a document; it may subsequently be abbreviated.

American Chemical Society	ACS
Council of Biology Editors	CBE
National Environment Research Council	NERC
National Institutes of Health	NIH
National Research Council of Canada	NRCC, usually NRC
World Health Organization	WHO
United States Geological Survey	USGS

The shortening of an official name can take several forms, including inversion, colloquialization, and truncation (in addition to the initialisms illustrated above). Once the entity has been fully named, subsequent references in 1 of these forms need not be capitalized, a style widely used in newspapers.

> [in a science newspaper referring to the US Geological Survey]
> After mapping Mars, the survey issued the first planetary atlas.

For some readerships and types of publication, however, notably documents of the named organization, the formal capitalization may have to be retained in the shortened or abbreviated form.

> [in a publication of the US Geological Survey]
> On 1 July, the Survey will release the *Geological Atlas of Mars and Venus.*

SCIENTIFIC NAMES

7·7 Many style conventions developed by scientific associations govern the formation of scientific names and the details of proper usage in scientific style. See Table 7·6 for a summary of the chapters relevant to the names of various scientific subjects.

GEOGRAPHIC NAMES

7·8 In general, a geographic entity should be referred to by the accepted name in the country where it is located. Consult an authority such as *Webster's New Geographical Dictionary* (MW 1988) or *Cambridge World Gazetteer: A Geographical Dictionary* (Munro 1990) to confirm spellings. Sometimes major features or populated places have an anglicized name that has become accepted through long use. Unless there is a risk of misinterpretation, these anglicized versions are preferable in English text.

> Florence [for "Firenze"]   Munich [for "München"]
> Cologne [for "Köln"]

In most contexts in scientific writing, short forms of country names and of names of constituent parts are sufficiently accurate for identification of areas and sites of observation or research. Note, however, that in some contexts (such as references to the government or its agencies) the full formal name of a country may be needed.

Table 7·6 Scientific names: relevant chapters in this manual

Subject	Chapter
Astronomical objects	27, Astronomical Objects and Time Systems
Archaeologic subjects	25, Human History and Society
Biologic kingdoms	
Kingdom Prokaryotae (Monera)	22, Bacteria
Kingdom Protista (Protoctista)	23, Plants, Fungi, Lichens, and Algae
Kingdom Fungi	23, Plants, Fungi, Lichens, and Algae
Kingdom Plantae	23, Plants, Fungi, Lichens, and Algae
Kingdom Animalia	24, Human and Animal Life
Chemical elements and names	15, Subatomic Particles, Chemical Elements, and Related Notations
	16, Chemical Names and Formulas
Disease names	24, Human and Animal Life
Drugs	19, Drugs and Pharmacokinetics
Ethnic groups	25, Human History and Society
Fossils	26, The Earth
Particle physics	15, Subatomic Particles, Chemical Elements, and Related Notations
Rock formations	26, The Earth
Soils	26, The Earth
Viruses	21, Viruses
Units of measure	11, Numbers, Units, Mathematical Expressions, and Statistics

the Islamic State of Afghanistan [for "Afghanistan"]

the Gabonese Republic [for "Gabon"]

the Hashemite Kingdom of Jordan [for "Jordan"]

the Kingdom of Spain [for "Spain"]

Many of these formal names can be found in *The Oxford English Encyclopedic Dictionary* (Hawkins and Allen 1991) in the definitions of country names serving as entries in their short common form. A comprehensive list of the short and the formal names can be found in "Annex 2" of the *WHO Editorial Style Manual* (WHO 1993), along with the appropriate form of the adjective representing the country and the term for its people.

See sections 13·5–13·12 for more information about geographic names and section 8·21 on capitalization.

DRUGS AND REAGENTS

7·9 When drugs and reagents are referred to in the text of a scientific article, the nonproprietary (generic) name must be given. If desired, the trade name (proprietary name) may be given within parentheses after the nonproprietary name; the trade name must be capitalized. In editorials and less formal writing, trade names are often used alone. *USAN and the USP Dictionary of Drug Names* (USPC [annual]) and the *Compendium of Pharmaceuticals and Specialties* (CPA [annual]) provide information about trade and nonproprietary names of drugs. *The Merck Index* (Budavari 1989) provides information about the names of chemical reagents. For more detailed guidelines on referring to drugs and for other sources for drug names, consult Chapter 19.

TRADE NAMES

See section 8·25 for a discussion of trade names and trademarks.

REFERENCES

Cited References

[AMA] American Medical Association. 1989. Manual of style. 8th ed. Baltimore: Williams & Wilkins.

Budavari S, editor. 1989. The Merck index: an encyclopedia of chemicals, drugs, and biologicals. 11th ed. Rahway (NJ): Merck.

[CBE] Council of Biology Editors, CBE Style Manual Committee. 1983. CBE style manual. 5th ed. Bethesda (MD): CBE.

[CMA] Canadian Medical Directory. [annual]. Don Mills (ON): Southam Business Information & Communications Group.

[CPA] Canadian Pharmaceutical Association. [annual]. Compendium of pharmaceuticals and specialties. Ottawa: CPA.

Di Lauro A, Brandon E. 1990. Manual for preparing records in microcomputer-based bibliographic information systems. IDRC-TS67e. Ottawa (ON): International Development Research Centre.

Hawkins JM, Allen R, editors. 1991. The Oxford encyclopedic English dictionary. Oxford: Clarendon Pr.

Mossman J. 1994. Acronyms, initialisms & abbreviations dictionary 1994. 18th ed. Detroit: Gale Research.

Munro D, editor. 1990. Cambridge world gazetteer: a geographical dictionary. Cambridge (UK): Cambridge Univ Pr.

[MW] Merriam-Webster. 1988. Webster's new geographical dictionary. Springfield (MA): MW.

[UCP] University of Chicago Press. 1993. The Chicago manual of style. 14th ed. Chicago: UCP.

[USPC] United States Pharmacopeial Convention. [annual]. USAN and the USP dictionary of drug names. Rockville (MD): USPC.

Walton J, Beeson PB, Bodley Scott R, editors. 1986. The Oxford companion to medicine. Oxford: Oxford Univ Pr.

[WHO] World Health Organization. 1993. WHO editorial style manual. Geneva: WHO.

Additional References

[AGPS] Australian Government Publishing Service. 1988. Style manual for authors, editors and printers. 4th ed. Canberra: AGPS Pr.

Ash M, Ash I. 1993. Chemical tradename dictionary. New York: VCH.

Department of the Secretary of State of Canada. 1985. The Canadian style: a guide to writing and editing. Toronto: Dundurn Pr.

Freelance Editors' Association of Canada: Burton L, Cragg C, Czarnecki B, Paine SK, Pedwell S, Phillips IH, Vanderlinden K. 1987. Editing Canadian English. Vancouver (BC): Douglas & McIntyre.

Huth EJ. Medical style & format. 1987. Philadelphia: ISI Pr. Available from Williams & Wilkins, Baltimore.

8 Capitalization

> . . . an admiral of the fleet is not necessarily identical with an Admiral of the Fleet, nor a foreign secretary with a Foreign Secretary.
>
> —H W Fowler, *A Dictionary of Modern English Usage,* 1965

Some capitalization conventions such as those for capitalizing the names of persons and cities are widely known and accepted; others such as those for position titles are often debated and will differ through the preferences of individual publishers. Capitalization can imply authority and importance, and it has been suggested that the British are "more favourably disposed toward authority than North Americans" (FEAC 1987; p 28) and therefore tend to use capitals more liberally than Americans.

This manual recommends a spare style, so the use of an initial capital letter is restricted to those instances where it is clearly warranted.

SYNTACTIC CAPITALIZATION

FIRST WORD OF A SENTENCE

8·1 The 1st word of every complete sentence should be capitalized except when the sentence is within parentheses or brackets in another complete sentence.

> We delineated 6 study plots measuring 3 × 3 m; no plots were adjoining.
>
> Subjects were assigned to groups on the basis of age and symptoms (sex was recorded for each person but was not a factor in determining group placement).
>
> Are there enough data to support the author's conclusions?

If the sentence begins with an accepted symbol that has an initial lowercase letter, it should be restructured, if possible, to reposition the symbol; otherwise, the lowercase form should be retained, and a capital should be used on the 1st appropriate word.

> Student *t*-test was applied to each set of data. [preferred to "*t*-Tests were used in the analyses."]
>
> The pH must be carefully controlled. [preferred to "pH must be carefully controlled."]

If the sentence begins with a chemical name that has a locant or other prefix, it should be restructured, if possible, to reposition the chemical name; otherwise, the root term is capitalized, not the prefix.

> α-Toluene can be used for this purpose.
>
> [rewrite to] The solvent α-toluene can be used for this purpose.

> crs5 Cells were maintained under these conditions for several months.
>
> [rewrite to] These conditions were suitable for maintaining crs5 cells for several months.

Multiplying prefixes (uni, bi, tri, and so on) are treated as integral parts of the words they modify and should be capitalized if they appear at the beginning of a sentence.

> Univalent molecules are created in this reaction.

> Bicarbonate levels in the blood were abnormally low.

If the 1st word of the sentence is a proper name that usually begins with a lowercase letter, it should be capitalized.

> We agree with the conclusion reached by du Pont.
>
> Du Pont argued this point in an earlier paper.

FIRST WORD AFTER A COLON

8·2 The 1st word after a colon should be capitalized if it begins a direct quotation; the word can be capitalized if what follows forms a complete sentence or independent clause that does not logically depend on the preceding clause.

> Dr Frost spoke for the committee: "We are concerned about the escalating costs and recommend an immediate inquiry." [capital required]
>
> When we examined the evidence, we were led to ask an unexpected question: Did the fish arrive in the lake through an underground connection with the river? [capital optional]

If what follows the colon does not form a complete sentence, a capital letter should not be used.

> Two alternatives for treatment were proposed: surgical resection or relief of the pain through medication.
>
> Each candidate will be given the same materials for the test: exam booklets, a calculator, and pencils.

QUOTATIONS

8·3 The 1st word of a direct run-in quotation that is complete in itself, formally introduced, and not grammatically joined to what comes before is capitalized.

> The program's director said, "We are now working to incorporate this new information into our budget."

If the quotation is a syntactic part of the sentence, a capital is not required.

> One professor stated her concern that "the quality of research is no longer the deciding factor".

See section 9·9, item 4, for an additional recommendation on the capitalization of quoted material.

FOOTNOTES

8·4 If a footnote is a complete sentence, the 1st word should be capitalized and the note should end with a period. For clarity, unless the footnotes in a group are all very short, it may be best to end each with a period.

a Radiobiology terms are defined in the glossary.

b NS = not significant.

PREFIXES

8·5 Prefixes such as "anti", "ante", "ex", "inter", "mid", "post", "pre", "semi", and "un" remain lowercase when connected to a proper noun that does not appear at the beginning of a sentence.

mid-July anti-American post-Christmas

In scientific terms the hyphen is often dropped and the prefix attached to the stem; see section 5·16. Some prefixes are capitalized in certain terms ("Pan-American"), others have been incorporated into a single capitalized ("Precambrian") or lowercased word ("transatlantic"). There are no rules for the words that have evolved in these ways; a dictionary should be consulted for a choice.

TITLES AND HEADINGS

8·6 The rules that follow for capitalization within a title or heading apply when the title appears at the beginning of the work or within running text. These rules apply to books, pamphlets, newspapers, magazines, poems, articles, and lectures and to the parts, chapters, and sections of a published work. This capitalization is recommended to help the reader more readily distinguish a title from the adjacent text.

TITLES

Multiterm Titles

8·7 The 1st and the last word of a title should be capitalized, regardless of the part of speech. All nouns, pronouns, verbs, adjectives, adverbs, and subordinating conjunctions should be capitalized. Articles (a, an, the), coordinate conjunctions (and, or, nor), and prepositions that do not appear at the beginning of the title and are not part of a phrasal verb should not be capitalized.

The Beginnings of Life is a monograph on the early stages of fetal development.

She suggested that new users consult the brochure *Knowing Where to Look.*

Beauty and the Dreadful Beast   To Tread Lightly on the Earth

Weight Loss without Dieting

A preposition that forms an integral part of a phrasal verb should be capitalized.

Doing Without the Extras: An Approach to Fiscal Management

A good introduction is Finwinder's article "Bubble Theory: Finding Their Way Up".

How I Did Without   My Life: What I Put Up With

In titles that have a marked break indicated by punctuation, the article, preposition, or conjunction that immediately follows the break should be capitalized (but not in references).

Hibernation in Canadian Mammals: A Review

Saturated Fats—Out of the Frying Pan

The word "to" in infinitives within a title should not be capitalized.

Learning to Read   To Move or to Stay: The Ethnic Dilemma

8·8 Locants and other similar prefixes of chemical names should not be capitalized in titles, although the root terms should be.

β-Agonists in Respiratory Medicine

L-Erythrose and Related Sugars ["L" is a small capital]

8·9 In book reviews and similar settings, capitalization of a title should follow that of the original work; this form can usually be found on the verso of the title page. Note, however, that original titles in all-capital letters may appear too conspicuous in running text if the original capitalization is followed. Therefore, such titles can be advantageously capitalized in accordance with the other rules in this section. Remember that titles in bibliographic references have capital letters only as the initial letter of the 1st word and of proper nouns; see section 8·14.

8·10 The 1st word of a hyphenated term should be capitalized. Both components of a 2-word hyphenated term should be capitalized if the term is a temporary compound (as in a unit modifier that would not otherwise be hyphenated) or a coordinate term.

Behavior of Well-Adjusted Children in the Classroom

Determining Acid–Base Status

["Acid–Base" is a coordinate term with en-dash linkage]

Nitrogen-Fixing Bacteria

The 2nd and subsequent components of a term that would normally be hyphenated should not be capitalized unless they are nouns or proper adjectives or their grammatical weight is the same as or greater than that of the 1st component.

How to Plant Forget-me-nots   Neo-Darwinian Approaches to Evolution

Attitudes toward Abortion among Non-Christian Women in New York

8·11 Scientific names should be capitalized as they would be in running text (see section 8·26); therefore, the specific epithet in the name of a species should not be capitalized.

The Metabolism of *Escherichia coli*   *Homo sapiens* and Predecessors

Single-Word Titles

8·12 Single words serving as titles are capitalized in accordance with the convention presented in section 8·7.

She [a novel]   Style [a guide to writing good prose]

Some trademarks that serve as titles of published material or organizational names have, in addition to the conventional initial capital letter, 1 or more internal capital letters.

WordPerfect [a word-processing program]

MedEdit [name of an organization]

These conventions should be observed, but note in section 8·9 the conversion of titles in all-capital letters to conventional capitalization. (See also section 8·7.)

PARTS OF WORKS

8·13 The titles of parts of works appearing in cross-references should be capitalized according to the principles in sections 8·6–8·12.

according to criteria outlined in Materials and Methods . . .

See the chapter "Daily Life with the Chimps".

. . . , who wrote the Preface for this volume, . . .

temperature was the controlling factor (Table 1, Figure 3).

If the part of the work is, however, not referred to in its original form (for example, if a modified form is used or if the original form is used as an adjective), capitalization is not required.

A complete description can be found in the methods section
[or "in the Materials and Methods"].

This discrepancy is analyzed further in the discussion section
[or "in the Discussion"].

in the chapter on living with the chimpanzees
[see the example in the preceding group]

The title of a chapter in a book should be capitalized, as already specified, and enclosed in quotation marks.

The last chapter, "Future Prospects", was originally published in *New Scientist.*

Nouns (or their abbreviations) used with a letter or a numeral representing their position in a series of similar parts of a book, chapter, or accessories to text are usually capitalized.

Volumes 2–4 Part A, Parts 3 and 4 Plate II, Plates I and II

Figure 6, Figures 9–12 Table 3, Tables 6 and 7

The words "line" and "page" are always set lowercase, not capitalized, and should not be abbreviated in running text.

The only error I could find is on page 37, line 10.

REFERENCE LISTS

8·14 See sections in Chapter 30 for recommendations on capitalization in reference lists and bibliographies: 30·23 (article titles), 30·24 (journal titles), 30·34 (book titles), and 30·43 (conference and subject titles)

INDEXES

8·15 Capitalization should be reserved for terms that would have an initial capital in running text, such as proper names. Users of indexes can thus be helped in distinguishing common from proper nouns, for example, in an index of microbiologic names, and nonproprietary drug names from trade names.

hexylresorcinol	[*US Pharmacopoeia* nonproprietary drug name]
Hibiclens	[Stuart trademark for its brand of chlorhexidine gluconate]
Histalog	[Eli Lilly trademark for its brand of betazole hydrochloride]
histidine	[International Nonproprietary Name (generic) for an amino acid]

Even if the editor chooses a style that calls for initial capitalization of every term, some terms should retain an initial lowercase letter.

1 Compound surnames in which the prefix begins with a lowercase letter (du Pont, von Willebrand disease).
2 The italicized prefix of a chemical compound (*p*-Aminobenzoic acid, *o*-Toluic acid).
3 A standard symbol or abbreviation that begins with a lowercase letter (pH, pK′, mRNA).

Also see section 31·36.

TABLES

8·16 See sections in Chapter 31 for recommendations on capitalization in tables: section 31·4 (table titles), 31·6 (column headings), 31·8 (row headings), 31·10 (text-field entries), and 31·16 (general considerations).

NAMES

PROPER NOUNS

8·17 An initial capital letter, or capital letters, should be used for all proper nouns, including the names of persons and places, the official names of organizations, institutions, and political entities, and some adjectives based on proper nouns.

Albert Einstein Vancouver, British Columbia

Council of Biology Editors

the Annual Meeting and Scientific Assembly of the Radiological Society of North America

Canadian weather patterns Roman civilization

Truncated terms that retain the specifying elements of a name may be capitalized.

> the National Library [for "the National Library of Medicine"]

A lowercase letter should be used for truncated terms that eliminate the specifying element or elements and contain only a generic term unless the lack of capitalization would produce ambiguity.

> the hospital the state the institute
>
> the Bank [for "The World Bank" in text also referring to other banks]

A publisher may decide to capitalize even these truncated forms or commonly used but unofficial forms, especially in publications intended for an internal audience or in news-type publications, when the distinctive importance of the institution is supported by capitalization.

> the University's policy will be . . .
> [when readers of the document wish to see the particularity of the institution symbolized by the capitalization]
>
> the Brundtland Commission
>
> Welcoming Dr Pitt to the Laboratory were . . .

In a plural construction, lowercase should be used for the generic part of the names that would be capitalized in the singular, unless it precedes the specific names.

> Harvard and Princeton universities
>
> the Departments of Botany, Forestry, and Zoology
>
> the Universities of California and British Columbia
>
> Toronto Western and Mount Sinai hospitals

The name of a discipline or specialty should not be capitalized except when it is part of a proper name, such as the name of a department.

> students interested in pursuing a career in astrophysics . . .
>
> The Department of Astrophysics will host a seminar . . .
>
> The discipline of internal medicine is attracting fewer medical graduates.
> [not "The discipline of Internal Medicine is attracting fewer medical graduates."]

For recommendations on capitalizing the names of racial, linguistic, tribal, religious, and other groupings within the human race and dates, consult sections 25·3–25·11.

EPONYMIC TERMS

8·18 The 1st letter of the proper name in an eponymic term for a virus, a disease, a syndrome, a named chemical reaction, or a named equation that incorporates a proper name referring to a theoretician, researcher, physician, patient, or place should be capitalized. Derivative or adjectival forms are not capitalized.

Bareggi reaction	Hodgkin disease
cesarean section	Hunter sore (hunterian chancre)
Down syndrome	linnaean system of classification
fallopian tube	duct of Müller (müllerian duct)
Gasser (gasserian) ganglion	Parkinson disease (parkinsonian tremor)
graafian follicle	Wilms tumor
Gram stain (gram-negative bacteria)	

For discussion of possessive forms of eponyms, see section 5·38.

Several kinds of terms should not be capitalized: noncapitalized terms established by an authoritative decision (such as names of units); terms derived from proper nouns that relate to objects, especially apparatuses; terms in well-established common usage.

ampere	congo red	joule	portland cement
angstrom	coulomb	lambert	roentgen
benday process	curie	london purple	roman type
bessemer steel	diesel fuel	merino sheep	rutherford
brussels sprouts	draconian	paris green	timothy grass
bunsen burner	gauss	petri dish	venturi tube
burley tobacco	india ink	plaster of paris	

Special rules should be applied when an eponymic adjective may be used in phrases with different contexts. An eponymic adjective should be capitalized when the phrase is used in a context directly and specifically concerned with the original historic and cultural meaning of the phrase and lowercased when the phrase is used in contexts not directly concerned with those origins, as in contexts concerned with present-day applications.

roman alphabet	[for present-day uses of letters having their origin in ancient Rome; the historic Roman alphabet was modified in the Middle Ages to its present form]
Roman alphabet	[in contexts dealing with the alphabet of ancient Rome]
Roman civilization	[in contexts referring directly to the civilization of Rome]

Roman numerals	[in contexts concerned with uses of numerals in ancient Rome]
roman numerals	[in contexts discussing present-day uses of numerals from ancient Rome]
roman type	[the Romans did not have type; the type was developed centuries later to represent the present roman alphabet]
Arabic medicine	[in contexts referring to medicine in Arabic culture]
arabic numerals	[numerals developed within historic Arabic culture but referred to in contexts dealing with present-day uses; present arabic numerals are not identical in form with those of their origin]

Verbs derived from proper nouns should be in lowercase.

italicize pasteurize

Terms such as "Anglophile" and "Francophobe" may be capitalized for some readerships but are preferably not capitalized because of their adjectival form.

English written by francophone authors may incorporate confused translations of "false cousin" terms.

TITLES OF PERSONS

8·19 A professional, civil, military, or religious title is capitalized when it immediately precedes the name of the person and is used to address the person as part of the name.

Professor Katherine Dobbs President David Green General Shaw
Archbishop Morland Dean Cablebein

If the person's title is set off from an institutional or organizational name by a comma, the title should be capitalized, because it represents the formal title of the position.

David Green, President, the University of Arizona

If the title follows the name, is descriptive of the person, is linked to an institutional or organizational name by "of", "for", "in", or "at", and is not necessarily the complete formal title, it should be in lowercase so that the term after the comma is not misconstrued as the formal title.

Katherine Dobbs, a professor of chemistry at Syracuse, published . . .

David Green, president of the University of Arizona, announced that . . .

Michael Shaw, general in the US Army

Peter Morland, archbishop of New York

If the position title precedes the name but is immediately preceded by a modifier, such as an article or an adjective, and is separated from the name by a comma, the 2 terms are said to be in nonrestrictive apposition (that is, the person's name or the title could be used alone, although using both provides more information). In this case, the position title need not be capitalized, although a publisher may decide to do so in certain documents.

a director of the board, Joseph Smith, wrote to . . .

the president of the corporation, Henrietta Cullingham, joined . . .

the chair of the committee, Melinda Jenkins, recommended . . .

ACADEMIC DEGREES AND HONORIFIC TERMS

8·20 Academic degrees and scholastic and military honors and decorations should be capitalized when they follow a proper name, whether they are written in full or abbreviated; see Table 7·2 for examples of abbreviations.

Bernice Cooper MD FRCSC Kerry Stanton PhD

Debra Kellington, Fellow of the Royal Geographic Society

Denton McLeod OBE

When a degree or honor is referred to as such, it may be lowercased. If the desire is to represent it by the formal title (as in a publication of the institution), it should be capitalized.

The university grants master of science degrees in many disciplines. [in a newspaper]

The university grants Master of Science degrees in 20 disciplines. [in the university's catalogue]

GEOGRAPHIC DESIGNATIONS

8·21 The names of the great divisions (zones) of the earth's surface and the names of distinct regions or districts of the earth should be capitalized; adjectives and some nouns based on these terms begin with a lowercase letter.

the Antipodes (in reference to Australia and New Zealand); antipodean

the Great Divide   the Tropic of Cancer, the tropics

North Temperate Zone, the temperate zone   the Equator, equatorial

Formal terms for the hemispheres of the earth and their inhabitants are capitalized.

Eastern Hemisphere, Eastern(er)   Western Hemisphere, Western(er)

The names of the points of the compass and derived adjectives are not usually capitalized.

northern Saskatchewan   an eastern exposure   traveled south . . .

western songbirds   visible to the north of . . .

They should be capitalized when they form part of a name that designates a specific area.

East Africa   the Midwest, the Middle West [but "midwesterner"]

Sherman's march to the sea drove a wedge through the South.

North Africa [but "central Africa"]

Southeast Asia, southeastern Asia, central Asia

The Arctic is the region of the earth north of the Arctic Circle. The word "arctic" is used adjectivally in several ways: when it refers to the geographic region, the adjective should be capitalized, as in "Arctic communities"; when it refers to very low temperatures, it should not be capitalized, as in "arctic gale". Established names of Arctic flora and fauna are usually lowercased, for example, "arctic char". Also see section 8·18 on capitalizing adjectives derived from proper nouns.

Many regions and localities have informal names that are recognized as representing a specified area; these should be capitalized.

the Levant [the countries bordering on the eastern Mediterranean Sea]

the Great Plains [plains region in Canada and the United States east of the Rockies]

the Bruce Peninsula [peninsula in Lake Superior, Ontario]

the National Capital [USA]   the National Capital Region [Canada]

the Maritimes [Nova Scotia, New Brunswick, Prince Edward Island; Canada]

Such terms should be used with care in scientific publications. If one is used to describe, for example, a collecting area, readers unfamiliar with the boundaries of the regions could be deprived of important information; furthermore, the definition of such an area may not be absolute. Locations should be specified when possible by geographic coordinates or, more rarely, by map-indicated political boundaries (depending on the context) to eliminate any possibility of confusion.

The article "the" should be capitalized when it is part of a geographic name; to determine whether "the" is part of the name, consult a comprehensive atlas or dictionary such as *Webster's New Geographical Dictionary* (MW 1988) and *Cambridge World Gazetteer: A Geographical Dictionary* (Munro 1990).

The Gambia   The Hague   the Black Forest   the Bronx
the Netherlands

Generic geographic terms (such as "lake", "river", "ocean", "mountain") that form part of a proper name should be capitalized, unless the generic noun and the word "the" precede the proper noun.

Atlantic Ocean   Canadian Shield   Canoe Lake   Fraser River
Grouse Mountain   Mount Everest   Nile River [but "the river Nile"]

A generic term that follows a capitalized generic term should not be capitalized.

the Fraser River valley [but "the Fraser Valley"]   the Rio Grande valley

In a plural construction, lowercase may be used for the generic noun that would be capitalized in the singular unless the common noun precedes a group of proper nouns.

Great Slave and Lesser Great Slave lakes
Vancouver and Saltspring islands   Lakes Ontario and Huron

Some geographic names contain foreign words that are the equivalents of generic terms (for example, "rio" means "river"; "mauna", "sierra", and "yama" mean "mountains"); therefore, the generic English term should not be added.

Fujiyama [or] Mount Fuji [not "Fujiyama Mountain" or "Mount Fujiyama"]
Mauna Loa [not "Mauna Loa Mountain" or "Mount Mauna Loa"]
Sierra Madre [not "Sierra Madre Mountains"]
Rio Grande [not "Rio Grande River"]   the Sahara [not "Sahara Desert"]

For recommendations on selection of geographic names and on their capitalization, see sections 13·7–13·9, and on capitalization in addresses, section 13·1.

GEOLOGIC NAMES

8·22 The accepted names for geologic and stratigraphic time units are usually capitalized in formal usage, but the 2nd element may be lowercase in running text (or omitted altogether when the 1st term is preceded by "the").

Archean Eon   Cenozoic Era   Devonian Period

Pleistocene Epoch   disappeared during the Pleistocene and was not . . .

Names of geologic formations are capitalized only if they have been formally published according to the rules set out in the *North American Stratigraphic Code* (NACSN 1983). Lists of names that have been published acceptably can be found in the *Lexicons of Canadian Stratigraphy* (CSPG); other guidance on style can be found in the *Guide to Authors* (GSC 1979) of the Geological Survey of Canada.

Consult Chapter 26 for a discussion of geologic names and recommendations for style; Tables 26·1 and 26·2 give the names of geologic time.

ASTRONOMICAL NAMES

8·23 The proper names of planets, stars, and other astronomical bodies should be capitalized.

Mars   the North Star (Polaris)   the Milky Way   Halley's Comet

The words "earth", "moon", and "sun" are capitalized only when they are used in connection with other astronomical terms.

The sun provides energy for photosynthesis.

The eccentric orbit of Pluto takes it around the Sun only once every 248.5 years.

Gravitational effects between the moon and the earth influence the tides.

The Moon was the 1st destination for space travelers; Mars may be reached in our lifetime, but manned travel to Venus or Saturn is unlikely for various reasons.

The word "earth" is never capitalized when preceded by the article "the".

the 4 corners of the earth   the salt of the earth

The sequence of planets is Mercury, Venus, Earth, Mars . . .

See Chapter 27 for more detail on names of astronomical objects.

NAMES FOR ABSTRACTIONS

8·24 The names of abstract ideas or objects (including the seasons) can be capitalized when personified. Thus, "Mother Nature" is always capitalized, but "Nature" is capitalized only when the personification is clear.

> Three varieties are found in nature, but more have been developed in the laboratory.
>
> Living near the Arctic Circle, he learned that Nature can be a harsh taskmistress.

TRADEMARKS

8·25 A trademark is a proprietary name for a product legally registered by the firm or the person making the product. If the name must be used, the word or phrase must be capitalized. However, it is preferable to use a generic name whenever possible, because many firms insist on capitalization as well as a certain phrasing of the trademark name and a generic term (for example, "BAND AID Brand Adhesive Bandages") to protect their trademark in law.

> acetylsalicylic acid [instead of "Aspirin"] cola [instead of "Coca-Cola"]
> tissue [instead of "Kleenex"] clear plastic [instead of "Plexiglas"]
> synthetic fiber or polyester [instead of "Dacron"]
> petroleum jelly [instead of "Vaseline"]
> photocopy, photocopier [instead of "Xerox"]
> plastic wrap [instead of "Saran Wrap"]

Note that the trademark "Aspirin" is no longer protected in Australia, Argentina, Britain, France, India, Ireland, Japan, New Zealand, Pakistan, the Philippines, South Africa, and the United States. It continues to be protected in other countries and if used should be capitalized in publications produced in those countries. The following former trademarks are no longer protected and need not be capitalized.

> cellophane escalator kerosene lanolin nylon zipper

The Canadian Trade Index (CMA [annual]), an annual publication of the Canadian Manufacturers' Association, includes a list of the trade names effective in Canada. The *Trademark Checklist* (USTA 1990) provides the same kind of information for the United States. The legitimacy of a trademark in the United States can be ascertained by telephoning the International Trademark Association (formerly the US Trademark Association), 212 768-9887. Also see the compilations by Wood (1988, 1989).

In the United States certain marking conventions are required by law. For marks that have been officially registered ® is used, whereas ™ is appropriate for marks that have not been registered but which the manufacturer wishes to identify as its own. In Canada there are no legal marking requirements. Some Canadian writers may follow the US conventions, but use of ® in a Canadian publication for a nonregistered mark could lead the reader to think that the mark is registered. The French equivalents are MD (marque déposée) for ® and MC (marque de commerce) for ™.

When a drug is mentioned, the brand name may be of interest, especially in studies of characteristics of a particular product. If so, the capitalized proprietary name should appear within parentheses after the 1st use of the nonproprietary name. To determine if a drug name is proprietary or to find the nonproprietary name, consult *USAN and the USP Dictionary of Drug Names* (USPC [annual]) or the *Compendium of Pharmaceuticals and Specialties* (CPA [annual]); see sections 19·3, 19·4, 19·8, 19·14.

ORGANISMS

8·26 The Latin scientific name of a phylum, class, order, family, or genus or any of their subdivisions should be capitalized; specific epithets and the designations of subspecific taxons should not be capitalized, except where permitted by international codes. For example, the *International Code of Botanical Nomenclature* (IBC 1988) allows capitalization of the specific epithet if it is based on the name of a person (actual or mythical), a vernacular name (in any language), or a former generic name, but this is not common practice. The common or vernacular names of organisms are rarely capitalized. Adjectives and English nouns derived from scientific names should not be capitalized or italicized.

salmonid   orthopteran   pneumocystis pneumonia   ameba

A few terms have more than 1 meaning, which may create difficulties in determining whether they should be capitalized. "Metazoa" and "Protozoa" should be capitalized when they are used as the names of divisions. Their lowercase common-noun designations are "metazoon" and "protozoon" (plurals, "metazoa" and "protozoa"). When these words are used with the article "the" (as in the members of the division and not the division itself), they should be in lowercase. A similar problem can arise with "Primates", the name of an order, and "primates", the members of the order.

Gorillas, chimpanzees, and baboons are representatives of the order Primates.

the adaptations of primates to communal living . . .

For details on capitalization for the names of organisms and for representation of other biological elements such as chromosomes and genes, see Chapters 20–24 in Part 3.

REFERENCES

Cited References

[CMA] Canadian Manufacturers' Association. [annual]. The Canadian trade index. Volume 1. Toronto: CMA.

[CPA] Canadian Pharmaceutical Association. [annual]. Compendium of pharmaceuticals and specialties. Ottawa: CPA.

[CSPG] Canadian Society of Petroleum Geologists. Lexicons of Canadian stratigraphy. Calgary: CSPG. Volumes 1, 2, 4, and 6 now available; Volumes 3 and 5 not yet complete.

[FEAC] Freelance Editors' Association of Canada; Burton L, Cragg C, Czarnecki B, Paine SK, Pedwell S, Phillips IH, Vanderlinden K. 1987. Editing Canadian English. Vancouver (BC): Douglas & McIntyre.

[GSC] Geological Survey of Canada. 1979. Guide to authors. GSC Misc Rep No 29. A new edition is in preparation.

[IBC] International Botanical Congress. 1988. International code of botanical nomenclature. Regnum vegetabile 118. Königstein (Germany): Koeltz Scientific Books.

Munro D, editor. 1990. Cambridge world gazetteer: a geographical dictionary. Cambridge (UK): Cambridge Univ Pr.

[MW] Merriam-Webster. 1988. Webster's new geographical dictionary. Springfield (MA): MW.

[NACSN] North American Commission on Stratigraphic Nomenclature. 1983. North American stratigraphic code. Am Assoc Pet Geol Bull 67:841–75.

[USPC] United States Pharmacopeial Convention. [annual]. USAN and the USP dictionary of drug names. Rockville (MD): USPC.

[USTA] US Trademark Association. 1990. Trademark checklist. New York: USTA. A new edition of this checklist is scheduled to appear in 1994.

Wood D. 1988. International trade names dictionary. Detroit: Gale Research.

Wood D. 1989. Trade names dictionary 1989. 7th ed. Detroit: Gale Research.

Additional References

American Medical Association. 1989. Manual of style. 8th ed. Baltimore: Williams & Wilkins.

Ash M, Ash I. 1992. Industrial chemical thesaurus. New York: VCH.

Ash M, Ash I. 1993. Chemical tradename dictionary. New York: VCH.

Dodd JS, editor. 1986. The ACS style guide: a manual for authors and editors. Washington: American Chemical Soc.

Huth EJ. 1987. Medical style & format. Philadelphia: ISI Pr. Available from Williams & Wilkins, Baltimore.

[UCP] University of Chicago Press. 1993. The Chicago manual of style. 14th ed. Chicago: UCP.

9 Type Styles, Excerpts, Quotations, and Ellipses

> Typography is the efficient means to an essentially utilitarian and only accidently minor aesthetic end, for enjoyment of patterns is rarely the reader's chief aim.
>
> —Stanley Morison, *First Principles of Typography,* 1936

> Typography is . . . a craft by which the meanings of a text . . . can be clarified and honored . . . .
>
> —Robert Bringhurst, *The Elements of Typographic Style,* 1992

Various styles of typefaces are often applied in scientific publishing to represent specific meanings of text. Particular arrangements of type on pages and sequences of punctuation in quotations are additional devices that help guide the reader to a clear understanding of texts.

TYPE STYLES FOR STYLE CONVENTIONS

9·1 Virtually all scientific literature in English is printed with a typeface of the roman variety whether it be a serif face like that used for the text of this manual or a sans serif face (such as Helvetica). Variants of basic faces are used for specific conventions in prose style or for specific symbolic meaning. The 2 main variants are italics and boldface. Other conventions call for capital (uppercase) letters, small capitals, superscript characters (raised above the line of type), or subscript characters (depressed below the line). See Chapter 32 for more detail on characteristics of type.

ITALIC TYPE

9·2 Italic type can be represented directly in manuscripts prepared with word-processing programs. Note that some of the "italic" typefaces in these programs are not true italic faces by typographers' criteria but represent "slanted type", a computer-produced slanted version of an upright serif or sans serif typeface. Nevertheless, "italic typefaces" thus produced are acceptable representations for editorial offices and publishers.

In manuscripts prepared on typewriters or written by hand, italic type is indicated by underlining the characters to be typeset in the italic version of the main typeface. If underlining is intended in the published text (a rare usage), the author must write a marginal note adjacent to the beginning or end of the relevant line indicating "Underlining, not italics". See section 32·5 and Table 32·1.

General Uses

Seven uses of italic type are widely applied in published texts, whether general or scientific.

1 In running text for the title of a book, journal, or other complete document. Some publishers also use italics for the full or abbreviated journal titles in bibliographic references as a presumed aid to readers in picking out the title. Quotation marks can be used to distinguish titles of book sections (such as chapters) from titles of whole books (italicized) and of journal articles from journal titles; see section 4·18.

> His report, "The Gene Responsible for Multiple Sclerosis", was published in *The New England Journal of Medicine.*
>
> *Suggestions to Authors* is an authoritative style manual for geologists; the chapter "Geographic Names" will be useful to writers in other fields.

2 For many non-English words and phrases and their abbreviations, but not for non-English proper names. This rule does not apply to widely used phrases such as "a priori", "in vitro", amputated phrases (like "post hoc" representing *post hoc ergo propter hoc* as in "a post

hoc argument"), or abbreviations (such as "et al." for *et alii*) that are now considered standard English in science.

3 For a letter or number in text or a legend referring to the corresponding character in an illustration.

We could not account for the outlier *h* that is conspicuous in Figure 3.

4 For a word or phrase when introduced in text for definition, explanation, or discussion.

The concept of *stratum* took a long time to develop in the evolution of geology.

5 For a word or phrase that must be represented as such.

Typeface unfortunately has given way in desktop publishing to *font*.

Note that this convention and that of item 4 above are not used in this manual because of the need to confine the use of italics as much as possible to scientific conventions; see section 9·3. Terms representing words, illustrative terms, and explanatory terms in this manual are enclosed by quotation marks; see section 1·4.

6 For explanatory words used within text that is in a roman typeface. This use is frequently needed in indexes.

NIH *for* National Institutes of Health
AIDS *for* acquired immunodeficiency syndrome

coronal suture 236
coronary heart disease, *see* ischemic heart disease
corticotrophin 479

7 For a short preface or an explanatory note by a publisher or editor that must be clearly distinguished from the text of the author.

[*The following short paper was submitted anonymously, but the editors felt it merited publication nonetheless. —The Editor-in-Chief*]

One night as I was about to turn in for the 1st sleep I would have had in 48 hours, the phone rang and I was called to the cancer ward to . . .

Scientific Uses

9·3 The scientific uses of italics itemized below should be applied in all scientific publications; other uses may differ in different disciplines.

1 For single-letter symbols standing for a quantity or a variable (known or unknown); see section 3·5 and Table 3·5. This rule applies throughout science. Multiletter symbols for quantities may have to be in a roman face; see Table 3·5. For applications in mathematics and statistics, see sections 11·14 and 11·24 and Tables 11·8–11·13; in chemical kinetics, sections 17·4, 17·11; in pharmacokinetics, section

19·17 and Table 19·2; in cardiovascular and respiratory physiology, sections 24·13, 24·27 and Tables 24·3, 24·6.

2 For single-letter modifiers (usually subscript) of symbols when the modifier itself is a symbol for a quantity; see Table 3·5.

3 For the scientific name of a genus, species, subspecies, or variety. The name of a higher taxon is usually not italicized. For the conventions in virology, see section 21·3; in bacteriology, section 22·6; for plant sciences, sections 23·4, 23·26, 23·31 and Tables 23·1–23·3; for zoology, section 24.2.

4 For letters and numerals designating mutant viruses; see section 21·8.

5 For letters and numerals used as a symbol for a gene or allele in most systems of gene symbolization; see sections 20·23–20·48 and Tables 20·4–20·14.

6 For a prefix to a chemical name identifying the position of a labeled element; see section 16·32.

7 For a prefix identifying a configurational relationship in an organic compound; see sections 16·13, 16·14, 16·24.

8 For the symbols of elements used as locants in the names of organic compounds; see section 16·13.

9 For some crystallographic symbols; see section 26·12.

Note that scientific text that should be in italic type (see item 3 above) may be shifted to roman type when the surrounding text must be in italic type for some reason of design.

> Sablefish, *Anoplopoma fimbria,* is sensitive to handling but . . .
> [changes to] *Sablefish,* Anoplopoma fimbria, *is sensitive to handling but . . .*

In general, design requirements of this kind should be avoided if possible in scientific publishing, but such design may be needed for a level of text heading when many levels are needed.

BOLDFACE TYPE

9·4 Boldface type is marked in typed or handwritten manuscript with wavy underlining (see section 32·5 and Table 32·1. There are no standard uses of boldface in general literature. Some publications assist readers by using boldface for some elements in indexes (such as indicating page numbers of the main text on a subject) and in bibliographic references (such as volume or year numbers). Boldface can be used to indicate letters in terms that are the basis for an acronym or abbreviation, especially when the letters used are not the initial letters in the words they stand for.

> The abbreviation AIDS represents "**a**cquired **i**mmuno**d**eficiency **s**yndrome".

A few scientific uses of boldface type are or were well established.

1 Symbols for vectors in mathematical expressions; see section 11·18.
2 Names of virus families and subfamilies and of virus genera that were formerly in italic boldface are no longer bold; see section 21·3.

CAPITAL (UPPERCASE) LETTERS

9·5 Capital letters are usually typed as such in manuscripts and need not be specifically marked. In manuscript preparation and proof correction, capitalization for existing lowercase letters can be indicated by triple underlining; see sections 32·5 and Table 32·1.

General Uses

The uses of initial capital letters are described in detail in Chapter 8. These uses are summarized below.

1 For the 1st word of a sentence unless the sentence is within parentheses or square brackets within an enclosing sentence; see section 8·1.

> The survey was launched (we had the funds) but soon abandoned.

2 For the 1st word after a colon if the following words form a complete independent clause not logically dependent on the preceding clause; see sections 4·10 and 8·2.

> The Faculty Senate established only 1 rule: Do not plagiarize.

3 For trade names; see section 8·25.

> Xerox Naldegesic Nikon F Listerine

4 For certain words in the titles of articles, journals, books, and other documents: all words except conjunctions (such as "and", "or"), prepositions (such as "between", "for", "to", "within"), and articles ("a", "the") when these exceptions are not the initial word of a title; see sections 8·7–8·11.

> *Within the Amazon Jungle without Permission*
> *A Solution to a Problem*
> *Important Historic Sites between London and Oxford*

5 For the initial letter of the 1st word and for all proper nouns and proper adjectives in titles of articles and books in bibliographic references, whether abbreviated or not; see sections 30·23, 30·34.

Jones J. A study of Hispanic demographics in Boston. Ann Epidemiol Res 1998;12:15–8.

6 For a professional, civil, military, or religious title that immediately precedes a personal name; see section 8·19.

President Harding   Colonel Jones   Monsignor Smith

7 For the official name of a private or government organization or institution; see section 8·7.

Stanford University   US Geological Survey   the City of Tucson

8 For a generic geographic name that is part of a proper name; see sections 8·21 and 13·8.

Lake Michigan   Columbia River   Red Sea

Capital letters are widely used for contraction abbreviations whether the words represented are capitalized or not; see section 10·2.

NIH [for "National Institutes of Health"]
DNA [for "deoxyribonucleic acid"]

Nouns, adjectives, and verbs derived from proper names are usually not capitalized when they have entered common usage; see section 8·18.

sousaphone   petri dish   bunsen burner   pasteurize

Scientific Uses of Capitalization

9·6

1 For the scientific (taxonomic) name of a phylum, class, order, family, or genus and their subdivisions, but not of a specific or subspecific taxon except where permitted by international codes; see sections 21·3, 22·6, 23·4, and 24·2.

2 For many gene, chromosome, blood-group, cardiovascular, and respiratory physiology symbols, and other symbols; see section 3·5, Table 3·5, and sections specifically relevant to a symbol group as noted in Table 3·4.

3 For the name of a formal historical epoch, geologic age or stratum, zoogeographic zone, or other term used for convenience in classification; see sections 25·4–25·8, 25·10, 25·11, 26·3, 26·5, and 26·16.

4 For the proper name of a star or other astronomical body. But note that "earth", "moon", and "sun" are capitalized only when used with names of other bodies in the Solar System; see sections 8·23 and 27·2.

Some organizations in some disciplines recommend capitalizing common names of organisms of interest to them.

5 The complete vernacular or common name of a species of bird in accordance with the checklist of the American Ornithologists' Union; see section 24·4.
6 The common name of an insect in accordance with the list approved by the Entomological Society of America; see section 24·4.

The generally accepted rule is that common names of organisms are not capitalized except for proper-name components of common names.

American black currant   Arizona white oak   Hawaiian bud moth
Virginia pine sawfly

SMALL CAPITAL LETTERS

9·7 Small capital letters (often called "small caps") were generally designed to be the height of the letter x in a particular typeface and font. For typefaces lacking small capital fonts, they are often produced by an arbitrary reduction of capital letters. Small capital letters are marked in manuscript with double-underlining. See section 32·5 and Table 32·1.

General Uses

Small capital letters have often been used for typographic variation judged by a designer to aid the reader in distinguishing some parts of text from each other. Examples of such use include modifying abbreviations like those for academic degrees following personal names and those modifying dates.

Robert Hutchins MD [rather than "Robert Hutchins MD"]

AD 1066 [rather than "AD 1066"]

10:07 AM [rather than "AM" or "a.m."]

This type convention does not convey meaning beyond that carried by capital letters in such text and hence is not recommended for scientific style.

A suggested general use of small capitals (Dong 1993) is to set surnames (family names) in initial capital letters and small capital letters in author bylines of articles and author designations in references to make clear which part of the name is the surname.

NGUYEN Van Cu   DONG Geng   SZENT-GYORGI Albert

Adam SMITH

Table 9·1 Examples of superscript symbols

Item represented	Type of symbol	Symbol	Section or table in which described
Celestial coordinate, right ascension	Alphabetic	$14^{h}17^{m}13^{s}$	27·3
Citation by reference number	Numeric	. . . paper^{12} in . . .	30·5
Degrees, geographic	Graphic	75° 00′ 15″ W	13·13
Degrees, temperature	Graphic	50 °C	11·10
Electric charge of a particle	Graphic	p^{+}	15·4
Electronic spin multiplicity	Numeric	$^{2}P_{1/2}$	15·8
Integration limit, upper	Numeric	$\int^{\infty}$	11·14
Ionic charge	Graphic	Na^{+}	15·6
Ionic charge	Numeric and graphic	Al^{3+}	15·6
Mass number	Numeric	^{14}C	15·6, 16·31
Oxidation number	Roman numeric	Mn_{VII}	15·6
Phenotype, wild-type	Graphic	Tol^{+}	Table 20·4
Power function	Numeric	50–250 W/m^{2}	11·20
Power function	Alphabetic	e^{x}	Table 11·8
Thermodynamic state	Graphic	ΔG°	17·5

Scientific Uses

Small capital letters have been designated in several conventions to indicate particular kinds of meaning.

1 Configurational prefixes (D, L) for carbohydrates and amino acids; see sections 16·21, 16·16, and 16·24.
2 Modifiers indicating the anatomic site of a gas phase for symbols of respiratory function; see Tables 24·3 and 24·6.

SUPERSCRIPT AND SUBSCRIPT CHARACTERS

9·8 Alphabetic, numeric, and graphic characters in a superscript position (above the main line of type) are widely used in scientific notation as qualitative and quantitative modifiers of symbols on the line. These should usually be in a smaller font than that used for the main line of type. Some superscripts may themselves have superscripts. Table 9·1 illustrates many of the most common kinds of superscripts and cites the sections of this manual that describe their use in detail.

Alphabetic and numeric characters in a subscript position (below the

Table 9·2 Examples of subscript symbols

Item represented	Type of symbol	Symbol	Section or table in which described
Atomic number	Numeric	$_{7}N$	15·6
Constant designation	Alphabetic	K_i	Table 17·4
Derivative specified	Alphabetic	$D_x y$	Table 11·9
Gas phase specified	Alphanumeric	$P\mathrm{A}_{\mathrm{CO}_2}$	Table 24·6
Integration limit, lower	Alphabetic	$\int_x$	11·14
Member of chemical group	Numeric	vitamin D_3	16·59
Number of atoms	Numeric	H_2O	15·7
Number of molecular units	Numeric	$\alpha_2\beta_2$	24·17
Ring conformation, carbohydrate	Numeric	$^{2,5}H_5$	16·21
Scalar designation	Alphabetic	F_x,F_y	11·15
Thermodynamic change, type	Alphabetic	$\Delta_f H$	17·6
Thermodynamic quantity (heat capacity), state for	Alphabetic	C_p	Table 17·1

main line of type) are widely used in scientific notation as qualitative and quantitative modifiers of symbols on the line. These should usually be in a smaller font than that used for the main line of type. Table 9·2 illustrates many of the most common kinds of subscripts and cites the sections of this manual that describe their use in detail.

Superscript and subscript characters that serve as symbols for quantities when used alone (see sections 3·5 and 9·3), should also be in italics. Superscript and subscript characters serving in these positions as qualitative (descriptive) modifiers should be in roman type.

EXCERPTS AND QUOTATIONS

Short passages taken from another text are usually efficiently presented as quotations with relevant punctuation in the line of the text quoting them; such quotations are sometimes called "run-in quotations". Long passages may be more effectively presented as excerpts distinguished from the text into which they are inserted by special typographic devices, which accounts for their often being called "block quotations" or "set-off quotations".

EXCERPTS (BLOCK QUOTATIONS, EXTRACTS)

9·9 Various typographic devices can be applied to identify a block of text as quoted text. One pair of devices is indenting the quoted text from the left margin of the quoting text and using a smaller type font.

> Physicians in some fields may wonder why they should understand the fundamental principles of the management of fluid and electrolyte disturbances. Elkinton and Danowski made clear in their classic monograph, *The Body Fluids*[12], that any physician, no matter his or her specialty, will sooner or later have to deal with such problems.
>
>> Many disease states which are otherwise unrelated have certain features in common such as starvation, dehydration, vomiting, diarrhea, sweating, and renal dysfunction. These symptoms can exert profound influences upon the volume and composition of body fluids and solutes.
>
> But despite this point, many physicians still feel that . . .

If the text preceding the excerpt does not make clear its source, the excerpt should end with a parenthetic indication of source placed after the closing punctuation mark of the excerpt.

> and composition of body fluids and solutes. (Elkinton and Danowski, *The Body Fluids*[12])

The selection of such devices is usually best made with the advice of a typographer or designer. A journal can have a standing style for excerpts, but a particular style may have to be developed for a particular book or report.

The excerpt should maintain the structure, typographic style, and content of the passage in its original form. Note the following rules.

1 If the original opened as an indented 1st sentence of a paragraph, the excerpt should include the indent.

> The second paragraph of Lewin's text[17] goes directly to the heart of the matter.
>
>> The gene is the unit of genetic information. The crucial feature of Mendel's work, a century ago, was the realization that the gene is a distinct entity. The era of molecular biology began in 1945 when Schrödinger developed the view that the laws of physics might be inadequate to account for . . . its stability during innumerable generations of inheritance.

2 If the excerpt is from within a paragraph and opens with a complete sentence, the excerpt should not be indented.

3 If the excerpt begins with a sentence fragment, it need not have an added indent or initial capital letter, and ellipsis marks are not necessary (see section 9·14).

4 The initial letter of the excerpt should follow the original's style: capitalization for the beginning of a sentence, lowercase for the word following an ellipsis (unless it is properly capitalized, as for a proper noun).

5 Original content omitted within the excerpt must be indicated by ellipsis marks. If original content has been omitted and replaced with substituted equivalent text, the replacement must be within square brackets.

> These symptoms can . . . influence . . . body fluids and solutes.
>
> These [disorders] can [disturb] body fluids and solutes.
>
> The cases came from only 3 cities. → The [patients] came from only 3 cities.

6 Excerpts in a non-English language should preserve the conventions of the original, including diacritical marks and similar devices such as guillemets ("chevrons") used in French as quotation marks. A translation within the excerpt of a word, phrase, or longer elements of text should appear within square brackets after the translated matter.

Some exceptions to the general rule of excerpting the original exactly may be reasonable.

1 Citations in the original may be omitted, but the omission must be treated as an ellipsis.

2 An obviously typographic error in the the original can be corrected.

> the survey of New Yorke was . . . → . . . survey of New York was . . .

Idiosyncratic spelling and phrasing in historical sources should be preserved. If such idiosyncrasies might be interpreted by a reader as the quoting author's errors rather than the original form, they can be followed by *sic* and the correct form, both within square brackets.

> and the physiological analyses of Barnard [*sic,* Bernard] are often pointed to as . . .

3 Words or terms in boldface for emphasis in the original can be presented in the normal weight of the typeface. But note that typographic devices used for specific scientific meaning, such as boldface for vectors in mathematical and physics texts and italics for species names, must be preserved in the excerpt. If an author italicizes some of the excerpted (or quoted) text for emphasis, that change should be explained by a phrase within square brackets.

> and Schmidt *proved* [italics mine] that Cohen's conjecture was . . .

QUOTATIONS

9·10 Run-in quotations must be marked with quotation marks and separated from the other text by appropriate punctuation so as to make clear the quoted matter and its relation to the rest of the text. Styles for run-in quotations differ between American usage and British usage. There are 2 main differences.

1. American style uses double quotation marks for a primary quotation and single quotation marks for a quotation within a quotation. It places the closing quotation mark at the end of the quoting sentence after its closing stop (period, exclamation mark); this illogical practice appears to have arisen from American printers' aesthetic distaste for the space beneath the closing quotation mark and the following period.
2. British style uses single quotation marks for a primary quotation and double marks for a quotation within a quotation. Closing quotation marks are positioned before the closing stop of the quoting sentence except when the quotation is a complete sentence.

In general, US and Canadian publications (USGPO 1984; DSSC 1985; FEAC 1987; UCP 1993) follow the American style, while UK and Australian publications (OUP 1983; AGPS 1988) follow the British style. The British style is more logical (and akin to the equally logical French style of quotation), but the American style is deeply embedded in North American publishing practices.

Both styles are described here. This manual recommends using double quotation marks for the primary marks, then single marks (American style) for a quotation within a quotation but positioning them in accordance with British practice; for the British style for placement of quotation marks in relation to other punctuation marks, see section 9·12.

Quotations Within Sentences

9·11 The quoted text is enclosed by primary quotation marks.

[American style] Nininger described[7] “tiny droplets of melted country rock,” and he felt he had tangible proof of an explosion.

[British style] Nininger described[7] ‘tiny droplets of melted country rock’, and he felt he had tangible proof of an explosion.

[CBE style]	Nininger described[7] “tiny droplets of melted country rock”, and he felt he had tangible proof of an explosion.

An internal quotation is indicated by secondary quotation marks.

[American style]	It is worth noting that “Nininger noticed tiny ‘bombs’ of what he described as yellow-green-brown slag,” and we must recall that . . .
[British style]	It is worth noting that ‘Nininger noticed tiny “bombs” of what he described as yellow-green-brown slag’, and we must recall that . . .
[CBE style]	It is worth noting that “Nininger noticed tiny ‘bombs’ of what he described as yellow-green-brown slag”, and we must recall that . . .

An internal quotation within a block quotation is treated as it would be in a within-sentence quotation.

A tertiary quotation (a quotation within a secondary quotation) should be enclosed by the primary quotation marks of either style, but tertiary quotations can rarely be justified in scientific writing.

If the quotation is a syntactical part of the quoting sentence, its 1st letter should be lowercase even if the quoted text began with a capital. If the quotation is not a syntactical part, the initial capital should be kept.

> Peabody pointed out that “one of the essential qualities of the clinician is interest in humanity”, and he went on with his famous aphorism about the care of the patient.
>
> Peabody made a strong point, “One of the essential qualities of the clinician is interest in humanity”, and then he stated his famous aphorism. [Note the British style for the relation of the closing quotation mark and the comma.]

Quotation Marks and Other Punctuation Marks

9·12 This manual recommends the British style for positioning of quotation marks in relation to other punctuation marks. In the British style (OUP 1983), “All signs of punctuation used with words in quotation marks must be placed *according to the sense.* If an extract ends with a point or exclamation or interrogation sign, let that point be included before the closing quotation mark; but not otherwise.” (This 2nd quoted sentence ends in the source with a period.) *Hart’s Rules for Compositors and Readers,* quoted above (OUP 1983), summarizes the British rules with this maxim: “Place punctuation according to sense.”

> He saw the ecological effects as 'an utter disaster'.
>
> His brief note was 'A disaster!'.

But if the quoted passage is a complete sentence ending with a period, the period is not repeated after the closing quotation mark.

> At his retirement dinner he said, 'My work is finished.'

If the quotation is an internal, not a closing, quotation, the full stop (period) is omitted even if it was in the original.

> He ended his address with the statement 'I am finished with this project' and sat down.

9·13 In the American style, the following rules apply to the placement of closing quotation marks in relation to other punctuation marks.

1 After a comma or period, even if the punctuation mark is not part of the quotation.

> He saw the ecological effects as "an utter disaster."
>
> He commented, "An utter disaster," and left immediately.

2 Before a semicolon or colon.

> He listed elements of "an adequate study design": statement of question, definition of . . .

3 After any other punctuation mark if it is part of the quotation.

> "A disaster!" was his brief note.

Additional details on the American style can be found in *The Chicago Manual of Style* (UCP 1993); on the British style, in the section "Punctuation" of *Hart's Rules* . . . (OUP 1983).

ELLIPSES

9·14 In a run-in or block quotation any omission from the quoted text must be represented by ellipsis marks, which are 3 on-the-line dots (periods) separated from each other and adjacent characters by single spaces. Two systems for their application are in general use and are summarized in detail in *The Chicago Manual of Style* (UCP 1993).

> He reported, "Laboratory findings . . . were similar for 3 patients."

Punctuation preceding or following an omission can be retained if the passage's sense is made clearer.

> This is what he concluded, but only after much field work: the crater was probably not meteoric.
>
> This is what he concluded . . . : the crater was probably not meteoric.

An omission following a complete sentence can be indicated by 4 dots, the 1st representing the sentence's closing period; the 1st is not spaced from the preceding word (the last word of the sentence).

> There are mutations that affect the ability of *E. coli* cells to engage in DNA repair. . . . The major known pathways are the *uvr* excision-repair system and the *dam* replication mismatch-repair system.

Dashes or asterisks should not be used to indicate omissions.

REFERENCES

Cited References

[AGPS] Australian Government Publishing Service. 1988. Style manual for authors, editors and printers. 4th ed. Canberra (Australia): AGPS Pr.

Dong G. 1993. The name problem. Eur Sci Editing; May(49):6.

[DSSC] Department of the Secretary of State of Canada. 1985. The Canadian style: a guide to writing and editing. Toronto: Dundurn Pr.

[FEAC] Freelance Editors' Association of Canada: Burton L, Cragg C, Czarnecki B, Paine SK, Pedwell S, Phillips IH, Vanderlinden K. 1987. Editing Canadian English. Vancouver: Douglas & McIntyre.

[OUP] Oxford University Press. 1983. Hart's rules for compositors and readers at The University Press Oxford. 39th ed. Oxford: OUP.

[UCP] University of Chicago Press. 1993. The Chicago manual of style. 14th ed. Chicago: UCP.

[USGPO] United States Government Printing Office. 1984. United States Government Printing Office style manual 1984. Washington: USGPO.

Additional Reference

Bringhurst R. 1992. The elements of typographic style. Vancouver (BC): Hartley & Marks.

10 Abbreviations

> Although present-day abbreviation in English descends from such [classical forms as *SPQR* for *Senatus Populusque Romanus*], its more immediate origin was in the practices of scribes, among whom short forms were mnemonic and a means of economizing on parchment, effort, and time.
> —Tom McArthur, *The Oxford Companion to the English Language,* 1992

English does not have formal and consistent rules for forming abbreviations, and many variations in abbreviations representing the same terms can be found throughout the history of English prose. Nor has the punctuation of abbreviations been standardized. A useful summary of the history and current practice of abbreviating can be found in *The Oxford Companion to the English Language* (McArthur 1992).

FORMATION OF ABBREVIATIONS

10·1 English-language abbreviations are generally of 3 types, with minor variations within each: contraction, suspension, and hybrid. Punctuation of abbreviations has frequently differed in American and British usage.

1 Initial letter or letters (2 or more) of the word (true abbreviations, also known as contraction abbreviations).

species → sp. nitrogen → N Massachusetts → Mass.

2 Initial and 2nd, or another internal, letter of the word (contraction abbreviations).

natrium [sodium] → Na plumbum [lead] → Pb

3 Initial letters of 2 or more words, occasionally with 1 or more internal letters, usually capitalized (contraction abbreviations).

acquired immunodeficiency syndrome → AIDS

flavin adenine dinucleotide → FAD

National Institutes of Health → NIH

ethylenediaminetetraacetic acid → EDTA

also known as → a.k.a. soluble ribonucleic acid → sRNA

left ventricular end-diastolic pressure → LVEDP

cardiac output → CO

Some abbreviations of this type may be called "acronyms" if they serve as pronounceable words; they usually cease to be capitalized.

laser [for "**l**ight **a**mplification by **s**timulated **e**mission of **r**adiation"]

4 Initial and terminal letters of the word (suspension abbreviations).

numero → no. number → nr Doctor → Dr

Pennsylvania → PA

Note that "no." for "*numéro*" must have the period to avoid ambiguity with the word "no", the negative statement.

5 Initial, internal, and terminal letters of the word (hybrid abbreviations).

Mistress → Mrs Pennsylvania → Penna

undetermined → undetd

For plural forms of abbreviations see section 5·29. Note that abbreviations formed from multiword terms may conceal the possibility that the term can be treated grammatically as a plural.

The NIH have agreed to combine their efforts . . .
[NIH for "National Institutes of Health"]

Note also that "abbreviation" and "symbol" are not mutually exclusive terms. A symbol may be formed by abbreviation ("Na" for "natrium") but need not be an abbreviation; many abbreviations do not serve as symbols but simply represent short forms of words or terms. See section 3·4 for further discussion.

PUNCTUATION AND TYPOGRAPHY

10·2 In recent years American practice has generally been to place a period (full stop) after an abbreviation of a single word (Dr. for "Doctor", "undetd." for "undetermined") and to omit periods (full stops) in abbreviations of multiword terms ("USA" for "United States of America", "NAS" for "National Academy of Sciences"). British practice has tended to be the reverse ("Dr" for "Doctor", "I.U.P.A.P." for "International Union of Pure and Applied Physics", "N.I.H." for "National Institutes of Health"). Some periods are required by nomenclatural systems or to prevent misunderstanding, but eliminating unneeded periods in abbreviations would bring American and British usage into convergence or identity, save authors' and editors' time, and eliminate keystrokes. This change has been recommended by various authorities, including *The Chicago Manual of Style* (UCP 1993).

> It is often an open question whether or not periods should be used with particular abbreviations. The trend now is strongly away from the use of periods with all kinds of abbreviations that have carried them in the past. . . . this is to the good: anything that reduces the fussiness of typography makes for easier reading.

A British authority on usage commented 30 years ago in his detailed manual on punctuation, *You Have a Point There: A Guide to Punctuation and Its Allies* (Partridge 1953):

> from the United States has come a practice that is rapidly growing and that could advantageously become universal. If it did, it would merely fall in line with the very general abandonment of [periods] in chemistry, physics, electricity. . . . For the initials of all organizations, the omission of [periods] would be—for Americans it already is—an excellent thing. . . . If ever there was—who doubts that there is?—a strong case for mankind v. useless conventions, the discarding of all but clarificatory [periods] constitutes such a case.

In view of these several considerations, the following rules are recommended.

1 Abbreviations should not, in general, be followed by or include periods (full stops).

 Dr [not "Dr."] PM [not "P.M."]

 PhD [not "Ph.D."] IUPAP [not "I.U.P.A.P."]

2 Periods should be retained if they are required by a nomenclatural document of a scientific organization, for example, the International Code of Nomenclature of Bacteria.

 species → sp. ("spp.", plural) [not "sp" ("spp", plural)]

3 Periods may be retained with initial-letter abbreviations of personal names if the publication must maintain the distinction between abbreviated personal names and single-letter elements of names properly presented without a period.

Franklin D. Roosevelt Jr   Charles C Thomas   Harry S Truman

[The "C" in "Charles C Thomas" and the "S" in "Harry S Truman" are not abbreviations but given letters.]

4 Periods after abbreviated personal names must be omitted in formats requiring their omission, such as formats for references specified in Chapter 30.

Zink MC, Narayan O. 1988. Host interaction in caprine arthritis-encephalitis. Ann NY Acad Sci 540:634–5.

5 Commas should not be used before or after abbreviations representing parts of personal names. Abbreviations for academic degrees and honorific titles that are readily recognized as such by readers need not be set off by commas; for some readerships, commas may be needed (see section 7·3).

Franklin D Roosevelt Jr and his sister Anna . . .

The members included Russell Tomar PhD and Thomas Smith MD to represent . . .

and the eminent English zoologist Herbert Graham Cannon FRS was widely known . . .

To simplify conventions further, 2 additional rules are recommended.

6 For abbreviations (acronyms or initialisms) of multiword terms, the initial letter or internal letters of words forming the term should be capitalized even if the term is not normally capitalized in running text.

AM [for "*ante meridiem*"]

AIDS [for "acquired immunodeficiency syndrome"]

7 Abbreviations should be set in roman (upright) type and in lowercase and capital letters unless the abbreviation must follow a formal nomenclatural convention that is different. Specifically, small capitals need not be used in abbreviations; italicized abbreviations should be used for properly italicized terms.

Observations were scheduled to begin at 12:01 AM on . . .
[not "at 12:01 AM"]

The dominant bacteria were *S. aureus* and . . . [not "were S. *aureus* and"]

USAGE

COMMON FORMS

10·3 The following general rules are recommended; they govern usage in this manual.

1 Abbreviations, except those widely and formally used in particular contexts, should not be used in running text.

> and the nomenclatural commission was headed by Ralph Schowitz PhD and his colleague . . .
>
> Observations were scheduled to begin at 12:01 AM on the 1st day of each month.
>
> He noted morning changes [not "a.m. changes"] in the ape's eating habits.
>
> among the changes, for example, was the decline in . . .
> [not "among the changes, e.g., was the decline in . . ."]

2 When an abbreviation is needed, as in addresses and tables, and formally standardized and widely recognized forms are available, they are preferred over nonstandard abbreviations frequently seen in variant forms.

> Philadelphia PA [not "Philadelphia, Penna."]

3 When an abbreviation is needed to save space, as in a table, it should be one derived from an English term rather than from Latin or another non-English language.

> nr [for "number", not "no.", derived from the French *numéro* or the Italian "*numero*"]

Also see section 10·5 and Table 10·2 for Latin-term abbreviations.

Editorial offices should maintain a list of abbreviations frequently appearing in their publications. When various forms are correct, the list should carry only 1 form selected from among them and additionally specify the contexts of appropriate and inapproriate use. A comprehensive dictionary of abbreviations such as De Sola's *Abbreviations Dictionary* (De Sola 1992), *Acronyms, Initialisms and Abbreviations Dictionary* (Mossman 1994), and *The Oxford Dictionary of Abbreviations* (OUP 1992) may supply different forms for selection. The list should also indicate the forms not to be used. The list should be the basis for a list of acceptable abbreviations in a journal's information-for-authors sheet or a publisher's style manual supplied to authors.

SCIENTIFIC USAGE

10·4 The enormous, accelerated growth of scientific knowledge in the past half century has generated great numbers of new terms, notably multiword terms. Their length and the pressures on use of journal space have led to frequent and widespread coinages of abbreviations. Some of these are coined anew in papers and do not represent established usage; others are well known only in relatively narrow disciplines. These abbreviations make for difficult reading for all but scientists thoroughly familiar with the topic.

Editors must control the use of scientific abbreviations in journals and books so as to balance readability of texts for nonspecialist readers against the need for economical use of their pages, while not impairing readability for specialists. For guidance, the following recommendations are offered; also see section 6·11.

1 Abbreviations widely known throughout science, such as "DNA" for "deoxyribonucleic acid", can be used in titles, abstracts, and text without definition. A helpful criterion for decisions on such use is whether the abbreviation has been accepted into thesauri and indexes widely used for searching major bibliographic databases in the scientific field, for example, the medical-subject-headings (MeSH terms) thesaurus of the National Library of Medicine (NLM [annual]).
2 Abbreviations not widely known throughout science and not carried in the index to a major bibliographic database relevant to the subjects and scope of the journal should not be allowed in titles. If they are well established in the discipline represented by a specialized journal, they may be used in its abstracts and text without explanation; these abbreviations should be indicated in the journal's information-for-authors pages as standard for the journal.
3 Abbreviations not acceptable by criteria 1 and 2 above must not be used in titles and abstracts but may be used in text, tables, and illustrations if they are parenthetically defined at 1st use. If, however, they are not needed more than a few times (perhaps 3 to 5 times) in the following text, they should not be allowed; the terms they represent should be spelled out, or the text in which they appear should be rewritten to eliminate the need for abbreviations.
4 Editorial offices should maintain lists of abbreviations allowed by criterion 3 and of the terms of their origin so that their use can be standardized in the journal and variant forms can be avoided.
5 Care should be taken to see that abbreviations allowed by criterion 3 are not used, notably adjectivally, where the meaning would be clear without the adjectival abbreviation and with the context.

Consider the following example and the revised version.

> [original] The publication of expressed sequence tags (ESTs) derived from randomly selected clones from commercially obtained brain complementary DNA libraries[1,2] has received much attention. . . . Initially we examined one EST of interest to us (EST01828, similar to the *Drosophila* homeobox gene *otd*), and found that it appears to contain an intron. Subsequently, it seemed that there were several ESTs reported by . . .
>
> [revised version] The publication of expressed sequence tags (ESTs) derived from randomly selected clones from commercially obtained brain complementary DNA libraries[1,2] has received much attention. . . . Initially we examined one tag of interest to us (EST01828, similar to the *Drosophila* homeobox gene *otd*), and found that it appears to contain an intron. Subsequently, it seemed that several tags were reported by . . .

For abbreviations applicable in specific fields and designated in this manual, see Table 10·1.

ABBREVIATIONS OF LATIN TERMS

10·5 Many Latin terms of scholarly reference have been used in abbreviated forms, especially in footnotes, endnotes, and parenthetic statements in text. These carryovers from the centuries when all scholarly literature was in Latin are not needed with present forms of citation and reference in scientific literature. Such Latin abbreviations as "etc." (for *et cetera*) and "et al." (for *et alii*) are readily replaced in text by English equivalents. For a compilation of many of these abbreviations and the Latin terms and phrases they represent, see Table 10·2. A more extensive compilation of abbreviations (English and Latin) widely used in the literature of the arts and the humanities can be found in *The MLA Style Manual* (Achtert and Gibaldi 1985).

Table 10·1 Abbreviations with specific applications as described in this manual

Subject	Section	Table
Academic degrees and honors	7·2–7·4	7·2, 7·3
Astronomical abbreviations	27·7	27·5
Astronomical catalogues	27·1, 27·2, 27·7	27·1
Astronomical objects and terms	27·1, 27·10, 27·11, 27·14	27·2, 27·5
Astronomical organizations	27·24	27·6
Astronomical time	27·15, 27·21–27·23	27·5
Author names, bacterial nomenclature	23·5	
Bacterial culture collections	22·16	22·2
Chromatographic methods	18·2, 18·3	18·1
Chronometric and radiometric dating	25·11, 25·12	
Circulation, hemodynamics	24·13	24·3
Constellations	27·7	27·2
Days of the week	12·6	12·1
Electrocardiographic recordings	24·14	
Eras	12·10	12·3
Hepatitis and hepatitis viruses	24·33	
Hormones	24·23	24·4
Journal-title words	Appendix 1	A1·1
Military titles and organizations	7·5	7·4, 7·5
Mineral terms	26·9	
Months	12·7	12·2
Organization names	7·6	
Plant taxonomic terms	23·4	23·1, 23·2
Postal code abbreviations	13·1–13·4	13·1
Professional designations	7·4	7·2
Publisher names	Appendix 2	
Radio frequency ranges	14·1	
Renal structures	24·26	
Respiration	24·27	24·6
Soils	26·23	
Spectroscopic terms	18·4	18·2, 18.3
State and province names	13·1	13·1
Taxonomic ranks	23·4	23·1, 23·2
Thermal regulation	24·28	24·7
Thyroid hormones	24·24	24·4, 24·5
Time, time zones	12·1–12·5	
Virus mutants	21·8	
Viruses	21·6	21·2
Viruses, satellite	21·1	
Vitamin B-6 and related compounds	16·58	
Wavelengths	14·1–14·5	14·2

Table 10·2 Scholarly Latin abbreviations, the corresponding terms and phrases, and their English equivalents

Abbreviation	Latin term	English equivalent
c., ca.	*circa*	about, approximately (with reference to a date or quantity)
cf.	*confer*	compare [not "see"]
ed. cit.	*editio citata*	edition cited
e.g.	*exempli gratia*	for example
et al.[a]	*et alii*	and others
etc.	*et cetera*	and so forth, and so on
et seq.	*et sequens*	and the following
et seqq.	*et sequentes*	and the following
	et sequentia	and the following
fl.	*floruit*	flourished (used before a date representing a historical figure for whom exact birth and death dates are unknown)
ib., ibid.	*ibidem*	in the same place (the work cited in the immediately preceding note)
i.e.	*id est*	that is
infra[b]	*infra*	below
loc. cit.	*loco citato*	in the place (passage) cited, the same passage indicated in a preceding reference
NB	*nota bene*	take notice
ob.	*obiit*	he, or she, died
op. cit.	*opere citato*	in the work cited
q.v.	*quod vide*	which see
r.	*recto*	righthand page
supra[b]	*supra*	above
s.v.	*sub verbo*	under the word, under the heading
	sub voce	under the word, under the heading
v.	*verso*	lefthand page
	vide	see
	versus	against (in contrast to)
viz.	*videlicet*	namely
vs.	*versus*	against (in contrast to)

[a] This manual recommends that, in general, English equivalents be used; see section 10·5. Specifically, "and others" is recommended instead of "et alii" or "et al.".

[b] Not abbreviated.

REFERENCES

Cited References

Achtert WS, Gibaldi J. 1985. The MLA style manual. New York: Modern Language Assoc.

De Sola R. 1992. Abbreviations dictionary. 8th ed. Boca Raton (FL): CRC Pr.

McArthur T. 1992. The Oxford companion to the English language. Oxford: Oxford Univ Pr.

Mossman J. 1994. Acronyms, initialisms and abbreviations dictionary 1994. 18th ed. Detroit: Gale Research.

[NLM] National Library of Medicine. [annual]. Medical subject headings—annotated alphabetic list. Bethesda (MD): NLM.

[OUP] Oxford University Press. 1992. The Oxford dictionary of abbreviations. Oxford: OUP.

Partridge E. 1953. You have a point there: a guide to punctuation and its allies. London: Routledge & Kegan Paul. p 42.

[UCP] University of Chicago Press. 1993. The Chicago manual of style. 14th ed. Chicago: UCP. p 460.

Additional References

[AGPS] Australian Government Publishing Service. 1988. Style manual for authors, editors and printers. 4th ed. Canberra (Australia): AGPS Pr.

Buttress FA, Heaney HJ. 1988. World guide to abbreviations of organizations. 8th ed. Detroit: Gale Research.

Department of the Secretary of State of Canada. 1985. The Canadian style: a guide to writing and editing. Toronto: Dundurn Pr.

Hensyl WR, editor. 1992. Stedman's abbreviations, acronyms & symbols. Baltimore: Williams & Wilkins.

Jablonski S. 1992. Dictionary of medical acronyms and abbreviations. 2nd ed. Philadelphia: Hanley & Belfus.

[OUP] Oxford University Press. 1991. The Oxford dictionary for scientific writers and editors. Oxford: OUP.

11 Numbers, Units, Mathematical Expressions, and Statistics

> Is it not wonderful that man's reason should be made a judge over Gods works, and should measure, and weigh, and calculate, and say at last 'I understand I have discovered—It is right and true'.
>
> —James Clerk Maxwell, *Inaugural Lecture at Marischal College, Aberdeen*, 3 November 1856

The following guidelines for representing numbers may occasionally collide with the realities of some documents. When common sense or editorial judgment says a guideline is a poor choice for a specific document, follow sense or judgment. Perfect consistency in representing numbers and numeric terms is impossible; choose the form that is best suited to your document and follow that form as consistently as is reasonable.

The international general standard for expressing quantities and units is *International Standard ISO 31-0-1992, Quantities and Units — Part*

Table 11·1 Values of roman numerals

Roman numeral[a]	Arabic equivalent
I, i	1
V, v	5
X, x	10
L, l	50
C, c	100
D, d	500
M, m	1000

[a]A bar over a letter multiples its value by 1000; for example, $\overline{\text{V}}$ = 5000

0: General Principles (ISO 1992a). The standard for mathematical symbols in the physical sciences and technology is *Part 11, ISO 31/11-1992* (ISO 1992b); the 1978 version of this standard can be found as *ISO 31/11-1978* in *ISO Standards Handbook 2: Units of Measurement* (ISO 1982).

NUMBERS

EXPRESSING NUMBERS IN TEXT

A broad and useful discussion of numeric presentations in text is found in "Writing about Numbers" by Mosteller, Chapter 20 in *Medical Uses of Statistics* (Bailar and Mosteller 1992).

11·1 Numerals are the symbols that are used alone or in combination to represent numbers. There are 10 arabic numerals: 0, 1, 2, 3, 4, 5, 6, 7, 8, and 9. A number is the representation in numeric or word form of a count, an enumeration, or a measurement: for example, 547.2, 6 million, 1.54×10^6.

Numbers are formed in roman numerals by combinations of 7 letters, whose values are shown in Table 11·1. When a numeral with a smaller value follows one of equal or higher value, the values are added. When a numeral of smaller value precedes one of higher value, the smaller value is subtracted from the larger.

XXI = 21 XXIV = 24 XXVI = 26 XLIV = 44

The pagination of the front matter of books and reports is generally in lowercase roman numerals. When roman numerals are used in numbering tables, figures, or sections, capitals are more common.

NUMERALS OR WORDS

11·2 In scientific text, arabic numerals should be used in preference to words when the number designates anything that can be counted or measured (exceptions are given below).

3 hypotheses 7 samples 52 trees 328 amino acids

Ordinal numbers are treated in the same manner as cardinal numbers.

3rd 7th 52nd 328th

Numerals are also used to designate mathematical relationships such as ratios and multiplication factors.

5:1 1000 times (or "1000×") 4-fold

One of the exceptions to the use of numerals is that numerals are not used to begin a sentence. If logic calls for a number to begin the sentence, the solution is to spell it out, reword the sentence, or, if appropriate, join the sentence to the previous sentence in the text.

Twenty milligrams is the desired amount, but 15 mg is enough.

The desired amount is 20 mg, but 15 mg is enough.

The drug is administered in a single dose; 20 mg is the desired amount, but 15 mg is enough.

The 2nd exception is when 2 numeric expressions are adjacent in a sentence. The number easiest to express in words should be spelled out and the other left in numeric form, or the sentence should be recast to separate the numbers. In general it is preferable to retain the numeric form with units of measurement.

The sample was divided into eight 50-g aliquots.

The sample was divided into 8 aliquots of 50 g each.

Finally, when numbers are not directly associated with a count or measurement, words may be used in text.

When a letter of smaller value follows one of equal or higher value . . . ["one letter" is implied]

Note the difference, however, between the number "one" and the pronoun "one", which is always spelled out.

In supporting scientific ethics, one is obliged to . . .

FORMAT OF NUMBERS

11·3 For numbers consisting of 2 to 4 digits, the numerals are run together, that is, set close, with no extra spaces.

27 368 1000 2568

In numbers of more than 4 digits, British and American practice has been to mark off groups of 3 digits, starting at the decimal point, with commas. In the European convention (ISO 1992a), however, the comma serves as the decimal point, and periods or spaces are used to group the digits of large numbers.

[traditional British and American]	1,000,523	47,938.275
[European]	1.000.523	47 938,275

To prevent confusion in an international readership, thin spaces should be used to separate groups of 3 digits in either direction from the decimal point. A single digit remaining after those to the right of the decimal point have been grouped is added to the final group.

[recommended] 1 000 523 47 938.275 1.527 4304

In the United States, some journals and organizations have adopted the practice, but many have not. The use of spacing is not recommended if thin spaces are not available and full spaces must be used between groups of digits. In using word-processing or typesetting programs with full justification of lines, the spacing within large numbers may be lost when the program decreases interword spacing because a line is very full. Spacing is not used in some numbers with more than 4 digits such as US postal (ZIP) codes, patent numbers, and telephone numbers in some countries.

An initial zero (0) should always be used before the decimal point for numbers smaller than 1.0. The initial zero removes any ambiguity about a possibly erroneous omission of a digit before the decimal point; it also improves the readability of numbers in tables. Also see section 11·24 for this requirement in statistical reporting. An integer is never terminated by a decimal point except at the end of a sentence.

0.497 [not ".497"] $P = 0.05$ [not "$P = .05$"] 74 [not "74."]

RANGES OF NUMBERS

11·4 When expressing ranges in text, the word "to" or "through" connects the numbers.

Samples 42 to 153 were scanned at wavelengths from 240 to 350 nm.

When the ranges are numbers of several digits, no digits from the 2nd number in the range should be omitted.

1938 through 1954 [not "1938–54"]

pages 1466 to 1472 [not "pages 1466 to 72"]

In references, duplicate numbers are omitted in specifying pages, and an en-dash is used instead of the word "to"; see section 30·27.

If a range of values begins a sentence in the text, both numbers in the range, the span word ("to" or "through"), and the accompanying units are spelled out.

Twenty-three to forty-seven kilovolts was the test range.

[not "Twenty-three to 47 kV was the test range."]

If the resulting written expression is cumbersome, the sentence should be recast.

Ranges of numbers and their accompanying units are expressed with a single unit symbol following the 2nd number of the range, except when the symbol is set close without spacing.

23 to 47 kV   50 to 250 W/m^2 0.7 to 0.9 kg   10% to 100%

In documents containing many ranges or in tables where space is at a premium, an en-dash may be preferable as a substitute for the word "to". To avoid confusion, however, the en-dash should not be used with ranges in which either (or both) of the values is a negative number. Note also that the en-dash represents only the word "to" or "through"; when the text phrasing is "between x and y," the en-dash cannot be used. Finally, in a "from . . . to" phrase (such as "increments of from 15 to 70 V"), the en-dash should not be substituted for "to".

SCIENTIFIC NOTATION

11·5 In scientific writing, very large or very small numbers are usually expressed in powers of 10 (scientific) notation.

2.6×10^4 [rather than "26 000"]

4.23×10^8 [rather than "423 000 000"]

7.41×10^{-6} [rather than "0.000 007 41"]

When expressing a range of values in scientific notation, either spell out both limits of the range in full or enclose the range within parentheses.

2.6×10^4 to 9.7×10^4 [or] $(2.6 \text{ to } 9.7) \times 10^4$ [not "2.6 to 9.7×10^4"]

If the limits of the range have different exponents, both limits must be shown in full.

3.7×10^2 to 5.9×10^6

A value of 7.4×10^3 with an error of $\pm\ 0.4 \times 10^3$ should be shown as

$7.4 \times 10^3 \pm 0.4 \times 10^3$ [or as] $(7.4 \pm 1.4) \times 10^3$
[not as "$7.4 \pm 0.4 \times 10^3$"]

When numbers are large and are not expressed with a high degree of precision (that is, they have trailing zeros, such as 3 000 000), the text form may be a combination of numerals and words, particularly when the numbers do not represent experimental quantities or precise quantitative measurements. Such expressions are never used with scientific units of measure.

3 million people $13.9 million

1.5×10^6 km [not "1.5 million km"]

The combination of numerals and words can cause confusion because of differing meanings of the terms beyond "million". For example, "billion" means 1000 million (1 000 000 000) in the United States but still usually means 1 million million (1 000 000 000 000) in the United Kingdom (Anonymous 1992). In any document that may have an international readership, either write out the number or use scientific notation if appropriate.

PERCENTAGES

11·6 Percentages are expressed in scientific writing using numerals and the percent sign (%) with no space between them: 38%. If the percentage starts a sentence, the number and "percent" are written out in words.

Fifty-seven percent of the samples were contaminated.

[note the plural verb "were"; the subject of the sentence is not a singular "percent" but the implied "Fifty-seven samples of 100 samples"]

A range of percentages is expressed with a percent symbol following each value.

The success rate was 15% to 47%.

The rates of change were –13% to +24%.

Use of the 2 symbols eliminates any ambiguity as to whether the 1st number represents simply a number or a percentage.

MONETARY UNITS

11·7 In scientific text, amounts of currency should be written out with numerals preceded by the symbol or abbreviation for the unit of currency. Very large amounts can be represented by numerals preceding a numeric unit (such as "million") in its spelled-out form.

> $245 million [or] $245 000 000

The units of currency should not be mixed with the prefixes of the SI system or with scientific notation to designate thousands, millions, or larger amounts.

> $58 000 [not "$58k" or "$5.8 × 10^5"]

Tables, illustrations, and other places calling for abbreviated forms should maintain this distinction.

> $, millions [or] $, 000 000 [or] millions of $
> [not "$M" or "$, M" or "M$"]

For currency units that are represented by symbols, the symbol is set close to the numerals. For currency units represented by abbreviations, a space separates the abbreviation from the numeral.

> $749   £749   F 749   DM 749

Because many nations use "dollar" as the name of the unit of currency, the dollar symbol should be prefixed with the appropriate abbreviation if several dollar-based currencies could fit the context.

> A$749 [Australian dollars]   NZ$749 [New Zealand dollars]
> Can$749 [Canadian dollars]   US$331 [US dollars]

The pound sterling symbol (£) may be similarly prefixed to designate the pound currencies of various countries. Note that other currencies (such as the peso) use the dollar symbol for their currency. For symbols for many national currency units, see Table 11·2. A more extensive compilation of units and their symbols can be found in Chapter 18 of the *United States Government Printing Office Style Manual 1984* (USGPO 1984).

An international standard, ISO 4217 : 1987 (E/F), which can be found in *ISO Standards Handbook 1: Documentation and Information* (ISO 1988), specifies 3-letter alphabetic codes for specific representation of currencies. The standard does not specify the position of the code in relation to the number representing the amount. Because the usual practice is to place the currency symbol before the number, that location is recommended.

AUD 749 [Australian dollars]	NZD 321 [New Zealand dollars]
USD 453 [US dollars]	FRF 749 [French francs]
IEP 579 [Irish pounds]	CHF 561 [Swiss francs]
GBP 798 [pounds sterling of the United Kingdom]	

These codes can be adapted by truncations for standard designations with the conventional symbol.

AU$ 749   CA$ 749   IE£ 579

These codes are not widely used in scientific literature; the more widely used designations are those represented in Table 11·2.

UNITS AND SYSTEMS OF MEASURE

SYSTÈME INTERNATIONAL (SI)

11·8 In scientific writing, metric measure is the accepted form of expression for physical and chemical quantities. Several metric systems have been used in the past; the currently recommended system is the Système international d'unités (the International System of Units, abbreviated SI) (ASTM Committee E-43 on Metric Practice 1989; Taylor 1991). SI provides a coherent system of measure constructed from 7 base units plus 2 supplementary units (Table 11·3). From these base units are derived all other units for physicochemical quantities, some of which are expressed in terms of mathematical manipulation of the base unit and others of which have special names and their own symbols; for examples, see Table 11·4; for more complete lists of units, see ASTM D380–89a (ASTM Committee E-43 on Metric Practice 1989; Taylor 1991) or Jerrard and McNeill (1992). Several Canadian guides to SI usage are available (CSA 1980, 1989; ACN 1990).

SI PREFIXES

11·9 Standard prefixes are used with the SI units to designate quantities much larger or smaller than a given unit (Table 11·5). Normal practice is to express units using the prefixes for multiples of 10^3 or 10^{-3} such that the number accompanying the unit is less than 1000. Prefixes are combined with the symbol for the base or derived SI unit. (The exception is the kilogram, because this base unit for mass is already prefixed. In this case, the prefixes are attached to the unit stem "gram" rather than being added to "kilogram".)

Table 11·2 Symbols for selected currencies[a]

Country	Monetary unit	Symbol	Footnote
Australia	Australian dollar	A$, $A	b
Bahamas	Bahamian dollar	B$	
Belgium	franc	BF	
Bermuda	Bermuda dollar	Bd$	
Brazil	cruzeiro	Cr$	
Canada	Canadian dollar	Can$	
Chile	peso	Ch$	
China	yuan	¥	
Colombia	peso	Col$	
Costa Rica	Costa Rican colón	₡	
Denmark	krone	DKr	
Dominican Republic	Dominican peso	RD$	
Ecuador	sucre	S/	
Finland	finnmark	Fimr	
France	franc	F	
Germany	deutsche mark	DM	
Guyana	Guyana dollar	G$	
Hungary	forint	Ft	
India	rupee	Rs	
Israel	shekel	I£	
Italy	lira	Lit	
Japan	yen	¥	
Mexico	peso	Mex$	
New Zealand	dollar	NZ$, $NZ	b
Norway	krone	NKr	
Peru	sol	s/	
Poland	zloty	Zl	
Sweden	krona	SKr	
Switzerland	franc	SwF	
United Kingdom	pound sterling	£, £ stg.	c
United States of America	dollar	$, US$	d

[a]A much more extensive "Foreign Money" table is in Chapter 18, "Useful Tables", of the *United States Government Printing Office Style Manual 1984* (USGPO 1984), the source of Table 11·2.

[b]The 1st form is recommended for consistency in style among the many symbols for different national currencies based on a dollar unit. The form $A is recommended by *Style Manual for Authors, Editors and Printers* (AGPS 1988), the manual of the Australian Government Publishing Service.

[c]The 2nd form is uncommon.

[d]The 1st form is recommended for use in US publications having mainly a domestic readership; the 2nd form is recommended for US documents that refer to several dollar-currencies, whether they have a domestic or an international readership.

Table 11·3 SI base units and symbols

Quantity	Name	Symbol
Base units		
length	meter (metre)	m
mass	kilogram	kg
time	second	s
electric current	ampere	A
thermodynamic temperature	kelvin	K
amount of substance	mole	mol
luminous intensity	candela	cd
Supplementary units		
plane angle	radian	rad
solid angle	steradian	sr

Table 11·4 Examples of SI-derived units, including some with special names

	SI Unit		
Quantity	Name	Symbol	In terms of other units
Activity (of a radionuclide)	becquerel	Bq	s^{-1}
Acceleration			m/s^2
Capacitance	farad	F	C/V
Current density			A/m^2
Electric charge, quantity of electricity	coulomb	C	s·A
Electric potential, electromotive force, potential difference	volt	V	W/A
Energy, work, quantity of heat	joule	J	N·m
Energy density			J/m^3
Force	newton	N	$(m \cdot kg)/s^2$
Frequency	hertz	Hz	s^{-1}
Heat capacity, entropy			J/K
Illuminance	lux	lx	lm/m^2
Luminance			cd/m^2
Luminous flux	lumen	lm	cd·sr
Magnetic flux	weber	Wb	V·s
Moment of force			N·m
Power, radiant flux	watt	W	J/s
Pressure, stress	pascal	Pa	N/m^2

Table 11·5 SI unit prefixes

Term		Multiple	Prefix	Symbol
10^{24}	=	1 000 000 000 000 000 000 000 000	yotta	Y
10^{21}	=	1 000 000 000 000 000 000 000	zetta	Z
10^{18}	=	1 000 000 000 000 000 000	exa	E
10^{15}	=	1 000 000 000 000 000	peta	P
10^{12}	=	1 000 000 000 000	tera	T
10^{9}	=	1 000 000 000	giga	G
10^{6}	=	1 000 000	mega	M
10^{3}	=	1000	kilo	k
10^{2}	=	100	hecto	h
10^{1}	=	10	deka[a]	da
		1 unit	—	—
10^{-1}	=	0.1	deci	d
10^{-2}	=	0.01	centi	c
10^{-3}	=	0.001	milli	m
10^{-6}	=	0.000 001	micro	μ
10^{-9}	=	0.000 000 001	nano	n
10^{-12}	=	0.000 000 000 001	pico	p
10^{-15}	=	0.000 000 000 000 001	femto	f
10^{-18}	=	0.000 000 000 000 000 001	atto	a
10^{-21}	=	0.000 000 000 000 000 000 001	zepto	z
10^{-24}	=	0.000 000 000 000 000 000 000 001	yocto	y

[a]Outside the United States, the spelling "deca" is often used.

Prefixes are set in roman type with no space between the prefix and unit symbol.

ng   kV   mJ   MN

Both the prefixes and the units they modify are either abbreviated or spelled out.

kV or kilovolt [not "kvolt" or "kiloV"]

Only 1 prefix may be used per unit symbol, and a prefix is never used alone.

8 ng [not "8 mμg"] 10^6/s [not "M/s"]

When a prefix is required for a derived unit, the appropriate prefix is attached to the term in the numerator; for example, 4 000 000 N/m^2 becomes 4 MN/m^2, not 4 N/mm^2. An acceptable alternative for expressing large numbers with units is to use scientific notation, for example, 4×10^6 N/m^2. The choice between an SI prefix on the units and scientific notation is made for clarity and convenience for the particular topic and for the particular situation in a document. Whatever choice is made, that usage should be followed consistently throughout the document.

SI UNITS AND THEIR SYMBOLS

11·10 When used in text without an accompanying numeric value, the names of units are spelled out.

> The measurements were recorded in kilojoules.

In tables or figures (illustrations), unit symbols may be used without accompanying numeric values to save space.

When numeric values are given, the unit symbols are always used, with a space between the number and the symbol. If, however, the value begins a sentence, both the number and the unit are spelled out. The symbols for SI units are printed in lowercase roman type except for the units with a name derived from a proper name, in which case the 1st letter of the symbol is capitalized (uppercase): for example, watt, W; tesla, T; hertz, Hz. Unit symbols are identical for singular and plural representations, and they are not followed by periods.

> 1 min   15 min   45 °C   386 K
> 17.6 W   37 lx   0.076 Pa   0.59 cd

11·11 A compound unit that is the product of SI units is indicated by a raised period (centered dot) in US practice (ANSI 1982) or by a space between the symbols.

> V · s [or] V s

When the names of such units are spelled out, the words are separated by a space or a hyphen, but not by a centered dot (ANSI 1982).

> volt second [or] volt-second [not "volt·second"]

A compound unit that includes a quotient should have the word "per" spelled out when the unit name is written in text.

> was measured in kilojoules per hour
> [not "was measured in kilojoules/hour"]

When the unit symbol is used, the form should be presented with a slash (/) or in negative exponent form rather than a letter-based abbreviation.

> kJ/h [or] $kJ \cdot h^{-1}$ [not "kph" for "kilojoules per hour"]

When an expression has more than 1 unit in the denominator, only 1 slash may be used. Multiple slashes in a mathematical expression are ambiguous. For example, consider "8/2/4"; (8/2)/4 = 1 but 8/(2/4) = 16. Thus, a unit "kilograms per milligram per hour" must be represented without ambiguity.

Table 11·6 Non-SI units (names and abbreviations) and their status in relation to SI

Other units used with SI	Units in temporary use with SI	Units deprecated[a]
day (d, = 24 h)	angstrom (Å, = 10^{-10} m)	atmosphere, standard (atm, = 101 325 Pa)
degree (°, = [π/180] rad)	are (a, = 100 m^2)	calorie (cal, = 4.18 J)[b]
hour (h, = 60 min)	bar (bar, = 10^5 Pa)	carat, metric (= 2 × 10^{-4} kg)
liter (L, or l, = 1 dm^3)[c]	barn (b, = 10^{-28} m^2)	fermi (fm, = 10^{-15} m)
minute (min, = 60 s)	curie (Ci, = 3.7 × 10^{10} Bq)	gamma (γ, = 10^{-9} T)
minute (′, = [π/10 800] rad)	gal (Gal, = 10^{-2} m/s^2)	gamma (γ, = 10^{-9} kg)
angular second (″, = [π/648 000] rad)	hectare (ha, = 10^4 m^2)	kilogram-force (kgf, = 9.8067 N)
tonne, or metric ton[d] (t, = 10^3 kg)	knot (kn, = 1 nautical mi/h)	lambda (λ, = 10^{-6} L)
	nautical mile (= 1852 m)	micron (μ, = 10^{-6} m)
	rad (rad, = 10^{-2} Gy)[e]	stere (st, = 1 m^3)
	rem (rem, = 10^{-2} Sv)[f]	torr (= 133.322 Pa)
	roentgen (R, = 2.58 × 10^{-4} C/kg)	

[a]CGS (centimeter, gram, second) units that are not recommended for use with SI units include the erg, dyne, poise, stokes, gauss, oersted, maxwell, stilb, and phot.

[b]The SI equivalent of the calorie varies beyond the 2nd decimal place depending on the definition of "calorie" being used.

[c]In the United States, the symbol L is generally used and the spelling "liter" is preferred to "litre" (ANSI 1982).

[d]The official SI name is "tonne"; "metric ton" is used in the United States.

[e]The unit is "rad", not "radian". "Gy" = "gray", the unit of absorbed radiation dose.

[f]"Sv" = "sievert", the unit of radiation dose equivalent.

kg/(mg·h) [or] kg·mg^{-1}·h^{-1} [not "kg/mg/h"]

Some units still being used in various fields are not part of the SI system. Some of these are recognized to be of value indefinitely, some are recognized but their use is considered temporary, and the use of some is deprecated; for examples, see Table 11·6.

OTHER MEASUREMENT SYSTEMS

11·12 Other systems of measurement are still in use despite the support for the SI system of units in almost all countries. Parts of the avoirdupois, apothecaries, and troy weight systems are still in use, particularly in the United States. Some of such units of weights and measures in common use in the United States are listed in Table 11·7. For further discussion of metric and common systems of measure, see Jerrard and McNeill

(1992). An extensive set of unit conversion factors was published in 1986 (Horvath 1986).

MATHEMATICAL SYMBOLS AND EXPRESSIONS

11·13 How mathematical expressions are edited depends on the means available for producing a document. Sophisticated typesetting systems allow more flexibility in choosing styles than simpler systems. The remaining sections of this chapter deal with the more general and commonly encountered aspects of editing and typesetting mathematical copy. These recommendations are suitable for the types of word-processing or document-preparation systems commonly available today.

The editing of mathematical expressions and text presents difficulties beyond those in the editing of general scientific and technical prose. Some guidelines for working on mathematical copy may be helpful.

1 An editor who does not understand the mathematics and does not know what to do, should ask someone who does.
2 Mathematics calls for particularly careful proofreading. There is no rational way to tell whether 1.457 should have been "spelled" 1.547.
3 Higher mathematics often contains idiosyncratic notations of the author. The author should be queried liberally on proposed editorial changes to mathematical symbols and be shown all changes for approval before printing.
4 Editing and typesetting take longer for pages of mathematical copy than the same number of pages of prose. Editors should plan for the extra time and expense.

TYPESETTING CONVENTIONS

Symbols

11·14 Mathematical expressions use arabic numerals. Scalar variables and constants represented by a single letter are set in italics in equations and text (for example, A, M, x, y). Abbreviations or symbols of several letters are set in roman type, as are abbreviations for mathematical functions (such as "d" for "derivative"); see examples in Table 11·8.

The symbols for many mathematical quantities and operations take distinguishing accessory marks (embellishments) above or below the character. Many of these marks are difficult and expensive (and sometimes impossible) to set in type, so alternatives to some of the most common marks have been developed. For example, vectors, traditionally rep-

Table 11·7 Selected non-SI units of weights and measures in use in the United States and the United Kingdom and their metric (SI) equivalents

Common	Metric equivalent	Common	Metric equivalent
acre	4047 m^2	ounce (oz)	
board foot	0.00236 m^3	apothecaries	31.10 g
bushel		avoirdupois	28.35 g
US	0.0352 m^3	troy	31.10 g
imperial	0.0364 m^3	US fluid	29.57 mL
calorie (cal)[a]	4.18 J	UK fluid	28.41 mL
carat, metric	0.200 g	peck, dry	
cord	3.625 m^3	US	8810 cm^3
drachm (UK)		UK	9092 cm^3
dry[b]	3.888 g	pint (pt)	
fluid	3.552 mL	US dry	550.6 cm^3
dram (US)		US liquid	0.4732 L
dry[b]	3.888 g	UK	568.3 cm^3
liquid	3.697 mL	pound	
fathom	1.829 m	apothecaries ($\overline{\text{lb}}$)	373.2 g
foot (ft)	30.48 cm	avoirdupois (lb)	453.6 g
furlong	201.2 m	troy (lb)	373.2 g
gallon (gal)		quart (qt)	
US, liquid	3.785 L	US liquid	0.9464 L
UK	4.546 L	US dry	1101 cm^3
grain (gr)[c]	0.065 g	UK	1137 cm^3
inch[d]	2.54 cm	rod	5.029 m
knot (kn)	0.514 m/s	ton (ton)	
mile (mi)		long	1016 kg
statute	1.609 km	short	907.2 kg
nautical	1.852 km	yard	0.9144 m

[a]The value of a calorie in SI units beyond the 2nd decimal place depends on the definition of "calorie" being used.

[b]Apothecaries drachm or dram. The avoirdupois dram in both the United States and the United Kingdom is equal to 1.772 g.

[c]The grain is a unit in both the avoirdupois and the apothecaries systems of measure.

[d]The unit "inch" should be spelled out. If it is abbreviated, a period must be used to distinguish the abbreviation "in." from the preposition "in".

resented by a right arrow above the variable, $\vec{v}$, are represented in print as a boldface italic letter "***v***" instead. As another example, sigma-class symbols often have boundaries represented by indices that are traditionally placed directly above and below the symbol, and this is the normal practice in display. When these symbols are needed in text, the indices can be placed adjacent to the symbol as subscripts and superscripts, which allows a better fit in text lines. For example,

Table 11·8 Mathematical functions

Symbol	Meaning	Remarks
$\exp x$ [or] e^x	exponential of x	
$\log_a x$	logarithm to the base a of x	When the subscript is omitted, the base 10 is assumed.
$\ln x$ [or] $\log_e x$	natural (Napierian) logarithm of x	
$\sin x$, $\cos x$, $\tan x$, $\cot x$, $\sec x$, $\csc x$	trigonometric functions of x	
$\sinh x$, $\tanh x$, [and so on]	hyperbolic functions of x	
$\arcsin x$ [or] $\sin^{-1} x$, arctan x, arcsec, x, etc.	inverse trigonometric functions of x	
arcsinh x [or] $\sinh^{-1} x$, arctanh x, arcsech x, [and so on][a]	inverse hyperbolic functions of x	
$\lim_{x \to a} y$	the limit of y as x approaches a	$\overline{\lim}$, least upper limit $\underline{\lim}$, greatest lower limit

[a]Note that for the inverse hyperbolic functions, IUPAP (Cohen and Giacomo 1987) recommends dropping the letter c of "arc"; the symbols thus become "arsinh", "arsech", and "arcosh", and so on.

$$\sum_{i=1}^{n} \text{ becomes } \textstyle\sum_{i=1}^{n} \qquad \displaystyle\int\limits_{\pi}^{\infty} \text{ becomes } \int_{\pi}^{\infty}$$

Although occasionally 2 levels (or more!) of embellishments are used by an author, for example, $\overline{\overline{\text{P}}}$, every effort should be made to secure the author's consent to replace such notation with alternatives. More examples of alternative embellishments can be found in *Mathematics into Type* (Swanson 1979) and in *A Manual for Authors of Mathematical Papers* (AMS 1990).

Some mathematical presentations use roman letters in a script-type face (for example, $\mathfrak{R}$) or non-English characters. Greek letters are commonly used, but occasionally Fraktur (German black letter) or hebraic letters are needed; see sections 3·2 and 3·3. For further discussions of such uses see Burton (1992) and *Mathematics into Type* (Swanson 1979).

11·15 The symbols for vectors are set in bold italics; the components of a vector are scalars, which are set in the same typeface without bolding. Tensors are sometimes represented by bolding, but this can be misread as the notation for vectors. A clearer form is for tensors to be distinguished by setting their symbols in bold italics of a sans serif typeface (Cohen and Giacomo 1987).

Scalar components of vector: F_x, F_y

Vector: $\boldsymbol{F}$

Tensor: $\boldsymbol{\mathsf{S}}$

This representation of tensors is not universal in the mathematics and physics literature; exceptions and other symbols are common.

11·16 Symbols used in calculus are shown in Table 11·9, and Table 11·10 contains some of the common notation for set theory. More complete lists and definitions of mathematical symbols can be found in specialized books (Beyer 1987; James 1992). Mathematical symbols used in physics can be found in the IUPAC recommendations, 1987 revision (Cohen and Giacomo 1987). *Markup of Mathematical Formulas* (AAP 1989) includes not only detailed directions for markup of manuscripts in Standard Generalized Markup Language (SGML; see sections 32·7–32·11) but also extensive tables of mathematical symbols, entity references for markup, and verbal descriptions of the symbols.

11·17 Matrices and determinants are arrays of elements in columns and rows. In text, an overall symbol for the matrix may be used (for example, matrix **A**) or general symbols representing the array may be shown between double vertical bars (or alternatively, in parentheses or braces). Double vertical bars or brackets (and sometimes large parentheses) are used to enclose displayed matrices (see Table 11·11).

[in text] [in display]

$$\mathbf{A} = \|a_i b_i c_i\| = \begin{bmatrix} a_1 & b_1 & c_1 \\ a_2 & b_2 & c_2 \\ a_3 & b_3 & c_3 \end{bmatrix} = \begin{Vmatrix} a_1 & b_1 & c_1 \\ a_2 & b_2 & c_2 \\ a_3 & b_3 & c_3 \end{Vmatrix}$$

A determinant is represented by a similar array of elements set between single vertical bars. Because vertical bars are also used to indicate the absolute value of a real number or the modulus of a complex number, the text notation for a determinant should be of the form "det **A**" rather than |**A**|.

$$\det \mathbf{B} = \det (x_{ij}) = \begin{vmatrix} x_{1,1} & x_{1,2} & x_{1,3} & x_{1,4} \\ x_{2,1} & x_{2,2} & \cdots & \\ x_{3,1} & x_{3,2} & \cdots & \\ x_{4,1} & x_{4,2} & x_{4,3} & x_{4,4} \end{vmatrix}$$

Table 11·9 Calculus symbols

Class	Symbol	Meaning	Remarks
Sigma-class			Sigma-class symbols are often used with the upper and lower limits displayed.
	Σ	summation of terms	
	Π	product of terms	
	$\int$	integration of terms	
	$\oint$	curvilinear integration	
Other			
	Δ	delta; a finite increment of a function	As in Δx
	d	an infinitesimal increment of a function; the derivative symbol[a]	As in $\mathrm{d}x$
	∂	a variation in a function; the partial derivative symbol	
	lim	limit	
	$\mathrm{d}y/\mathrm{d}x$ [or] $\mathrm{D}_x y$	derivative of y with respect to x	Where $y = f(x)$
	$\partial u/\partial x$ [or] $\mathrm{D}_x u$	partial derivative of u with respect to x	Where $u = f(x,y)$
	$\partial^2 u/\partial x \partial y$ [or] $\mathrm{D}_y(\mathrm{D}_x u)$	the 2nd partial derivative of u, the 1st with respect to x and the 2nd with respect to y	Where $u = f(x,y)$
	∇	del [or] nabla	Del is used as an operator on vector functions or, with superscript 2, as the Laplacian operator.
	grad f [or] ∇f	gradient of f	
	div A [or] $\nabla \cdot A$	divergence of A	
	$\nabla \times v$	curl of v	
	∇^2 [or Δ]	the Laplacian operator	

[a]The derivative symbol should be in roman type in accordance with the recommendation of the International Union of Pure and Applied Physics (Cohen and Giacomo 1987); the italic form is occasionally used in the United States, but this style should be discouraged. The derivative symbol represents a function, not a variable or quantity.

Operators

11·18 Table 11·12 contains common mathematical operators. Minus symbols (signs) should be represented by an en-dash (–) rather than by a hyphen (-).

In the multiplication of vectors and tensors, some of the traditional mathematical symbols, such as centered dots and multiplication signs, have special meanings.

Table 11·10 Notations used in set theory

Symbol	Meaning	Remarks
$\in$	is an element of	$x \in M$; x is an element of set M
$\notin$	is not an element of	$y \notin M$; y is not an element of set M
$\ni$	contains as an element	$M \ni z$; set M contains z as an element
$\supset$	contains as a proper subclass	$M \supset N$; set M contains set N as a proper subclass
$\subset$	is contained as a proper subclass within	$M \subset N$; set M is contained as a proper subclass within set N
$\supseteq$	contains as a subclass	$C \supseteq E$; set C contains set E as a subclass
$\subseteq$	is contained as a subclass within	$C \subseteq E$; set C is contained within set E as a subclass
$\cup$	union or sum of	$A \cup B$; the union of set A and set B
$\cap$	intersection of	$A \cap B$; the intersection of set A and set B
Λ or $\emptyset$	the empty (or null) set	A set containing no members

Table 11·11 Matrix notations

Symbol	Meaning
$\mathbf{A}$ or (a_{ij}) or $\|a_{ij}\|$ [in physics, A]	symbol for a matrix
$\mathbf{A}^{-1}$	inverse of matrix **A**
$\mathbf{A}'$ [or] $\mathbf{A}^{\mathrm{T}}$ [in physics, $A^{\sim}$]	transpose of matrix **A**
$\bar{\mathrm{A}}$ [in physics, A^*]	complex conjugate of matrix **A**
$\mathbf{A}^{\mathrm{H}}$ [in physics, $A^{\dagger}$; do not use $\mathbf{A}^{+}$]	Hermitian conjugate of matrix **A**
det **A** [or] $\|\mathbf{A}\|$	determinant of matrix **A**
tr **A** [in physics, Tr A]	trace of matrix **A**

$\boldsymbol{A} \cdot \boldsymbol{B}$	scalar product of 2 vectors
$\boldsymbol{A} \times \boldsymbol{f}$	vector product of 2 vectors
$\boldsymbol{Af}$	dyadic product of 2 vectors
$\boldsymbol{P} \cdot \boldsymbol{B}$	product of a tensor and a vector
$\boldsymbol{P} : \boldsymbol{R}$	scalar product of 2 tensors
$\boldsymbol{P} \cdot \boldsymbol{T}$	tensor product of 2 tensors

A slash (/) through an equality, identity, or congruency symbol (sign) produces the negative of that symbol ("not identical to"). Wavy lines are used to indicate approximate rather than exact relationships. The addi-

Table 11·12 Common operators in arithmetic, algebra, and number theory[a]

Symbol	Meaning	Remarks
$+$	plus	
$-$	minus	
$\times$ [or] $\cdot$	times	$x \times y$ [or] $x \cdot y$ [also shown by juxtaposition of the quantities] $2axz$
/ [or] $\div$	divided by	x/y [or] $\frac{x}{y}$ [or] xy^{-1} [also shown as] $x \div y$
$=$	equals, is equal to	
$\neq$	does not equal, is not equal to	
$\equiv$	is identical with, identically equal to	
$\cong$	approximately equal to, congruent	Plane geometry; physics uses $\approx$
$\simeq$	asymptotically equal to	Physics
$\sim$	is similar to	Plane geometry
$\sim$	is equivalent to	Matrix calculus
$>$	is greater than	Double symbols, $>>$, mean "much greater than" in physics
$<$	is less than	Double symbols, $<<$, mean "much less than" in physics
$\geq$ [or] $\geqq$	is greater than or equal to	
$\leq$ [or] $\leqq$	is less than or equal to	
$\propto$	is proportional to	
!	factorial	Example: 4! represents $1 \times 2 \times 3 \times 4$

[a]Some mathematical symbols are frequently called "signs", but this manual recommends not using "sign" for operators; see section 3·7.

tion of such lines to symbols sometimes has specific meaning (for instance, $\simeq$ in physics, meaning "asymptotically equal to").

The symbol ~ is often used without particular care in prose to mean "about" or "approximately". The preferred indicator of approximation in such text usage is the appropriate word spelled out.

> The temperature of the system was approximately 45 °C.
>
> [not "The temperature of the system was ~ 45 °C."]

The symbol ~ should be retained in its specific mathematical usages in plane geometry and matrix calculus (see Table 11·12). The abbreviation for *circa,* "ca.", should not be used in mathematical expressions.

Enclosures

11·19 Enclosures such as parentheses and brackets are called "fences" in mathematics and are used to aggregate groups of symbols. The order of use of the common fences generally is the reverse of their use in nonmathematical prose.

[mathematics] $\{[(\,)]\}$ [nonmathematical prose] $([\{\,\}])$

If more levels of fences are needed, larger parentheses, brackets, and braces should be used in the order shown.

Other types of fences include angle brackets and single and double vertical bars, which have special meanings and should not be used to extend the basic sets shown above. Angle brackets, such as in $<a + b>$, are sometimes used for aggregation, but caution is needed because 1) such brackets may also be used as an alternative to the horizontal bar above a symbol to mean an average value; 2) they may be used to represent an ordered set of objects, as in $<x, y, z>$; and 3) the angle brackets can be mistaken for the "less than" or "greater than" operators. Furthermore, fences are normally used in matching pairs, but in physics and higher mathematics, unlike members may be used as pairs, for example, $|z>$ for the state z of a system, and occasionally only the left-hand fence is used.

Spacing within Mathematical Symbols

11·20 No space is left between quantities multiplied together when the multiplication operator is not shown;

$2b$ ac $6yx$b

between fences and the variables on either side of them;

$(a - 1)y$ $(4p - 4bc)(1 - a)$ $a|x|$

between terms and their subscripts or superscripts, or between subscripts or superscripts and the following terms;

$\cos^3 y$ $(a - 1)y^3 z$ $c^{x-2}d$

or between the symbols for plus, minus, or plus or minus when they are used to designate positive or negative values of numbers or variables.

$-2x$ . . . the values $+13$, -7, or ± 2.

A thin space (1/6th em-quad) is used before and after all of the operator symbols listed in Table 11·12 when they are used to represent mathematical operations. If a thin space is not available, a full space should be used in preference to no space.

$x = -4y - 1$ $0 < y < zw$ $(x + p)a \geq y^3 z(1 - 3r)$

Thin spaces are also used on either side of symbols for trigonometric functions, logarithms, and exponential and limit functions. No space is left, however, if 1) the quantities preceding or following these symbols are enclosed by fences, 2) the function carries a superscript or subscript, or 3) the function itself is part of a superscript, subscript, or limit of a sigma-class symbol.

$b \sin x \qquad (ac)\sin^3 2y \qquad \log x \qquad \exp(a + 2b) \qquad y^{\sin x}$

Superscripts and subscripts on the same variable are generally set close to the variable, with the superscript directly above the subscript. However, in higher mathematics (for example, tensors), spaces may be put between the variable and either index. Such notations should not be altered without querying the author.

[usual] m_n^{2r} [alternatives] $m_n{}^{2r}$ $m^{2r}{}_n$

An ellipsis in a mathematical expression is indicated by 3 centered dots; also see section 4·6, item 2.

$x_1 + x_2 + \cdots x_n$

MATHEMATICS IN TEXT AND DISPLAY

Guidelines for Text

11·21 When mathematical expressions are set within lines of text, every effort should be made to limit the vertical dimensions of fractions to maintain the spacing and appearance of those lines. This means limiting the fractions, complex exponents, and large symbols used in the expressions. Fractions and other quantities should be displayed using slashes or exponents rather than the vertically stacked form; graphic symbols such as for "square root" should be converted to exponential notations.

a/b or ab^{-1} $\left[\text{rather than } \frac{a}{b}\right]$

$(b - d)^{1/2}$ $\left[\text{rather than } \sqrt{b - d}\right]$

Note that only 1 slash can be used in any expression.

$\frac{a/b}{c}$ [or] (a/b)/c [but not] a/b/c

Exponents containing more than 1 level become unwieldy within text lines. The exponential form should be converted to the "exp" form for use in text (although large exponents, such as the 2nd example below, may even then require a display format).

e^{x^2-1} $\exp(x^2 - 1)$

$e^{\frac{a-b}{c+d}}$ $\exp\left(\frac{a - b}{c + d}\right)$

11·22 When an equation set in the text line will carry over from a line of text to the next line, choosing where to break the equation requires a knowledge of mathematics. The most basic guidelines are given here; for a more complete treatment, see *Mathematics into Type* (Swanson 1979).

In text, break equations in the following order of preference.

1 Before or after an equals symbol (sign).

> All variables in this equation should be set in italics: $x - 2 =$ $3by^3z - 4m$.

2 Before or after an addition or subtraction symbol (sign), but not if these symbols occur within a set of fences.

> For our goal of approximating the solution, $b_st_r = -K[(u/y) + (v/r)]^n +$ $ba[(u_m/y) + (v_m/r)]$.

3 Before or after a multiplication symbol (sign), or between sets of fences when the multiplication symbol "×" is not displayed (in the latter case, make the × explicit between the fences).

> The authors represented this complex equation as $R_e = (d^n u^{2-n} \rho / 8^{n-1} K) \times$ $(4n/3n + 1)^n$

4 Before a sigma-class symbol.

Breaks after operators are preferred so the reader can recognize that more of the expression is still to come. Such breaks are more important when the text lines are ragged right than when the text on the page is justified both left and right. Breaks within a pair of fences should be avoided whenever possible.

Display Guidelines

11·23 In general, the guidelines for breaking displayed equations are the same as for text equations, but in display, breaks are before, not after, operators. In particular, the alignment of displayed equations is a concern not addressed when equations are within lines of text.

For a sequence of equations in which the left-hand side is unchanged, align the "=" symbol in each line.

$$\begin{aligned} 2u_0v_0 &= u_0^2 + v_0^2 - (u_0 - v_0)^2 \\ &= k - (u_0 - v_0)^2 \end{aligned}$$

For continued expressions in which the left side is long, align the "=" symbol with the 1st operator in the 1st line.

$$[(a_1 + ia_2) + (a_{11}s_1 + a_{21}s_2)]/[(b_1 + ib_2) + (b_{11}s_1 + b_{21}s_2)]$$
$$= f(x)g(y) + \ldots$$

For expressions in which the right side is long, align the continuing operator with the 1st term to the right of the "=" symbol.

$$f(x) = 2k(a^2 + 5b_1)(3c - b^2c)$$
$$+ 4ac\{a_1b_1 + [4 - b^2)^2(ab + 4ck + b_2c)]\}$$

If you must split within fences, align the continuing operator with the 1st symbol within the enclosed group.

$$f(x)g(x) = \sin ab[R(2k \cos b) - 2R_0(2k \cos b)$$
$$+ R_1(b \sin ab) + \cos b]$$

STATISTICS

11·24 The international standard for statistical definitions and symbols is ISO 3534-1977 (E/F) available in *Statistical Methods: ISO Standards Handbook 3 1979* (ISO 1981). Several aspects of statistical usage not covered in that standard merit comment here.

For commonly used symbols see Table 11·13. Additional tables of statistical symbols can be found in Bailar and Mosteller (1992) and the ISO standard cited above. Nonstandard symbols for variables not represented in the ISO recommendations have been suggested in a paper by Dong (1992).

The recommended abbreviation for "confidence interval" is "CI". The numeric limits represented by a confidence interval should be connected by "to" rather than an en-dash so as to avoid possible confusion of a dash with a minus symbol.

with a mean value of 4.23 (CI, –2.13 to 10.33), which is greater than . . .

The term "confidence interval" is conceptually more desirable than "confidence limits" because "interval" implies a span of values. The term "confidence limits" implies, on the other hand, 2 discrete points rather than a span of values. Some authors may, however, prefer the representation by "confidence limits".

Table 11·13 Symbols used in statistics[a]

Population symbol	Sample symbol	Explanation
	F	variance ratio [F-test]
H_0		null hypothesis
N		number of subjects (population or lot size)
	n	number of subjects
	P	probability of wrongly rejecting the null hypothesis
	R	coefficient of multiple correlation, range of a sample
ρ	r	coefficient of correlation
σ	s	standard deviation[b]
σ^2	s^2	variance
	$s_{\bar{x}}$	standard error of the mean[c]
	t	statistic derived in Student t-test
$\bar{X}$	$\bar{x}$	arithmetic mean
	α	probability of a type I error; significance level
	β	probability of a type II error
	χ^2	statistic derived in chi-square test
	v	number of degrees of freedom

[a]For additional symbols, see section 5, "Symbols" in the ISO standard cited in section 11·24.

[b]"SD" is not recommended as an abbreviation for "standard deviation".

[c]"SEM" is not recommended as an abbreviation for "standard error of the mean".

with a mean value of 4.23 (CL, –2.13, 10.33), which is greater than . . .

Numbers in statistical presentations should be styled in accordance with the recommendations in sections 11·1–11·4. Note especially that values for P (probability) should follow the general rule in section 11·3 that an initial zero (0) must precede the decimal point for numeric values smaller than 1.

and the mean differed from that of the control group by 17.8 ($P = 0.23$), which . . .
[not "by 17.8 ($P = .23$)"; a P value of 1.00 is extremely unlikely but not conceptually impossible]

Note that designations of "±" values following values of means must be accompanied by a notation indicating whether the ± value presented is a standard deviation (s) or a standard error of the mean ($s_{\bar{x}}$). The notation "SD" is not recommended for "standard deviation", and "SEM" is not recommended for "standard error of the mean".

and in the infected population, the mean value was 14.3 ± 2.5 (s).

A preferable notation does not use the ± symbol.

[for standard deviation]
and for the infected population, the mean value was 14.3, $s = 2.5$.
[for standard error of the mean]
and for women, the mean value was 25.7, $s_{\bar{x}} = 1.2$.

ROUNDING NUMBERS

11·25 In reporting a quantity, the number of significant digits must be commensurate with the precision of the method of measurement. If the quantity must be converted to SI units (see section 11·8), the quantity must be multiplied by the exact conversion factor and then rounded to the appropriate number of significant digits. The following rules and examples illustrate the rounding of a number in which 4 significant digits are to be retained.

1 If the digit to the right of the 4th digit is less than 5, leave the 4th digit unchanged.

4.1282 rounds to 4.128

2 If the digit to the right of the 4th digit is greater than 5, increase the 4th digit by 1.

4.1286 rounds to 4.129

3 If the digit to the right of the 4th digit is exactly 5, is followed only by zeros or nothing, and the 4th digit is even, leave the 4th digit unchanged.

4.1285 rounds to 4.128
4.12850 rounds to 4.128

If the 4th digit is odd, increase the 4th digit by 1.

4.1275 rounds to 4.128
4.12750 rounds to 4.128

4 If the digit to the right of the 4th digit is 5 and there is at least 1 digit other than zero to the right of the 5, increase the fourth digit by 1.

4.12851 rounds to 4.129
4.12751 rounds to 4.128

For an alternative system, consult "Guide to the Rounding of Numbers", Annex B in *International Standard ISO 31-0* (ISO 1992a).

REFERENCES

Cited References

[AAP] Association of American Publishers. 1989. Markup of mathematical formulas. Version 2. 0. Rev ed. Dublin (OH): AAP.

[ACN] l'Association Canadienne de normalisation. 1990. Guide canadien de familiarisation au système métrique. Rexdale (Toronto): ACN. Norme nationale du Canada: CAN/CSA-Z234.1–89.

[AGPS] Australian Government Publishing Service. 1988. Style manual for authors, editors and printers. 4th ed. Canberra: AGPS Pr.

[AMS] American Mathematical Society. 1990. A manual for authors of mathematical papers. Providence (RI): AMS.

[Anonymous]. 1992. Billion bites the dust. Nature 358 (Jul 2):2.

[ANSI] American National Standards Institute. 1982. ANSI/IEEE Standard 268–1982, Metric practice. New York: ANSI.

ASTM Committee E-43 on Metric Practice. 1989. Standard practice for use of The International System of Units (SI): the modernised metric system. ASTM E380-89a. Philadelphia: American Soc for Testing and Materials.

Bailar JC, Mosteller F. 1992. Medical uses of statistics. 2nd ed. Boston: NEJM Books.

Beyer WH. 1987. CRC standard mathematical tables. 28th ed. Boca Raton (FL): CRC Pr.

Burton BW. 1992. Dealing with non-English alphabets in mathematics. Tech Commun 39:219–25.

[CSA] Canadian Standards Association. 1980. Metric editorial handbook. Rexdale (Toronto): CSA. CSA special publication Z372-1980.

[CSA] Canadian Standards Association. 1989. Canadian metric practice guide. Rexdale (Toronto): CSA. National standard of Canada: CAN/CSA-Z234.1-89.

Cohen ER, Giacomo P. 1987. Symbols, units, nomenclature and fundamental constants in physics: 1987 rev. Physica 146A(1&2):v-67. Document IUPAP-25, 1987 (SUNAMCO 87–1).

Dong G. 1992. Conventions in statistical symbols and abbreviations. CBE Views 15:95–6.

Horvath AL. 1986. Conversion tables of units in science & engineering. New York: Elsevier.

[ISO] International Standards Organization. 1981. ISO standards handbook 3 1979: statistical methods. Geneva: ISO.

[ISO] International Organization for Standardization. 1982. ISO standards handbook 2: units of measurement. 2nd ed. Geneva: ISO.

[ISO] International Organization for Standardization. 1988. ISO standards handbook 1: documentation and information. 3rd ed. Geneva: ISO. p 810–36.

[ISO] International Organization for Standardization. 1992a. Quantities and units—Part 0: general principles: ISO 31-0 (E). Geneva: ISO.

[ISO] International Organization for Standardization. 1992b. Quantities and units—Part 11: mathematical signs and symbols for use in the physical sciences and technology: ISO 31/11-1992 (E). Geneva: ISO.

James R. 1992. Mathematics dictionary. 5th ed. New York: Van Nostrand Reinhold.

Jerrard HG, McNeill DB. 1992. A dictionary of scientific units: including dimensionless numbers and scales. 6th ed. London: Chapman and Hall.

Swanson E. 1979. Mathematics into type: copy editing and proofreading of mathematics for editorial assistants and authors. Rev ed. Providence (RI): American Mathematical Soc. Reprinted with corrections, 1986.

Taylor BN. 1991. The international system of units (SI). NIST Special Publication 330. Washington: National Inst of Standards and Technology.

[USGPO] United States Government Printing Office. 1984. United States Government Printing Office style manual. Washington: USGPO.

Additional References

[AIP] American Institute of Physics. 1990. AIP style manual. 4th ed. New York: AIP.

Baron DN. 1988. Units, symbols, and abbreviations: a guide for biological and medical editors. 4th ed. London: Royal Soc Med.

Cook JL. 1991. Conversion factors. New York: Oxford Univ Pr.

Darton M, Clark J. 1994. The Macmillan dictionary of measurement. New York: Macmillan.

Hansen WR. 1990. Suggestions to authors of the reports of the United States Geological Survey. 7th ed. Washington: US Government Printing Office.

Lapedes DN. 1978. McGraw-Hill dictionary of mathematics and physics. New York: McGraw-Hill.

Marriott FHC. 1990. A dictionary of statistical terms. 5th ed. Burnt Mill, Harlow, Essex, (UK): Longman.

McCoubrey AO. 1991. Guide for the use of the International System of Units: the modernized metric system. National Institute of Standards and Technology Special Publication 811. Washington: US Government Printing Office.

Monteith JL. 1984. Consistency and convenience in the choice of units for agricultural science. Exp Agric 20:105–17.

12 Time and Dates

> Absolute, true, and mathematical time, of itself, and from its own nature, flows equably without relation to anything external, and by another name is called duration: relative, apparent, and common time, is some sensible and external (whether accurate or unequable) measure of duration by the means of motion, which is commonly used instead of true time; such as an hour, a day, a month, a year.
>
> —Isaac Newton, *Philosophiae Naturalis Principia Mathematica,* 1687

Measuring the passing of time and the seasons is among the oldest activities of the human mind. Early in human societies men and women were spurred to find ways to record what they measured. As science has grown, it has increasingly needed ever more precise recording of time in smaller and smaller increments. Those needs have led to present-day notations. The present comprehensive standard for time and date representations is *Data Elements and Interchange of Formats — Information Exchange — Representation of Dates* (ISO 1988); also note the ISO standard for quantities and units of space and time (ISO 1992). The recommendations in the 1988 standard are not yet in wide use.

TIME

UNITS AND SYMBOLS

12·1 Units of time are generally spelled out in text when they are not represented by standard symbols in statements including scientific units of measurement; see sections 11·10 and 11·11 and Tables 11·3 and 11·6.

day d   hour h   minute min   second s

month mo   year y

The year can be measured accurately to a millionth of a second.

If the value is for only 1 unit, insert a space between the number and the symbol. But if the value calls for more than 1 unit, set the numbers and symbols close, with the symbols either on the line or superscripted.

1 y   14 mo   27 h   22h3min   2h15min4s   5min14s

The current position of the north galactic pole is near right ascension 12^h52^m, declination 27°8′ and the galactic center is located at about right ascension 17^h45^m, declination –28°56′.

Values with fractions of seconds should be written with decimal fractions.

3.21 s   10min14.6s   3h5min37.5s   1.4 h
45min5.8s   14h18min0.7s

For values of time, do not use the symbols for minute (′) and second (″) that are correctly used for geographic coordinates (latitude and longitude).

CLOCK TIME

12·2 Two systems are used to designate the hour of the day. In the 12-hour system the hours are divided into two 12-hour portions: 1 through 12 before noon (AM), and 1 through 12 midnight (PM). In the 24-hour system, the hours are numbered consecutively, 1 through 24 (24 being midnight).

The 12-hour system requires a distinction between the 2 halves of the day, for example, to distinguish 12:01 AM from 12:01 PM. Minutes are separated from hours by a colon, and a 0 is added as needed to make the minutes a 2-digit number. A period should be used only for a decimal fraction of the unit.

12:27 AM   7:57 AM   12:27 PM   7:57 PM

12.3 h = 12 and 3/10 h [or] 12h18min [not "12 h and 30 min"]

8.5 h = 8 and 1/2 h [or] 8h30min [not "8 h and 5 min"]

The abbreviations AM (*ante meridiem* = before noon) and PM (*post meridiem* = after noon) may be in capital letters, small capitals, or lowercase, but capital letters are recommended.

With the 12-hour system, it is correct to express time with the AM and PM notation or the informal and ambiguous "o'clock" but not with both.

10 PM or 10 o'clock [but not "10 o'clock PM"]

12·3 The 24-hour system used throughout Europe and the US military and by many scientists obviates ambiguities as to which part of the day is meant. Time is expressed as a 4-digit number (0's added as appropriate) without punctuation.

0602 = 6:02 AM   1802 = 6:02 PM
0028 = 28 minutes past 12 midnight   1228 = 28 minutes past 12 noon

The day begins at 0000 (midnight) and ends at 2359; 2400 of 1 day = 0000 of the next. The abbreviation "h" should not be used following the 4 digits because the last 2 digits are minutes and not decimal fractions of an hour. Do not use a period or colon between hours and minutes.

If the context of a time designation in the 24-hour format would produce ambiguity, add "hours".

At 1530 hours we stopped counting. [not "At 1530 we stopped counting": number or time?]

The ISO standard (1988) presents a format "hours minutes seconds" (hhmmss) with decimalization for fractions of seconds. The 2-digit representations of the values of each unit can be separated by colons; the comma is recommended by ISO for the decimal point, but this manual recommends a period (full-stop symbol) for American and British usage. The format can be truncated for less precise representations.

232050.5 [or] 23:20:50.5
[the hhmmss.s formats for "23 hours, 20 minutes, and 50.5 seconds"]
2320 [or] 23:20 [the hhmm formats for "23 hours and 20 minutes"]

TIME ZONES

12·4 International time zones were established legally in the 19th century. The standard time system, fixed in 1883, partitioned the natural continuous time (which gains 1 minute every 22.4 km [14 miles] traveled from east to west) into 24 international time zones in increments of 15° of longitude. (One degree of latitude covers about 111 km [69 miles].) Greenwich, UK, is at 0°, the prime meridian. Mean solar time is determined by this meridian: Greenwich mean time (GMT). GMT is equivalent to coordinated universal time (also known as universal time coordinated, UTC) and is the time scale available from broadcast time signals;

also see section 27·15. Each time zone differs from Greenwich mean time by a whole number of hours, or in some instances half hours, except where some political considerations prevail.

12·5 In the conterminous United States, 4 meridians are designated for standard time: 75° (eastern), 90° (central), 105° (mountain), and 120° (Pacific) west of Greenwich. Alaska and Hawaii are in the Alaska-Hawaii time zone (150° west of Greenwich) and the Aleutian Islands are in the Bering time zone (165° west of Greenwich). In addition to the 4 meridians for the conterminous United States, Canada contains the 135° meridian (Yukon), the 60° meridian (Atlantic), and the Newfoundland zone within the Atlantic zone, which encircles the island of Newfoundland and is 0.5 h later than Atlantic time. When the time in a zone is advanced in the spring by 1 hour to lengthen evening time that is in daylight, the term for the time zone replaces "standard" with "daylight".

The names of time zones are not capitalized when written out (except for proper nouns and in contexts with formal reference to time systems); their abbreviations are capitalized without periods when they immediately follow the time statement.

Greenwich mean time (GMT) eastern standard time (EST)
central daylight time (CDT) central standard time (CST)
mountain standard time (MST) Pacific standard time (PST)

The abbreviation for the zone is not set off from the time by commas.

When it is noon CDT, it is noon EST, which is 11:00 AM CST.

When it is 0100 in British Columbia (Pacific standard time), it is 0500 in New Brunswick (Atlantic standard time) but 0530 in Newfoundland (Newfoundland standard time).

DATES

DAYS

12·6 In English, the names of the days of the week are capitalized and written out in text but commonly abbreviated to their first 3 letters in tables and other locations where short forms are needed. They are named after planets and gods in terms of Anglo-Saxon origin; see Table 12·1.

MONTHS

12·7 Months are written out in text but may be abbreviated to their first 3 letters in tables and graphs; see Table 12·2. Using the 3-letter abbrevia-

Table 12·1 Days of the week: names, abbreviations, and derivations

English	Abbreviation	Anglo-Saxon[a]	Latin	French
Sunday	Sun	Sun's day	*dies solis* (Solis)	dimanche
Monday	Mon	Moon's day	*dies lunae* (Moon)	lundi
Tuesday	Tue	Tiw's day	*dies martis* (Mars)	mardi
Wednesday	Wed	Woden's day	*dies mercurii* (Mercury)	mercredi
Thursday	Thu	Thor's day	*dies jovis* (Jupiter)	jeudi
Friday	Fri	Frigg's day	*dies veneris* (Venus)	vendredi
Saturday	Sat	Saturne's day	*dies saturni* (Saturn)	samedi

[a]Modern English equivalent

Table 12·2 Months: names and abbreviations

English	French	Abbreviations
January	janvier	Jan, janv.
February	février	Feb, févr.
March	mars	Mar, mars
April	avril	Apr, avr.
May	mai	May, mai
June	juin	Jun, juin
July	juillet	Jul, juill.
August	août	Aug, août
September	septembre	Sep, sep.
October	octobre	Oct, oct.
November	novembre	Nov, nov.
December	décembre	Dec, déc.

tion in dates avoids any ambiguity inherent in using a numeric equivalent; also see section 12·9.

> 5 Jan 1993 [rather than "5/1/93" (European style) or "1/5/93" (US style)]

YEARS

12·8 Years are expressed in numerals. Only in informal writing are they abbreviated to 2 digits, for example, "class of '91", "the spirit of '76".

> In the 1980s (1984, to be exact) Lindow Man was found; about the 1st century AD he was ritually killed and deposited in a pond, which by the time we discovered him in the 20th century had become a peat bog.
>
> the 1880s and 1890s [not "the 1880s and '90s" or "the '80s and '90s"]

SEQUENCE OF DATE ELEMENTS

12·9 In scientific communications, dates should be written in the sequence of year, month, and day (common in astronomical and aeronautical literature and in references) or day, month, and year, common in Europe and Canada and in US military organizations. The order is the same for abbreviated dates or dates written out.

1992 April 23	1992 Apr 23	92 Apr 23	1992/4/23	92/4/23
23 April 1992	23 Apr 1992	23 Apr 92	23/4/1992	23/4/92

The sequence of month, day, and year, common in the United States, should be avoided in scientific writing, but if it must be used, the year should be set off by commas.

> The total solar eclipse of July 11, 1991, passed directly over the world's largest telescopes on Mauna Kea volcano, Hawaii.

The day of the month unaccompanied by the year is an ordinal and is so pronounced when spoken, but American usage usually expresses it as a cardinal number. British usage applies the ordinal.

> [American] 18 April or April 18   [British] 18th April or April 18th

If only the month and year are used in text, they are not separated by punctuation.

> An eclipse will pass over Uruguay and the South Atlantic Ocean in June 1992.
>
> The July 1991 eclipse darkened the earth as far east as Mexico City.

The relevant ISO standard (ISO 1988) recommends a numeric "calendar year–month–day" (CCYYMMDD) format and variations on it (including week designations).

> 19850412 [for "12 April 1985"]   1985–05 [for "May 1985"]
>
> 85W155 [for "1985, 15th week, 5th day of that week"]

ERAS

12·10 Numerals are used for year numbers followed or preceded by era designations, if necessary, in capitals; some typographers prefer small capitals when they are available in the typeface. "AD", "AC", and "AH" precede

Table 12·3 Eras: names and abbreviations

Era	Abbreviation
after Christ	AC
anno Domini (in the year of our Lord)	AD
anno hegirae (in the year of the [Muhammad's] Hegira AD 622) or *anno Hebraico* (in the Hebrew year)	AH
anno mundi (in the year of the world)	AM
anno salutis (in the year of salvation)	AS
ab urbe condita (from the founding of the city [Rome] in 753 BC)	AUC
before Christ	BC
before the common (Christian) era (= BC)	BCE
before present ("present" = 1950)	BP
of the common (Christian) era (= AD)	CE
millions of years before present	MaBP

the year; the others follow the year. See Table 12·3 for common designations of eras.

AD 1492 [or] AD 1492 2050 BC [or] 2050 BC

Note that there is no year 0.

Also see the sections 25·11–25·13 in Chapter 25, "Human History and Society", on dating and sections 26·1–26·3 on geologic time.

REFERENCES

Cited References

[ISO] International Standards Organization. 1988. International standard ISO 8601:1988(E): Data elements and interchange of formats—information exchange—representation of dates. Geneva: ISO.

[ISO] International Standards Organization. 1992. International standard ISO 31–1: Quantities and units, Part 1—Space and time. Geneva: ISO.

Additional References

American Society of Agronomy. 1988. Publications handbook and style manual. Madison (WI): American Soc of Agronomy, Crop Science Soc of America, and Soil Science Soc of America.

Huth EJ. 1987. Medical style & format: an international manual for authors, editors, and publishers. Philadelphia: ISI Pr. Available from Williams & Wilkins, Baltimore.

O'Malley M. 1990. Keeping watch: a history of American time. New York: Viking.

University of Chicago Press. 1993. The Chicago manual of style. 14th ed. Chicago: Univ Chicago Pr.

[USGPO]. US Government Printing Office. 1984. US Government Printing Office style manual 1984. Washington: USGPO.

13 Addresses and Geographic Descriptions

This is the place. —Brigham Young, 1847 [Leonard Arrington. 1985. *Brigham Young: American Moses*. New York: Knopf. p 459.]

With increasing populations and expectations of speedier communication, delivery of mail has become commensurately complex. Most countries have been officially divided into areas denoted by letters or numbers or both (postal codes), and many postal agencies will no longer deliver mail without the appropriate code. Furthermore, each country has its own spelling of its name, which probably differs from one's own country's spelling of that country's name. In recent years some countries have redrawn their boundaries and political affinities and sometimes changed their names and capital cities.

ADDRESSES

13·1 Addresses should include specific elements.

1 Name (for personal addresses).
2 Department or unit.
3 Institution or street number and name.

4 City or town, state (province for Canada; or other subnational unit if applicable), and postal code. The name of the state or province should be represented by its 2-letter postal abbreviation; see Table 13·1.

5 Country name if the letter is mailed outside of the country of the addressee. In general, the country name should not be abbreviated and should be in capital letters; if the proper form of the name is not known with certainty, the spelling specified by the post office of the origin of the mailing should be used.

Marco Polo PhD
Research & Exploration
National Geographic Society
17th and M Streets, NW
Washington DC 20036
UNITED STATES OF AMERICA
[or "USA"]

Marco Polo MD
Department of Anatomy
University of Pennsylvania
33rd and Spruce Streets
Philadelphia PA 19104

An office or suite number should follow the name of the institution or office building in which it is located, either after a comma and a space or within parentheses. Also note that the city name may be in capital letters and not followed by a comma, to facilitate computer scanning of the last line of the address. The US Postal Service requests that the entire address on mailing labels be printed in block capital letters.

POSTAL CODES

United States

13·2 US postal ZIP (for "Zone Improvement Plan") codes consist of 5- or 9-digit numbers (hyphenated between the first 5 and last 4 digits), placed 2 spaces after the 2-letter state postal code abbreviation.

American Association for the
Advancement of Science
1333 H Street NW
WASHINGTON DC 20005
USA

National Geographic Society
PO Box 1111
Washington DC 20013-9990

The 1st digit represents one of 10 large groups of states: 0 for New England states and New Jersey, 1 for New York and Pennsylvania, and so on to 9 for the Pacific Coast states, Alaska, and Hawaii. The 2nd and 3rd digits represent areas within the larger group; the 4th and 5th, local delivery areas. ZIP code numbers are available in the *National Five-Digit ZIP Code® and Post Office Directory* (USPS [annual]), which also carries

Table 13·1 US and Canadian governmental units and their postal abbreviations

US unit	Abbreviation	US unit	Abbreviation
Alabama	AL	Montana	MT
Alaska	AK	Nebraska	NE
American Samoa	AS	New Hampshire	NH
Arizona	AZ	New Jersey	NJ
Arkansas	AR	New Mexico	NM
California	CA	New York	NY
Colorado	CO	North Carolina	NC
Connecticut	CT	North Dakota	ND
Delaware	DE	Northern Mariana Islands	CM
District of Columbia	DC	Ohio	OH
Florida	FL	Oklahoma	OK
Georgia	GA	Oregon	OR
Guam	GU	Pennsylvania	PA
Hawaii	HI	Puerto Rico	PR
Idaho	ID	Rhode Island	RI
Illinois	IL	South Carolina	SC
Indiana	IN	South Dakota	SD
Iowa	IA	Tennessee	TN
Kansas	KS	Texas	TX
Kentucky	KY	Trust Territory	TT
Louisiana	LA	Utah	UT
Maine	ME	Vermont	VT
Maryland	MD	Virginia	VA
Massachusetts	MA	Virgin Islands	VI
Michigan	MI	Washington	WA
Minnesota	MN	West Virginia	WV
Mississippi	MS	Wisconsin	WI
Missouri	MO	Wyoming	WY

Canadian unit	Abbreviation	Canadian unit	Abbreviation
Alberta	AB	Nova Scotia	NS
British Columbia	BC	Ontario	ON
Manitoba	MB	Prince Edward Island	PE
New Brunswick	NB	Quebec	QC [or] PQ
Newfoundland	NF	Saskatchewan	SK
Northwest Territories	NT	Yukon	YT

an explanation of the ZIP code system and abbreviations approved for addresses needing compression to fit computerized address fields.

Canada

13·3 Canadian postal codes contain 6 letters (L) and numbers (N), the 1st letter of which stands for a region, in the form: L-N-L N-L-N. The last 3 units further narrow the location down to the appropriate side of the street in a single block. A major building has its own code, for example, each building on a college campus. The letters O and I are not used, but the numerals 0 and 1 are.

In Canadian addresses for domestic use, the postal code should go on a line by itself or after the 2-letter province abbreviation, but if the mail is sent from outside Canada, the postal code goes after the country name (CPC [annual]):

Paul K Anderson PhD	Paul K Anderson PhD	Paul K Anderson PhD
Department of Biology	Department of Biology	Department of Biology
University of Calgary	University of Calgary	University of Calgary
Calgary, AB	Calgary, AB T2N 1N4	Calgary, AB
T2N 1N4		CANADA T2N 1N4

Other Countries

13·4 Other countries apply their own standards. In Europe, the postal code often goes before the city, and sometimes a regional abbreviation goes after the city.

Eckart Ehlers
Institut für Wirtschaftsgeographie
der Universität Bonn
Mechenheimer Allee 166
D-5300 Bonn 1
GERMANY

Nadia Leal
83, rue de Paris
89000 Auxerre
FRANCE

Eugenio Ozhogin
Pasaje Villasis 1, 4E
41003 Sevilla
SPAIN

Prof V L S Bhimasankaram
Faculty of Science
Osmania University
Hyderabad - 500 007
INDIA

Henry W Esbenshade
International Tree Crops Institute
Nedlands 6009
AUSTRALIA

Andrzej Mochon
Institute of Geography
Pedagogical University Kielce
Kielce 25–409
POLAND

A David Cooper
Luton College of Higher Education
Luton LU1 3JU
UNITED KINGDOM

Hans Elsässer
Institut de Géographie
Université de Zurich
8057 Zurich
SWITZERLAND

A & G Marco
Via Fortezza 27
20126 Milano
ITALY

Lin Zhenyao
Institute of Geography
Chinese Acedemy of Sciences
Beijing 100101
PEOPLE'S REPUBLIC OF CHINA

Sergei B Rostotsky
Institute of Geography
Russian Academy of Sciences
Moscow 109017
RUSSIA

José O Moncada Maya
Institute of Geography
Universidad Nacional Autonoma de México
04510 México DF
MEXICO

GEOGRAPHIC NAMES

13·5 The spelling of many geographic names can be verified in the *Cambridge World Gazetteer* (CUP 1988) and *New Geographical Dictionary* (MW 1988). The rest of this section and section 13·10 describe sources of names not represented in these suggested sources for verification.

US AND CANADIAN NAMES

The US Board on Geographic Names formulates the principles and policies that determine the governmental use of domestic and foreign names and names of undersea and extraterrestrial features. The Domestic Names Committee at US Geological Survey, Reston, Virginia, maintains the Geographic Names Data Base, managed by the Geographic Names Information System (GNIS), which contains more than 2 million name records. Official names are identified as such; the database also includes records of named features that do not fall within the purview of the Board on Geographic Names. Inquiries should be directed to US Geological Survey, National Mapping Division, Domestic Names, Mail Stop 523, Reston VA 22092, USA.

In Canada, the official body that accepts or rejects geographic names is of the province or territory in which the feature lies, except for federally administered lands. The Canadian Permanent Committee on Geographical Names (CPCGN), a federal–provincial–territorial committee administered by a secretariat in Ottawa, acts as a clearinghouse and central registry for all approved names in Canada. The Secretariat enters all official, and some unofficial, names into the National Toponymic Data Base, a computer file from which the official names are drawn for gazetteers, topographic maps, and responses to inquiries. Most new names are generated from the general public. Sources of published names include the *Gazetteer of Canada* (CPCGN), the *Guide toponymique du Québec* (CTQ 1990), and the *Gazetteer of Undersea Feature Names* (CPCGN 1983). Inquiries should be addressed to Secretariat, Geographical Names, Room 650, 615 Booth Street, Ottawa, ON, CANADA K1A OE9.

GEOGRAPHIC NAMES IN TEXT

States and Similar Divisions

13·6 In text, names of states, territories, possessions, and provinces should be spelled out. A comma should follow the name of a city when it is followed by the state name, and a comma follows the state name unless it is the last word of the sentence.

> The Titan missile would be designed and developed in Denver, Colorado, at the Martin Company's new plant.
>
> Evidence of a new fault line was found 5 miles south of Turkington, Missouri.

Where abbreviations of state names are needed, as in tables and bibliographic references, the 2-letter postal code (see Table 13·1 and section 30·35) should be used.

Streets

13·7 Elements in a street address in text, including avenue, boulevard, street, road, north, south, should be spelled out and capitalized. Direction designations such as NW, NE, SW, or SE used as part of the street name are exceptions to this rule. Numbered streets and avenues should be designated with ordinal numbers. "Post office box" can be represented by the abbreviations "PO Box" or "POB".

> 3420 16th Street, NW   108 5th Street, NE   PO Box 32

In French addresses, these elements are set in lowercase type whether they are part of the address or stand alone. Building numbers are usually followed by a comma.

le boulevard Saint-Germain   13, rue des Beaux-Arts
la place de l'Opéra   le carrefour de Buci

Elements of an address that constitute a formal name are always spelled out. Generic terms are not capitalized.

the Fourteenth Street Bridge turnoff   the Fifth Avenue Hospital grounds

Regions and Geographic Features

13·8 A geographic term is capitalized if it is part of a place-name.

Mississippi River [but "the river" in a subsequent text reference]
Great Dismal Swamp [but "the swamp" in a subsequent text reference]

When the geographic term is plural and applied to 2 or more proper names it is not capitalized.

Mississippi and Missouri rivers

Descriptive terms (for example, "east", "west", "lower", "blue", "zone") are capitalized only when they are consistently used to denote a defined region (and therefore constitute a proper name).

West Coast   Upstate New York   Middle East   Blue Nile
Rust Belt   Corn Belt   the Southwest   central Manitoba
[but] western California   northern New York

PUNCTUATION

13·9 North American place-names seldom include punctuation.

Richardsons Creek [not "Richard's son's creek"]
Jamestown [not "James' town"]
Baileys Crossroads [not "Bailey's Crossroads"]

"Martha's Vineyard" is an exception.

NUMBERS

A place-name that includes a number is spelled out, and the number is not hyphenated.

Two Rivers   Three Mile Island   Fourteen Mile Point
Three River Stadium

An exception is cities named with dates.

20 de Junio, Argentina   25 de Diciembre, Peru

Such names are alphabetized as if they were spelled out in English.

MULTIWORD NAMES

All elements except prepositions and articles within multiword names are capitalized.

Fond du Lac   Point of Rocks   Rock of Ages
Truth or Consequences

PREFIXES

Prefixes of most geographic names (county, fort, point, port) should be spelled out.

Point Lobos, Port Arthur, San Diego

"Saint" and "Mount" may be abbreviated, respectively, to "St" and "Mt".

St Louis, Mt Washington [but "Mount St Helens"]

"Saint", "Sainte", and "Mont" should not be abbreviated for French place-names unless absolutely necessary. If the word is abbreviated (St, Ste, Mt), the 2 parts of the word are hyphenated.

Saint-Lawrence   St-Lawrence   rue de Ste-Geneviève
Sainte-Foy   Ste-Foy   Saint-Augustin-de-Demaures   Saint-Henri
Sainte-Anne-de-la-Pérade

FOREIGN GEOGRAPHIC NAMES

13·10 Place-names outside the United States and its possessions are handled in the United States by the Foreign Names Committee of the Board on Geographic Names, maintained at the Defense Mapping Agency, Washington, DC. The names are published in more than 100 gazetteers available from US Geological Survey libraries (Reston, Virginia; Denver, Colorado; Menlo Park, California) and in more than 600 federal, state, university, and local libraries. For some foreign place-names, the board has approved optional names in anglicized form.

Jordan River [instead of "Mahr al Urdunn"]

Rome [instead of "Roma"]

Danube River
[instead of "Donau" (Austria, Germany) or "Duna" (Hungary)]

Vatican City [instead of "Città del Vaticano"]

The Board's *Romanization Guide, Gazetteer of Conventional Names,* and additional information (including information on undersea and extraterrestrial features and Antarctica) is available from the Executive Secretary, Foreign Geographic Names, US Board on Geographic Names, Defense Mapping Agency, Washington DC 20305, USA.

If an author feels strongly about the spelling of a place-name and prefers the original national spelling, disregarding that opinion is perhaps insensitive when the only reason is that the US Department of Defense prefers a different spelling. For instance, non-Malawian authors may spell the African country "Malawi", but its residents point out that the official language is Chichewa and that Malawi has a board to determine official spelling. "Laŵi" means a flame; "Malaŵi" means flames, and without the circumflex over the "w" the word is meaningless, and hence not accepted nationally. Editorial offices may reject an author's legitimate preference of this kind because creating a character not represented in English may be difficult with standard equipment, but increasingly word-processing programs can produce a wide range of characters not used in English.

Regions

13·11 Regions are multinational areas that have some affinity but are not always strictly defined. For example, North America includes all the land from, and including, Panama north to the Arctic; the term includes the West Indies (Caribbean islands) and Greenland.

Middle America is composed of Mexico, Central America, and the West Indies (Caribbean islands).

Central America includes Guatemala, Belize, El Salvador, Honduras, Nicaragua, Costa Rica, and Panama (not including the Caribbean islands).

The Middle East has no precise definition; it is generally thought of as including Turkey, Cyprus, Syria, Lebanon, Israel, Jordan, Egypt, Iraq, Iran, and the countries of the Arabian Peninsula. At its maximum it would extend from Morocco to Bangladesh and would include North Africa, the Horn of Africa, Greece, and Bulgaria. "Near East" is a dated term for the same area that may be used in historical references.

Table 13·2 Countries that made up the former Union of Soviet Socialist Republics

Nation	Capital	Nation	Capital
Armenia	Yerevan	Lithuania	Vilnius
Azerbaijan	Baku	Moldova	Chişineu
Belarus	Minsk	Russia	Moscow
Estonia	Tallinn	Tajikistan	Dushanbe
Georgia	Tbilisi	Turkmenistan	Ashgabat
Kazakhstan	Alma-Ata	Ukraine	Kiev
Kyrgyzstan	Bishkek	Uzbekistan	Tashkent
Latvia	Riga		

Countries

13·12 For countries whose names are properly preceded by "the", the article should not be capitalized, except for "The Bahamas" and "The Gambia".

"United Kingdom" and "United States of America" as nouns are spelled out, but "UK" and "US" may be used as adjectives.

"Russia" is not a synonym for the Commonwealth of Independent States (CIS) or for the former Soviet Union (USSR). "Russian" is a term for a citizen of the Russian Federation or the former Russian Soviet Federative Socialist Republic. The peoples of other parts of the Commonwealth of Independent States or the former Soviet Union have specific names; for example, Ukrainian, Kazakh, Uzbek. In reference to the communist period, "Soviets" (in the plural) is acceptable for the government or its people. "Soviet" is the preferred adjectival form for the former USSR. The 3 Baltic states have become independent nations; they were followed by the 12 former Soviet republics; see Table 13·2.

At the end of 1992, Czechoslovakia separated into the Czech Republic (capital, Prague) and the Slovak Republic (capital, Bratislava).

Additional changes of names have taken place in recent years; see Table 13·3.

GEOGRAPHIC COORDINATES

LATITUDE AND LONGITUDE

13·13 Latitude is the distance north or south of the equator, designated by parallels and measured in degrees, minutes, and seconds, beginning with 0° at the equator and progressing to 90° north or south of the equator. Longitude is the distance east or west of the prime meridian Greenwich, designated by meridians and measured in degrees, minutes, and seconds from 0° to 180°. Both are spelled out when they appear alone in text, but

Table 13·3 Recent changes of country names[a]

Current Name	Former Name
Belize	British Honduras
Benin	Dahomey
Burkina Faso	Upper Volta
Cambodia	Kampuchea
Central African Republic	Central African Empire
Côte d'Ivoire	Ivory Coast
Czech Republic[b]	Czechoslovakia[b]
Germany	East Germany, West Germany
Madagascar	Malagasy Republic
Myanmar	Burma
Namibia	South West Africa
Slovakia[b]	Czechoslovakia[b]
Sri Lanka	Ceylon
Thailand	Siam
Vanuatu	New Hebrides
Yemen	North Yemen, South Yemen
Zaire	Congo
Zimbabwe	Rhodesia

[a]Most of these countries have longer formal names.
[b]Czechoslovakia is now divided into 2 countries.

are abbreviated when given as part of a coordinate. Latitude is given 1st, then, after a comma, longitude; the abbreviation of each precedes its coordinate, whose numbers (2 digits) are written without spaces.

lat 43°15′09″N, long 116°40′18″E   lat 04°59′17″S, long 01°02′03″W

In compilations of latitude and longitude for remote sensing and computers where the machinery cannot handle the symbols and extra space, the format may be simplified by omitting the abbreviations for "latitude" and "longitude" and the symbols for degrees, minutes, and seconds; latitude is always 1st.

4315091164018   045917010203

If breaking the coordinate at the end of a line is unavoidable, the break should come after the symbol for a coordinate unit and be indicated with a hyphen.

long 116°-   lat 45°29′-
40′18″E   14″S

Latitude–longitude coordinates for a very large number of US geographic entities are available in the volumes of the *National Gazeteer of the United States of America* (USGS 1990).

OTHER COORDINATE SYSTEMS

13·14 Some other coordinate systems (for example, plane rectangular coordinates, US Geological Survey topographic map coordinates) are not based on the latitude–longitude system. But any coordinate system must be based upon a known reference point, stated at the outset if necessary.

Some older literature in geography and history may specify a small area or locale with designations based on the US Public Lands Survey inaugurated in 1785. This system has as reference points a number of principal north–south meridians and east–west base lines established by the Survey. The intersection of a meridian and a base line is designated an "initial point". Townships (6 miles square) are numbered north and south of the baseline with a township number and east and west of the meridian with a range number. Townships are divided into 36 square sections (1 mile square) that are numbered from 1 to 36 beginning with the northeastmost section as section 1. Sections are divided into quarter sections. Topographic maps of the US Geological Survey carry symbolization relevant to identifying tracts covered by the system of public-lands subdivisions. Further details on style for identifying such tracts are in section 26·26.

Some maps of the British Ordnance Survey carry grid-square alphanumeric identifiers based on the British National Grid that enable one to specify a location in Great Britain to the nearest 100 meters; for example, the grid reference for Treglossik, a village in Cornwall, is SW 787 236. The SW is the designation for the grid square covering a large fraction of Cornwall, and the numbers represent with their first 2 digits small squares within SW and their 3rd digits easting and northing estimates from the grid markings on the Ordnance Survey map.

REFERENCES

Cited References

[CPC] Canada Post Corporation. [annual]. Canada postal guide. Ottawa: CPC.

[CPCGN] Canadian Permanent Committee on Geographical Names. Gazetteer of Canada. Otttawa: Department of Energy, Mines, and Resources. This is a series of 11 volumes, 1 for each province except Quebec and 1 each for Yukon Territory and the Northwest Territories. There is no single date of publication; each volume has its own year of publication and some are now in 2nd edition. Each volume is named "Gazetteer of Canada: [province name]".

[CPCGN] Canadian Permanent Committee on Geographical Names, Advisory Committee on Undersea Feature Names. 1983. Gazetteer of undersea feature names 1983. Ottawa: Department of Fisheries and Oceans.

[CTQ] Commission de toponymie, Québec [Province]. 1990. Guide toponymique du Québec. 2nd ed. Québec: Les Publications du Québec.

[CUP] Cambridge University Press. 1988. Cambridge world gazetteer: a geographical dictionary. Cambridge (UK): CUP.

[MW] Merriam-Webster. 1988. New geographical dictionary. Springfield (MA): MW.

[USGS] US Geological Survey. 1990. National gazetteer of the United States of America, Professional Paper P1200. Reston (VA): USGS.

[USPS] US Postal Service. [annual]. National five-digit ZIP code and post office directory. Washington: USPS.

Additional References

Hansen WR, editor. 1991. Suggestions to authors. 7th ed. Reston (VA): US Geological Survey.

Huth EJ. 1987. Medical style & format: an international manual for authors, editors, and publishers. Philadelphia: ISI Pr. Available from Williams & Wilkins, Baltimore, MD.

[NGS] National Geographic Society. 1988. Historical atlas of the United States. Washington: NGS.

[NGS] National Geographic Society. 1992. Atlas of the world. 6th ed. Washington: NGS.

[UCP] University of Chicago Press. 1993. Chicago manual of style. 14th ed. Chicago: UCP.

[USGPO] US Government Printing Office. 1984. US Government Printing Office style manual 1984. Washington: USGPO.

PART

3 Special Scientific Conventions

14 The Electromagnetic Spectrum

I have reason to believe that the magnetic and luminiferous media are identical . . . —James Clerk Maxwell, *Letter to William Thomson,* 1861

The electromagnetic spectrum is a continuum of radiated energy that is defined by the relation $c = \lambda\nu$, where c is the speed of light (3×10^8 m/s), λ is the wavelength of the radiation in meters (m), and ν is the frequency of the radiation in hertz (Hz). As the value of the frequency in hertz increases, the energy of the radiation quanta increases.

The practical bounds of the electromagnetic continuum covered in this chapter are from extremely low-frequency radio waves (approximately 10 Hz, or wavelengths of approximately 3×10^7 m) to the gamma rays produced by primary cosmic rays (approximately 3×10^{22} Hz [30 ZHz], wavelength approximately 10^{-14} m). Given that the spectrum is a continuum, none of the boundaries between segments of the spectrum is exact; all of the segments necessarily overlap with their neighboring segments (see Table 14·1).

THE LONGER WAVELENGTHS

14·1 The long wavelengths at the low-energy end of the spectrum include bands of radiation used for radio and television transmission as well as microwave bands. The segments of this end of the spectrum and their corresponding abbreviations, frequencies, and wavelengths are given in Table 14·2.

The longest-wavelength radio bands (ELF, SLF, ULF, VLF, and LF) are generally used for maritime communications. The medium-frequency (MF) wavelengths carry amplitude modulation (AM) radio broadcasts, and the high-frequency (HF) wavelengths carry shortwave radio. The VHF band carries frequency modulation (FM) radio and tele-

Table 14·1 The electromagnetic spectrum

Region	Frequencies	Wavelengths
Radio broadcast	3 Hz to 30 MHz	100 Mm to 10 m
Television broadcast	30 MHz to 900 MHz	10 m to 0.3 m
Microwave transmission	900 MHz to 30 GHz	0.3 m to 10 mm
Infrared	300 GHz to 400 THz	1 mm to 760 nm
Visible light	400 THz to 750 THz	760 nm to 400 nm
Ultraviolet	750 THz to 3 PHz	400 nm to 100 nm
X-ray	30 PHz to 6 EHz	10 nm to 50 pm
Gamma ray	6 EHz to 600 EHz	50 pm to 500 fm

Table 14·2 Radio frequency segments of the electromagnetic spectrum

Name	Abbreviation	Frequency range	Wavelength range
Extremely low-frequency	ELF	3 to 30 Hz	100 to 10 Mm
Superlow-frequency	SLF	30 to 300 Hz	10 to 1 Mm
Ultralow-frequency	ULF	300 Hz to 3 kHz	1 Mm to 100 km
Very low-frequency	VLF	3 to 30 kHz	100 to 10 km
Low-frequency	LF	30 to 300 kHz	10 to 1 km
Medium-frequency	MF	300 kHz to 3 MHz	1 km to 100 m
High-frequency	HF	3 to 30 MHz	100 m to 10 m
Very high-frequency	VHF	30 to 300 MHz	10 to 1 m
Ultrahigh-frequency	UHF	300 MHz to 3 GHz	1 m to 100 mm
Superhigh-frequency	SHF	3 to 30 GHz	100 to 10 mm
Extremely high-frequency	EHF	30 to 300 GHz	10 to 1 mm
Micrometer waves	—	300 GHz to 30 THz	1 mm to 10 μm

vision broadcasts, and the UHF band also carries television. The superhigh-frequency (SHF) band, also known as microwaves, is used for communication and radar. The band of wavelengths shorter than EHF contains micrometer waves (which are not the same as microwaves), and this band merges into the longest wavelengths of the thermal infrared region of the electromagnetic spectrum.

THE OPTICAL WAVELENGTHS

14·2 The electromagnetic radiation having frequencies of between 3×10^{14} Hz (300 THz, $\lambda = 1$ mm) and 3×10^{15} Hz (3 PHz, $\lambda = 100$ nm) is termed "optical radiation". This range includes the infrared (IR), visible, and ultraviolet (UV) regions of the spectrum. Optical radiation is more often referred to by wavelength than by frequency.

Infrared radiation lies between wavelengths of 1 mm and 760 nm (3×10^{11} Hz and 4×10^{14} Hz). Three divisions of the IR spectrum are labeled C, B, and A.

IR-C	1 mm (1 000 000 nm) to 3000 nm
IR-B	3000 to 1400 nm
IR-A	1400 to 760 nm

14·3 Visible light occupies the segment of the electromagnetic spectrum having wavelengths of 760 to 400 nm. These wavelengths produce the familiar optical spectrum of colors from red to violet. Some scientific fields change the boundaries of this region to better match their needs; for example, photosynthetically active radiation is taken to be from 700 to 400 nm. The colors of the visible spectrum occupy (approximately) the following wavelengths.

red	700 to 630 nm
orange	630 to 590 nm
yellow	590 to 530 nm
green	530 to 480 nm
blue	480 to 440 nm
violet	440 to 400 nm

14·4 Ultraviolet radiation occupies the segment of the electromagnetic spectrum between visible light and the X-ray region, the boundaries being approximately 400 nm on the visible light side and approximately 100 nm on the X-ray side. The UV region has traditionally been subdivided in physics into 4 ranges.

near UV	400 to 300 nm
middle UV	300 to 200 nm
far UV	200 to 100 nm
extreme UV	below 100 nm

The short-wavelength limit of extreme UV is variously cited as between 40 and 10 nm. Because the wavelengths below 200 nm are strongly absorbed in air, studies of this region must be conducted in another gas or in a vacuum, resulting in the alternative name, "vacuum ultraviolet". The region from 185 to 120 nm is also called the Schuman region, after its 1st investigator.

For biology (McKinlay 1986), the ultraviolet spectrum has conventionally been subdivided differently into 3 regions.

near UV	400 to 315 nm
actinic UV	315 to 200 nm
vacuum UV	less than 200 nm

Other biologically oriented subdivisions have been designated by the International Commission on Illumination (CIE 1991) on the basis of the interactions of the wavelengths with biological materials. These subdivisions were labeled A, B, and C, with the original wavelength boundaries listed below.

UV-A	400 to 315 nm
UV-B	315 to 280 nm
UV-C	280 to 100 nm

In practice, the boundary between UV-A and UV-B has been regarded by many researchers as 320 nm for a number of years.

Although the symbols for the ultraviolet spectrum segments are often written as 3 unspaced capital letters, the use of a hyphen (as shown above) facilitates recognition of the standard abbreviation for "ultraviolet". Because of the possible misreading as subdivisions of the UV region, the symbols UVR and UVL to designate "ultraviolet radiation" and "ultraviolet light" should be avoided; the preferred terms are "UV radiation" and "UV light".

THE SHORTER WAVELENGTHS

14·5 Beyond the UV portion of the spectrum lie X-rays, gamma rays, and cosmic rays. The term "X-ray" is always hyphenated and the "X" always capitalized (IUPAP 1987); "gamma ray" (or "γ ray") is unhyphenated except when used as an adjective.

X-rays occupy the spectrum segment from approximately 3×10^{16} Hz (30 PHz, $\lambda = 10$ nm) to perhaps 6×10^{18} Hz (6 EHz, $\lambda = 50$ pm). X-rays of lower frequencies (3×10^{16} Hz to approximately 3×10^{18} Hz) are known as "soft" X-rays; those of shorter wavelengths are "hard" X-rays. X-rays are distinguished from the previous forms of electromagnetic radiation by having enough energy to produce ionization upon passing through matter (ionizing radiation).

X-rays are produced by transitions of electrons from outer to inner atomic orbits within ionized atoms, the energy of the X-ray being equal to the energy difference between the orbital levels. When the transition is to the *K* shell of an atom, the series of spectral lines produced by these X-rays is called the *K* series (*K* lines); other series are named after the appropriate orbital shell, also using italic capital letters (IUPAP 1987): *L, M, N, O,* and *P.* The *K* series of virtually all elements consists of 4 major lines traditionally named γ (or β_2), ß (consisting of β_1 and β_3), α_1, and α_2. In the traditional designation for a particular line, the series capital is followed without space by the line name: $K\alpha_2$ or $L\beta_3$. A 1991 IUPAC document (Jenkins and others 1991) recommends a new system using the starting and ending shell letters: $K\alpha_2$ becomes $\mathrm{K}\text{-}\mathrm{L}_2$ and $L\beta_3$ becomes $\mathrm{L}_1\text{-}\mathrm{M}_3$; roman capital letters are used instead of italic capitals.

Gamma rays occupy the next segment of the electromagnetic spectrum. Soft gamma rays share an ambiguous boundary with hard X-rays, starting near 6×10^{18} Hz ($\lambda = 50$ pm) and extending to perhaps 6×10^{19} Hz ($\lambda = 5$ pm); hard gamma rays extend to frequencies of approximately 6×10^{20} Hz (600 EHz; $\lambda = 500$ fm), which is roughly the lower limit of energy for the gamma rays produced from atmospheric collisions of primary cosmic rays.

REFERENCES

Cited References

[CIE] Commission internationale de l'Eclairage [International Commission on Illumination]. 1991. International lighting vocabulary. 3rd ed. Publication CIE 17 nr (E-1.1) Paris: CIE. Cited in Duchêne AS, Lakey JRA, Repacholi MH, editors. IRPA guidelines on protection against non-ionizing radiation. Oxford (UK): Pergamon.

[IUPAP] International Union of Pure and Applied Physics. 1987. Symbols, units, nomenclature and fundamental constants in physics. Physica 146A:1–68.

Jenkins R, Manne R, Robin R, Senemaud C, IUPAC Commission on Spectrochemical and Other Optical Procedures for Analysis. 1991. Nomenclature, symbols, units, and their usage in spectrochemical analysis: VIII. Nomenclature system for X-ray spectroscopy. X-ray Spectroscopy 20:149–55.

McKinlay AF. 1986. Ultraviolet radiation: potential hazards. In: McAinsh TF, editor. Physics in medicine & biology encyclopedia, Volume 2. Oxford (UK): Pergamon.

Additional References

[ICRU] International Commission on Radiation Units and Measurements. 1980. Radiation quantities and units. Washington: ICRU. ICRU report 33.

Kathren RL, Petersen GR. 1989. Units and terminology of radiation measurement: a primer for the epidemiologist. Am J Epidemiol 130:1076–87.

Lide DR, editor. 1993. CRC handbook of chemistry and physics. Boca Raton (FL): CRC Pr.

[NCRP]. National Council on Radiation Protection and Measurements. 1985. SI units in radiation protection and measurements. Bethesda (MD): NRCP. NRCP report 82.

15 Subatomic Particles, Chemical Elements, and Related Notations

> There cannot be any atoms or parts of matter which are indivisible of their own nature. . . . For though God has rendered the particle so small that it was beyond the power of any creature to divide it, He could not deprive himself of the power of division, because it was absolutely impossible that He should lessen His own omnipotence. . . .
>
> —René Descartes, *Principia Philosophiae,* 1644

The authorities for nomenclature, symbolization, and other kinds of notation for subatomic particles and chemical elements are documents issued by the International Union of Pure and Applied Physics (Cohen and Giacomo 1987) and the International Union of Pure and Applied Chemistry (IUPAC 1990, IUPAC 1993). Definitions in this chapter are drawn or paraphrased from these documents, which are also the sources of many of the examples.

ELEMENTARY PARTICLES

15·1 The smallest units of matter, the elementary particles, are grouped into the designations “gauge bosons”, “leptons”, and “quarks”. The particles composed of quarks are designated “mesons” and “baryons”. Their symbols are greek and roman alphabetic characters, some with superscript

Table 15·1 Elementary particles: examples of symbols

Particles	Symbols[a]	Charge
Elementary particles		
Gauge bosons	γ, W, Z, g	
Leptons		
electron neutrino	ν_e	0
electron	e	−1
muon neutrino	ν_μ	0
muon	μ	−1
tau neutrino	ν_τ	0
tau (also "tauon")	τ	−1
Quarks [section 15·3]		
Particles having quarks and antiquarks as constituents		
Mesons		
Only u and d quarks	π^0, π^+, π^-, η	
"Strange" quarks	K^+, K^0, ϕ	
"Charm" quarks	$D^\pm$, D^0, D_S, D^*	
"Bottom" quarks	B^0, $B^\pm$, B_S, B_c	
The common baryons, composed of 3 quarks	p, n, Λ_c^+ Λ, Σ^+, Σ^0, Σ^- Ξ^0, Ξ^-, Ξ_c^0, Ξ_c^- Ω^-	

[a]The superscripts ^{0}, $^+$, and $^-$ indicate charge; the superscript * indicates "excited state". The subscripts, s, c, and b indicate "strange", "charm", and "bottom" quarks. Distinct antiparticles are indicated by a bar (overline).

or subscript modifiers. The IUPAP authority (Cohen and Giacomo 1987) for these symbols specifies that they be upright letters. Table 15·1 illustrates typical symbols.

15·2 For additional illustrations of particle symbols, consult the publications of the Particle Data Group (1992a, 1992b). The "Review of Particle Properties" published biennially in *Physical Review D* (Particle Data Group 1992a) specifies that these symbols should be "italic (slanted) characters" except for nonalphabetic superscripts and subscripts. This recommendation is based on a desire to be able to distinguish particle symbols readily in text, but it is not consonant with the IUPAP recommendations, which call for upright letters for symbols for particles, chemical elements, and nuclides. Note that p and e stand, respectively, for "proton" and "electron"; *p* and *e* stand, respectively, for "pressure" and "elementary charge". Nevertheless, the convention of slanted characters specified by the Particle Data Group is used in some physics journals and in Chemical Abstracts Service (CAS) indexes; authors must be

Table 15·2 Quarks: names (symbols), informal names, and charges

Name (symbol)	Informal name	Charge
d	"down"	−1/3
u	"up"	+2/3
s	"strange"	−1/3
c	"charmed"	+2/3
b	"bottom"; also "beauty"	−1/3
t	"top"; also "truth"	+2/3

aware of a particular journal's preference. Particle symbols may be used as variables in field theory.

QUARKS

15·3 The names of the quarks (the constituents of protons, neutrons, and related composite particles called hadrons [mesons and baryons]) are the symbols themselves; the informal names, such as "up", "down", "top", should be considered mnemonics. Table 15·2 gives their names (symbols), informal names, and charges.

For each charged lepton and quark there is an antilepton or antiquark with the same mass but the opposite electric charge.

NUCLEAR PARTICLES

15·4 The symbols for the nuclear particles that are projectiles or products in nuclear reactions are greek and roman alphabetic characters (Cohen and Giacomo 1987); see Table 15·3. The general designation of a heavy ion is HI (to be used only when ambiguity is not likely).

The electric charge of particles may be indicated by adding the superscripts +, −, or 0; examples are π^0, π^+, π^-, p^+, n^0, e^-. If the symbols p and e are used without a charge, they refer to the positive proton and negative electron, respectively. A bar above the symbol for a particle is used to indicate the corresponding antiparticle (both $\bar{e}$ and e^+ are commonly used for the positron [the positive electron] but "e^+" is preferred); for the antiproton, $\bar{p}$. A superimposed tilde is used to indicate a supersymmetric particle, for example, $\tilde{e}$.

A bound state of an electron and a positron (e^-, e^+) is known as "positronium". A hydrogen-like atom in which the proton is replaced by a positive muon, for example a μ^+ with an e^- bound to it, is known as "muonium" and given the notation "Mu".

Table 15·3 Nuclear particles

Name	Symbol and comment
photon (also "gamma")	γ
neutrino	ν, ν^{e}, ν_{μ}, ν_{τ}
electron	e, β [do not confuse "e" with "*e*", the symbol for elementary charge]
positron	e^{+}
muon	μ [may be positively or negatively charged]
tau (also "tauon")	τ [may be positively or negatively charged]
pion	π [may be neutral or positively or negatively charged]
nucleon	N [often used either with mass within parentheses after it for the resonances, for example, "N(1440)", or with other particle symbols when indicating the products of a reaction, for example, "$\pi\pi$N"; it should be used by itself only when there is no possibility of confusion with "N" for "nitrogen"]
neutron	n
proton	p [$^{1}H^{+}$]
deuteron	d [$^{2}H^{+}$]
triton	t [$^{3}H^{+}$]
helion	h [$^{3}He^{2+}$]
α particle	α [$^{4}He^{2+}$]

The notation for a nuclear reaction should have the following unspaced sequence of symbols: initial nuclide (incoming particle or photon, outgoing particle[s] or photon[s]) final nuclide.

$^{14}N(\alpha,p)^{17}O$

CHEMICAL ELEMENTS

15·5 The symbols for the chemical elements are in general derived from their latin or greek names and consist of 1, 2, or 3 letters (for example: H from "hydrogen"; K from "kalium"; Na from "natrium"; Ca from "calcium"; Unh from "unnilhexium", element 106). They must be printed in roman type with the initial letter capitalized. The 3-letter symbols are used for elements of atomic number greater than 103 until internationally accepted names and corresponding symbols are adopted for them. The names for the elements should be written in lowercase letters except at the beginning of a sentence or for other appropriate reasons such as capitalization in book titles. For the names and symbols for the chemical elements through atomic number 109, see Table 15·4.

The following roots have been selected for the formation of interim systematic names of the elements with atomic numbers above 103.

Table 15·4 Atomic numbers, names, and symbols of the 109 chemical elements

Atomic number	Name	Symbol	Atomic number	Name	Symbol
1	hydrogen	H	46	palladium	Pd
2	helium	He	47	silver	Ag
3	lithium	Li	48	cadmium	Cd
4	beryllium	Be	49	indium	In
5	boron	B	50	tin	Sn
6	carbon	C	51	antimony	Sb
7	nitrogen	N	52	tellurium	Te
8	oxygen	O	53	iodine	I
9	fluorine	F	54	xenon	Xe
10	neon	Ne	55	cesium	Cs
11	sodium	Na		caesium[a]	
12	magnesium	Mg	56	barium	Ba
13	aluminum	Al	57	lanthanum	La
	aluminium[a]		58	cerium	Ce
14	silicon	Si	59	praseodymium	Pr
15	phosphorus	P	60	neodymium	Nd
16	sulfur	S	61	promethium	Pm
17	chlorine	Cl	62	samarium	Sm
18	argon	Ar	63	europium	Eu
19	potassium	K	64	gadolinium	Gd
20	calcium	Ca	65	terbium	Tb
21	scandium	Sc	66	dysprosium	Dy
22	titanium	Ti	67	holmium	Ho
23	vanadium	V	68	erbium	Eb
24	chromium	Cr	69	thulium	Tm
25	manganese	Mn	70	ytterbium	Yb
26	iron	Fe	71	lutetium	Lu
27	cobalt	Co	72	hafnium	Hf
28	nickel	Ni	73	tantalum	Ta
29	copper	Cu	74	tungsten	W
30	zinc	Zn	75	rhenium	Re
31	gallium	Ga	76	osmium	Os
32	germanium	Ge	77	iridium	Ir
33	arsenic	As	78	platinum	Pt
34	selenium	Se	79	gold	Au
35	bromine	Br	80	mercury	Hg
36	krypton	Kr	81	thallium	Tl
37	rubidium	Rb	82	lead	Pb
38	strontium	Sr	83	bismuth	Bi
39	yttrium	Yt	84	polonium	Po
40	zirconium	Zr	85	astatine	At
41	niobium	Nb	86	radon	Rn
42	molybdenum	Mo	87	francium	Fr
43	technetium	Tc	88	radium	Ra
44	ruthenium	Ru	89	actinium	Ac
45	rhodium	Rh	90	thorium	Th

Table 15·4 (*cont.*)

Atomic number	Name	Symbol	Atomic number	Name	Symbol
91	protactinium	Pa	104	unnilquadium[b]	Unq
92	uranium	U		[rutherfordium[c]]	[Rf]
93	neptunium	Np		[kurchatovium[c]]	[Ku]
94	plutonium	Pu	105	unnilpentium[b]	Unp
95	americium	Am		[hahnium[c]]	[Ha]
96	curium	Cm	106	unnilhexium[b]	Unh
97	berkelium	Bk		[seaborgium[c]]	[Sg]
98	californium	Cf	107	unnilseptium[b]	Uns
99	einsteinium	Es		[nielsbohrium[c]]	[Ns]
100	fermium	Fm	108	unniloctium[b]	Uno
101	mendelevium	Md		[hassium[c]]	[Hs]
102	nobelium	No	109	unnilennium[b]	Une
103	lawrencium	Lr		[meitnerium[c]]	[Mt]

[a]British spelling.

[b]Interim systematic names; see section 15·5.

[c]As of mid-1994, the names and symbols proposed by national groups for elements 104, 105, 106, 107, 108, and 109 awaited official action by the Inorganic Nomenclature Committee of the International Union of Pure and Applied Chemistry.

1 un	4 quad	7 sept	0 nil
2 bi	5 pent	8 oct	
3 tri	6 hex	9 enn	

The symbols for the elements are formed by joining the roots corresponding to the digits of the atomic number and adding the ending "ium"; the initial letter is capitalized. For example, the symbol for element 106 is formed thus: un [1] + nil [0] + hex [6] = Unh = unnilhexium.

Detailed information on each element, including historical data and tabulations of key isotopes (see section 15·6), can be found in *The Elements* (Emsley 1991).

NUCLIDES

15·6 The symbols for the elements can be modified with additional symbols to indicate atomic number, mass number, charge or oxidation number, numbers of atoms per molecule, or other information. "E" represents the element symbol; the positions represented below the modifying symbols indicate their meaning.

$^{b}_{a}E^{c}_{d}$ a = atomic number; b = mass number (nucleon number, baryon number); c = charge number (or oxidation number, or other information); d = number in molecular formulas

When no left superscript is listed, the symbol is read as including all isotopes in natural abundance. Ionic charge is denoted by a right superscript consisting of a number and the appropriate sign (the number is omitted when equal to 1).

Na^{+} Al^{3+} S^{2-}

Oxidation numbers are indicated by positive or negative roman numerals or by zero.

Mn^{VII} O^{-II} Pt^{0}

The term "nuclide" implies an atom of specified atomic number (proton number) and mass number (nucleon number). "Isotopic nuclides" ("isotopes") are nuclides that have the same atomic number but different mass numbers (for example, $_{6}C$, but ^{12}C, ^{13}C, ^{14}C). "Isobaric nuclides" or "isobars" are nuclides that have the same mass number but different atomic numbers (for example, ^{14}C, ^{14}N, but $_{6}C$, $_{7}N$).

CHEMICAL FORMULAS

15·7 Chemical formulas are used to represent entities composed of more than 1 atom (molecules, complex ions, groups of atoms).

N_2 $CaSO_4$ $SO_4{}^{2-}$ $Fe_{0.91}S$ CH_3OH

Note the sequence $_4{}^{2-}$ in $SO_4{}^{2-}$, which is superior to a stacking sequence ($^{2-}_{4}$, 2– directly over the 4) in that it is easier to set and also conveys more accurate information. Consider the representation $I_3{}^{-}$: the entity I_3 bears a single negative charge. In contrast, $I^{-}{}_{3}$ would represent 3 I^{-} entities. That distinction would be lost in a superscript directly over a subscript, or stacked, representation.

Parentheses, brackets, and braces may be needed to identify complex groups and to remove ambiguity. The specific enclosure is often mandated by the type of compound, complex, or group being represented. Those recommendations are given in detail in IUPAC 1990.

SYMBOLS FOR ATOMIC STATES

15·8 The electronic states of atoms are labeled by the value of the quantum number *L* for the state. The *L* values 0, 1, 2, 3, 4, 5, 6, 7, . . . , are represented by roman capital letters S, P, D, F, G, H, I, K, . . . , respectively. The electronic spin multiplicity is indicated as a left-hand superscript to the letter, with the value of the total angular momentum being right-hand subscripts.

$^{2}\mathrm{P}_{1/2}$ $^{3}\mathrm{P}_{0}$ $^{4}\mathrm{S}$

The electronic configuration of an atom is indicated by giving the occupation of each 1-electron orbital as in the examples below; note that s and p (and d, f, . . .) are set in roman type.

boron	$(1\mathrm{s})^2(2\mathrm{s})^2(2\mathrm{p})^1$ [$1\mathrm{s}^2 2\mathrm{s}^2 2\mathrm{p}^1$ is acceptable]
carbon	$(1\mathrm{s})^2(2\mathrm{s})^2(2\mathrm{p})^2$ [$1\mathrm{s}^2 2\mathrm{s}^2 2\mathrm{p}^2$ is acceptable]
nitrogen	$(1\mathrm{s})^2(2\mathrm{s})^2(2\mathrm{p})^3$ [$1\mathrm{s}^2 2\mathrm{s}^2 2\mathrm{p}^3$ is acceptable]

SYMBOLS FOR MOLECULAR STATES

15·9 The electronic states of molecules are labeled by the symmetry species label of the wavefunction in the molecular point group. These labels should be roman or greek upright capital letters.

The electronic configuration is indicated in a manner analogous to that for atoms.

ground state of OH $(1\sigma)^2(2\sigma)^2(3\sigma)^2(1\pi)^3$

REFERENCES

Cited References

Cohen ER, Giacomo P. 1987. Symbols, units, nomenclature and fundamental constants in physics: 1987 revision. Physica 146A(1&2):v–67. Document IUPAP-25, 1987 (SUNAMCO 87-1).

Emsley J. 1991. The elements. 2nd ed. Oxford (UK): Oxford Univ Pr, Clarendon Pr.

[IUPAC] International Union of Pure and Applied Chemistry, Commission on the Nomenclature of Inorganic Chemistry. 1990. Nomenclature of inorganic chemistry, recommendations, 1990. Oxford (UK): Blackwell Scientific.

[IUPAC] International Union of Pure and Applied Chemistry, Physical Chemistry Division. 1993. Quantities, units and symbols in physical chemistry. Oxford (UK): Blackwell Scientific.

Particle Data Group. 1992a. Particle properties data booklet: June 1992. College Park (MD): American Institute of Physics. A condensed pocket-version of the document referenced immediately below.

Particle Data Group. 1992b. Review of particle properties. Physical Rev D 45(11 Pt 2):I.1–XI.8.

Additional References

International Union of Pure and Applied Chemistry, Inorganic Chemistry Division, Commission on Atomic Weights and Isotopic Abundances. 1992. Atomic weights of the elements 1991. Pure Appl Chem 64:1519–34.

Lide DR, editor. 1993. CRC handbook of chemistry and physics. 74th ed. Boca Raton (FL): CRC Pr.

16 Chemical Names and Formulas

> . . . chemistry, like mathematics, possesses a particularly intimidating obstacle in its language and symbolism, which potentially obscures what are usually quite simple theoretical ideas and experimental techniques.
>
> — W H Brock, *The Norton History of Chemistry,* 1993

Most of the nomenclature rules and publication styles for inorganic and organic chemistry are established by committees of the International Union of Pure and Applied Chemistry (IUPAC) and the International Union of Biochemistry and Molecular Biology (IUBMB). Definitions in this chapter and many of the examples have been taken from the cited references. Many nomenclature documents of particular importance in biochemistry are available in a single source (IUBMB 1992a). Two widely used reference handbooks, *CRC Handbook of Chemistry and Physics* (Lide 1993) and *Lange's Handbook of Chemistry* (Dean 1992), have detailed sections on chemical nomenclature and symbolization.

INORGANIC COMPOUNDS

The nomenclature of inorganic chemistry has considerable flexibility. Sections 16·1–16·3 briefly summarize the most commonly used principles (IUPAC 1990).

NOMENCLATURE

The Binary Approach

16·1 The binary approach to nomenclature bases names on the ionic nature of the constituents, with electropositive ions being listed 1st, followed by electronegative ions. When neutral segments are inherent components of the compound in question, those segments follow the positive ions and precede the negative ions. All terms for ions and neutral segments should be separated by spaces.

sodium chloride   magnesium potassium fluoride
zinc hydroxide iodide   platinum diammine dichloride

The Coordination Approach

16·2 In coordination nomenclature, compounds are regarded as having central atoms to which other ions or neutral segments are attached. It is an additive system (associated ligands added to a central atom). In this system, the central atom is cited last in the name segment; the groups attached are cited in alphabetic sequence.

diamminedichloroplatinum(II)   pentacyanonitrosylferrate(2–)
sodium pentacyanonitrosylferrate(2–)

The Substitutive Approach

16·3 The substitutive approach to naming inorganic compounds is borrowed from the principles of nomenclature of organic chemistry, in which compounds of carbon are regarded as containing a maximum number of hydrogen atoms unless indicated otherwise (by prefix or suffix). Because names for saturated hydrocarbons contain the ending "ane" attached to the hydrocarbon stem, these inorganic hydrides are patterned similarly: "ane" attached to a stem.

trichlorophosphane	[3 chloro groups substituting for the 3 hydrogen atoms implied by the stem "phosph" and the ending "ane"]
trichlorophosphorane	[3 chloro groups substituting for 3 of the 5 hydrogen atoms implied by the stem "phosphor" and the ending "ane"]
pentachlorophosphorane	[5 chloro groups substituting for the 5 hydrogen atoms implied by the stem "phosphor" and the ending "ane"]

FORMULAS

16·4 Empirical formulas are the simplest possible formulas for expressing composition; the sequence of symbols is generally alphabetic.

ClHg   OSi

Molecular formulas are formulas corresponding to the relative molecular masses; the order of citation is based on relative electronegativities, with the more electropositive constituents being cited 1st. If the compound contains more than 1 electropositive or electronegative constituent, the sequence within each class is the alphabetic order of their symbols.

Hg_2Cl_2 [not "HgCl"] $IBrCl_2$

The numbers of atoms or groups, designation of oxidation states, and indication of ionic charge follow the recommendations in Chapter 15. The following examples illustrate these recommendations and also indicate some of the patterns encountered with enclosing marks.

$[Co(NH_3)_6]_2(SO_4)_3$ $[\{Fe(CO)_3\}_3(CO)_2]^{2-}$

$[P^{V}{}_2Mo_{18}O_{62}]^{6-}$ $Pb^{II}{}_2Pb^{IV}O_4$

Cu^+ Cu^{2+} $[PCl_4]^+$ $[Fe(CN)_6]^{4-}$

FREE RADICALS

16·5 IUPAC (1990) recommends that the term "radical" be restricted to species conventionally termed free radicals. A radical may be indicated in formulas by a dot as a right superscript to the symbol of the element or group. If a charge is present, the dot(s) precede the indication of charge.

$H^{\cdot}$ $(NH_3)^{\cdot +}$ $(N_2O)^{\cdot\cdot 2+}$ [or] $(N_2O)^{2\cdot 2+}$ $[FeCl_4]^{4\cdot 2-}$

ORGANIC COMPOUNDS

GENERAL PRINCIPLES

16·6 The nomenclature of organic chemistry is a carbon- and hydrogen-based nomenclature (IUPAC 1979). Names are formulated by considering the number of carbon atoms in a compound and assuming that the valences of those carbon atoms are satisfied by hydrogen atoms unless explicitly stated otherwise.

ALICYCLIC COMPOUNDS

16·7 Names of unsubstituted alicyclic (straight-chain) carbon compounds have a stem and a characteristic ending; the stem denotes the number of carbon atoms and the ending the degree of saturation.

[name]	[stem + ending]	[number of carbon atoms; degree of saturation]
methane	meth + ane	1 carbon atom; fully saturated
ethane	eth + ane	2 carbon atoms; fully saturated
propane	prop + ane	3 carbon atoms; fully saturated
butane	but + ane	4 carbon atoms; fully saturated
pentane	pent + ane	5 carbon atoms; fully saturated
hexane	hex + ane	6 carbon atoms; fully saturated
heptene	hept + ene	7 carbon atoms; 1 double bond
octadiene	oct + (a)diene	8 carbon atoms; 2 double bonds
nonatriyne	non + (a)triyne	9 carbon atoms; 3 triple bonds
decadienyne	dec + (a)dienyne	10 carbon atoms; 2 double bonds, 1 triple bond

Note the following points.

1. Trivial (nonsystematic) stems are preserved for compounds containing 1–4 carbon atoms.
2. An "a" is inserted between the stem and the characteristic ending (for euphony) when the stem ends in a consonant and the ending begins with a consonant.
3. An "e" in endings denoting the state of saturation is elided when followed by a vowel ("decadienyne" = "dien(e) + yne", with the designated "e" being elided).
4. When unsaturation is present, locants are often necessary to indicate the carbon atoms involved.

butene [not clearly defined name]

1-butene [or] but-1-ene $CH_2{=}CHCH_2CH_3$
1 2 3 4

2-butene [or] but-2-ene $CH_3CH{=}CHCH_3$
1 2 3 4

decadienyne [not clearly defined name]

3,5-decadien-1-yne [or] deca-3,5-dien-1-yne

$CH{\equiv}CCH{=}CHCH{=}CH(CH_2)_3CH_3$
1 23 4 5 6 7–9 10

There are no firm rules for representation of single and double bonds in typescript but the following rules, based on *The ACS Style Guide: A Manual for Authors and Editors* (Dodd 1985), are suggested.

1. For linear formulas in text, single bonds need not be represented.
2. If bonds must be represented, a hyphen (-) can be used for a single bond and the equals symbol (=) for a double bond. Triple bonds may

have to be represented by a hand-drawn symbol or an equals symbol with a superscripted hyphen.

3 For some publications, a double bond may be satisfactorily represented by a colon (:).

CYCLIC COMPOUNDS

Names of cyclic compounds fall into 2 principal categories: compounds that are saturated or partially saturated and compounds that are aromatic (maximally unsaturated). Heterocyclic compounds (compounds that contain other atoms in addition to carbon) form an important subset of both groups. Names and locants of known ring systems can be looked up by ring analysis order in the Chemical Abstracts Service *Ring Systems Handbook* (CAS 1993) and its supplements.

Saturated and Partially Saturated Compounds

16·8 The prefix "cyclo" is attached to the name of the corresponding straight-chain compound ("bicyclo", "tricyclo", "tetracyclo", and so on, and "spiro" are other possible prefixes; if these prefixes are used, additional locants are necessary).

cyclobutane

cyclopentene [carbons containing the double bond are numbered 1 and 2]

cyclohexadiene [locants are necessary to specify multiple bond location]

cyclooctatrienyne [locants are necessary to specify multiple bond location]

bicyclo[3.2.1]octane [note the pattern of brackets and the numbers within, necessary for complete description of the compound; see IUPAC 1979 for further details]

spiro[2.5]hepta-1,4-diene [note the pattern of brackets and the numbers within, necessary for complete description of the compound; see IUPAC 1979 for further details]

Large cyclic carbon molecules with a spheroidal structure have been named "fullerenes" and, informally, "buckyballs". The C_{60} molecule is called "buckminsterfullerene" after Buckminster Fuller, the designer of the geodesic domes and spheres. This molecule can be formed through coalescence of cyclo(30)carbon (*cyclo*-C_{30}) molecules. Smaller cyclo carbons, *cyclo*-C_{18} and *cyclo*-C_{24}, can form the C_{70} molecule. At high temperatures a gap may develop in a fullerene molecule that permits entry of an atom of an element into its hollow center, such as an atom of helium, lithium, or neon. This kind of structural variant has been represented with an "at" symbol connecting the element symbol and the "C_{60}".

He@C_{60} Li@C_{60}

Aromatic Compounds

16·9 Aromatic compounds are cyclic compounds based on benzene, C_6H_6, that contain a maximum number of noncumulative double bonds. Their characteristic ending is "ene", attached to a trivial name stem. This ending is also used for the class of fused-ring compounds, similarly maximally unsaturated. The nomenclature of this class of compounds is beyond the scope of this manual; see IUPAC 1979 for further details.

[examples of aromatic compounds] benzene, toluene, styrene

[examples of fused-ring names; note the bracket and italic patterns]

benz[*a*]anthracene phenanthrene

1*H*-benzo[*a*]cyclopent[*j*]anthracene

dibenz[*a,j*]anthracene 9*H*-dibenzo[*de,rst*]pentaphene

Heterocyclic Compounds

16·10 Heterocyclic compounds contain atoms in addition to carbon as integral parts of the skeleton. Many heterocyclic compounds are important in biochemistry and in medicinal and pharmaceutical chemistry. These heterocyclic compounds have primarily trivial names, some of which are incorporated into specialized nomenclature systems. Typical examples follow.

pyrrole [in nomenclature of tetrapyrroles and corrinoids]
pyridine
purine [in nucleoside and nucleotide nomenclature]
pyrimidine [in nucleoside and nucleotide nomenclature]

Derivatives of heterocyclic compounds are named in accord with the principles of substitutive nomenclature.

IONS AND RADICALS

16·11 Suffixes and substituent prefix endings for describing radical and ionic centers have been established in IUPAC 1993a, which deals with classical valence structures and does not deal with delocalization or with concepts such as paired and unpaired electronic configurations.

[operation]	[suffix, example]	[substituent prefix ending, example]
addition of H+	-onium, "sulfonium"	-oniumyl, "sulfoniumyl"
	-ium, "methanium"	-iumyl, "methaniumyl"
loss of H+	-ide, "phosphanide"	-idyl, "phosphanidyl"
	-ate, "methanesulfonate"	-ato, "sulfonato"
	-ate, "methanethiolate"	-ido, "sulfido"
addition of H^-	-uide, "boranuide"	-uidyl, "boranuidyl"
	-ylium, "methylium"	-yliumyl, "methyiumyl"
loss of H^-	-yl, "methyl"	-ylo . . . yl, "ylomethyl"

Zwitterionic parent hydrides are named by combining appropriate operational prefixes at the end of the name of a neutral parent hydride in the order "ium", "ylium", "ide", and "uide".

1,2,2-trimethylhydrazin-2-ium-1-ide

OTHER NOMENCLATURE SYSTEMS

16·12 Although substitutive nomenclature is the basis for formulating systematic names for organic compounds, other systems are also used. One of the most common is functional group nomenclature, in which compounds are named by focusing on a class of compounds, such as alcohols, amines, ethers, or ketones, retaining the class name, and identifying the specific compound by using the appropriate form of the hydrocarbon name.

methyl alcohol triethylamine
diethyl ether ethyl methyl ketone

Note the spaces with "alcohol", "ether", and "ketone". Each of these names is a class name representing a functional group with no hydrogens being replaced: –OH, –O–, –CO–. The name "amine", however, is regarded as a special name for ammonia, which has 3 replaceable hydrogens; the names of its derivatives are written as 1 word.

SPECIAL SYMBOLS AND CAPITALIZATION

16·13 The nomenclature of organic chemistry is replete with special symbols indicating site of substitution and stereochemical configuration and orientation. The list is extensive, the use of punctuation is significant, and capitalization is important. The following rules should be observed.

1 The "cis" and "trans" in compound names should be set in italic type.

9-*cis*-retinal

2 Symbols for atoms being substituted should be set in italic type.

N-methylethanamine *S*-ethylcysteine

3 *E*, *Z*, *R*, and *S* symbols indicating stereochemistry should be set in italic type and should be placed, with the appropriate locant when necessary, within parentheses.

(*Z*)-2-butene (2*E*,5*Z*)-2,5-decadiene

(2*S*,5*R*)-2,5-dimethylheptan-1-ol

4 Italic representations should be ignored in capitalizing.

9-*cis*-Retinal . . . [at the beginning of a sentence]

o-Methylphenol . . . [at the beginning of a sentence]

Note the importance of not altering the correct pattern for stereochemical representation, illustrated by the preceding examples. Capitalizing the "*o*" in the last example would convert the ortho-substituted compound into an oxygen-substituted compound.

POLYMER NOMENCLATURE

Polymers are characterized by multiple repetitions of atoms or groups of atoms linked to each other in amounts sufficient to provide a set of properties that would not differ significantly with the addition or deletion of 1 or a few of the repeating units.

SOURCE-BASED NOMENCLATURE

16·14 Polymers have long been named by attaching the prefix "poly" to the name of the real or assumed monomers from which they are derived (IUPAC 1991). When the name of the monomer consists of 2 or more words, it should be enclosed by parentheses.

poly(adenylic acid)

Source-based nomenclature for polymers is misleading in that the chemical structure of the monomer is different from the chemical structure of the monomeric unit in the polymer (for example, a monomer CHX= CHX may form a polymer with the sequence -CHX-CHX-).

1 Copolymers: polymers derived from more than 1 species of monomer.

poly(adenosine-*co*-guanosine) copoly(adenosine/guanosine)

2 Copolymer types:	*-co-*	[unspecified]
	-stat-	[statistical]
	-ran-	[random]
	-alt-	[alternating]
	-per-	[periodic]
	-block-	[block]
	-graft-	[graft]

STRUCTURE-BASED NOMENCLATURE

16·15 For regular organic polymers (polymers that have only 1 species of repeating group), names are formulated by using the pattern "poly(constitutional repeating unit)", with the constitutional repeating unit being the name of a bivalent organic group (IUPAC 1991, 1993b). The constitutional repeating unit is always placed within enclosures.

poly(methylene) [structure-based name]
polyethylene [source-based name]
poly(propylene) [structure-based name]
polypropene [source-based name]
poly(1-acetoxyethylene) [structure-based name]
poly(vinyl acetate) [source-based name]

COMPOUNDS OF BIOCHEMICAL IMPORTANCE

AMINO ACIDS AND PEPTIDES

16·16 Amino acids are a series of compounds containing both an amino group and a carboxylic acid group. Peptides are compounds in which amino acids are bonded by an amide linkage (a so-called peptide bond) between the carboxylic acid group of 1 amino acid and the amino group of a 2nd amino acid. Proteins are polypeptides.

A detailed nomenclature of amino acids and peptides is presented in IUPAC–IUB 1972 and 1984 and in IUBMB 1992a. Several important points merit summarizing.

An α-amino acid is an amino acid in which the amino group is substituted on the carbon immediately adjacent to the carboxylic acid group—the α-carbon. Most naturally occurring amino acids are α-amino acids.

The simplest amino acid is glycine, H_2NCH_2COOH (2-aminoethanoic acid, aminoacetic acid). All other α-amino acids, for ex-

Table 16·1 Names and symbols of α-amino acids[a]

Trivial name	Symbols		Systematic name
	3-letter	1-letter	
alanine	Ala	A	2-aminopropanoic acid
arginine	Arg	R	2-amino-5-guanidinopentanoic acid
asparagine	Asn	N	2-amino-3-carbamoylpropanoic acid
aspartic acid	Asp	D	2-aminobutanedioic acid
cysteine	Cys	C	2-amino-3-mercaptopropanoic acid
glutamic acid	Glu	E	2-aminopentanedioic acid
glutamine	Gln	Q	2-amino-4-carbamoylbutanoic acid
glycine	Gly	G	aminoethanoic acid
histidine	His	H	2-amino-3-(1*H*-imidazol-4-yl)propanoic acid
isoleucine	Ile	I	2-amino-3-methylpentanoic acid
leucine	Leu	L	2-amino-4-methylpentanoic acid
lysine	Lys	K	2,6-diaminohexanoic acid
methionine	Met	M	2-amino-4-(methylthio)butanoic acid
phenylalanine	Phe	F	2-amino-3-phenylpropanoic acid
proline	Pro	P	pyrrolidine-2-carboxylic acid
serine	Ser	S	2-amino-3-hydroxypropanoic acid
threonine	Thr	T	2-amino-3-hydroxybutanoic acid
tryptophan	Trp	W	2-amino-3-(1*H*-indol-3-yl)propanoic acid
tyrosine	Tyr	Y	2-amino-3-(4-hydroxyphenyl)propanoic acid
valine	Val	V	2-amino-3-methylbutanoic acid
unspecified amino acid	Xaa	X	

[a]The trivial name refers to the L, D, or DL amino acid. For those that are chiral, only the L form is used for protein biosynthesis.

ample, alanine, $CH_3CH(NH_2)COOH$ (2-aminopropanoic acid, α-aminopropionic acid), contain chiral centers—the α-carbon atom is asymmetric—and thus have 2 optically active forms, D and L (small capitals) forms. All naturally occurring α-amino acids (with the exception of glycine, which has no chiral center) are L (small capital) forms.

Names of Common α-Amino Acids

16·17 The trivial names, traditional and well-known, of the common α-amino acids were, in general, given to them by their discoverers. The trivial names of the α-amino acids commonly found in proteins, together with their symbols and systematic names, are given in Table 16·1. For the acyl names of α-amino acids see Table 16·2.

Principles of Forming Names for Amino Acids and Derivatives

Semisystematic names of substituted α-amino acids are formed according to the general principles of organic nomenclature, by attaching

Table 16·2 Acyl names of α-amino acids

Animo acid	Radical name	Structure
alanine	alanyl	$CH_3CH(NH_2)CO-$
arginine	arginyl	$H_2NC(:NH)NH(CH_2)_3CH(NH_2)CO-$
asparagine	asparaginyl	$H_2NCOCH_2CH(NH_2)CO-$
aspartic acid	aspartoyl	$-COCH_2CH(NH_2)CO-$
	α-aspartyl	$HOOCCH_2CH(NH_2)CO-$
	β-aspartyl	$HOOCCH(NH_2)CH_2CO-$
cysteine	cysteinyl[a]	$HSCH_2CH(NH_2)CO-$
glutamic acid	glutamoyl	$-COCH_2CH_2CH(NH_2)CO-$
	α-glutamyl	$HOOCCH_2CH_2CH(NH_2)CO-$
	γ-glutamyl	$HOOCCH(NH_2)CH_2CH_2CO-$
glutamine	glutaminyl	$H_2NCOCH_2CH_2CH(NH_2)CO-$
glycine	glycyl	H_2NCH_2CO-
histidine	histidyl	$(N_2C_3H_2)CH_2CH(NH_2)CO-$
isoleucine	isoleucyl	$CH_3CH_2CH(CH_3)CH(NH_2)CO-$
leucine	leucyl	$(CH_3)_2CHCH_2CH(NH_2)CO-$
lysine	lysyl	$H_2N(CH_2)_4CH(NH_2)CO-$
methionine	methionyl	$CH_3SCH_2CH_2CH(NH_2)CO-$
phenylalanine	phenylalanyl	$C_6H_5-CH_2CH(NH_2)CO-$
proline	prolyl	$\overline{NHCH_2CH_2CH_2C}HCO-$
serine	seryl	$HOCH_2CH(NH_2)CO-$
threonine	threonyl	$CH_3CH(OH)CH(NH_2)CO-$
tryptophan	tryptophyl	$(C_8H_6N)CH_2CH(NH_2)CO-$
tyrosine	tyrosyl	$4-HOC_6H_4CH_2CH(NH_2)CO-$
valine	valyl	$(CH_3)_2CHCH(NH_2)CO-$

[a]An exception to the rule for names of radicals to differentiate it from the cysteic acid radical.

the name of the substituent group, with the appropriate numerical locant, to the trivial name of the amino acid.

trans-4-hydroxy-L-proline

S-(D-2-amino-2-carboxyethyl)-D-homocysteine
[synonym, "D-cystathionine"]

Peptides and Polypeptides

16·18 Peptides are compounds containing amino acids joined through a peptide linkage (–CO–NH–) between the carboxylic acid group of 1 amino acid and the amino group of another. The name of any peptide with 1 or more peptide linkages may be formed by combining the acyl forms of the names of all amino acids (Table 16·2), in sequence from left (the amino terminal) to right (the carboxy terminal), until the carboxy-terminal amino acid is reached; the trivial name of that amino acid is retained.

L-valyl-L-alanylglycyl-D-phenylalanyl-L-tryptophyl-D-proline

[Note the omission of the hyphen between L-alanyl and glycyl, the only amino acid without a chiral carbon atom. The example displayed is a 6-peptide (a peptide containing 6 amino acid residues)]

When the sequences of proteins (which are longer chain polypeptides, many having trivial names) are being represented, it is customary in text and tabular material to use the 3-letter symbols for the amino acids. The 3-letter and 1-letter representations of the 6-peptide illustrated above are as follows.

L-Val-L-Ala-Gly-D-Phe-L-Trp-D-Pro

LVal-LAla-Gly-DPhe-LTrp-DPro

[The hyphens between the configurational prefix and the amino acid symbol may be omitted for brevity.]

V A G F W P

[One-letter symbols should be used sparingly, never in text; configurations must be annotated in text and tables. A blank between letters indicates the sequence was identified experimentally.]

The following conventions, shown by examples, are important.

	[a hyphen between amino acids represents a peptide linkage]
-Cys	[loss of H from the NH_2]
Cys-	[loss of OH from the COOH of the amino acid (the acyl form of the amino acid)
-Cys-	[loss of H from the NH_2 and loss of OH from the COOH]
| Lys [or] Lys |	[linkage through the side chain functional group (the amino group at C-6 with lysine; with cysteine, the SH group would be involved; with serine, the OH group; with glutamic and aspartic acids, the 2nd carboxylic acid group)]
|5 Lys	[substitution at C-5 of lysine]
ψ(xxxx)	[replacement of the peptide bond (CO–NH) by another linkage (used with peptide analogues)]

Substitutions involving side chains may also be presented by placing the symbol of the substituent within parentheses immediately after the symbol of the amino acid being substituted.

Cys(Et)	*S*-ethylcysteine
Cys-OEt	cysteine ethyl ester

Ser(Ac) O^3-acetylserine

Ac-Glu(OEt)-OMe O^5-ethyl methyl *N*-acetylglutamate
[or] O^5-ethyl O^1-methyl *N*-acetylglutamate

Lack of substitution at the amino and carboxy groups may be emphasized by applying the principles stated above.

H-Ala Ala-OH

Modification of Named Peptides

16·19 Amino acids in named peptides are numbered starting from the amino terminal. When 1, or more, of these amino acids is replaced, the name of the replacement amino acid preceded by the number of the amino acid being replaced is placed within square brackets immediately preceding the name of the peptide.

[8-citrulline]vasopressin [Cit^8]vasopressin

[5-isoleucine,7-alanine]angiotensin II [Ile^5,Ala^7]angiotensin II

Replacement of a peptide bond is indicated similarly; note, however, that parentheses are used in the symbol.

[$^3\Psi^4$,CH_2-S]oxytocin . . . -Ψ(CH_2-S)- . . .

Extensions of the peptide chain are named, for the amino terminal, by preceding the name of the peptide with the acyl form of the extending amino acid (Table 16·2) and, for the carboxy terminal, adding "yl" to the name of the peptide and attaching that "yl" form to the name of the extending amino acid.

valylvasopressin Val-vasopressin

(angiotensin II)ylglycine angiotensin II-Gly

Insertion and deletion of amino acid residues are indicated by the terms "endo" and "des".

endo-4a-tyrosine-oxytocin endo-Tyr^{4a}-oxytocin

des-7-proline-oxytocin des-Pro^7-oxytocin

Polymerized Amino Acids: Definitions, Symbols, and Names

16·20 The following definitions and examples illustrate the variety of conventions for polymerized amino acids.

1 Linear polymer: all amino acid residues are linked in an unbranched chain.

2 Simple homopolymer.

poly(Ala) [or] $(Ala)_n$, polyalanine
[no parentheses with single, simple word]

poly(D-Ala) [or] $(D\text{-}Ala)_n$ [or] $(DAla)_n$, poly(D-alanine)
[parentheses with the expanded name, a more complex form]

poly(Glu) [or] $(Glu)_n$, poly(glutamic acid)

3 Linear copolymer, unknown sequence, composition not specified.

poly(Ala,Lys) [or] $(Ala,Lys)_n$

4 Linear copolymer, regular alternating sequence.

poly(Ala-Lys) [or] $(Ala\text{-}Lys)_n$

5 Linear sequence of unknown order.

poly(AlaGluLys) [or] $(AlaGluLys)_n$

6 Linear sequence of unknown order but known composition.

poly($DLAla^{22}DGlu^{12}LGlu^{24}LLys^{42}$)

7 Block: a polymer that forms a distinct part of a larger polymer.
8 Block polymer: 2 or more blocks linked to form a larger linear polymer.

poly(Glu-Lys-Tyr-Tyr)-poly(Ala) [or] $(Glu\text{-}Lys\text{-}Tyr\text{-}Tyr)_{n}\text{-}(Ala)_m$

9 Graft polymer: 1 or more blocks linked to the functional groups of a linear polymer, thus creating a branch or branches.

poly(Ala,Lys)- -poly(Ala)-poly(Tyr)
[the 2 hyphens indicate the point of attachment is not known]

poly(Ala)-poly(Tyr)——┐
poly(Ala,Lys)
[attachment to the ε-amino group of lysine]

CARBOHYDRATES

16·21 The principles of naming carbohydrates are elaborated in the references cited. The following examples illustrate representative types, names, and symbols (IUPAC–IUB 1971, 1981a–c, 1982b,c, 1983c,d; IUBMB 1992a).

Aldoses (carbohydrates containing an aldehyde functional group).

triose [3 carbon atoms]:	glyceraldehyde
tetrose [4 carbon atoms]:	erythrose, threose

pentose [5 carbon atoms]:	arabinose, lyxose, ribose, xylose [for "*arabino*-pentose", "*lyxo*-pentose", and so on]
hexose (6 carbon atoms):	allose, altrose, galactose, glucose, gulose, idose, mannose, talose [for "*allo*-hexose", "*altro*-hexose", and so on]

Ketoses (carbohydrates containing a keto functional group).

D-fructose (D-*arabino*-2-hexulose)

Note the following points.

1 Trivial names (ribose, for example) are set in roman type.
2 Systematic names (D-*ribo*-pentose, for example) have 4 components: the configurational symbol D or L (small capitals), the configurational prefix (representing the relative positions of the hydroxyl groups), the stem name (indicating the number of carbon atoms), and the characteristic ending (indicating the type of carbohydrate: aldose, ketose, or combination thereof). Carbohydrates may be substituted, forming derivatives and larger carbohydrates.

The following names are examples.

[1] methyl ribofuranoside

[2] ethyl (methyl 3,4-di-*O*-methyl-α-L-*ribo*-2-hexulopyranosid)onate

[3] 4-*O*-β-D-galactopyranosyl-α-D-glucopyranose

EXAMPLE 1: The "anomeric" hydroxyl group of the sugar (the OH on the aldehyde or ketone carbon) is substituted. That type of substitution is indicated by changing the ending "ose" to "oside" and preceding the name of the original sugar with the name of the substituent group, separated by a space. The infix "furan" indicates that the sugar is in a 5-membered ring formation.

EXAMPLE 2: Note the space patterns, the use of parentheses, the use of italics, and the following infixes or suffixes: "pyranosid(e)", "onate" (the e of pyranoside being elided before another vowel). The 1st space is that associated with esters (the ethyl ester of the ". . . onic" acid). The parentheses associate the group preceding the 2nd space with the appropriate part of the compound (for example, methyl 2-hexulopyranoside). The infix "pyran" indicates that the sugar is in a 6-membered ring formation.

EXAMPLE 3: In the 3rd name, note the absence of spaces. The parent compound, glucose (in the pyranose ring form, as opposed to the straight-chain or the furanose ring form), is substituted through the O at C-4.

In addition to configuration, symbols are used to describe conformation, the shape of the ring form: chair, boat, skew, half-chair, envelope, and twist; and location of hydroxyl groups.

$^{2,5}B$ [boat conformation, with the hydroxyl groups at C-2 and C-5 above the reference plane]

$^{4}H_{5}$ [half-chair conformation, with the hydroxyl group at C-4 above the reference plane and the hydroxyl group at C-5 below the reference plane]

Because the systematic names of substituted carbohydrates are quite long, abbreviations are used extensively. The following examples represent extended and condensed forms.

D-Glc*p*, DGlc*p*

["Glc" is used for "glucose"; "*p*" represents a pyranose ring; thus, these are symbols for "D-glucopyranose"]

L-Rib*f*, LRib*f*

["Rib" is used for "ribose"; "*f*" represents a furanose ring; thus, these are symbols for "L-ribofuranose"]

β-D-Xyl*p*-(1→4)-α-D-Gal*p*-(1→6)-α-D-Man*p*-(1→2)-β-D-Fru*f*

DXyl*p*-(β1–4)DGal*p*(α1–6)DMan*p*(α1–2β)DFru*f*

CAROTENOIDS

16·22 Carotenoids, a class of hydrocarbons called carotenes and their oxygenated derivatives (xanthophylls), consist of 2 groups of 4 isoprenoid units joined at their ω carbon atoms such that the methyl groups of one 4-unit group are in opposite configurational relationships to the methyls of the 2nd 4-unit group (IUPAC–IUB 1975b, IUBMB 1992a).

Specific names are constructed by adding 2 greek-letter prefixes to the stem name "carotene". The appropriate greek letter depends on the type of the C_9 end groups (positions 1–6, 16–18, 1′–6′, and 16′–18′).

[type]	[prefix]	[formula]
acyclic	ψ	C_9H_{15}
cyclohexene	β, ε	C_9H_{15}
cyclopentane	κ	C_9H_{17}
aryl	φ, χ	C_9H_{11}
methylenecyclohexane	γ	C_9H_{15}

Substituents on the carbon chains are named according to the general rules of organic chemistry nomenclature.

2,2′-dinor-β,β-carotene

2,3-seco-ε,ε-carotene

3-hydroxy-3′-oxo-β,ε-caroten-16-oic acid

CORRINOIDS

16·23 Corrinoids are a group of compounds containing 4 reduced pyrrole rings linked to form a macrocycle. Three of the linkages are through -CH= groups; the 4th is a direct ring-to-ring linkage involving Cα-Cα bonds. B-12 vitamins and their derivatives are important members of the corrinoid family (IUPAC–IUB 1976, IUBMB 1992a).

The following names are representative of the cobalamin series of corrinoids (the oxidation number of the cobalt is given when it is significant).

vitamin B-12	cyanocobalamin	*Co*α-[α-(5,6-dimethylbenz-imidazolyl)]-*Co*β-cyanocobamide
	cyanocob(III)alamin	
vitamin B-12r	cob(II)alamin	
vitamin B-12s	cob(I)alamin	
vitamin B-12a	aquacobalamin	*Co*α-[α-(5,6-dimethylbenz-imidazolyl)]-*Co*β-aquacobamide
vitamin B-12b	hydroxocobalamin	*Co*α-[α-(5,6-dimethylbenz-imidazolyl)]-*Co*β-hydroxocobamide
vitamin B-12c	nitritocobalamin	*Co*α-[α-(5,6-dimethylbenz-imidazolyl)]-*Co*β-nitritocobamide
coenzyme B-12	adenosylcobalamin	*Co*α-[α-(5,6-dimethylbenz-imidazolyl)]-*Co*β-adenosylcobamide

CYCLITOLS

16·24 Cyclitols are cycloalkanes that contain 1 hydroxyl group on each of 3 or more ring atoms. Because cyclitols and related compounds have been extensively studied and because they possess stereochemical features characteristic of their class, special methods of designating that stereochemistry are used. In other than stereochemical respects, their nomenclature should follow the general rules of organic chemistry (IUPAC–IUB 1974a, IUBMB 1992a).

A series of trivial names is recommended for cyclitols, with

1,2,3,4,5,6-cyclohexanehexols being called "inositols". Individual inositols are distinguished by the italicized prefixes.

cis- (1,2,3,4,5,6/0) *epi*- (1,2,3,4,5/6) *allo*- (1,2,3,4/5,6)
neo- (1,2,3/4,5,6) *myo*- (1,2,3,5/4,6) *muco*- (1,2,4,5/3,6)
chiro- (1,2,4/3,5,6) *scyllo*- (1,3,5/2,4,6).

The numbers within parentheses represent the positions of the hydroxyl groups. Numbers preceding the virgule indicate that those hydroxyl groups are above the cyclohexane plane; those following the virgule are below the plane.

Several unusual features occur in cyclitol nomenclature. Some are common to carbohydrate nomenclature. Among these features the following should be noted.

1 Because the principal functional group is most commonly hydroxyl, variations are generally emphasized.

(2,3,4/1(COOH),5)-2,3,4,5-tetrahydroxycyclopentanecarboxylic acid

2 Because most carbon atoms have hydroxyl substituents that are in turn substitutable, replacement of hydrogen atoms directly attached to carbon atoms is generally emphasized.

1-*C*-methyl-*myo*-inositol

3 When a hydroxyl group is replaced, the italic symbol for the replacing group may be appended to the locant.

(1,2*N*,3,5/4,6)-2-amino-1,3,5,4,6-cyclohexanepentol

4 Configurational D and L symbols (small capitals) are often preceded by a locant.

1L-1-*O*-methyl-6-mercapto-6-deoxy-*chiro*-inositol

5 Whereas the recommended rules use italic configurational prefixes, older names in the field (traditional names) deviate from these recommendations (for example, "D-*chiro*-inositol" or "(+)-chiroinositol"; note that on occasion the locants in the recommended rules may differ from those in the standard traditional names, and as such, error may be introduced in attempts to edit the latter to the former).

ELECTRON-TRANSFER PROTEINS

16·25 Electron-transfer proteins are proteins involved in the transfer of reducing equivalents of hydrogen atoms or electrons from an initial donor to a final acceptor. Although many of these proteins satisfy the general definition of an enzyme, it is undesirable to give the subunit of an enzyme a separate enzyme name. As a result, recommendations addressing these specific types of carrier as well as more general metal-containing electron transport proteins have been formulated (IUB 1991). Detailed recommendatons for flavoproteins, cytochromes, non-heme iron proteins including iron–sulfur proteins, copper proteins, molybdenum proteins, nickel proteins, vanadium proteins, quinoproteins, and metal-substituted metalloproteins are presented.

ENZYMES

16·26 Enzymes are proteins with catalytic activity. Multienzymes are proteins with more than 1 catalytic function contributed by distinct parts of a polypeptide chain ("domains") or by distinct subunits.

Most enzyme names end in "ase" and are formulated according to the recommendations in *Enzyme Nomenclature: Recommendations 1992,* a compilation prepared and updated by the Internaional Union of Biochemistry and Molecular Biology (formerly the International Union of Biochemistry) (IUB 1989; IUBMB 1992a, 1992b). The long-established names of proteinases such as chymotrypsin, thrombin, coagulation factor Xa, and subtilisin are permitted as exceptions.

Classification

16·27 Enzymes are classified in 6 main divisions, each of which may have subclasses. These classifications serve as the basis for assigning 4 code numbers to each enzyme.

1 A number representing the 6 main divisions (classes) to which the enzyme belongs.

1 oxidoreductases	4 lyases
2 transferases	5 isomerases
3 hydrolases	6 ligases

2 A number representing the subclass.
3 A number representing the sub-subclass.
4 A serial number.

Below are 2 representative names and numbers; the systematic names of these enzymes (given below the recommended names) and of all enzymes form the basis for their classification.

adenosinetriphosphatase, EC 3.6.1.3	[recommended name]
ATP phosphohydrolase	[systematic name]

[EC is the abbreviation for "Enzyme Commission". Adenosinetriphosphatase is a hydrolase (class 3) that catalyzes the hydrolytic cleavage of acid anhydrides (subclass 6 of hydrolases) in phosphorus-containing hydrides (sub-subclass 1). It was the 3rd enzyme thus categorized.]

asparagine synthase (glutamine-hydrolyzing), EC 6.3.5.4 [recommended name]

L-asparatate:L-glutamine amido-ligase [AMP-forming] [systematic name]

[Asparagine synthase (glutamine-hydrolyzing) is a ligase (class 6) that catalyzes the formation of carbon–nitrogen bonds (subclass 3) involving an amido-N donor (sub-subclass 5). It was the 4th enzyme thus categorized.]

The table of contents in *Enzyme Nomenclature* (IUBMB 1992b) gives a complete list of the subclasses and sub-subclasses.

Symbolism

16·28 Catalytic domains (part of a polypeptide chain with a catalytic function) are indicated by capital letters from early in the alphabet (A, B, C . . .); substrate–carrier domains are indicated by capital letters from late in the alphabet (R, S, T . . .). Regulatory domains are given lowercase letters from early in the alphabet (a, b, c . . .). Domains in the same polypeptide chain are placed within the same pair of parentheses. Thus, "(ABC)" represents a multienzyme polypeptide, and "(A)(BC)" represents a multienzyme complex.

Curly brackets {} may be used to indicate stable association. For example, tryptophan synthase from *Escherichia coli,* "$(A)_2(B)_2$", portrayed as "$\{(A)_2(B)_2\}$" indicates that the association is stable, but portrayed as "$\{(A)(B)\}_2$" indicates that each A chain binds 1 B chain tightly, but that the 2 {(A)(B)} units are more loosely associated.

FOLIC ACIDS AND RELATED COMPOUNDS

16·29 Folates are a group of heterocyclic compounds in which 1 or more L-glutamic acid units are conjugated with the 4-[(pteridin-6-ylmethyl)amino]benzoic acid skeleton (pteroic acid; 4-[[(2-amino-3,4-dihydro-4-oxopteridin-6-yl)methyl]amino]benzoic acid). "Folate" and "folic acid" are synonyms for "pteroylglutamate" and "pteroylglutamic acid" (IUPAC–IUB 1987d).

In the glutamate conjugates, the acid group of the pteroic acid is involved in an amide linkage with the α-amino group of glutamic acid. If more than 1 glutamic acid residue is present, it is assumed that these are linked by peptide bonds through the γ-carboxyl group. Names such as "pteroylpentaglutamate" represent a pteroyl derivative of a pentapeptide; "pteroylhexaglutamate", of a hexapeptide.

Symbols may be used to indicate substitution and state of reduction. The following symbols are representative.

	folate [never abbreviated]
Pte	pteroate [or "pteroic acid" or "pteroyl"]
$PteGlu_4$	pteroyltetraglutamate
$H_2folate$	dihydrofolate
$H_4PteGlu$	tetrahydropteroylglutamate
(6*S*)-5-CHO-H_4folate	(6*S*)-5-formyltetrahydrofolate
(6*R*)-5,10-CH_2-H_4folate	(6*R*)-5,10-methylenetetrahydrofolate

GLYCOPROTEINS, GLYCOPEPTIDES, AND PEPTIDOGLYCANS

16·30 Glycoproteins and glycopeptides are compounds containing carbohydrates covalently linked to proteins and peptides (IUPAC–IUB 1987a, IUBMB 1992a). Proteoglycans are a subclass in which the carbohydrate units are polysaccharides that contain amino sugars; these polysaccharides are also known as glycosaminoglycans. Peptidoglycans are glycosaminoglycans formed by alternating residues of D-glucosamine and either muramic acid or L-talosaminuronic acid.

ISOTOPICALLY MODIFIED COMPOUNDS

16·31 An isotopically modified compound has a composition such that the isotopic ratio of nuclides for at least 1 element deviates measurably from that occurring in nature. It is either an isotopically substituted or an isotopically labeled compound.

Isotopically Substituted Compounds

An isotopically substituted compound is one in which essentially all the molecules of the compound have only the indicated nuclide at each of the designated positions. Formulas for these compounds are written with the appropriate nuclide symbol. Names are written by including the nuclide symbol in parentheses, with locants if necessary, before the name of the compound or the group being substituted (IUPAC 1979).

$^{14}CH_4$ (^{14}C)methane

CH_3CH^2HOH (1-^2H_1)ethanol

Specifically Labeled or Selectively Labeled Compounds

16·32 An isotopically labeled compound may be either specifically labeled or selectively labeled. A specifically labeled compound is one in which a unique isotopically substituted compound is added to the analogous isotopically unmodified compound. A selectively labeled compound is one in which a mixture of isotopically substituted compounds is added to the isotopically unmodified compound in such a way that the position(s) but not necessarily the number of each labeling nuclide is defined. Formulas and names for isotopically labeled compounds both use square brackets (IUPAC 1979, IUBMB 1992a). Note the pattern shown.

	[formula as labeled]	[name indicating labeling]
[specifically labeled]	$[^{14}C]H_4$	[^{14}C]methane
	$CH_2[^2H_2]$	[$^2H_2$]methane
	$CH_3CH[^2H]OH$	[1-$^2H_1$]ethanol
	$CH_3CH_2[^{18}O]H$	[^{18}O]ethanol
	$CH_3CH_2O[^2H]$	ethan[^{2}H]ol
[selectively labeled]	$[^2H]CH_4$	[^{2}H]methane
	$[1\text{-}^{14}C,^{18}O]CH_3CH_2OH$	[1-^{14}C,^{18}O]ethanol

A nonselectively labeled compound is one in which both the position(s) and the number of the labeling nuclide(s) are undefined.

$[^{14}C,^2H]CH_3CH_2OH$ [^{14}C,^{2}H]ethanol

Some examples of specifically labeled compounds follow. Note that the label indication may precede the name of the compound (in which case a locant is often mandatory) or the group designation.

1-(amino[^{14}C]methyl)cyclopentanol

ethan[^{2}H]ol [or] [*O*-^{2}H]ethanol

N-(6,7-[6-^{131}I]diiodofluoren-2-yl)acetamide

[*O*-^{2}H]acetic acid or acetic [^{2}H]acid

cyclohexane[^{2}H]carboxylic acid or [*O*-^{2}H]cyclohexanecarboxylic acid

2-([2,2-$^2H_2$]ethyl)-4-ethyl-1-hexanol

Isotopically Deficient Compounds

An isotopically deficient compound is one in which 1 or more of the nuclides are present in less than the natural ratio.

[*def*^{13}C]$CHCl_3$ [*def*^{13}C]chloroform

General and Uniform Labeling

16·33 In general labeling, all positions of the designated element are labeled, but not necessarily in the same isotopic ratio.

[G-^{14}C]pentanoic acid   D-[G-^{14}C]glucose

In uniform labeling, all positions or all specified positions of the designated element are labeled in the same isotopic ratio.

[U-^{14}C]pentanoic acid   D-[U-^{14}C]glucose   D-[U-1,3,5-^{14}C]glucose

LIPIDS

Fatty Acids, Neutral Fats, Long-chain Alcohols, and Long-chain Bases

16·34 "Fatty acid" is the term given to the aliphatic monocarboxylic acids obtained by hydrolysis of naturally occurring fats and oils. Neutral fats are esters of glycerol and fatty acids and are appropriately termed "monoacylglycerol", "diacylglycerol", and "triacylglycerol"; use of the terms "glyceride", "diglyceride", and "triglyceride" is discouraged. "Long-chain alcohol" refers to alcohols whose carbon chain length is greater than 10. "Long-chain base" refers to any base containing a long-chain aliphatic radical. Sphinganine (D-*erythro*-2-amino-1,3-octadecanediol) and its homologs, stereoisomers, and derivatives are called "sphingoid" or "sphingoid bases".The following examples are representative (IUPAC–IUB 1977a, IUBMB 1992a).

4D-hydroxysphinganine	[also]	(2*S*,3*S*,4*R*)-2-amino-1,3,4-octadecanetriol
		phytosphingosine
sphingosine	[also]	(4*E*)-sphingenine
		trans-4-sphingenine
		(2*S*,3*R*,4*E*)-2-amino-4-octadecene-1,3-diol
tristearoylglycerol	[also]	tri-*O*-stearoylglycerol
		glycerol tristearate
		glyceryl tristearate
glycerol 2-phosphate	[also]	2-phosphoglycerol
sn-glycerol 1-phosphate	[also]	L-(glycerol 1-phosphate)
		D-(glycerol 3-phosphate
sn-glycerol 3-phosphate	[also]	D-(glycerol 1-phosphate)
		L-(glycerol 3-phosphate)

Phospholipids

16·35 "Phospholipid" is a term applicable to any lipid containing phosphoric acid as a monoester or diester. The prefix "phospho" may be used as an infix to designate phosphodiester bridges, as in glycerophosphocholine; this prefix is used as a contracted form of either "phosphono" to denote the $-P(O)(OH)_2$ group or as a contracted form of "phosphinico" to denote (HO)P(O)= in phosphodiester bridges. "Phosphono" is never used to denote phosphodiester bridges. Some examples of phospholipids include the glycerophospholipids, sphingophospholipids, and inositolphospholipids.

Phosphatidic acids are derivatives of glycerol phosphate in which both of the hydroxyl groups of glycerol are estified with fatty acids. The common glycerophospholipids are named as derivatives of phosphatidic acids, for example, "3-*sn*-phosphatidylcholine" (trivial name "lecithin"; systematic name "1,2-diacyl-*sn*-glycero-3-phosphocholine") and "2-phosphatidylserine" (IUPAC–IUB 1977b, IUBMB 1992a).

Glycolipids

16·36 Glycolipids are compounds in which 1 or more monosaccharide residues are linked through a glycosyl group to a lipid group. The structures are often complex, and the use of abbreviations is prevalent (but "acyl" should not be abbreviated). Examples follow.

1,2-diacyl-3-β-D-galactosyl-*sn*-glycerol [or]
1,2-diacyl-*sn*-glycerol 3-β-D-galactoside
[mucotriaosylceramide] $McOse_3Cer$ [or] Gal(β1–4)Gal(β1–4)Glc(1–1)Cer

NEURAMINIC ACID

16·37 Neuraminic acid, symbol "Neu", is the specific compound 5-amino-3,5-dideoxy-D-*glycero*-D-*galacto*-nonulosonic acid. The term "sialic acid" is used to designate *N*-acylneuraminic acid and its oxygen derivatives.

NUCLEIC ACIDS, NUCLEOSIDES, NUCLEOTIDES, AND POLYNUCLEOTIDES

Nucleic Acids

16·38 Nucleic acids are compounds consisting of nucleotide units, purine or pyrimidine bases attached to a ribosyl or deoxyribosyl group and a phosphoric acid residue, joined by the phosphoric acid residue in ester linkages between hydroxyls of the ribosyl groups.

Table 16·3 Purines, pyrimidines, and nucleosides: recommended symbols

Purines and pyridimines	3-letter symbol	Nucleosides	3-letter symbol	1-letter symbol
adenine	Ade	adenosine	Ado	A
guanine	Gua	guanosine	Guo	G
		inosine	Ino	I
		thioinosine	Sno	
xanthine	Xan	xanthosine	Xao	X
hypoxanthine	Hyp			
unknown purine	Pur	a purine nucleoside	Puo	R
thymine	Thy	ribosylthymine	Thd	T
cytosine	Cyt	cytidine	Cyd	C
uracil	Ura	uridine	Urd	U
		thiouridine	Srd	
		pseudouridine	ψrd	ψ
orotate	Oro	orotidine	Ord	O
unknown pyrimidine	Pyr	a pyrimidine nucleoside	Pyd	Y
		a nucleoside	Nuc	N
unknown base	Base			

ABBREVIATIONS AND SYMBOLS

The 5′-phosphates (mono, di, tri) of the common nucleosides may be represented by commonly used abbreviations of the form (for adenosine) “AMP”, “ADP”, “ATP”. For single nucleoside and nucleic acid representations a 3-letter formulation is preferred (“ATP” becomes “Ado-5′*PPP*”, for example), whereas for chains single-letter symbols are used, with connecting hyphens or the lowercase letter “p” representing phosphate linkages. See Table 16·3 for the recommended symbols (IUPAC–IUB 1970, 1983d, IUBMB 1992a).

GENERAL PRINCIPLES FOR DESCRIBING CHAINS

16·39 **CHAIN DIRECTION:** The conventional representation of a polynucleotide chain is from the 5′ terminus to the 3′ terminus.

A nucleotide, the repeating unit of a polynucleotide chain, is composed of 3 parts: the D-ribose or 2-deoxy-D-ribose sugar ring, the phosphate group, and the purine or pyrimidine base, with the sugar ring and the phosphate forming the backbone of the chain. The unit is defined by the sequence of atoms from the phosphorus atom at the 5′-end to the oxygen atom at the 3′-end of the sugar. The following are equivalent representations for a 6-nucleotide unit.

GAATTC [or] G-A-A-T-T-C [or] pGpApApTpTpC

NUMBERING: Specific nucleotide units are designated by letter or number in parentheses, starting with the 1st nucleotide residue in the sequence.

A(1)   pU(5)   C(10)p   pG(*i*)p

Atoms in each of the constituents follow the standard numbering practice.

C2, C2(5)   N3, N3(10) O^5 P(*i*+1)   H1   H2′   O5′H

GENERAL PRINCIPLES FOR DESCRIBING INCOMPLETELY SPECIFIED BASES IN NUCLEIC ACID SEQUENCES

16·40 Nucleotide sequences code for specific amino acids. Some amino acids, however, may be expressed by several nucleotide sequences. The following symbols are used in representing sequences where variations are possible (IUB 1985).

[symbol]	[base component of the nucleoside]
A	adenine
C	cytosine
G	guanine
T	thymine
R	purine (adenine or guanine)
Y	pyrimidine (cytosine or thymine)
W	adenine or thymine
M	adenine or cytosine
S	guanine or cytosine
K	guanine or thymine
H	adenine or thymine or cytosine (thus, not guanine)
B	guanine or cytosine or thymine (thus, not adenine)
V	adenine or guanine or cytosine (thus, not thymine)
D	adenine or guanine or thymine (thus, not cytosine)
N	adenine or guanine or cytosine or thymine

For example, the triplets CGN and AGR are codes for arginine. Variability in the 1st and last unit may be indicated by substituting M (either C or A), giving rise to MGN (N includes R).

PEPTIDE HORMONES

16·41 Many peptide hormones have well-established trivial names. Some of these trivial names are, however, so long that the hormones are known mainly by abbreviations. To preclude the proliferation of additional abbreviations and to decrease the use of many of the abbreviations employed, IUPAC–IUB (IUPAC–IUB 1975c, IUBMB 1992a) proposed alternative names based on 3 principles.

1 New names for hormones of the adenohypophysis should end with "tropin".

corticotropin [for "adrenocorticotropic hormone", ACTH]

2 Hypothalamic releasing factors should end with "liberin".

corticoliberin [for "corticotropin-releasing factor", CRF]

3 Hypothalamic release-inhibiting factors should end with "statin".

somatostatin [somatotropin release-inhibiting factor]

(The abbreviations ACTH and CRF are not recommended for biochemical literature; they are given here only for identification.)

PHOSPHORUS-CONTAINING COMPOUNDS OF BIOCHEMICAL IMPORTANCE

16·42 The nomenclature of phosphorus-containing compounds in inorganic and organic chemistry is well defined. Strict application of the rules thereof to biochemically important compounds would, however, result in complicated names inconvenient for most biochemists and biologists. The following principles are recommended for use in biochemical, biological, and medicinal fields (IUPAC–IUB 1977b).

1 Phosphoric esters, RO-PO$(OH)_2$, are named as *O*-substituted phosphoric acids or as substituted alcohols.

glycerol 1-phosphate glycero-1-phosphate
[the 2nd, condensed, form is acceptable]

sn-glycerol 3-phosphate *sn*-glycero-3-phosphate
[note the use of "*sn*" representing "*s*tereospecific *n*umbering"]

choline *O*-(dihydrogen phosphate)
["phosphate" may be used instead of "dihydrogen phosphate"]

O-phosphonocholine
["phosphono" may be contracted to "phospho"; "phosphoryl" is incorrect for this group]

The prefixes "bis", "tris", and so on should be used to indicate 2 or more independent phosphoric acid residues; the prefixes "di", "tri", and similar prefixes representing higher numbers are used to denote phosphate chains.

fructose 1,6-bis(phosphate) [not "fructose 1,6-diphosphate"]
adenosine diphosphate

2 Phosphodiesters, -PO(OH)-, are named by using the infix "phospho".

glycerophosphocholine

3 Nucleoside triphosphate analogues (methylene groups, imido groups, or sulfur atoms replace an oxygen atom bridging 2 phosphorus atoms) may be named by indicating, within brackets (implying replacement), the locants of the phosphorus atoms being bridged followed by the name of the replacement group.

adenosine 5′-[α,β-methylene]triphosphate
[or] adenosine 5′-α,β-μ-methylenetriphosphate

PRENOLS

16·43 Prenols are a group of alcohols containing 1 or more isoprene units. Prenols are, with their esters, the biological precursors of the isoprenoids, a variety of compounds including terpenes and steroids that contain much of the carbon skeleton intact. Prenols are also related to the simplest juvenile hormones.

The term "prenol" (from iso*pre*noid alcoh*ol*) is recommended (IUPAC–IUB 1987b) to describe the structure shown below. The carbon adjacent to the hydroxyl group is numbered 1. The carbon of the methyl group attached to C-3 is numbered 3^1 (note this unusual representation; it occurs in many natural products containing rings and chains with fixed numbering systems). The repeating C_5H_8 unit (enclosed within parentheses) is called an isoprene unit or an isoprene residue, and compounds containing this unit are called isoprenoids.

$$\begin{array}{c} \quad CH_3 \\ \quad | \\ H\text{-}(CH_2\text{-}C{=}CH\text{-}CH_2)_n\text{-}OH \\ \;\; 4 \quad 3 \;\; 2 \;\; 1 \end{array}$$

If n is known, the compounds and their derivatives are named accordingly.

[n = 6]	hexaprenol
[n = 7, ester of diphosphoric acid]	heptaprenyl diphosphate

Stereochemistry

16·44 The stereochemistry of unsubstituted prenols is associated with the double bond and is indicated by the terms "*cis* "and "*trans*". How the terms "*cis*" and "*trans*" are used in names (for example, "*tritrans*", "*polycis*") is unusual in organic nomenclature. The following list gives some common patterns, names, and symbols (W, ω-residue; T, trans-residue; C, cis-residue; S, saturated-residue; *POP,* diphosphate group).

W-*POP*	dimethylallyl diphosphate [or] 3,3-dimethylallyl diphosphate [or] 3-methyl-2-butenyl hydrogen diphosphate
WC-*POP*	neryl diphosphate
	cis-diprenyl diphosphate
WCC-*POP*	*cis,cis*-farnesyl diphosphate
	2-*cis*,6-*cis*-farnesyl diphosphate
	dicis-triprenyl diphosphate
WTTT-*POP*	geranylgeranyl diphosphate
	2-*trans,*6,10-ditrans-geranylgeranyl disphosphate
	all-trans-tetraprenyl diphosphate
WTTC-*POP*	2-*cis,*6,10-*ditrans*-geranylneryl diphosphate
	2-*cis,*6,10-*ditrans*-tetraprenyl diphosphate
$WTTTT_n$-*POP*	*all-trans*-polyprenyl diphosphate
$WTTTC_n$-*POP*	*tritrans,polycis*-polyprenyl diphosphate

Juvenile Hormones

16·45 Farnesol ($n = 3$) is the prenol that corresponds to the carbon skeleton of the simplest juvenile hormone. Other juvenile hormones have ethyl instead of methyl groups at C-3, and so on These may be named as methyl derivatives of farnesol.

ditrans-11^1-methylfarnesol   *ditrans*-7^1,12-dimethylfarnesol

Dolichols

Dolichols are a group of prenol derivatives in which the residue containing the hydroxyl group is saturated. The name "dolichol" is retained, but alternative names, for example, "2,3-dihydropolyprenol", may be used.

PROSTAGLANDINS

16·46 Prostaglandins are eicosanoids, derivatives of long-chain fatty acids having 20 carbon atoms. Structurally they have a cyclopentane ring with 2 stereochemically defined adjacent *trans* alkyl side chains of 7 and 8 carbon atoms, 1 of which terminates in a COOH group. Prostaglandins have been divided into A, B, C, D, E, F, G, H, and I families. The families are distinguished by the presence and position of unsaturation and oxo and hydro groups giving rise to forms such as postaglandin A_1 (PGA_1) and prostaglandin A_2 (PGA_2). The F family is subdivided further into α and β groups to distinguish between 2 stereochemical forms; a typical designation is "prostaglandin $F_{2\alpha}$". The subscript numbers with family letter-designations designate particular double-bond configurations.

[subscript number]	[double-bond configuration]
1	13*E*
2	5*Z* and 13*E*
3	5*Z*, 13*E*, and 17*Z*

Because of the complexity of prostaglandin compounds and the cumbersome quality of their systematic names derived by the rules for naming organic compounds, proposals have been presented that use the trivial names for common prostaglandin compounds as bases for further substitution and modification (Nelson 1974). The following names are examples.

ent-prostanoic acid (19*R*)-19-hydroxyprostaglandin B_1

2,3,4,5-tetranorprostaglandin E_1 (18*Z*)-18,19-didehydroprostaglandin E_1

8-*iso*-prostaglandin E_1 11,15-anhydro-11-*epi*-prostaglandin E_1

4,5,6-trinor-3,7-*inter-m*-phenylene-3-oxaprostaglandin A_1

Note the italicizing of *ent, epi, inter,* and *iso* (*ent* = inversion of all stereochemical features; *epi* = inversion of the normal configuration of a particular substituent at the numbered position; *inter* = replacement of the carbons at the numbered positions by the specified infix; *iso* =inversion of the normal chirality at the numbered center).

PROTEINS

16·47 In contrast to the standardized nomenclature for enzymes (see sections 16·26–16·28), there is no broadly applicable system for naming nonenzyme proteins. The Joint Commission on Biochemical Nomenclature (JCBN) has recognized this lack to be "an outstanding example of a problem that is in need of a solution" (IUPAC–IUBMB 1992). The goal of a solution would be a systematic nomenclature for proteins, analogous to that for enzymes.

Current standardization generally involves specific families of proteins, for example, the kallikrein and annexin families (Berg and others 1992, Moss and others 1991). Such nomenclatures arise after a consensus has formed on the membership of individual proteins in a particular family. The designation of a protein family is based on common attributes such as functions and structures within a family. A family may then be subclassed from more specific criteria about the relationships among members; the annexin family has been thus subclassed: annexin I–annexin XII. Before a nomenclatural system emerges for a protein family, a plethora of trivial designations may call for some expertise within the interested subdiscipline to recognize any relationships that may exist among proteins of the emerging family. The annexin family is a good example of this situation. Trivial designations for individual proteins in the annexin family include "p35", "lipocortin", "chromobindin", "calelectrin", "calphobindin", and other coinages. Until a nomenclatural system broadly applicable to all nonenzyme proteins is codified, this situation will continue.

On studying a novel protein, one immediately faces the question of what to name it. The name usually depends on what is initially known about the protein. The name may evolve as more observations are made, starting from physical characteristics such as molecular weight, and going to more functional characteristics such as a biochemical role in cellular or physiological processes. Some trends in the naming of proteins before their becoming established as members of a protein family with an associated standardized nomenclature are apparent in the biochemical literature.

Protein Designations Based on Open Reading Frames

16·48 Gene sequencing has led to identifying potential protein-coding regions referred to as open reading frames (ORFs). Existence of the translated protein may or may not be experimentally shown. An identified protein may be designated as "protein ORF" with this designation including the number of amino acids in the protein or their sequential numbering. For example, "protein ORF216" would indicate a protein of 216 amino acids in length. This system of designation has a dual purpose of referring to the DNA region containing the open reading frame and to the protein resulting from the translation coded by the DNA region.

Protein Designations Based on Molecular Weight

16·49 One of the earliest experimental findings for a protein is the molecular weight; it may be used as the basis for the name of the protein. For example, "p21" and "p75" identify proteins of molecular weights of 21 and 75 kilodaltons, respectively; posttranslationally modified forms of these proteins may be referred to as "gp21" and "pp75", indicating a glycoprotein of 21 kilodaltons and a phosphoprotein of 75 kilodaltons.

Designations by molecular weight may be further modified by including the gene symbol as a superscript. For example, "gp160^{env}" is the designation for the 160 kilodalton glycoprotein encoded by the *env* gene in the human immunodeficiency virus (HIV). This form in turn may be used to identify proteins that are processed to smaller molecular-weight forms encoded by a single gene, such as to processing of HIV's gp160^{env} to gp120^{env} and gp41^{env}.

Molecular weight designations may also be used in combination with the class of the protein, especially if the protein has multiple isoforms or multiple subunits. The class is given as an acronym in superscript following the molecular-weight designation. For example, "p75^{NGFR}" is a 75 kilodalton isoform of a nerve growth-factor receptor.

Protein Designations Based on Gene Names

16·50 Many protein names are derived from the symbols for the corresponding gene names. Variation of this system can be found throughout the literature. In systems where the gene symbols are in lowercase italic letters, a protein is designated by capitalizing the initial letter, or all letters, of the relevant gene symbol and printing the resulting designation in roman letters. For organisms for which nonmutant gene symbols are fully capitalized, as for yeast, a protein is sometimes indicated by the gene symbol in upright characters (all capitals or only the initial letter) followed by a "p" suffix.

[gene symbol]	[the encoded protein]
ras	Ras
myc	Myc
NPL3	NPL3p [or] Npl3p

Note the importance of distinguishing between these closely related symbol pairs by using italic characters for gene symbols and upright characters for protein designations.

Names for proteins representing a mutant that is characterized by replacement of a single amino acid are also sometimes derived from the corresponding gene symbol. For example, an amino acid replacement with valine at position 12 of the Ras protein may be indicated by Ras$^{\mathrm{Val12}}$.

RETINOIDS

16·51 Retinoids are a class of compounds consisting of 4 isoprenoid units (IUPAC–IUB 1983b) . "Vitamin A" is a generic descriptor for retinoids having the biological activity of retinol, a retinoid with a terminal hydroxy functional group. Recommended names are based on 3 defined parent compounds: retinol, retinal, and retinoic acid. The stereoparent name implies that the polyene chain has the trans configuration around all double bonds unless the contrary is indicated:

13-*cis*-retinol	neovitamin A

STEROIDS

16·52 Steroids are compounds with the skeleton of cyclopenta[*a*]phenanthrene or a skeleton derived from it by bond scissions or ring expansions or contractions. The absolute stereochemistry of some chiral centers of steroids is defined by the name of the parent; that of other chiral centers is indicated by the use of α, β, *R*, and *S*, with ξ (xi) being used when the configuration is not known.

The names of steroid derivatives are based on the standard principles of the nomenclature of organic chemistry (IUPAC–IUB 1989). Many steroid derivatives have special names. The following is a list of special names, with IUPAC names given for comparison and illustrating the principles of substitutive nomenclature.

aldosterone	11β,18-epoxy-18ξ,21-dihydroxypregn-4-ene-3,20-dione
androsterone	3α-hydroxy-5α-androstan-17-one
bufalin	3β,14-dihydroxy-5β-bufa-20,22-dienolide
i-cholesterol	3α,5-cyclo-5α-cholestan-6β-ol
cholic acid	3α,7α,12α-trihydroxy-5β-cholan-24-oic acid
cycloartane	4,4,14-trimethyl-9,19-cyclo-5α,9β-cholestane
digitoxigenin	3β,14-dihydroxy-5β-card-20(22)-enolide
estrone	3-hydroxyestra-1,3,5(10)-trien-17-one
lanostane	4,4,14-trimethyl-5α-cholestane
progesterone	pregn-4-ene-3,20-dione
sarsasapogenin	(25*S*)-5β-spirostan-3β-ol
scillarenin	3β,14-dihydroxybufa-4,20,22-trienolide
spironolactone	7α-acetylthio-3-oxo-17α-pregn-4-ene-21,17-carbolactone
strophanthidin	3β,5,14-trihydroxy-19-oxo-5β-card-20(22)-enolide
testosterone acetate	3-oxoandrost-4-en-17β-yl acetate

16·53 A group of prefixes can indicate particular structural features associated with modifications of ring size or stereochemistry.

abeo	Bond migration (another way of indicating both ring expansion and contraction; original numbering is retained); for example, "10(5→6)*abeo*-6a(H)-androstane".
des	Removal of a terminal ring; for example, "des-*A*-androstane"; "des-*D*-5β-estrane".
ent	Inversion of all asymmetric centers.
homo	Expansion of a ring or lengthening of a side chain by insertion of 1 or more (dihomo) methylene groups. The numbering of the original parent nucleus is retained. The position of the new carbon atom is denoted by adding a lowercase letter a, b, c, . . . to the locant of the highest numbered nonbridgehead carbon of the expanded ring or of the side-chain where the expansion occurs.
nor	Ring or side-chain contraction. The numbering of the original parent nucleus is retained, with the "nor" prefix associated with the highest numbered nonbridgehead carbons of the contracted ring or that side-chain carbon where the contraction occurs.
rac	Racemic mixture; for example "*rac*-testosterone".
seco	Ring fission; for example, "(5*Z*,7*E*)-(3*S*)-9,10-secocholesta-5,7,10(19)-trien-3-ol"; trivial name "cholecalciferol" or "calciol".

The D vitamins and their derivatives are seco steroids. See section 16·59.

TETRAPYRROLES

16·54 Tetrapyrroles are members of a class of compounds containing 4 pyrrole rings linked by single-atom bridges between the alpha positions of the rings. Macrocyclic tetrapyrroles (as in the porphyrins) and linear tetrapyrroles (as in the bile pigments) are the most common examples, but the nomenclature system has been extended to include dipyrrole systems as well (IUPAC–IUB 1987c).

Earlier nomenclature in the field was based on a large number of trivial names. A more formal system based upon traditional IUPAC recommendations is available. The trivial names for the porphyrins follow.

coproporphyrin I	cytoporphyrin	deuteroporphyrin	etioporphyrin I
hematoporphyrin	mesoporphyrin	phylloporphyrin	protoporphyrin
pyrroporphyrin	rhodoporphyrin	uroporphyrin I	phytoporphyrin

Reduced Porphyrins, Including Chlorins

16·55 The most common reduced porphyrins are dihydroporphyrins in which saturated carbon atoms are located at the nonfused carbon atoms of 1 of the pyrrole rings. The parent compound of this series is called chlorin. The following names are representive examples, including tetrahydroporphyrins and porphyrins with the saturation involving the bridgehead carbons.

chlorin [2,3-dihydroporphyrin]
bacteriochlorin [7,8,17,18-tetrahydroporphyrin]
isobacteriochlorin [2,3,7,8-tetrahydroporphyrin]
porphyrinogen [5,10,15,20,22,24-hexahydroporphyrin]

The chlorophylls are close relatives of substituted reduced porphyrins. The chlorophylls are metallated; the corresponding demetallated compounds are called pheophytins (if the ester of the 17-propionic acid group is phytyl) or pheofarnesins (if the ester of the 17-propionic acid group is farnesyl). Compounds that are demetallated and also possess a free propionic acid group or acrylic acid group at C-17 are called pheophorbides. Thus, chlorophyll *a,* minus the metal plus hydrolysis of the C-17 ester, becomes pheophorbide *a.*

Linear Tetrapyrroles

16·56 The fundamental systems containing 4 tetrapyrrole groups linked by methylene (-CH_2-) or methene (=CH-) bridges are called bilane (trivial name bilinogen) and bilin, respectively. (Bilin has a further double bond; it is a bilatriene.) When the linkages are both methylene and methene, the compounds are called bilenes, with attached italic letters denoting where the methene is located (for example, bilene-*a*), or biladienes, with a combination of letters attached denoting where the 2 methenes are located (for example, biladiene-*bc*). The rest of the rings are maximally unsaturated.

Biliverdins (for example, biliverdin IXb, biliverdin IIIa) are substituted bilins. Note the hyphenation and italicization patterns of the names above: "chlorophyll", no hyphen, italic lower case letter; "bilene", hyphen, italic lower case letter; "biliverdin", no hyphen, roman numeral.

Relatives of Linear Tetrapyrroles

Tripyrrin and dipyrrin (formerly dipyrromethene or pyrromethene) are the names given to the 3- and 2-pyrrole systems linked by methene bridges.

Metal Coordination Complexes

16·57 The common structural pattern for metal coordination compounds of tetrapyrroles is for the metal to be coordinated, via the pyrrole nitrogens, to dianionic (through loss of the hydrogens at those nitrogens), roughly planar, tetrapyrrole ligands. In addition, other ligands may also be attached, in axial (perpendicular to the plane of the tetrapyrrole) position, to the metal. The rules of coordination nomenclature apply, with the axial ligands being designated α and β (the α and β are relative; what is correct with the semisystematic name may be reversed in a strictly systematic name because of the change in numbering of the ring). Other coordination patterns (for example, metals associated with the delocalized double bonds rather than the pyrrole nitrogens; metals associated with the nitrogens as well as delocalized double bonds) may occur.

VITAMINS

Vitamin A

See "Retinoids", Section 16·51.

Vitamins B-6 and Related Compounds

16·58 The term "vitamin B-6" is used as the generic description for all 3-hydroxy-2-methylpyridine derivatives with pyridoxine-type activity (IUPAC–IUB 1974b). Pyridoxine (3-hydroxy-4,5-bis(hydroxymethyl)-2-methylpyridine) is 1 of the B-6 vitamins, and the term "pyridoxine" should not be used as a synonym for "vitamin B-6". The following trivial names are accepted; the abbreviations are not recommended for text use.

[trivial names]	[abbreviations]
pyridoxal	PL
pyridoxal-*P* (*P*, phosphate ester)	PLP
pyridoxamine	PM
pyridoxamine-*P*	PMP
pyridoxine	PN
pyridoxine-*P*	PNP
pyridoxyl	Pxy- [acceptable in formulas]
pyridoxylidene	Pxd= [acceptable in formulas]

Vitamins B-12

See “Corrinoids”, Section 16·23.

Vitamin D Compounds

16·59 The D vitamins and their derivatives are seco steroids. Because the seco names can be cumbersome, trivial names have been adopted. The following is a list of representative trivial names and their recommended forms (IUPAC–IUB 1982a).

[trivial name]	[recommended trivial name]
1α,24*R*,25-trihydroxycholecalciferol [the 1,3,24,25-tetrol]	calcitetrol
cholecalciferol	calciol or cholecalciferol
25-hydroxycholecalciferol [the 3,25-diol]	calcidiol [D_3]
1α,25-dihydroxycholecalciferol [the 1,3,25-triol]	calcitriol [D_2]
ergocalciferol	ercalciol [or] ergocalciferol
previtamin D_3 [or] precalciferol	(6*Z*)-tacalciol
tachysterol$_3$	tacalciol
isovitamin D_3	(5*E*)-isocalciol

It is recommended that the configurations of chiral centers in ring A be denoted by *R* or *S*, and the geometry of double bonds at C-6 or C-7 be denoted by *E* or *Z*.

Vitamin E: Tocopherols and Related Compounds

16·60 “Vitamin E” is the term used as the generic descriptor for all tocol and tocotrienol derivatives having the biological activity of α-tocopherol (IUPAC–IUB 1982d, 1983a). “Tocol” is the trivial name for “2-methyl-2-(4,8,12-trimethyltridecyl)chroman-6-ol”, and “tocopherol” is the generic descriptor for monomethyl-, dimethyl-, and trimethyl-substituted tocols.

The following are examples of recommended trivial names and some corresponding semisystematic names.

[recommended trivial names]	
RRR-α-tocopherol	(2*R*,4′*R*,8′*R*)-α-tocopherol
2-*epi*-α-tocopherol	(2*S*,4′*R*,8′*R*)-α-tocopherol
2-*ambo*-α-tocopherol	mixture of the 2*R*,4′*R*,8′*R* and 2*S*,4′*R*,8′*R* forms

4′-*ambo*-8′-*ambo*-α-tocopherol	mixture of 4 isomers

Vitamin K

16·61 Compounds with vitamin K activity have side chains consisting of prenyl groups substituted on quinone ring systems, which themselves have various substitution patterns. The following are representative trivial names (IUPAC–IUB 1975a).

[recommended name]	[trivial name]	[vitamin K name]
phylloquinone	phytylmenaquinone	vitamin K_1(20)
menaquinone-*n*	prenylmenaquinone-*n*	vitamin K_2
	n = 10	vitamin K_2(50)
	$n = 7$	vitamin K_2(35)
	$n = 6$	vitamin K_2(30)

BIOTECHNOLOGY

16·62 A glossary of biotechnology terms (IUPAC 1992) defines and explains, as needed, over 230 terms frequently used in publications representing the multidisciplinary field of biotechnology. The glossary was developed to help facilitate communication among chemists, chemical engineers, biologists, and bioengineers and to make biotechnology and its methods more accessible to the chemical profession. The range of terms includes microbiology, genetic engineering, biochemistry, molecular biology, biochemical engineering, bioprocessing, and general concepts of biotechnology. [This summary was adapted from the document (IUPAC 1992).]

CLINICAL CHEMISTRY, QUANTITIES AND UNITS

16·63 Recommendations on specific usage in clinical chemistry build on standard IUPAC practice (IUPAC–IFCC 1984). The specific needs of this field are addressed in nomenclature documents too extensive, however, to summarize here. Included are definitions of fundamental concepts, types of quantities and units typically encountered, and proposed new derived kinds of quantities, such as mean catalytic activity rate.

REFERENCES

Cited References

Berg T, Bradshaw RA, Carretero OA, Chao J, Chao L, Clements JA, Fahnestock M, Fritz H, Gauthier F, MacDonald RJ, Margolius HS, Morris BJ, Richards RI, Sciti AG. 1992. A common nomenclature for members of the tissue (glandular) kallikrein gene families. Agents Actions Suppl 38:19–25.

[CAS] Chemical Abstracts Service. 1993. Ring systems handbook. Columbus (OH): American Chemical Soc, CAS.

Dean JA, editor. 1992. Lange's handbook of chemistry. New York: McGraw-Hill.

Dodd JS. 1985. The ACS style guide: a manual for authors and editors. Washington: American Chemical Soc.

[IUB] International Union of Biochemistry, Nomenclature Committee. 1985. Nomenclature for incompletely specified bases in nucleic acid sequences: recommendations 1984. Eur J Biochem 150:1 N5.

[IUB] International Union of Biochemistry, Nomenclature Committee (NC-IUB). 1989. Nomenclature for multienzymes: recommendations 1989. Eur J Biochem 185:485–6.

[IUB] International Union of Biochemistry, Nomenclature Committee (NC-IUB). 1991. Nomenclature of electron-transfer proteins: recommendations 1989. Eur J Biochem 200:599–611.

[IUBMB] International Union of Biochemistry and Molecular Biology, Nomenclature Committee. 1992a. Biochemical nomenclature and related documents: a compendium. 2nd ed. London: Portland Pr.

[IUBMB] International Union of Biochemistry and Molecular Biology, Nomenclature Committee. 1992b. Enzyme nomenclature: recommendations 1992. San Diego (CA): Academic Pr.

[IUPAC] International Union of Pure and Applied Chemistry, Organic Chemistry Division, Commission on the Nomenclature of Organic Chemistry. 1979. Nomenclature of organic chemistry, Sections A, B, C, D, E, F, and H. 4th ed. Oxford (UK): Pergamon.

[IUPAC] International Union of Pure and Applied Chemistry. 1990. Nomenclature of inorganic chemistry: recommendations 1990. 3rd ed. Oxford (UK): Blackwell Scientific.

[IUPAC] International Union of Pure and Applied Chemistry, Macromolecular Division, Commission on Macromolecular Nomenclature.

1991. Compendium of macromolecular nomenclature. Oxford (UK): Blackwell Scientific.

[IUPAC] International Union of Pure and Applied Chemistry, Applied Chemistry Division, Commission on Biotechnology. 1992. Glossary for chemists of terms used in biotechnology: recommendations 1992. Pure Appl Chem 64:143–68.

[IUPAC] International Union of Pure and Applied Chemistry, Organic Chemistry Division, Commission on Nomenclature of Organic Chemistry. 1993a. Revised nomenclature for radicals, ions, radical ions and related species: recommendations 1993. Pure Appl Chem 65:1357–1455.

[IUPAC] International Union of Pure and Applied Chemistry, Macromolecular Division, Commission on Macromolecular Nomenclature. 1993b. Nomenclature of regular double-strand (ladder and spiro) organic polymers: recommendations 1993. Pure Appl Chem 65:1561–80.

[IUPAC–IFCC] International Union of Pure and Applied Chemistry; International Federation of Clinical Chemistry. 1984. Physicochemical quantities and units in clinical chemistry with special emphasis on activities and activity coefficients: recommendations 1983. Pure Appl Chem 56:567–94.

[IUPAC–IUB] IUPAC–IUB Commission on Biochemical Nomenclature (CBN). 1970. Abbreviations and symbols for nucleic acids, polynucleotides and their constituents: recommendations 1970. Eur J Biochem 15:203–8.

[IUPAC–IUB] IUPAC Commission on the Nomenclature of Organic Chemistry (CNOC) and IUPAC–IUB Commission on Biochemical Nomenclature (CBN). 1971. Tentative rules for carbohydrate nomenclature. Biochemistry 10:3983–4004; correction 10:4995.

[IUPAC–IUB] IUPAC–IUB Commission on Biochemical Nomenclature (CBN). 1972. Abbreviated nomenclature of synthetic polypeptides (polymerized amino acids): revised recommendations (1971). Biochemistry 11:942–4.

[IUPAC–IUB] International Union of Pure and Applied Chemistry–International Union of Biochemistry, Commission on Biochemical Nomenclature. 1974a. Nomenclature of cyclitols: recommendations 1973. Pure Appl Chem 37:283–97.

[IUPAC–IUB] International Union of Pure and Applied Chemistry–International Union of Biochemistry, Commission on Biochemical

Nomenclature (CBN). 1974b. Nomenclature of vitamins B-6 and related compounds: recommendations 1973. Biochemistry 13:1056–8.

[IUPAC–IUB] International Union of Pure and Applied Chemistry–International Union of Biochemistry, Commission on Biochemical Nomenclature (CBN). 1975a. Nomenclature of quinones with isoprenoid side-chains: recommendations 1973. Eur J Biochem 53:15–8.

[IUPAC–IUB] International Union of Pure and Applied Chemistry–International Union of Biochemistry, Commission on Biochemical Nomenclature (CBN). 1975b. Nomenclature of carotenoids: recommendations 1974. Pure Appl Chem 41:405–31.

[IUPAC–IUB] International Union of Pure and Applied Chemistry–International Union of Biochemistry, Commission on Biochemical Nomenclature (CBN). 1975c. The nomenclature of peptide hormones: recommendations (1974). Biochemistry 14:2559–60.

[IUPAC–IUB] International Union of Pure and Applied Chemistry–International Union of Biochemistry, Commission on Biochemical Nomenclature (CBN). 1976. Nomenclature of corrinoids: rules approved 1975. Pure Appl Chem 48:495–502.

[IUPAC–IUB] International Union of Pure and Applied Chemistry–International Union of Biochemistry, Commission on Biochemical Nomenclature. 1977a. Nomenclature of lipids: recommendations 1976. Lipids 12:455–68.

[IUPAC–IUB] International Union of Pure and Applied Chemistry–International Union of Biochemistry, Commission on Biochemical Nomenclature (CBN). 1977b. Nomenclature of phosphorus-containing compounds of biochemical importance: recommendations 1976. Proc Natl Acad Sci USA 74:2222–30.

[IUPAC–IUB] International Union of Pure and Applied Chemistry–International Union of Biochemistry, Joint Commission on Biochemical Nomenclature (JCBN). 1981a. Nomenclature of unsaturated monosaccharides: recommendations 1980. Eur J Biochem 119:1–3.

[IUPAC–IUB] International Union of Pure and Applied Chemistry and International Union of Biochemistry, Joint Commission on Biochemical Nomenclature (JCBN). 1981b. Nomenclature of branched-chain monosaccharides: recommendations 1980. Eur J Biochem 119:5–8.

[IUPAC–IUB] International Union of Pure and Applied Chemistry and International Union of Biochemistry, Joint Commission on Biochemical Nomenclature (JCBN). 1981c. Conformational nomenclature for five- and six-membered ring forms of monosaccharides and their derivatives. Pure Appl Chem 53:1901–5.

[IUPAC–IUB] International Union of Pure and Applied Chemistry and International Union of Biochemistry, Joint Commission on Biochemical Nomenclature (JCBN). 1982a. Nomenclature of vitamin D: recommendations 1981. Pure Appl Chem 54:1511–6; Eur J Biochem 124:223–7.

[IUPAC–IUB] International Union of Pure and Applied Chemistry and International Union of Biochemistry, Joint Commission on Biochemical Nomenclature (JCBN). 1982b. Abbreviated terminology of oligosaccharide chains: recommendations 1980. Pure Appl Chem 54:1517–22.

[IUPAC–IUB] International Union of Pure and Applied Chemistry and International Union of Biochemistry, Joint Commission on Biochemical Nomenclature (JCBN). 1982c. Polysaccharide nomenclature. Pure Appl Chem 54:1523–6.

[IUPAC–IUB] IUPAC–IUB Joint Commission on Biochemical Nomenclature (JCBN). 1982d. Nomenclature of tocopherols and related compounds: recommendations 1981. Eur J Biochem 123:473–5.

[IUPAC–IUB] IUPAC-IUB Joint Commission on Biochemical Nomenclature (JCBN), Nomenclature Committee. 1983a. Newsletter 1983. Eur J Biochem 131:1–3.

[IUPAC–IUB] International Union of Pure and Applied Chemistry and International Union of Biochemistry, IUPAC–IUB Joint Commission on Biochemical Nomenclature (JCBN). 1983b. Nomenclature of retinoids: recommendations 1981. Pure Appl Chem 55:721–6; Eur J Biochem 1982;129:1–5.

[IUPAC–IUB] International Union of Pure and Applied Chemistry and International Union of Biochemistry, Joint Commission on Biochemical Nomenclature (JCBN). 1983c. Symbols for specifying the conformation of polysaccharide chains: recommendations 1981. Pure Appl Chem 55:1269–72.

[IUPAC–IUB] International Union of Pure and Applied Chemistry and International Union of Biochemistry, Joint Commission on Biochemical Nomenclature (JCBN). 1983d. Abbreviations and symbols for the description of conformations of polynucleotide chains: recommendations 1982. Pure Appl Chem 55:1273–80; Eur J Biochem 131:9–15.

[IUPAC–IUB] IUPAC–IUB Joint Commission on Biochemical Nomenclature (JCBN). 1984. Nomenclature and symbolism for amino acids and peptides: recommendations 1983. Eur J Biochem 138:9–37.

[IUPAC–IUB] IUPAC–IUB Joint Commission on Biochemical Nomenclature (JCBN). 1987a. Nomenclature of glycoproteins, glycopeptides and peptidoglycans: recommendations 1985. J Biol Chem 262:13–8.

[IUPAC–IUB] International Union of Pure and Applied Chemistry and International Union of Biochemistry, Joint Commission on Biochemical Nomenclature (JCBN). 1987b. Nomenclature of prenols: recommendations 1986. Pure Appl Chem 59:683–9.

[IUPAC–IUB] International Union of Pure and Applied Chemistry and International Union of Biochemistry, Joint Commission on Biochemical Nomenclature (JCBN). 1987c. Nomenclature of tetrapyrroles: recommendations 1986. Pure Appl Chem 59:779–832; Eur J Biochem 1988;178:277–328.

[IUPAC–IUB] IUPAC–IUB Joint Commission on Biochemical Nomenclature (JCBN). 1987d. Nomenclature and symbols for folic acid and related compounds: recommendations 1986. Pure Appl Chem 59:833–6; Eur J Biochem 168:251–3.

[IUPAC–IUB] International Union of Pure and Applied Chemistry and International Union of Biochemistry, Joint Commission on Biochemical Nomenclature (JCBN). 1989. Nomenclature of steroids: recommendations 1989. Pure Appl Chem 61:1783–822.

[IUPAC-IUBMB] Nomenclature Committee of IUBMB (NC-IUBMB) and IUPAC–IUBMB Joint Commission on Biochemical Nomenclature (JCBN). 1992. Newsletter 1992. Arch Biochem Biophys 294:322–5.

Lide DR, editor. 1993. CRC handbook of chemistry and physics. 74th ed. Boca Raton (FL): CRC Pr.

Moss SE, Edwards HC, Crumpton MJ. 1991. Diversity in the annexin family. In: Heizman C, editor. Novel calcium-binding proteins. Berlin: Springer Publishing. p 533–66.

Nelson NA. 1974. Prostaglandin nomenclature. J Med Chem 17:911–8.

Additional References

American Thyroid Association Nomenclature Committee. 1987. Revised nomenclature for tests of thyroid hormones and thyroid-related proteins in serum. Clin Chem 33:2114–9.

Ash M, Ash I. 1992 Industrial chemical thesaurus: Volume 1, chemical to tradename reference; Volume 2, tradename to chemical cross reference. 2nd ed. New York: VCH.

Block BP, Powell WH, Fernelius W. 1990. Inorganic nomenclature: principles and practice. Washington: American Chemical Soc.

Budavari S, editor. 1989. The Merck index: an encyclopedia of chemicals, drugs, and biologicals. 11th ed. Rahway (NJ): Merck.

Connolly JD, Hill RA. 1992. Dictionary of terpenoids. New York: Chapman & Hall.

Elks J, Ganelin CR. 1990. Dictionary of drugs: chemical data, structures and bibliographies. New York: Chapman & Hall.

Emsley J. 1991. The elements. 2nd ed. Oxford (UK): Oxford Univ Pr, Clarendon Pr.

Fasman GD. 1989. CRC practical handbook of biochemistry and molecular biology. 3rd ed. Boca Raton (FL): CRC Pr.

Glatz JFC, van der Vusse GJ. 1990. Nomenclature of fatty acid-binding proteins. Mol Cell Biochem 98:231–6.

Gold V, Loening KL, McNaught AD, Sehmi P, editors. 1987. Compendium of chemical terminology: IUPAC (International Union of Pure and Applied Chemistry) recommendations. Oxford (UK): Blackwell Scientific.

Hill RA, Kirk DN, Makin HLJ, Murphy GM. 1991. Dictionary of steroids: chemical data, structures and bibliographies. New York: Chapman & Hall.

Howard PH, Neal M. 1992. Dictionary of chemical names and synonyms. Boca Raton (FL): Lewis Publishers.

Larson LL, Kenaga EE, Morgan RW. 1985. Commercial and experimental organic insecticides. Lanham (MD): Entomol Soc America.

Macintyre JE. 1992. Dictionary of inorganic compounds. New York: Chapman & Hall.

Marczenko Z, Newman E, Thorburn-Burns DT, Townshend A. 1993. Dictionary of analytical reagents. New York: Chapman & Hall.

Raina AK, Gäde G. 1988. Insect peptide nomenclature. Insect Biochem 18:785–7.

Wiggins G. 1991. Chemical information sources. New York: McGraw-Hill.

17 Chemical Kinetics and Thermodynamics

Every system in stable chemical equilibrium submitted to the influence of an exterior force which tends to cause variation either in its temperature or its condensation (pressure, concentration, number of molecules in the unit of volume) in its totality or only in some one of its parts can undergo only those interior modifications which, if they occur alone, would produce a change of temperature or of condensation, of a sign contrary to that resulting from the exterior force. —Henri Louis Le Chatelier, *A General Statement of the Laws of Chemical Equilibrium,* 1884

The key to a knowledge of enzymes is the study of reaction velocities, not of equilibria. —J B S Haldane, *Enzymes,* 1930

Kinetic studies are important in interpreting scientific reactions and in extrapolating the results of scientific studies to in vivo and in vitro systems. Because of the long history of accurate notation in the physical sciences, consistency in terminology in the field of kinetics is much more

Table 17·1 Symbols and units for some thermodynamic quantities[a]

Quantity name	Quantity symbol	SI unit	SI unit symbol
volume	V	cubic meter	m^3
force	F	newton	$N = m\ kg\ s^{-2}$
density	ρ	kilograms per cubic meter	$kg\ m^{-3}$
pressure	p	pascal	$Pa = N\ m^{-2}$
viscosity	η	pascal second	Pa s
energy	E	joule	J = N m
heat	q, Q	joule	J
work	w, W	joule	J
internal energy	U	joule	J
enthalpy	H	joule	J
Gibbs energy	G	joule	J
Helmholtz energy	A	joule	J
entropy	S	joule per kelvin	J/K [or] $J\ K^{-1}$
power	P	watt	W
heat capacity			
at constant pressure	C_p	joule per kelvin	J/K [or] $J\ K^{-1}$
at constant volume	C_v	joule per kelvin	J/K [or] $J\ K^{-1}$
osmotic pressure	Π	pascal	Pa
chemical potential of B	μ_B	joule per mole	J/mol [or] $J\ mol^{-1}$
absolute activity of B	λ_B	[dimensionless]	
activity coefficient			
mole fraction basis	f_B	[dimensionless]	
concentration basis	y_B	[dimensionless]	
osmotic coefficient	ϕ	[dimensionless]	

[a]The symbols for physical quantities should be printed in italic (slanted) type for the greek or roman letters (and underlined in typescripts); the symbols for units should be in roman (upright) type.

evident than in the life sciences (IUB 1979, 1982; IUPAC 1981, 1982). The need for greater systemization and consistency across fields resulted in a multicommission effort in biothermodynamics, with subsequent recommendations for presentation of data and results (IUPAC–IUPAB–IUB 1976, 1985).

PHYSICAL QUANTITIES, SI UNITS, AND THEIR SYMBOLS

17·1 A physical quantity is the product of a numerical value and a unit. For a discussion of SI units and symbols, see sections 11·8–11·11. Specific quantities used in thermodynamics are given in Table 17·1.

UNITS IN BIOTHERMODYNAMICS

MASS

17·2 The term "specific" preceding the name of a quantity means "divided by mass" (the SI unit kilogram). The term "molar" preceding the name of a quantity means "divided by amount of substance" (the SI unit mole).

Thermodynamic quantitites should, where possible, be reported in terms of molar quantities. In research with macromolecules, molecular mass or molar mass may be reported. The values of molecular mass and molar mass are numerically identical, but the units differ.

1 Molecular mass, u (atomic mass unit) or Da (dalton).
2 Molar mass, g mol^{-1}.

The name "dalton" is a special name for the atomic mass unit, u, an SI unit. Note that neither the name "dalton" nor the symbol Da has been approved by CGPM (General Conference on Weights and Measures), although recent IUPAC documents (IUPAC 1993, among others) include "dalton" and Da in lists of acceptable symbols and units. "Dalton" should not be used in expressions such as "the molecular weight of subunit A is 76 000 daltons". The term "molecular weight" is considered to be a synonym for "relative molecular mass", which is unitless. It is acceptable to use the following alternatives.

The molecular weight of subunit A is 76 000.
The molecular mass of subunit A is 76 000 Da.

The molecular mass of subunit A is 76 000 u.
The molar mass of subunit A is 76 000 g mol^{-1}.

VOLUME, TEMPERATURE, ENERGY, AND DENSITY

17·3 The SI unit for volume is the cubic meter. In the expressions for most units, the use of submultiples, including "deci" and "centi", is discouraged. The value of the cubic centimeter (cm^3; not "cc") in scientific notation is recognized. In addition, "liter" (symbol "L" or "l" [lowercase letter "ell"]) as a specialized name for the cubic decimeter, with its submultiples the milliliter and the microliter, is acceptable and may be retained.

In thermodynamic calculations and kinetic studies, temperature should be expressed as its SI base unit, the kelvin (K), although it is acceptable to express experimental temperatures in degrees Celsius (°C).

Joule is the SI unit for energy; the use of "calorie" is generally discouraged. Its usefulness in the field of nutrition is nevertheless recognized.

Whereas the SI unit for density is kilograms per meter cubed (kg m^{-3}), a more convenient unit is grams per centimeter cubed (g cm^{-3}).

NOTATIONS FOR VARIABLES, STATES, AND PROCESSES

VARIABLES

17·4 Symbols for variables are set in italic (slanted) type for letters of the greek or roman alphabets (IUPAC–IUPAB–IUB 1985). They may be appended within parentheses to the symbol for the thermodynamic function.

$\mu_b{}^{ss}(T)$

[the standard chemical potential of substance b at temperature *T;* ss, "standard state"]

STATES

17·5 Superscripts to symbols for themodynamic functions are used to denote particular states, for example, $\Delta G°$. The superscript symbols for the most frequently encountered states are the 7 below.

standard	°(or $^{\ominus}$)	apparent	′
pure substance	*	excess	E
infinite dilution	∞	activated complex	‡
ideal	id		

Symbols for the state of aggregation should be given within parentheses after the symbol for the property, for example, V^*(cr), the volume of a substance in its pure crystalline state. The symbols below specify the states of aggregation.

gas	g	crystalline solid	cr
liquid	l	amorphous solid	am
solid	s	vitreous substance	vit
fluid	fl	solution	sln
liquid crystal	lc	aqueous solution	aq

Below are examples of representations with these symbols.

C_V(fl)	constant volume heat capacity of a fluid
NaCl(sln)	sodium chloride in solution
NH_3 (aq)	ammonia in aqueous solution

PROCESSES

17·6 Thermodynamic changes are indicated by the symbol Δ before the corresponding thermodynamic symbol. The type of change is indicated by a subscript immediately after the Δ; for example, $\Delta_f H$, change in enthalpy of formation. Symbols for the most frequently occurring processes and their symbols are shown below.

vaporization (evaporation)	vap	solution (dissolution)	sol
sublimation (evaporation)	sub	reaction (except combustion)	r
melting (fusion)	fus	combustion	c
transition (solid-solid)	trs	formation	f
mixing of fluids	mix		

Below are examples of representations with these symbols.

$\Delta_f S°(HgCl_2$, cr, 298.15 K)	[change in (molar) standard entropy due to the formation of crystalline mercuric chloride from its elements at a temperature of 298.15 K]
$\Delta_r G°$(1000 K)	[change in (molar) standard Gibbs energy due to a chemical reaction at a temperature of 1000 K]

MICROBIAL PROCESSES

17·7 IUPAC 1992 presents a detailed list of terms important in the biotechnology of microbial processes. Many of the terms, definitions, and symbols are identical with those in other IUPAC documents (for example, IUPAC 1993) or have been brought into agreement with those recommendations. General quantities, intensive quantitites, rate quantitites, and concentrations and amounts are all described and defined, with symbols and units being given. Both SI units and units in common usage (such as atmosphere, bar, hour [instead of second]) are given. Table 17·2 shows some representative examples.

Table 17·2 Selected terms, symbols, and units related to microbial processes

Quantity	Symbol	SI unit	Customary unit
pressure	p	Pa	atm, bar
molar activation energy	E	J mol^{-1}	
for specific growth rate	E_{μ}		kJ mol^{-1}
for cell biomass yield	E_{Y}		kJ mol^{-1}
area per volume	a	m^{-1}	cm^{-1}
yield of cell biomass per amount of ATP produced in cells	Y_{ATP}	kg mol^{-1}	g mol^{-1}
gas hold-up (volume of gas divided by volume of liquid)	ε	1	1
doubling time, biomass	t_{d}	s	min, h, d
growth rate, colony radial (rate of extension of biomass colony on a surface)	K_{r}	m s^{-1}	m h^{-1}
specific mass metabolic rate	q	s^{-1}	h^{-1}
mutation rate	w	s^{-1}	h^{-1}

ION MOVEMENT

17·8 Research in physiology involves aspects of solute transport studied by convection, diffusion, and permeation across membranes (ion movement and transport). A standard set of symbols for practitioners of 1 method often differed from standard sets in other areas. In an attempt to improve communication in research in ion transport, recommendations for terminology were published by Bassingthwaighte and others (1986). Use of the terminology is encouraged by the American Physiological Society but has not been made mandatory. Table 17·3 presents some of the important symbols.

SOLUTIONS

17·9 A solution is a liquid or solid phase containing more than 1 component. The term "solvent" is generally used for the substance in largest concentration, with the other components called "solutes". In biological solutions, however, "solvent" is often used to refer to a fixed mixture of components (for example, water plus buffer components), with all other substances being "solutes".

Notations for the composition of solutions, recommended symbols, and appropriate units follow.

Table 17·3 Selected symbols for ion mass transport and exchange[a]

Symbol	Definition or description
Principal symbol	
a	activity, molar
A	area of indicator concentration—time curve excluding recirculation
C	concentration, mol/L
CV	coefficient of variation, dimensionless
D	diffusion coefficient, $cm^2\ s^{-1}$
E	electric potential, V
ECF	extracellular fluid
F	flow, $cm^3\ s^{-1}$
F_B	blood flow to an organ, $cm^3\ g^{-1}\ min^{-1}$
$h(t)$	transport function
Hct	hematocrit, the fraction of the blood volume that is erthycrocytes, dimensionless
ISF	interstitial fluid, the extravascular extracellular fluid
J	flux
P	permeability coefficient for a solute traversing a membrane, $cm\ s^{-1}$
RD	relative dispersion, dimensionless
Subscript	
A	arterial
B	blood
C or cap	capillary, or the region of blood-tissue exchange
cell	cell
ECF	extracellular fluid
F	flow or filtration
i, j	indices in series, or in summations, or in elements of arrays
in [or] i	into [or] inside [or] inflow
ISF or I	interstitial fluid space, the extravascular, extracellular fluid
m	membrane
out [or] o	out of [or] outside [or] outflow
p	plasma
RBC	red blood cell
R	reference, nonpermeant tracer
S	solute
T	total
V	venous
W	water

[a]From Bassingthwaighte and others (1986).

amount of substance B	n_B	mol
concentration of solute substance B	c_B	$mol\ dm^{-3}$
mass concentration of substance B	ρ_B	$g\ dm^{-3}$
molality of solute substance B	m_B	$mol\ kg^{-1}$
mole fraction of substance B	x_B	[dimensionless]
mass fraction of substance B	w_B	[dimensionless]
volume fraction of substance B	ϕ_B	[dimensionless]

Table 17·4 Enzyme kinetics: A selection of recommended kinetic terms, symbols, and their units

Term	Symbol	Customary unit[a]
concentration of substrate A	[A]	mol dm^{-3}
concentration of inhibitor I	[I]	mol dm^{-3}
concentration of product Y	[Y]	mol dm^{-3}
forward and reverse rate constants	k_i, k_{-i}	as k
catalytic constant	k_0	s^{-1}
rate constant of any order n	k	$(mol\ dm^{-3})^{1-n}\ s^{-1}$
inhibition constant	K_i	mol dm^{-3}
Michaelis constant	K_m	mol dm^{-3}
Michaelis constant for substrate A	K_{mA}	mol dm^{-3}
time	t	s
rate (or velocity) of reaction	v	mol dm^{-3} s^{-1}
initial rate of reaction	v_0	mol dm^{-3} s^{-1}
rate of conversion	$\dot{\xi}$	mol s^{-1}

[a]In all cases, "dm^3" may be replaced identically with "L" or "l" (liter).

ENZYME KINETICS

17·10 Practices in the field of enzyme kinetics follow closely those in chemical kinetics (IUB 1982; IUPAC 1981). However, the rigorous detail associated with all variables in chemical kinetics is generally not necessary with enzyme-catalyzed reactions, which most frequently occur in a liquid phase at constant pressure.

KINETIC EQUATIONS

Definitions of consumption and formation rates, reaction rates, and elementary and composite reactions are given in IUB 1982. Of special interest are the kinetic equations describing Michaelis–Menten and non-Michaelis–Menten relationships.

1 Michaelis–Menten. The relationship between the rate of an enzyme-catalyzed reaction and the substrate concentration is of the following form.

$$v = V[A]/(K_{mA} + [A])$$

2 Non-Michaelis–Menten. The following form is an example.

$$v = V'[A]/(K'_{mA} + [A] + [A]^2K_{iA})$$

Table 17·4 gives a selected list of symbols for kinetic reactions (IUB 1982).

ENZYME ACTIVITY

The catalytic activity of an enzyme is the property measured by the increase in the rate of conversion (that is, the rate of reaction expressed as an extensive quantity measured as the increase in amount-of-substance per unit of time) of a specified chemical reaction that the enzyme produces in a specific assay system (IUB 1982). It has been expressed by the unit "katal" (symbol "kat"), although use of the SI unit, mol sec^{-1}, is preferred. A katal is the catalytic activity that will raise the rate of reaction by 1 mole per second in a specified assay system (IUB 1979).

PRESENTATION OF RESULTS

17·11 In text, tables, and figures, all quantities reported should be given with units and include all necessary information about important experimental considerations (for example, temperature, concentration, pH). Tables and figure captions should use standard scientific notation. For designation of units in tabular-column heads, see section 31·17.

REFERENCES

Cited References

Bassingthwaighte JB, Chinard FP, Crone C, Goresky CA, Lassen NA, Reneman RS, Zierler KL. 1986. Terminology for mass transport and exchange. Am J Physiol 250(4 Pt 2):H539–45.

[IUB] International Union of Biochemistry, Nomenclature Committee (NC-IUB). 1979. Units of enzyme activity: recommendations 1978. Eur J Biochem 97:319–20.

[IUB] International Union of Biochemistry, Nomenclature Committee (NC-IUB). 1982. Symbolism and terminology in enzyme kinetics: recommendations 1981. Eur J Biochem 128:281–91.

[IUPAC] International Union of Pure and Applied Chemistry, Physical Chemistry Division, Subcommittee on Chemical Kinetics. 1981. Symbolism and terminology in chemical kinetics. Pure Appl Chem 53:753–71.

[IUPAC] International Union of Pure and Applied Chemistry, Physical Chemistry Division, Commission on Thermodynamics. 1982. Manual of symbols and terminology for physicochemical quantities and units: appendix IV, notation for states and processes, significance of the word *standard* in chemical thermodynamics, and remarks on commonly tabulated forms of thermodynamic functions. Pure Appl Chem 54:1239–50.

[IUPAC] International Union of Pure and Applied Chemistry, Applied Chemistry Division, Committee on Biotechnology. 1992. Selection of terms, symbols and units related to microbial processes: recommendations 1992. Pure Appl Chem 64:1047–53.

[IUPAC] International Union of Pure and Applied Chemistry, Physical Chemistry Division. 1993. Quantities, units and symbols in physical chemistry. Oxford (UK): Blackwell Scientific.

[IUPAC–IUPAB–IUB] IUPAC–IUPAB–IUB Interunion Commission on Biothermodynamics. 1976. Recommendations for the presentation of biochemical equilibrium data. J Biol Chem 251:6879–85.

[IUPAC–IUPAB–IUB] IUPAC–IUPAB–IUB Intercommission on Biothermodynamics. 1985. Recommendations for the presentation of thermodynamic and related data in biology (1985). Eur J Biochem 153:429–34.

Additional References

Dean JA, editor. 1992. Lange's handbook of chemistry. New York: McGraw-Hill.

Lide DR, editor. 1993. CRC handbook of chemistry and physics. 74th ed. Boca Raton (FL): CRC Pr.

Ulicky L, Kemp T, editors. 1993. Comprehensive dictionary of physical chemistry. Englewood Cliffs (NJ): Prentice Hall.

Wiggins G. 1991. Chemical information sources. New York: McGraw-Hill.

18 Analytical Methods

> Progress in science depends on new techniques, new discoveries and new ideas, probably in that order. —Sydney Brenner, *Nature,* 1980

Definitions of terms and descriptions of techniques in many areas of analytical chemistry are given in IUPAC 1985a, 1985b, 1987a, 1987b, 1988a–d, 1989, 1991a–f, 1993. Many of the terms and techniques are widely known and followed. Whenever a paper can report the use of established methods, the experimental details may be presented in summary form: for example, "... was determined by Lineweaver–Burk methodology as modified by Smith et al. (1985)". If new techniques are used or important modifications of basic methods are introduced, the experimental procedures should be presented in sufficient detail so that they could be duplicated by an experienced research scientist.

Definitions in this chapter have been taken directly, or paraphrased, from the nomenclature documents cited as references.

CHROMATOGRAPHY

18·1 A document incorporating all chromatographic terms and definitions used in the major chromatographic methods has been published (IUPAC 1993). Methods such as gas, liquid, and supercritical-fluid chromatography; column and planar chromatography; and partition, adsorption, ion-exchange, and exclusion chromatography are included; an index of terms and lists of symbols and acronyms used in chromatography are included. The following sections are representative of the scope and content.

CLASSIFICATION

18·2 Chromatographic methods are often classified according to the shape of the chromatographic bed, the physical state of the mobile phase, the mechanism of separation, or special methods.

[bed]	column chromatography planar chromatography
[physical state]	gas–liquid chromatography (GLC) liquid–liquid chromatography (LLC) gas–solid chromatography (GSC) liquid–solid chromatography (LSC)
[mechanism]	adsorption chromatography partition chromatography ion-exchange chromatography exclusion chromatography affinity chromatography
[special methods]	gradient elution post-column derivatization reversed-phase chromatography

TERMS, SYMBOLS, AND ABBREVIATIONS

18·3 Definitions and symbols (when applicable) for terms such as "column", "pore radius", "split injection", "chamber saturation", "standard deviation", "diffusion velocity", "retention volume", and "detector sensitivity" are covered by IUPAC 1993. Special terms used in ion exchange and exclusion chromatography (such as "sorption isotherm", "ion exchangers", "resin matrix", "peak resolution", "plate number", and "plate height") are defined.

Recommended abbreviations are shown in Table 18·1. They should be defined in a footnote or paragraph.

SPECTROSCOPY

18·4 Terms encountered in and definitions of the various spectroscopic techniques are given in IUPAC 1985a, 1987b, 1988b–d, 1989, 1991b, 1991c, 1991e. The abbreviations in Tables 18·2 and 18·3 are recommended for spectroscopy and surface science.

Table 18·1 Abbreviations of chromatography terms[a]

Term	Abbreviation
exclusion chromatography	EC
gas chromatography	GC
gas-liquid chromatography	GLC
gas-liquid partition chromatography	GLPC
gel-permeation chromatography	GPC
gas-solid chromatography	GSC
height equivalent to 1 theoretical plate	HETP
high-performance liquid chromatography	HPLC
ion-exchange chromatography	IEC
liquid chromatography	LC
liquid-liquid chromatography	LLC
liquid-solid chromatography	LSC
paper chromatography [or] planar chromatography	PC
porous-layer open-tubular (column)	PLOT
programmed-temperature vaporizer	PTV
relative retention time	RRT
support-coated open-tubular (column)	SCOT
supercritical-fluid chromatography	SFC
separation number	SN
thin-layer chromatography	TLC
wall-coated open-tubular (column)	WCOT

[a]Such abbreviations should be parenthetically defined at 1st use in text, defined in footnotes when they are used in tables, and defined in legends when they are used in figures.

Table 18·2 Abbreviations in electron, photoelectron, and related spectroscopic methods

Term	Abbreviation
Auger electron spectroscopy	AES
electron energy loss spectroscopy	EELS
high-resolution energy-loss electron spectroscopy	HREELS
vibrational energy-loss electron spectroscopy	VEELS
inelastic electron tunnelling spectroscopy	IETS
photo(n)electron spectroscopy	PS, PES
ultraviolet photoelectron spectroscopy	UPS, UPES
X-ray photoelectron spectroscopy	XPS, XPES
angle-resolved photoelectron spectroscopy	ARPS, ARPES

The IUPAC 1991c document suggests a sequence of letters that relate to descriptive adjectives or phrases, types of probes or particles involved, and the general type of technique, with the recommendation that the patterns suggested be followed when feasible. Many abbreviations have, however, been modified to form acronyms.

Table 18·3 Abbreviations for photon spectroscopy based on the use of electromagnetic radiation

Term	Abbreviation
electron paramagnetic resonance	EPR
electron spin resonance	ESR
far-infrared spectroscopy	FIR
microwave spectroscopy	MW
mid-infrared spectroscopy	MIR
near-infrared spectroscopy	NIR
nuclear magnetic resonance	NMR
nuclear quadrupole resonance	NQR
Raman spectroscopy	RS
resonance Raman spectroscopy	RRS
surface-enhanced Raman spectroscopy	SERS
ultraviolet spectroscopy	UV
vacuum ultraviolet spectroscopy	VUV
visible spectroscopy	VIS

COSY [for "correlation spectroscopy"]

NOESY [for "nuclear Overhauser enhancement spectroscopy"]

MISCELLANEOUS ANALYTICAL METHODS

18·5 Terms encountered in and definitions of other analytical methods are given in IUPAC 1985b, 1987a, 1991a, 1991f. Rules for naming derived quantities are described in IUPAC 1991a; the following terms are examples of its definitions.

[terms derived from denominator quantity]

areic ["divided by area"]
entitic ["divided by number of entities"]

lineic ["divided by length"]
massic ["divided by mass of the system"]

molar ["divided by amount of substance"]
rate ["divided by time"]

volumic ["divided by volume of the system"]

Definitions, classifications, conceptual aspects, and proposals for the determination of carry-over effects in clinical chemistry are presented in IUPAC 1991a.

Activities, activity coefficients, and pH are rigorously defined in IUPAC–IFCC 1984.

REFERENCES

Cited References

[IUPAC] International Union of Pure and Applied Chemistry, Analytical Chemistry Division, Commission on Spectrochemical and Other Optical Procedures for Analysis. 1985a. Nomenclature, symbols, units and their usage in spectrochemical analysis: V. Radiation sources: recommendations 1985. Pure Appl Chem 57:1453–90.

[IUPAC] International Union of Pure and Applied Chemistry, Analytical Chemistry Division, Commission on Electroanalytical Chemistry. 1985b. Recommended terms, symbols, and definitions for electroanalytical chemistry: recommendations 1985. Pure Appl Chem 57:1491–505.

[IUPAC] International Union of Pure and Applied Chemistry, Analytical Chemistry Division. 1987a. Compendium of analytical nomenclature: definitive rules 1987. 2nd ed. Oxford (UK): Blackwell Scientific.

[IUPAC] International Union of Pure and Applied Chemistry, Physical Chemistry Division, Commission on Molecular Structure and Spectroscopy. 1987b. A descriptive classification of the electron spectroscopies: recommendations 1987. Pure Appl Chem 59:1343–406.

[IUPAC] International Union of Pure and Applied Chemistry, Organic Chemistry Division, Commission on Photochemistry. 1988a. Glossary of terms used in photochemistry: recommendations 1988. Pure Appl Chem 60:1055–106.

[IUPAC] International Union of Pure and Applied Chemistry, Physical Chemistry Division, Commission on Molecular Structure and Spectroscopy. 1988b. Presentation of molecular parameter values for infrared and Raman intensity measurements: recommendations 1988. Pure Appl Chem 60:1385–8.

[IUPAC] International Union of Pure and Applied Chemistry, Analytical Chemistry Division, Commission on Spectrochemical and Other Optical Procedures for Analysis. 1988c. Nomenclature, symbols, units and their usage in spectrochemical analysis: VII Molecular absorption spectroscopy, ultraviolet and visible (UV/VIS): recommendations 1988. Pure Appl Chem 60:1449–60.

[IUPAC] International Union of Pure and Applied Chemistry, Analytical Chemistry Division, Commission on Spectrochemical and Other Optical Procedures for Analysis. 1988d. Nomenclature, symbols, units and their usage in spectrochemical analysis: X Preparation of materials for analytical atomic spectroscopy and other related techniques: recommendations 1988. Pure Appl Chem 60:1461–72.

[IUPAC] International Union of Pure and Applied Chemistry, Physical Chemistry Division, Commission on Molecular Structure and Spectroscopy. 1989. Recommendations for EPR/ESR nomenclature and conventions for presenting experimental data in publications: recommendations 1989. Pure Appl Chem 61:2195–200.

[IUPAC] International Union of Pure and Applied Chemistry, Clinical Chemistry Division, Commission on Automation and Clinical Chemical Techniques, and Analytical Chemistry Division, Commission on Analytical Nomenclature; International Federation of Clinical Chemistry, Subcommittee on Analytical Systems. 1991a. Proposals for the description and measurement of carry-over effects in clinical chemistry: recommendations 1991. Pure Appl Chem 63:301–6.

[IUPAC] International Union of Pure and Applied Chemistry, Analytical Chemistry Division, Commission on Spectrochemical and Other Optical Procedures for Analysis. 1991b. Nomenclature, symbols, units and their usage in spectrochemical analysis: VIII Nomenclature system for X-ray spectroscopy: recommendations 1991. Pure Appl Chem. 63:735–46.

[IUPAC] International Union of Pure and Applied Chemistry, Physical Chemistry Division, Commission on Molecular Structure and Spectroscopy. 1991c. English-derived abbreviations for experimental techniques in surface science and chemical spectroscopy: recommendations 1991. Pure Appl Chem 63:887–93.

[IUPAC] International Union of Pure and Applied Chemistry, Clinical Chemistry Division, Commission on Quantities and Units in Clinical Chemistry and International Federation of Clinical Chemistry, Scientific Division, Committee on Quantities and Units. 1991d. Nomenclature of derived quantities: recommendations 1991. Pure Appl Chem 63:1307–11.

[IUPAC] International Union of Pure and Applied Chemistry, Physical Chemistry Division, Commission on Molecular Structure and Spectroscopy, Subcommittee on Mass Spectroscopy. 1991e. Recommendations for nomenclature and symbolism for mass spectroscopy (including an appendix of terms used in vacuum technology): recommendations 1991. Pure Appl Chem 63:1541–66.

[IUPAC] International Union of Pure and Applied Chemistry, Commission on Analytical Radiochemistry and Nuclear Materials. 1991f. Glossary of terms used in nuclear analytical chemistry (provisional).

Available from the IUPAC Secretariat, Bank Court Chambers, 2–3 Pound Way, Templars Square, Cowley, Oxford OX4 3YF, UK.

[IUPAC] International Union of Pure and Applied Chemistry, Analytical Chemistry Division, Commission on Analytical Nomenclature. 1993. Nomenclature for chromatography: recommendations 1993. Pure Appl Chem 65:819–72.

[IUPAC–IFCC] International Union of Pure and Applied Chemistry and International Federation of Clinical Chemistry, Clinical Chemistry Division, Commission on Quantities and Units in Clinical Chemistry; International Federation of Clinical Chemistry, Scientific Committee, Analytical Section, Expert Panel on pH and Blood Gases. 1984. Physicochemical quantities and units in clinical chemistry with special emphasis on activities and activity coefficients: recommendations 1983. Pure Appl Chem 56:567–94.

Additional References

Marczenko Z, Newman E, Thorburn-Burns DT, Townshend A. 1993. Dictionary of analytical reagents. New York: Chapman & Hall.

Sheppard N, Willis HA, Rigg JC. 1987. Approved recommendation 1985 on names, symbols, definitions, and units of quantities in optical spectroscopy. J Clin Chem Clin Biochem 25:327–36.

Ulicky L, Kemp T, editors. 1993. Comprehensive dictionary of physical chemistry. Englewood Cliffs (NJ): Prentice Hall.

Wiggins G. 1991. Chemical information sources. New York: McGraw-Hill.

19 Drugs and Pharmacokinetics

> It is the object of a Pharmacopoeia to select from among substances which possess medicinal power, those the utility of which is most fully established and best understood. . . . It should . . . distinguish those articles by convenient and definite names, such as may prevent trouble or uncertainty in the intercourse of physicians and apothecaries.
>
> —Jacob Bigelow, *United States Pharmacopoeia,* 1820

Many details in nomenclature and notations for chemistry and chemical kinetics are relevant to some aspects of drug nomenclature and pharmacokinetics, but some needs special to pharmacology and related fields justify separate treatment.

DRUG NOMENCLATURE

19·1 Drugs are identified by various designations used alone or in combination in different contexts. A detailed summary of the various kinds of designations and their bases can be found in the preface to the dictionary issued annually by the United States Pharmacopeial Convention (USPC [annual]); see section 19·3, "United States Adopted Names (USAN)", below.

NONPROPRIETARY NAMES (GENERIC NAMES)

19·2 Most drugs have complex molecular structures. Hence the systematic chemical names of most of them are also complex. In literature not concerned specifically with chemical characteristics of drugs and in clinical medicine, shorter, more convenient names are used. This practice has given rise to national and international mechanisms for coining names that carry no implication of commercial ownership but make accurate comprehension of the pharmacologic and medical literature universally possible.

United States Adopted Names (USAN) and Other National Names

19·3 Nonproprietary (generic) names for drugs in the United States generally are proposed to the United States Adopted Names Council (USANC) by a company that has developed a therapeutic agent. Selected names and previously adopted names are published annually in *USAN and the USP Dictionary of Drug Names* (USPC [annual]) issued by the United States Pharmacopeial Convention (USPC) under the auspices of the USPC Drug Nomenclature Committee. Proposed names are expected to conform to the criteria set forth in "Guiding Principles for Coining U. S. Adopted Names for Drugs", published as an appendix to the *USAN and the USP Dictionary.* The "Principles" document gives rules for coining names and provides lists of approved stem terms and their meanings and of terms representing contractions for radicals (chemical names) and adducts. This document also provides guidelines for naming interferons, interleukins, somatotropins, colony-stimulating factors, erythropoietins, and monoclonal antibodies.

Most nonproprietary names are 1-word names, but those representing chelates, complexes, esters, or salts are 2-word terms.

rolitetracycline   rolitetracycline nitrate
salicylate meglumine   spiradoline mesylate

Note that radiopharmaceuticals are multiword terms; see section 19·11.

The standard source for US Adopted Names is the *USAN and the USP Dictionary* cited above. The *Dictionary* also carries information relevant to other designations for drugs; see Table 19·1 for a list, with examples, of the types of information. Note that in its entry terms the *Dictionary* capitalizes the initial letters of nonproprietary names, which should not be capitalized except as specified by the rules for capitalizing initial words in sentences and nouns in titles; see sections 8·1, 8·7, 8·25, and 19·8. New US Adopted Names are published between the annual issues of *USAN and the USP Dictionary* in issues of *USP DI Update,* a

Table 19·1 Kinds of information in entries of the *USAN and the USP Dictionary of Drug Names*[a], with examples[b]

United States Adopted Name (USAN)	mirincamycin hydrochloride
Year of publication of the USAN	[1963]
Status in the *United States Pharmacopeia* (USP)	**USP** (official in the USP)
Pronunciation guide	(sul fa dye′ a zeen)
Alternative names in other systems of nonproprietary names, such as the International Nonproprietary Names (INN) of the World Health Organization	[paracetamol is INN and BAN[c]]
Miscellaneous other names[d]	nonanoic acid
Molecular formula and weight	$C_{17}H_{31}NO$
Chemical name or names	ethanesulfonic acid, 2-mercapto-, monosodium salt
Chemical Abstracts Service Registry Number	*CAS-7660-71-1*
Pharmacologic and therapeutic activity	*Vasodilator; relaxant (smooth muscle)*
Brand name or names, manufacturer or manufacturers	Lidanar (Sandoz)
Code designation	*NSC-526280*
Indication that the USAN is also official in other systems	INN; BAN; DCF; MI[e]
Graphic formula	[Graphic molecular structure when known]

[a](USPC [annual]).

[b]Note that the examples are from different entries, not a single entry; the categories of information are listed in the sequence in which they appear in entries. Note that the *USAN and the USP Dictionary of Drug Names* capitalizes chemical and nonproprietary drug names that should not be capitalized except in titles and other settings that require capitalization of common nouns; see section 19·3. Such names are presented in this table properly uncapitalized rather than in the capitalized forms in the source.

[c]BAN = British Approved Name.

[d]As in the inverted *Chemical Abstracts* index name; see section 19·9.

[e]These abbreviations stand for, respectively, "International Nonproprietary Name", "British Approved Name", "Dénomination Commune Française", and *The Merck Index.*

periodical also published by the USPC; they are also published in a "New Names List" in *Clinical Pharmacology and Therapeutics.* Another valuable source of chemical information on drugs is *The Merck Index* (Budavari and others 1989).

Many other countries have similar means for establishing nonproprietary names, such as those of the British Pharmacopeia Convention and the French Codex Commission. These names are identified in a number of authoritative sources, for example, the *USAN and the USP Dictionary.* The Canadian authority for nonproprietary and brand names is *Compen-*

dium of Pharmaceuticals and Specialties (CPA [annual]). For the names established in European countries, comprehensive sources include *Martindale: The Extra Pharmacopoeia* (Reynolds 1993) and *Index Nominum* (SPS 1992).

International Nonproprietary Names (INN)

19·4 A committee of the World Health Organization (WHO) serves a function akin to that of the United States Adopted Names Council and establishes nonproprietary names. Newly accepted names are published from time to time in *WHO Drug Information* (WHO 1987), and they are identified in various authoritative sources such as the *USAN and the USP Dictionary.* Cumulative lists of international nonproprietary names (INNs) are published at about 5-year intervals; "Cumulative List 8" was published in 1992 (WHO 1992). In countries without their own national authority for nonproprietary names, those established by WHO should be used as properly established names.

Pharmacy Equivalent Names (PENs)

19·5 For convenient and brief designations of dosage forms containing 2 or more therapeutic substances, the US Pharmacopeial Convention (USPC 1991) recommends the use of pharmacy equivalent names (PENs) derived from stems in the separate drug names and with the prefix "co-", which indicates a combination dosage form. Specific short prefix and suffix units are suggested for the formation of the stem terms (those following "co-").

> co-trimoxazole [formed from "trimethoprim" and "sulfamethoxazole" with the prefix "co-"]
>
> co-codAPAP [representing "codeine phosphate" and "acetaminophen"]

The additional intent in recommending these names is to try to prevent the proliferation of trivial names and ad hoc abbreviations for these dosage forms. These names are not, however, official names in the *United States Pharmacopeia* and *The National Formulary* although they are entries in the *USAN and the USP Dictionary.* This system originated with the British Pharmacopeia Commission in 1988.

Abbreviations for Multidrug Regimens

19·6 Some drug treatment regimens may use 2 or more single drugs in a complex regimen. These have frequently been identified with abbreviations based on, usually, initial letters of the individual drug names (some of which may have been superseded by a later name).

MOPP [for]	**m**echlorethamine, **O**ncovin [vincristine sulfate], **p**rednisone, **p**rocarbazine hydrochloride
PVBMF [for]	cis**p**latin, **v**incristine, **b**leomycin, **m**ethotrexate, 5-**f**luorouracil

If the editorial policy is to allow such abbreviations without explanation in the title of an article, the explanatory elements should appear at least in its abstract; the explanation must be given at the 1st use of the abbreviation in the text.

Endocrinologic and Metabolic Drugs

19·7 Some drugs represent hormonal entities that in contexts other than diagnosis or treatment carry names preferred in endocrinology and nutrition.

seractide acetate = α^{1-39}corticotropin (human)

calcitriol = 1α,25-dihydroxycholecalciferol

PROPRIETARY NAMES (TRADE NAMES, TRADEMARKS)

19·8 Proprietary names are those established by manufacturers and vendors of drugs to represent their own products. Because such names are proper nouns and identify only specific products, they must be capitalized (see sections 8·17 and 8·25). Firms that have registered their proprietary names with the US Patent and Trademark Office may prefer to have those names include a suffix superscript ®. Proprietary names can be verified in sources such as the *USAN and the USP Dictionary* (USPC [annual]), *Compendium of Pharmaceuticals and Specialties* (CPA [annual]), the *American Drug Index* (Billups and Billups [annual]), and other sources such as those cited in section 19·3.

CHEMICAL NAMES

19·9 Most drugs are complex chemical compounds. The nonproprietary names cannot indicate the details of chemical structure; these details are properly indicated only by systematic chemical names developed by the principles set forth in various documents of the International Union of Pure and Applied Chemistry (IUPAC) and summarized in Chapter 16. The *USAN and the USP Dictionary* provides 2 chemical names for each nonproprietary name: the inverted *Chemical Abstracts* index name and an uninverted IUPAC-preferred chemical name sanctioned by WHO.

> [inverted form]
> hydrazinecarboximidamide, 2-[2-(2,6-dichlorophenoxy)ethyl]-, sulfate, (2:1)
> [uninverted form]
> 2-[2-(2,6-dichlorophenoxy)ethyl]hydrazinecarboximidamide sulfate (2:1)

The index (inverted) forms can readily be converted to the uninverted forms.

CHEMICAL FORMULAS

19·10 Chemical molecular formulas for drugs indicate the relative proportions of elements making up compounds and give little or no indication of chemical structure.

> [nonproprietary name] clemastine
> [chemical formula] $C_{21}H_{26}ClNO$

Structural formulas are illustrated in the 2 major reference sources on drugs, *The Merck Index* (Budavari 1989) and the *USAN and the USP Dictionary* (USPC [annual]).

RADIOPHARMACEUTICALS

19·11 The nonproprietary names of pharmaceuticals carrying radioisotopes do not follow the conventions applied elsewhere (see sections 16·31 and 16·32) but place the symbol for the radioactive isotope after the carrier name and then the number for its atomic weight on the line.

> iodopyracet I 125 albumin, chromated Cr 51 serum
> technetium Tc 99m pentetate calcium

CHEMICAL ABSTRACTS REGISTRY NUMBERS

19·12 The rapid handling of data in, for example, databases, may be facilitated by numbering systems for identifying complex chemical entities, including drugs. The most prominent of such systems is that of Chemical Abstracts Service (CAS). CAS registry numbers are randomly assigned to compounds and are unique for each compound. In the online Registry File of Chemical Abstracts Service (CAS REGISTRY File) available on STN International^{SM}, all the names for a single substance are brought together and may be displayed under its unique CAS Registry Number with an indication of the related bibliographic references to the chemical literature. Some entries in the *USAN and the USP Dictionary* carry more than 1 CAS registry number. For example, an anhydrous compound will have 1 number and its hydrate another.

theophylline monohydrate CAS 5967-84-0
theophylline [anhydrous] CAS 58-55-9

STN International^{SM} is the online Scientific and Technical Information Network operated in North America by Chemical Abstracts Service in Columbus, Ohio, a division of the American Chemical Society, with headquarters in Washington, DC.

CODE DESIGNATIONS

19·13 In developmental stages in the study and testing of potential new drugs, compounds are often assigned alphanumeric designations by the organization carrying out the work. The alphabetic component is usually derived from the organization's name.

NSC 26980
[the code designation for mitomycin assigned by the National Service Center, National Cancer Institute, National Institutes of Health]

USE OF NONPROPRIETARY (GENERIC) OR TRADE NAMES

19·14 Decisions on when to use nonproprietary names or trade names depend in part on the audience for the publication and on the context in which drug names are used. Because of pervasive drug advertising in medical journals, many physicians are more accustomed to trade names, especially those for relatively new drugs still available only from the originating manufacturer, than to nonproprietary names. There are, however, 2 important reasons for generally preferring nonproprietary names: they are more likely to represent the chemical characteristic of the drug, and trade names often differ greatly in different countries while nonproprietary names are more likely to be identical. Additionally, most indexing services index drug-related publications by nonproprietary names and not by trade names.

Note that when nonproprietary names are used, the specific drug used in the reported research should be identified, at least in the methods section of the paper, by trade name and manufacturer.

[Title] Sucralfate Compared with Placebo in the Treatment of Peptic Ulcer

[Methods section] The treatment group was prescribed sucralfate (Carafate, Marion Merrill Dow) at a dose of . . .

If the reported research involved comparing 2 drugs with the same nonproprietary name for some characteristic, for example, absorbability, the trade name of each should be used in the paper's title, abstract, and methods section along with the nonproprietary name.

DRUG AND OTHER RECEPTORS

19·15 Symbols for receptors of drugs and other humoral mediators have not been systematically standardized. Those in wide use do, however, represent a general pattern of alphanumeric symbols built of roman or greek alphabet characters and additional alphanumeric or numeric designators, which are usually subscript.

[alpha-adrenoreceptors] α_{1A} α_{1B} α_{1C} α_{1D} α_{2A} α_{2B} α_{2C}

[adenosine receptors] A_1 A_{2b} [bradykinin receptors] B_1 B_2

[cholecystokinin and gastrin receptors] CCK_A CCK_B

[dopamine receptors] D_1 D_2 D_4 D_5

[purine nucleotide receptors] P_{2x} P_{2u} P_{2t}

Additional symbols. along with previous symbolizations and other related information, can be found in the annually published *TiPS Receptor Nomenclature Supplement* of *Trends in Pharmacological Sciences* (Watson and Girdlestone 1993); also see Abbracchio and others (1993). Note that some of these symbols might be confused with other symbols (such as those for some vitamins), but they are unlikely to appear in potentially confusing contexts.

UNITS FOR DRUG CONCENTRATIONS IN BIOLOGICAL FLUIDS

19·16 Drug concentrations in biological fluids have generally been measured and reported in gravimetric units (such as milligrams/liter [mg/L]; nanograms/milliliter [ng/mL]). The Scientific Committee of the Association of Clinical Biochemists (United Kingdom) (Ratcliffe and Worth 1986) has recommended using SI units (see section 11·8), with the liter being the reference volume and millimole (mmol) or nanomole (nmol) the unit for amount of drug or metabolite measured; the resulting units would be mmol/L and nmol/L. The use of molar rather than gravimetric units has strong scientific rationales for it, but this change has not been widely adopted in the United States, largely because dosage units are still gravimetric.

PHARMACOKINETICS

19·17 Symbols for variables measured in pharmacokinetic studies are various combinations of capital and lowercase roman and greek letters (ACCP 1982, ASM 1985). Qualifiers specifically indicating sites of measurement, organs, and elimination routes, and routes of administration are subscripts; other modifiers for general conditions, such as "ss" for "steady-state", may be superscripts.

> Dosing rate = $\mathrm{CL}\cdot C^{\mathrm{ss}}$
> ["CL" stands for "clearance"; "C^{ss}" stands for "steady-state drug concentration"]

Note that single-letter symbols for variables (such as "*C*" for "concentration") should be italicized in accordance with the international style for symbols of quantities and for chemical kinetics; see sections 3·5, 9·3, and 17·4 and Table 3·5. Double-letter and triple-letter symbols are, however, in roman (upright) type rather than italic type, to remind readers that they represent single variables and not multiplied single-letter variables.

> CL = dose/AUC
> [where "CL" = "clearance" and "AUC" = "area under the curve"]

A useful convention for distinguishing multiletter variables from single-letter variables could be enclosure of the roman-letter characters within parentheses (Huth 1987), but this convention has not been generally accepted. See Table 19·2 for examples of symbols of variables and qualifiers.

Table 19·2 Examples of pharmacokinetic symbols and qualifiers[a]

Symbol	Term symbolized
Measured or calculated variables (italics if single letter; roman if 2- or 3-letter symbol)	
A^{SS}	amount of drug in the body at steady-state
$AUC^{0 \rightarrow \infty}$	area under plasma concentration-time curve from zero to infinity
C	drug concentration in plasma at any time t
C^{SS}	steady-state concentration of drug in plasma during infusion at a constant rate
CL_{NR}	nonrenal clearance of drug from plasma
CL_{CR}	creatinine clearance
E_H	hepatic extraction ratio
f_a	fraction of administered dose absorbed
k_m	rate constant (1st-order) for formation of metabolite
Q_R	renal blood flow; use qualifier for plasma flow
R_0	constant infusion rate (zero-order)
t_{max}	time to reach maximum or peak concentration after drug administration
$t_{1/2}$	elimination half-life associated with terminal slope of a semilogarithmic concentration-time curve
$t_{1/2a}$	absorption half-time
V_c	pharmacokinetic volume of central or plasma compartment
Specific modifiers representing sites of measurement (roman type, subscripted)	
b	blood
p	plasma
sal	saliva
t	tissue
u	unbound species
ur	urine
Specific modifiers representing organs or elimination routes (roman type, subscripted)	
e	excreted into urine
H	hepatic
m	metabolized
NR	nonrenal
R	renal
Specific modifiers representing routes of administration (roman type, subscripted)	
im	intramuscular
ip	intraperitoneal
iv	intravenous
oral	oral
po	peroral
pr	rectal
sc	subcutaneous
sl	sublingual
top	topical

[a]Based on *Manual of Symbols, Equations & Definitions in Pharmacokinetics* issued by the Committee for Pharmacokinetic Nomenclature, American College of Clinical Pharmacology (ACCP 1982). This source also indicates applicable units, previously used symbols, and recommended pharmacokinetic equations. Note that the international convention of italicizing single-letter symbols of variables has been applied in this table.

REFERENCES

Cited References

Abbracchio MP, Cattabeni F, Fredholm BB, Williams M. 1993. Purinoceptor nomenclature: a status report. Drug Dev Res 28:207–13.

[ACCP] American College of Clinical Pharmacology, Committee for Pharmacokinetic Nomenclature. 1982. Manual of symbols, equations & definitions in pharmacokinetics. J Clin Pharmacol 22:1S–23S.

[ASM] American Society for Microbiology. 1985. ASM style manual for journals and books. Washington: ASM.

Billups NF, Billups SM. [annual]. American drug index. St Louis (MO): Facts and Comparisons.

Budavari S, O'Neil MJ, Smith A, Heckelman PE. 1989. The Merck index: an encyclopedia of chemicals, drugs, and biologicals. 11th ed. Rahway (NJ): Merck.

[CPA] Canadian Pharmaceutical Association. [annual]. Compendium of pharmaceuticals and specialties. Ottawa: CPA.

Huth EJ. 1987. Medical style & format: an international manual for authors, editors, and publishers. Philadelphia: ISI Pr. Available from Williams & Wilkins, Baltimore, MD.

Ratcliffe JG, Worth HGJ. 1986. Recommended units for reporting drug concentrations in biological fluids. Lancet Jan 25;1(8474):202–3.

Reynolds JEF. 1993. Martindale: the extra pharmacopoeia. 30th ed. London: Pharmaceutical Pr.

[SPS] Swiss Pharmaceutical Society. 1992. Index nominum: international drug directory 1992–1993. 15th ed. Stuttgart: Medpharm Scientific Publications.

[USPC] United States Pharmacopeial Convention. [annual]. USAN and the USP dictionary of drug names. Rockville (MD): USPC. Also available as a database (USAN) on STN International^{SM}, the online Scientific & Technical Information Network operated in North America by Chemical Abstracts Service, a Division of the American Chemical Society.

[USPC] United States Pharmacopeial Convention, Drug Nomenclature Committee. 1991. Nomenclature policies and recommendations: II. pharmacy equivalent names for frequently prescribed combination drug dosage forms. Pharmacopeial Forum 17:1984–6.

Watson S, Girdlestone D. 1993. TiPS receptor nomenclature supplement 1993. Trends Pharmacol Sci 11(1 suppl):1–43.

[WHO] World Health Organization. 1987. WHO drug information. Geneva: WHO.

[WHO] World Health Organization. 1992. International nonproprietary names (INN) for pharmaceutical substances; Lists 1–65 of proposed INN and lists 1–31 of recommended INN; Cumulative list number 8. Geneva: WHO.

Additional Reference

Elks J, Ganellin CR, editors. 1990. Dictionary of drugs: chemical data, structures and bibliographies. London: Chapman & Hall.

20 Cells, Chromosomes, and Genes

> It has not escaped our attention that the specific pairing [of bases in the double helical structure] we have postulated immediately suggests a possible copying mechanism for the genetic material.
>
> —J D Watson and F H C Crick, Molecular Structure of Nucleic Acids: A Structure for Deoxyribose Nucleic Acid, *Nature,* 1953

The fields of cellular and molecular biology have grown rapidly in this century. The especially spectacular growth in our knowledge of molecular-level genetics in the past several decades has led to a parallel growth in the number of conventions for symbolizing chromosomes and their structure and components. The conventions presented in this chapter are representative and do not include all conventions for all organisms that have been studied with genetic and biochemical methods. Many conventions for different organisms fortunately represent only minor differences in symbolization and notation.

The initial formal system of genetic nomenclature and symbolization appears to have been the rules (Dunn and others 1940) on symbols for genes of the laboratory mouse. A later set of recommendations (CGSN 1957) for wider application influenced systems developed for a large number of organisms. Developments in symbolization and nomenclature for the structure and function of cells and their components will continue, especially in genetics; these should be called to the attention of the Council of Biology Editors committee responsible for this manual to aid it in better representing publication style for these topics; see section 1·5 for further information.

CELL LINES

20·1 When a cell line is isolated and 1st described, basic information on it should be provided in the report.

1 Procedures for isolation.
2 The species, strain (if known), age, and sex of the donor animal.
3 The organ or tissue of origin.
4 Normality of the tissue source: normal, neoplastic (benign or malignant).
5 Cell type.
6 Whether cloned; number of times cloned.
7 Specific properties or markers.
8 Tests for sterility.
9 If appropriate, number of population doublings accrued since isolation.

Cell cultures should be obtained, if possible, from recognized cell banks or the originator of a culture. The American Type Culture Collection (ATCC), 12301 Parklawn Drive, Rockville, Maryland 20852, USA, and the Coriell Institute for Medical Research (CIMR), 401 Haddon Avenue, Camden, New Jersey 08103, USA, provide many cell lines meeting desirable standards. Both ATCC and CIMR distribute catalogues. Detailed information on the ATCC cell lines and hybridomas is also available through the Microbial Strain Data Netwrok (MSDN). The CODATA Hybridoma Data Bank (HDB) is available through the MSDN, the National Institutes of Health (US), Institute for Physical and Chemical Research (RIKEN, Japan), and the Canada Institute for Scientific and Technical Information (CISTI).

Reports of research using established cell lines should identify the cell-line bank or the name of the originator (if the source) and indicate the number of passages since the last complete characterization of the line. The description should include the species and tissue of origin, the specific designation of the cell line, and the bank's designation.

rabbit kidney cells, LLC-RK_1, ATCC CCL 106
antihuman T-lymphocyte, T3–3A1, ATCC HB 2

There is no universally accepted system of notation for cell lines. In general, the designations identify in abbreviated form the laboratory of origin and include a numeric further identifier; some represent the species and tissue or person of origin. The Tissue Culture Association has recommended that the laboratory of origin be identified by the 1st element and with no more than 4 letters and that these be followed by an arabic-numeral designation indicating the sequence of isolation; in some designations the sequence identifier is the month, year, and sequence number.

NCTC 2071 IMR 381 NBL 6 NIH 10/81/33

The entire designation should be unique and unambiguous.

The ATCC catalogue (ATCC 1992) and catalogues from similar banks are reliable sources in which to verify notations for cell lines and hybridomas.

CELLULAR FUNCTIONS

For recommendations on reporting studies on cellular chemical kinetics and biothermodynamics and for the appropriate systems of notation, see Chapter 17.

CHROMOSOMES AND CHROMOSOMAL STRUCTURES

For some organisms, detailed systems for the symbolic representation of normal and abnormal chromosomes are available. The systems for other organisms follow, in general, the detailed systems represented in sections 20·2–20·18

Symbols for genes are summarized in sections 20·23–20·48.

NEMATODE (*Caenorhabditis elegans*)

20·2 Symbols representing chromosomal aberrations in the nematode *Caenorhabditis elegans* (Horvitz and others 1979) are similar to those for the fruit fly (see the following section, 20·3). The components of symbols are, in order, the laboratory mutation-name prefix (lowercase letters), the abbreviation for the aberration (as for the fruit fly), an arabic numeric designation, optionally the roman numeral (or numerals) for the affected linkage group (or groups)("*f*" for "free chromosome" or "fragmented chromosome"), and, finally, and optionally, the symbol for the affected gene or genes; all characters are italicized.

mnDp2 *sDf3(I)dpy-5* *mnDpI(X;;V)* *eDp6(III;f)*

Phenotypic characteristics should, in general, be described in words, but a nonitalicized abbreviation corresponding to a gene symbol can be used; the initial letter of the abbreviation should be capitalized. Abbreviations not corresponding to gene symbols must be explained at 1st use.

Dpy animals segregated both Dpy Unc and Dpy Tra progeny.

For further details, see Horvitz and others (1979). For gene symbols, see section 20·27.

FRUIT FLY (*Drosophila melanogaster*)

20·3 The chromosomes of *Drosophila melanogaster* are represented by their numbers and sex-chromosome letters: 1 [X], 2, 3, 4, Y. The symbols representing chromosomal aberrations and the rules for their use are summarized in *The Genome of* Drosophila melanogaster (Lindsley and Zimm 1992). This source includes a catalogue of symbols for chromosomal aberrations and their names, the inducing agents, alternative symbols or names or both, and additional details.

For gene symbolization, see section 20·28.

Chromosomal aberrations are classified as 6 types and assigned italicized symbols representing the descriptive terms.

Df [deficiencies, deletions]	*Dp* [duplications]	*In* [inversions]
R [rings]	*T* [translocations]	*Tp* [transpositions]

Aberration symbols do not have spaces or subscripts; note that superscripts are also italicized.

Deficiencies (Deletions)

The symbol *Df* is used for deficient chromosomes and for deficient genotypes. It is followed by a parenthetic indicator of the involved chromosomal arm and then by the symbol for the specific deficiency (an abbreviation of its unique name). Superscripts are not used with the specific designator as with allelic symbolization.

Df(2R)vg-B [not "$Df(2R)vg^{B}$"] *Df(3L)st3* [not "$Df(3L)st^{3}$"]

When a deficiency is not named after the deleted gene, the superscript is retained.

$Df(1)sc^{8}$ [not "*Df(1)sc8*"]

Duplications

A symbol for a duplication includes the following elements: *Dp* followed, within parentheses, by the symbol for the chromosome of origin of the duplicated segment, a semicolon, the symbol for the recipient chromosome, and then the specific designation. If the duplicated segment is carried as a free centric element, the symbol *f* (for *free*) follows the semicolon.

Dp(3;1)O5 $Dp(1;1)y^{bl}$ *Dp*(1;f)101

Inversions

A symbol for an inversion includes the following elements: *In;* within parentheses, the involved chromosome arm or arms; the specific designation. Superscripts *L* and *R* can identify the sources of the 2 ends of recombinations between similar inversions.

In(2L)Cy $In(2LR)bw^{V1}$ $In(2R)Cy^{L}bw^{VDe1R}$

Rings

The symbol *R* is followed by the symbols for the involved chromosome, within parentheses, and the specific designation.

R(1)1 [ring formed by rejoining of breaks in opposite arms of chromosome 1]

Translocations

The symbol *T* is followed within parentheses by the symbols of the participating chromosomes separated by semicolons (listed in the order 1 [X], Y, 2, 3, 4), and then the specific designation.

T(1;Y;3)127 $T(1;4)B^{S}$

Transpositions

In the parenthetic representation of the involved chromosomes, the symbol for the donor chromosome precedes the symbol for the recipient.

Tp(2;3)P   *Tp(3;1)ry*^{35}

More Complex Representations

When an aberration involves more than 1 type, the sequence for their representation is in this ranking: *T* >interchromosomal *Tp*>*R*>*In*>intrachromosomal *Tp*>*Dp*>Df. Combinations of aberrations may be shown individually and connected with a plus symbol.

In(1)sc^{8} + *In(1)dl-49*

Additional symbols representing gene content and sequence can follow after a comma with spacing.

In(1)dl-49, y w B   [for an X chromosome with the delta-49 inversion, the recessive markers *yellow* and *white* and the dominant marker *Bar*]

For recommendations on still more complex representations, see Lindsley and Zimm (1992) and *Drosophila: A Laboratory Handbook* (Ashburner 1989).

Centromeres

In symbols for chromosomes, the position of the centromere is indicated by a raised period, for example, *X·YL.*

MOUSE (*Mus* SPECIES)

20·4 The linkage group designations by roman numerals formerly used for mouse chromosomes have been replaced by arabic numerals and X and Y (CSGNM 1972).

Chromosome bands are designated by letters following chromosome numbers (Nesbitt and Francke 1973). Divisions of bands (subdivisions) are represented by numbers; further divisions of subdivisions are also numbered but following a period. A variant of a band may be designated by a superscript letter. Note that in referring to a specific chromosome, the "c" in "chromosome" is capitalized.

17B 17B1 17B2 17B1.1 17B1.2

17A2[s] [small A2 band in Chromosome 17]

As with symbolization for the fruit fly (see section 20·3), symbols for chromosome anomalies in the mouse (CSGNM 1989) begin with letter symbols for the types of anomalies.

Df [deficiency] Del [deletion] Dp [duplication]

Hc [pericentric heterochromatin]

HSR [homogeneous staining region]

Is [insertion] In [inversion]

Ms [monosomy] Rb [Robertsonian translocation]

T [translocation] Tp [transposition] Ts [trisomy]

These symbols are not italicized. "Deficiency" should be restricted to defining the unbalanced products of chromosome aberrations (such as malsegregation of reciprocal translocations); "deletion" refers to interstitial losses often, but not always, cytologically visible.

The symbols have this general pattern: symbol for the type of anomaly; within parentheses, a symbol (or symbols) for the chromosome (chromosomes) involved in the anomaly; followed by a series symbol made up of a number and a 2- or 3-letter abbreviation representing the laboratory or person reporting the anomaly. To avoid duplication, these laboratory registration codes must be registered with the central registry maintained by the Institute for Laboratory Animal Resources (ILAR), US National Academy of Sciences. The abbreviation may already have been assigned for inbred substrains or sublines or be coined because an abbreviation is not available in the standard list of abbreviations. The symbols are run together unspaced.

In(2)5Rk [inversion in Chromosome 2]

A semicolon separates the 2 chromosomes involved in a translocation or insertion; a period (point) represents the centromere in Robertsonian translocations.

T(4;X)37H Rb(9.19)163H

For insertions the number for the chromosome yielding the inserted part is placed 1st.

Is(7;1)40H [a part of Chromosome 7 inserted into Chromosome 1]

The symbols L and S represent the long and short arms of chromosomes; the L may be omitted if the meaning of the symbol is clear without it.

T(4S;5)99H

[a translocation, break in the short arm of Chromosome 4 and long arm of Chromosome 5]

Positions of breakpoints known relative to Giemsa bands are represented by the band numbers after the appropriate chromosome number or letter.

Is(7F1–7C;XF1)1Ct

[inverted insertion of segment 7F1–7C (Chromosome 7) into the X Chromosome at band F1]

In symbols for segmental trisomies and monosomies derived from reciprocal translocations, the 1st number within the parentheses refers to the derivation of the proximal end of the trisomic chromosome, and a superscript number designates the chromosome contributing the distal end; the series number indicates the translocation from which the trisomy is derived.

Ts(1^{13})89H

[trisomy for the proximal end of Chromosome 1 and the distal end of 13, derived from T(1;13)89H]

For additional details consult the source (CSGNM 1989) of these recommendations.

DEER MOUSE (*Peromyscus* SPECIES)

The chromosomes of the deer mouse are represented and rearrangements symbolized (CSCP 1977) by essentially the same conventions as for *Homo sapiens;* see sections 20·6–20·7.

DOMESTIC ANIMALS

20·5 Karyotypes of domestic animals are represented, in general, by the conventions (SCHCN 1978) for numbering and designating bands of human karyotypes; see section 20·6. The system for domestic animals had its origin at the 1976 Reading Conference (Ford and others 1980). It has been developed (Diberardino and others 1990) for standardization of cattle, goat, and sheep karyotypes with both G-banded ("G" for "Giemsa-stained") and R-banded ("R" for "reverse-Giemsa stained") karyotypes. Chromosomes are numbered beginning with chromosome 1; the sex

Table 20·1 Domestic and laboratory animals: chromosome numbers and letters[a]

Animal	Chromosomes
Cat (*Felis catus*)	A1, A2, A3, B1, B2, B3, B4, C1, C2, D1, D2, D3, D4, E1, E2, E3, F1, F2, X, Y
Cattle (*Bos taurus*)	1 to 29, X, Y
Deer mouse (*Peromyscus* species)	1 to 23, X, Y
Goat (*Capra hircus*)	1 to 29, X, Y
Horse (*Equus caballus*)	1 to 31, X, Y
Mouse (*Mus* species)	1 to 19, X, Y
Pig (*Sus scrofa*)	1 to 18, X, Y
Rabbit (*Oryctolagus cuniculus*)	1 to 21, X, Y
Sheep (*Ovis aries*)	1 to 26, X, Y

[a]Adapted in part from the report on the Reading conference of 1976 (Ford and others 1980).

chromosomes are designated X and Y (see Table 20·1). More specific recommendations have been published for the domestic sheep (CSKOA 1985), the domestic pig (CSKDP 1988), and the domestic silver fox (CSKVFD 1985).

HUMANS (*Homo sapiens*) AND OTHER PRIMATES

20·6 The basic document establishing nomenclature and symbolization for human chromosomes, normal and abnormal, is the 1978 report (SCHCN 1978) of the Standing Committee on Human Cytogenetic Nomenclature; this report is widely referred to as "ISCN 1978". Its "Historical Introduction" reviews and cites the preceding relevant documents.

The autosomal chromosomes are numbered from 1 to 22, generally in descending order of length; the sex chromosomes are designated X and Y. Normal females have 2 X chromosomes, normal males, an X and a Y. Historically they have been assigned to 7 alphabetically designated groups (A to G) determined by descending order of size and position of the centromere (the narrowed region of a chromosome joining its 2 sister chromatids), but these group-categories are no longer of importance.

The symbols p and q designate, respectively, the short and long arms of each chromosome. These and additional symbols are used to describe symbolically the normal and abnormal structures of chromosomes and to designate landmarks and regions on them (see Table 20·2).

Chromosome Bands, Landmarks, and Regions

20·7 Chromosomes stained with quinacrine show fluorescent bands called Q-bands. Some of these bands, the ends of the chromosome arms, and the

Table 20·2 Human chromosomes: symbols and their meaning[a]

Symbol	Meaning
A	Group A (chromosomes 1, 2, 3)
AI	1st meiotic anaphase
AII	2nd meiotic anaphase
ace	acentric fragment
→ [arrow]	from-to
* [asterisk]	mating [applied in designating interspecific animal hybrids]
B	Group B (chromosomes 4, 5)
b	break
C	Group C (chromosomes 6, 7, 8, 9, 10, 11, 12, X)
cen	centromere
chi	chimera
: [colon]	break [in detailed descriptions]
:: [double colon]	break and reunion [in detailed descriptions]
cs	chromosome
ct	chromatid
cx	complex
D	Group D (chromosomes 13, 14, 15)
del	deletion
der	derivative chromosome
dia	diakinesis
dic	dicentric
dip	diplotene
dir	direct
dis	distal
dit	dictyate
dmin	double minute
dup	duplication
E	Group E (chromosomes 16, 17, 18)
e	exchange
end	endoreduplication
= [equals symbol]	sum of
F	Group F (chromosomes 19, 20)
f	fragment
fem	female
G	Group G (chromosomes 21, 22, Y)
g	gap
h	secondary constriction
i	isochromosome
ins	insertion
inv	inversion
lep	leptotene
MI	1st meiotic metaphase
MII	2nd meiotic metaphase
mal	male
mar	marker chromosome
mat	maternal origin
med	median
min	minute

Table 20·2 (*cont.*)

Symbol	Meaning
− [minus symbol]	loss of
mn	modal number
mos	mosaic
oom	oogonial metaphase
p	short arm of chromosome
PI	1st meiotic prophase
pac	pachytene
( ) [parentheses]	[used to surround a symbol for a structurally altered chromosome]
pat	paternal origin
pcc	premature chromosome condensation
Ph	Philadelphia chromosome[b]
+ [plus symbol]	gain of
prx	proximal
psu	pseudo
pvz	pulverization
q	long arm of chromosome
qr	quadriradial
? [question mark]	questionable identification of chromosome or chromosome structure
r	ring chromosome
rcp	reciprocal
rea	rearrangement
rec	recombinant chromosome
s	satellite
sce	sister chromatid exchange
sdl	sideline, subline
; [semicolon]	separates chromosomes and chromosome regions in structural rearrangements involving more than 1 chromosome
sl	stemline
/ [slant line, slash, solidus]	separates cell lines in describing mosaics or chimeras
spm	spermatogonial metaphase
t	translocation
tan	tandem
ter	terminal [end of chromosome]
tr	triradial
tri	tricentric
[double underline]	[used to distinguish homologous chromosomes]
var	variable chromosome region
xma	chiasma[ta]
zyg	zygotene

[a]Adapted from *An International System of Human Cytogenetic Nomenclature (1978)* (SCHCN 1978).

[b]Other forms of the symbol have included Ph^1, Ph1, and Ph_1 (Sandberg and others 1985).

centromere serve as landmarks for chromosomal locations. A region is defined as a part of a chromosome lying between 2 adjacent landmarks.

Regions are numbered outwardly from the centromere; the region adjacent to the centromere is number 1 for each arm. A landmark band is treated as belonging to the region distal to the landmark and numbered 1.

A band is designated by the chromosome number, the arm symbol, the region number, and the band number, these symbols being run together in that sequence and without punctuation.

1p33 [chromosome 1, its short arm (p), region 3, and band 3 of that region]

Sub-bands are also numbered out from the centromere. A sub-band number follows a period placed after its band number; divisions of sub-bands are sequentially numbered in the same pattern but without periods.

1p33.1 [chromosome 1, its short arm (p), region 3 of the arm, band 3, sub-band 1]

Karyotype Symbolization

The general pattern of symbols for representing entire normal and abnormal karyotypes is the sequence of total chromosome number, comma, sex chromosomes, followed, if needed to designate abnormalities, by a comma and the additional symbols needed for description. These elements are not separated by spaces.

46,XX	[normal female karyotype]
46,XY	[normal male karyotype]
46,X,dic(Y)	[46 chromosomes, 1 X chromosome, dicentric Y chromosome]
47,XY,+14p+	[male karyotype, additional chromosome 14 with longer short arm]

The large number of possible symbolizations precludes describing all of them in text; see Table 20·3 for examples with explanations. Consult the ISCN 1978 report (SCHCN 1978) for additional guidance.

Great Apes

20·8 Karyotypes for the great apes (Pongidae: chimpanzee, *Pan troglodytes,* PTR; pygmy chimpanzee, *Pan paniscus,* PPA; gorilla, *Gorilla gorilla,* GGO; orangutan, *Pongo pygmaeus,* PPY) are represented by the conventions for human karyotypes.

Table 20·3 Symbolized human karyotypes and chromosomal aberrations: examples and explanations[a]

Represented feature	Example	Explanation
Numeric aberration	49,XXXY	49 chromosomes, XXXY sex chromosomes
Additional or missing whole chromosome	48,XXY,+G	48 chromosomes, XXY sex chromosomes, additional Group G chromosome
	45,XX,-?8	45 chromosomes, XX sex chromosomes, missing chromosome which is probably chromosome 8
Chromosome mosaic	mos45,X/46,XY	2 cell lines: 1 with 45 chromosomes, sex chromosome X; the other, with 46 chromosomes, XY sex chromosomes
Translocation	46,X,t(Xq+;16p-)	Reciprocal translocation between the q arm of the X chromosome and the p arm of chromosome 16
Ring chromosome	46,XX,r(16)	Female karyotype with 46 chromosomes; chromosome 16 is a ring chromosome
Deletion, interstitial	46,XX,del(1)(q21q31)	Deletion between bands 1q21 and 1q31 in the long arm (q) of chromosome 1
Marker chromosome	46,XX,t(12;?)(q15;?)	Karyotype with rearranged chromosome 12; the segment of the long arm (q) distal to band 12q15 not identified
Meiotic karyotype	M1,24,XY,+I(21)	Primary spermatocyte: 24 elements from a male with trisomy 21; the extra chromosome 21, univalent
Somatic cell hybrid	t(HSA+CGR)(1;9)	Translocation between a human chromosome 1 and a Chinese hamster chromosome 9

[a]Adapted from *An International System for Human Cytogenetic Nomenclature (1978)* (SCHCN 1978).

Phylogenetic homologies between human chromosomes and those of the great apes are represented by roman numerals. A chromosome can be represented by the ISCN 1978 conventions or the phylogenetic convention in which the species code (see the codes in the paragraph above) is followed by an arabic numeral for the chromosome; the presumptive homology is indicated within parentheses by the species compared and the relevant chromosome number. If the symbols for an aberration must include parentheses, the homology is represented within square brackets.

[conventional ISCN symbolization] PPY 3 [HSA inv(4)(p15q21)]

[phylogenetic symbolization] PPY IV [HSA inv(4)(p15q21)]

BARLEY (*Hordeum* SPECIES)

20·9 Linkage groups and the corresponding chromosomes for barley are represented (BGC 1972) by arabic numerals. Symbolic abbreviations represent chromosomal aberrations.

Df [deficiency] Dp [duplication] In [inversion]
T [translocation] Tp [transposition]

Extrachromosomal factors are placed within square brackets preceding the genic formula.

For gene symbolization see section 20·39.

COTTON (*Gossypium* SPECIES)

20·10 Chromosomes of cotton are designated (Kohel 1973) by arabic numerals (1 through 26) and linkage groups by roman numerals. Aberrations are designated by italicized letters representing the terms for aberrations followed by the symbols for the involved chromosomes. Translocations are represented by *T*. Different translocations in the same chromosomes are distinguished with lower-case letters.

T1–2a *T1–2b*

Primary monosomics and primary trisomics are designated by *mono* and *triplo* followed by the number of the chromosome.

mono-6 *triplo-6*

The same rules apply for inversions (*In*), deficiencies (*Df*), duplications (*Dp*), deletions (*Del*), and transpositions (*Tp*).

For gene symbolization, see section 20·40.

MAIZE (CORN, *Zea mays*)

20·11 Maize chromosomes are numbered from the longest, 1, to the shortest, 10. Short arms are symbolized by S, long arms by L. Linkage groups are represented beginning with position 0 (zero) at the end of the short arm, which is termed the “left” or “top” or “north” end on the linkage map. Reciprocal translocations are designated by T followed by the numbers of the rearranged chromosomes separated by a hyphen, followed by a letter or an isolate number.

T1–2a T1–2b T1–9(4995)

Translocations of A chromosomes with B chromosomes are designated by TB followed by a hyphen, the number of the A chromosome and the arm translocated, followed by a letter or an isolate number.

TB-1La TB-5Sc

Inversions are designated by Inv followed by the chromosome number and a letter or an isolate number.

Inv1c Inv2(8865)

Deletions are designated by Del, duplications by Dup.

For additional details, see "A Standard for Maize Genetics Nomenclature" (NS 1993). For gene symbolization, see section 20·41.

OAT (*Avena* SPECIES)

20·12 Recommendations for symbolizing chromosomes of oat were issued in 1978 (Simons and others 1978). Complete chromosomes are represented by arabic numerals. Aberrations are represented by abbreviations: Df (deficiency), Dp (duplication), In (inversion), T (translocation), and Tp (transposition). See Simons and others (1978) for additional details.

ONION (*Allium* SPECIES)

20·13 The nomenclature and notation for the chromosomes of *Allium cepa* L. (Kalkman 1984) represent the standard (De Vries 1990) for *Allium* species that cross-fertilize with *A. cepa* to yield F_1 interspecific hybrids.

The longest chromosome is at the left in an idiogram, and the successively shorter chromosomes are placed to the right, with the centromeres positioned at the same level. The chromosomes are numbered, beginning at the left end of the idiogram, with arabic numerals and an unspaced capital (uppercase) letter representing the specific epithet. Thus the chromosomes for *A. cepa* are numbered 1C, 2C, and so on to 8C.

RICE (*Oryza sativa*)

20·14 Chromosomes of rice (Kinoshita 1986) are numbered according to the length at the pachytene stage; the longest is chromosome 1 and the shortest, chromosome 12. Short arms are symbolized by S, long arms by L.

1S 2L

Structural changes are represented by a symbol representing the type of aberration (italicized letters) and the chromosome(s) (nonitalicized number[s]): *Dp* (duplication), *In* (inversion), *T* (translocation), and *Tp* (transposition). Lowercase letters (roman) following the chromosome number (nonitalicized) distinguish similar aberrations involving the same chromosome.

In(2)a *T*(1–2)b

Monosomics and trisomics are designated by the additional chromosome.

Mono-1 Triplo-2

For gene symbolization, see section 20·43.

RYE (*Secale* SPECIES)

20·15 The rye chromosomes are designated (Sybenga 1983) 1R through 7R; the capital letter L designates the long arm and S the short arm of a chromosome. When the *Secale* species must be designated to indicate the derivation of a chromosome arm, a 3-letter superscript is added to the chromosome symbol: $2R^{cer}L$ represents the long arm of chromosome 2R of *Secale cereale.*

For gene symbolization, see section 20·44.

SOYBEAN (*Glycine max*)

20·16 For the soybean, linkage groups and the corresponding chromosomes are represented (SGC 1991) by arabic numerals. Symbols for chromosomal aberrations are an abbreviation representing the aberration followed by the chromosome number or numbers (and a letter for additional aberrations on a chromosome).

Tran 1–2a [the 1st case of reciprocal translocations between chromosomes 1 and 2]

Tran 1–2b [the 2nd case of such translocations]

The symbols for deficiencies, inversions, and primary trisomics are also abbreviations: Def, Inv, and Tri, respectively.

Cytoplasmic factors are represented by 1 or more italicized letters following the hyphenated prefix *cyt-*.

cyt-G [cytoplasmic factor for maternal green cotyledons]

For gene symbolization see section 20·45.

TOMATO (*Lycopersicon esculentum*)

20·17 Chromosomes of the tomato are numbered according to the length, 1 for the longest, 12 for the shortest. Long arms are symbolized by L, short arms by S. Translocations are symbolized by T, inversions by In, and deficiencies by Df. Aneuploids with whole-arm interchanges are symbolized by the component arms.

1S·12L

[interchange between the short arm of chromosome 1 and the long arm of 12]

Additional details can be found in Clayberg (1970). For gene symbolization see section 20·46.

WHEAT (*Triticum* SPECIES)

20·18 Linkage groups and the corresponding chromosomes of wheat are designated (McIntosh 1983) by an arabic numeral followed by a capital (uppercase) roman letter representing the genome.

1A–7D [for the chromosomes of the hexaploid wheat of group *aestivum*]

Chromosome arms are designated L for "long" and S for "short". The designation of a particular band requires 5 symbol elements (Gill 1987): chromosome number, genome designation, arm symbol, region number, and the band number within the region.

1Bq21 [for "chromosome 1, genome B, long arm, region 2, band 1"]

Chromosomal aberrations are indicated by abbreviations: Df (deficiency), Dp (duplication), In (inversion), Tr for (translocation), and Tp (transposition). For a gene not in the standard chromosome position, the new chromosome designation can be given within parentheses following the gene designation.

Hp (Tp 6D)

[the introgressed "hairy neck" gene on chromosome 6D instead of the standard chromosome 4A]

For gene symbolization, see section 20·47.

DIMENSIONAL UNITS FOR CHROMOSOMES

Base Pairs

20·19 The length of base-pair sequences in DNA, for example, distances between loci on contig maps, can be designated by the number of base pairs encompassed. The symbol bp for "base pairs" can be used for relatively short sequences; for longer sequences or to indicate the total number of base pairs, the symbol b may be prefaced with k for "kilo" or M for "mega".

The total length of DNA sequenced from the clone is 255 bp.

The sequence-tagged site sY80 has a size of 1.075 kb.

The total amount of mapped sequence is now more than 10.3 Mb.

The Centimorgan

The genetic distance between 2 loci on a chromosome can be expressed in centimorgans (cM). The centimorgan numerically equals the statistically corrected recombination frequency expressed as a percentage; thus if the recombination frequency between 2 given markers is 15%, they are separated by 15 cM. The genetic distance of 1cM approximates, on average, the physical distance of 1 million base pairs (1 Mb), but this correlation varies throughout the genome, especially at telomeres, where recombination is more frequent. The centimorgan takes its name from Thomas Hunt Morgan (1866–1945), the American geneticist awarded the Nobel Prize for Medicine or Physiology in 1933 for his great contributions on the function of the chromosome in heredity, derived from his research on the fruit fly, *Drosophila melanogaster.*

Note that "morgan" in "centimorgan" is not capitalized, but the "M" in the symbol cM is capitalized. These are the conventions in the SI; see section 11·10.

CHROMOSOME COMPONENTS

Loci and Markers

20·20 A locus is the location on a chromosome (or plasmid or other kind of genetic molecule) of a gene, an anonymous DNA segment, a fragile site, a breakpoint, an insertion, or other distinguishable sequence by which the location can be specified. It is generally desirable to specify the character of the locus, for example, "gene locus", "RLFP locus". In a genome database, a locus entry is likely to include the cytogenetic location, the official name and symbol, the mapping method(s) used, and the homologous genes of a reference organism. Genes are represented by gene symbols; see sections 20·23–20·48 for gene symbols for viruses, prokaryotes, and eukaryotes.

A marker can be a gene locus with a particular, and usually readily identifiable, phenotype, or a subdivision of a gene, or an anonymous DNA sequence.

Anonymous DNA Sequences

20·21 A sequence of DNA with no known functional identity is an anonymous DNA sequence. These are represented for human chromosomes by unique D-number symbols (alphanumeric) assigned by the Human Gene Mapping Workshop's DNA Committee (Kidd and others 1988).

A DNA segment number has 3 components.

1 The letter D to designate an anonymous DNA segment.
2 The number of the chromosome on which the segment resides (1–22, X, Y; XY for sequences homologous on the X and Y chromosomes).
3 A 1-letter code to designate the type of sequence.

S for a single-copy sequence, followed by a sequential number unique for the chromosome on which the sequence resides.

F for a member of a family of homologous sequences on 1 or more chromosomes; the F is followed by a sequential number, unique in the genome, for the sequence family, then by the letter S and a sequential number, unique in the family, designating which member of that family is referred to.

Z for a repetitive sequence, followed by a sequential number unique for the chromosome on which the sequence resides; in some cases this sequential number is followed by another letter.

E if the sequence is expressed; this means that the sequence is likely to be a gene , but cannot be given a gene symbol because its function is not known.

X or Y for a sequence homologous on the X and Y chromosomes, designating on which chromosome the sequence lies.

Note below examples illustrating applications of these rules.

D20S103 [the 103rd single-copy anonymous sequence assigned to chromosome 20]

D9S220E [the 220th single-copy anonymous sequence assigned to chromosome 9; the sequence is expressed, but its function is unknown]

DYZ3	[the 3rd repetitive anonymous sequence assigned to the Y chromosome]
D14F23S2	[the 1st member of the 23rd anonymous-sequence family assigned in the genome; this member of the family is on chromosome 14]
D14F23S2	[the 2nd member of the 23rd family, also on chromosome 14]
D2F23S3	[the 3rd member of the 23rd family; this member is on chromosome 2]
DXYS7	[the 7th single-copy anonymous sequence assigned, which is homologous on the X and Y chromosomes]
DXYS7X	[the X chromosome copy of DXYS7]

Note that all D segments are also loci.

Probes

20·22 A "probe" in genetic usage is a DNA, RNA, protein, or other cellular constituent used to allow visualization of a genetic target object or structure. Probe names are expected to be distinct from gene symbols and names of loci and are assigned by the originating laboratories; superscripts and subscripts must not be used.

18K3-Alu   lambdaPKCB   Gsa   EPAcDNA

Lowercase-letter prefixes can indicate vectors: p for "plasmid", c for "cosmid", l for "lambdoid phage", and y for "yeast"; an example is "pHJ19".

Contigs

Overlapping DNA sequences that together constitute a continuous DNA segment of a genome are known as a contig. The components of the contig are generally represented by the symbols of the clones carrying the overlapping nucleotide sequences. Note that as of 1993 the symbolization of clone symbols was not standardized.

Sequence-Tagged Sites (STSs)

A sequence-tagged site (STS) is a short DNA segment that can be located by polymerase chain reaction (PCR) techniques and that identifies a unique location on a chromosome.

sY163 [sequence 163 on the Y chromosome]

As of 1993 the symbolization was not standardized. Note that STSs can be given D segment numbers.

GENES AND PHENOTYPES

BACTERIOPHAGES

20·23 There are no firm and formal rules for symbolizing bacteriophage genes. The customary conventions have evolved to represent the genes of bacteriophage lambda (phage l). A sample of symbols for phage λ can be found in the American Society for Microbiology style manual (ASM 1985) and in the compilation by Echols and Murialdo (1978).

In general, the symbols are 1, 2, or 3 letters in italics. The bacteriophage is specified with a prefix for the phage (for example, λ for bacteriophage lambda, Mu for bacteriophage Mu, P1 for bacteriophage P1), spaced or unspaced from the gene symbol proper. Because phages do not have metabolism outside bacterial cells, there are no phenotype symbols corresponding to those for genes; hence some authors prefer to disregard the italicization. Symbols may be combined, spaced or unspaced, to represent mutations. Superscripts indicate hybrid genomes.

λ*che22* Mu dII345 [bacteriophage Mu]
P1 *vir* [bacteriophage P1]

λcI857*int*2*red*114*susA*11 [mutations in genes *c*I, *int,* and *red;* a suppressible (*sus*) mutation in gene *A*]

λ att^{434} imm^{21} [hybrid of phage λ carrying the attachment (*att*) region of phage 434 and the immunity (*imm*) region of phage 21]

ONCOGENES

20·24 Symbols for oncogenes are 3-letter italicized symbols based on the retrovirus in which the oncogene was 1st identified. Nonitalicized (roman) single lowercase letter prefixes specify the homologue type; capital-letter prefixes indicate the the cellular localization.

v-*fes* [viral gene, from "feline sarcoma" virus]

c-*myc* [cellular, from "myelocytomatosis"]

N-*erb A* [nucleus localization, from "avian erythroblastosis"]

S-*sis* [secreted, from "simian sarcoma" virus]

HUMAN RETROVIRUSES

Various italicized 1-, 2-, or 3-letter and alphanumeric symbols have represented genes of the human immunodeficiency viruses (HIV-1 and HIV-2) and the human T-cell leukemia viruses, HTLV-1 and HTLV-2 (formerly designated HTLV-III, IV, I, and II, respectively).

p40x   *tel*   *tat-3*   *sor*   *X*

Gallo and colleagues (Gallo 1988) have proposed a consistent use of 3-letter italicized symbols, with letters reflecting gene functions.

tax_1 [transactivator]
rev [regulator of expression of virion proteins]
nef [negative factor]

BACTERIA

20·25 The document defining the basic rules for symbolizing bacterial genes and their phenotypes is *A Proposal for a Uniform Nomenclature in Bacterial Genetics* (Demerec and others 1966). The notations are similar to those for human and animal genes and phenotypes, but the Demerec conventions have been supplemented with subsequent additions. The main features of these conventions are illustrated in Table 20·4.

PLASMIDS

20·26 The notations for plasmid loci, genes, and alleles follow, in general, the rules set forth by Demerec and colleagues; see the paragraph above. A detailed set of recommendations was published by Novick and colleagues (1976) along with definitions of relevant terms, but symbols may take various degrees of completeness and spacing. In general, each newly described plasmid and newly isolated genotypic modification of a known plasmid is given a unique alphanumeric designation with the form pXY1234 in which "p" stands for "plasmid", "XY" are initials for the laboratory or the reporting scientist, and "1234" represents the laboratory's numeric designation. If the strain carries more than 1 plasmid, they are sequenced in the plasmid designation position. Italics are used only for species names.

pPL603 [PL for PS Lovett]   *E. coli* K-12(pML432)
E. coli K-12(pML432)(pml531)

The phenotypic notation for a plasmid gene should consist of a capital and a lowercase letter (or 2 letters if necessary) and should reflect the

Table 20·4 Examples of symbols for bacterial genes and phenotypes, with explanatory footnotes[a]

Feature	Convention	Example	Footnote
Genotype			
Wild-type	3-letter, lower-case italic symbol	*ara*	b
Locus designation	Capital italic letter	*araA*	
Phenotypically related	Same symbol with added italic capital letter	*araA araD*	c
Mutation site	Mutation indicated by adding an italic serial isolation number (allele number)	*araB1*	
Promoter site	Added italic p	*lacZp*	d
Terminator site	Added italic t	*lacAt*	
Operator site	Added italic o	*lacZo*	
Attenuator site	Added italic a	*lacZp*	
Phenotype	3-letter symbol, roman letters; 1st capitalized	FepA [the protein encoded by the *fepA* gene]	
Wild type	Superscript +	Tol^+	
Characteristic	Superscript lowercase letter	Str^r ["r" for "resistance"]	
Plasmids and episome	Symbol not italic; 1st letter capitalized; (Col E1) enclosed in parentheses when written as part of a bacterial genotype		

[a]Adapted from Demerec and others (1966) and the *ASM Style Manual for Journals and Books* (ASM 1985).

[b]Derived from *arabinose* and represents loci controlling the bacterium's response to this sugar as a source of carbon and energy. A superscript + may be added to indicate wild type, as in ara^+.

[c]Each capital (uppercase) letter must produce a unique symbol designating only 1 locus.

[d]Subscript numerals (not italicized) may be added to distinguish more than 1 site, for example, $glnAp_1$, $glnAp_2$.

phenotypic trait for which the gene is responsible. The genotypic notation should be in the form recommended by Demerec (see section 20·25 and Table 20·4). The complete genotypic identification includes as a prefix the name of the plasmid on which the gene was found.

Plasmid recombinants should be given new plasmid numbers. Deletions are indicated by a lowercase greek delta (δ) and identified by unique serial numeration and a list of deleted genes. Insertions, transpositions, and translocations should be indicated by a lowercase greek omega (ω) and identified by unique serial numeration and a list of translocated genes.

Note that prefix designations may be registered with the Plasmid Reference Center (Lederberg 1986), and new designations can be checked to avoid conflicts with already applied and registered prefixes.

NEMATODE (*Caenorhabditis elegans*)

20·27 Gene symbols for the nematode *Caenorhabditis elegans* are derived from gene names (representing broad phenotypic categories) represented in abbreviated form by 3 lowercase italicized letters.

che [for "abnormal chemotaxis"]

mec [for "mechanosensory abnormality"]

Different genes within a particular broad category are represented by hyphenated arabic numerals; symbols can include an italicized roman numeral indicating the linkage group of the gene.

unc-15　　*unc-54*　　*dpy-18 III*　　*lon-2 X*

Symbols for mutations have an italicized prefix-letter representing the laboratory designating the mutation followed by an italicized arabic numeral. Suffixes of lowercase nonitalicized letters (defined at 1st use) indicate characteristics of the mutation. Gene and mutation names can be used together, with the mutation name within parentheses after the gene name. A plus symbol (+) designates a wild-type allele.

*e1348*rl ["*e*" for MRC Laboratory of Molecular Biology, Cambridge, England; "rl" for "recessive lethal"]

unc-4 (e120) II　　*sma-2+*

These recommendations (Horvitz and others 1979) were expanded in 1987 (Riddle and others 1988) to cover additional needs for symbolizing complex alleles (including revertants), cloned DNA, transgenic strains, transposon names, transposon insertions in genes, DNA sequences not associated with specific genes (including restriction fragment length polymorphism names), rearrangment names, tandem duplications, crossover suppressors, hybrid strains of *C. elegans,* and nomenclature for *Caenorhabditis briggsae.*

Protein products of genes are represented by the same characters but in nonitalic (roman), capitalized (uppercase) form.

ANY-1　　[protein product of *any-1*]

FRUIT FLY (*Drosophila melanogaster*)

20·28 A symbol representing a gene locus in the fruit fly is the abbreviation of the name assigned to the locus. The symbol usually begins with the same letter as the name. The symbol is italicized and has no spaces or subscripts. Names are not italicized in text.

awd [for "abnormal wing disc"]   *R* [for "Roughened"]

In genotypes with several mutant genes, symbols are separated if the genes are on the same chromosome. If on homologous chromosomes, the symbols are separated by a slash; on nonhomologous chromosomes, by semicolons and spaces.

y w F B   *y w f/B*   *bw; e; ey*

Alleles are represented by the same name and symbol differentiated with italicized superscript alphabetic and numeric characters.

bw^{D} Hn^{r2} a^{badp}

Additional details on these and other conventions can be found in Lindsley and Zimm (1992) and Ashburner (1989).

FISH

20·29 A formal system published in 1990 (Shaklee and others 1990) established the nomenclature and symbolization for genes coding for enzymes and other proteins of fish. The system is similar to that for human genes (see section 20·32) and draws on the enzyme nomenclature of the International Union of Biochemistry (see section 16·27). Table 20·5 shows the main elements of the system.

MOUSE (*Mus* SPECIES)

20·30 The present system of gene symbolization for the mouse is described in Chapter 1, "Rules and guidelines for gene nomenclature", in *Genetic Variants and Strains of the Laboratory Mouse* (Lyon 1989). It represents the latest version of the system initiated in 1940 (Dunn and others 1940) and modified in 1963, expanded for biochemical variants in 1973, and revised in 1979; see Lyon's chapter (1989) for references to the relevant documents. Table 20·6 summarizes the main conventions; Lyon and others (1989) should be consulted for additional details, including specifications for symbolizing genes homologous with other species, gene complexes, pseudogenes, lethals, virus-expression-related genes, onco-

Table 20·5 Symbolization for protein-coding loci in fish[a], with explanatory footnotes

Feature	Convention	Examples	Footnote
Gene	Uppercase, italic; preferably ends with*; all characters on the line; no superscript or subscript characters	*IDHP* MPI**	b
Multiloci	Hyphen and arabic numeral (unknown relationships) or italic capital letter (established orthologies)	*IDHP-1* LDH-A**	
Isoloci	Comma between the 2 locus numbers	*GPI-B1,2**	
Regulatory locus suffix	Lowercase italic “r”	*LDH-Ar**	
Subcellular prefix	Lowercase-letter prefix	*sMDH-B**	c
Allele	Italic arabic numeral, lowercase letter, or relative electrophoretic mobility with a preceding asterisk	*MDH-B*1* *ADA-1*a* *EST-2*75*	d
Phenotype			
Enzyme	IUB-specified name and number	L-lactate dehydrogenase, 1.1.1.27	b
Enzyme symbol	Uppercase (capital) roman letter; with numbers, greek-letter stereochemical isomer symbols, or hyphens as necessary	LDH	e
	Lowercase prefix for subcellular localization	mIDHP	

[a]Based on Shaklee and others (1990). Additional details are available in this souce.

[b]Same alphanumeric symbols as the abbreviations for the coded proteins, which are derived from the IUB names (see section 16·27 and footnote e below). Italics and asterisk-marking distinguish the symbols from those for enzymes and proteins. The asterisk follows the locus symbol and precedes an allele designation.

[c]The prefix corresponds to that indicating the subcellular localization of the enzyme coded for by the gene: “l” for “lysosomal”, “m” for “mitochondrial”, “p” for “peroxisomal”, and “s” for “cytosolic” (“supernatant” or “soluble”). The prefixes should be ignored for alphabetization.

[d]The preferred convention is sequentially assigned number codes to designate alleles.

[e]See revised Table 1 in Shaklee and others (1990) (erratum in [1990. Trans Am Fish Soc 119(4)]) for a compilation of recommended enzyme and protein abbreviations. Hyphens, numbers, and greek letters can be used in enzyme abbreviations if needed to avoid ambiguity.

genes, antigenic variants, biochemical variants, loci for mouse lymphocyte antigens, globin genes, homoeobox-containing genes, and transgenic mice.

The nomenclature and the symbolization for mouse *H-2* genes were substantially modified in 1990 (Klein and others 1990).

Table 20·6 Symbols for mouse genes[a]

Feature	Convention	Examples	Footnote
Gene	2-, 3-, or 4-letter abbreviation in italic letters of the gene name; numerals if needed	*Hbb* [hemoglobin β-chain *G6pd* [glucose-6-phosphate dehydrogenase]	b
Member of a series	Same symbol for all plus a hyphen and a numeral	*H-1 H-2* *Es-1 Es-2*	c
Allele	Gene symbol and italic 1 or 2 letters (lowercase) superscript	Hbb^{d} Mi^{wh}	d
Wild-type	Plus symbol (not italic) and superscript italicized gene symbol	$+^{pe}$	
Reversion to wild-type	Symbolization reversed from the wild-type symbol	pe^{+}	
Phenotype	Same elements as for gene symbols but with capital, not italic letters and superscripts lowered to the type line	GPI-1A	

[a]Adapted from Lyon (1989).

[b]The initial letter is capitalized and the following letters are lowercase. The initial letter is lowercase for a recessive mutation that was the basis for identification of the locus. Numerals may be used if they occur in the name or abbreviation on which the symbol is based; roman numerals and greek letters should not be used. Many symbols run up to 9 characters, and this convention has been modified to allow up to 10 characters.

[c]As this book went to press, the International Committee for Standardized Genetic Nomenclature in Mice was considering deleting hyphens from all symbols except those where hyphens are needed to avoid confusion (for example, to separate numbers).

[d]When superscript symbols are not available (as in computer printouts), the superscript letter(s) may be replaced by an asterisk or angle brackets and the letter(s), for example, *Hbb**d* or *Hbb<d>* rather than Hbb^{d}.

DEER MOUSE (*Peromyscus* SPECIES)

The conventions applicable to the mouse are generally applied to gene representations for the deer mouse; specific rules have not been published. A list of genetic loci was published in 1968 (Rasmussen 1968); updated gene lists are published biannually in the *Peromyscus Newsletter*, an informal periodical.

RABBIT

20·31 The gene symbols for the rabbit follow conventions similar to those for the mouse, for example, *a* representing "agouti", *Acp-2* stands for the gene coding for acid phosphatase-2, *Grs* for the gene coding for glutathione reductase.

Table 20·7 Symbols for human genes and phenotypes, with explanatory footnotes[a]

Feature	Convention	Examples	Footnote
Gene	Uppercase roman letters or a combination of uppercase letters and arabic numerals; italicized or underlined (in typescript); all elements on the line	*ACADS AMY1 G6PD HPRT*	b
	Symbols for enzyme and protein genes use the conventions stated above; greek letters, roman numerals, superscripts, and subscripts are not acceptable	βGAL-1 [becomes] *GLB1*	c
Allele	As for genes; allele characters (4 or fewer) separated from gene symbol by an asterisk	*ADA*1 HBB*6V TG*A1*	b
Gene family	The character @ added to the relevant gene symbol	*HBB@* [beta hemoglobin gene family]	d
Genotype	Chromosomal homologs: Allele symbols separated by a horizontal line or if on the same line by a slash (slant line)	*ADA*1/ADA82 G6PD*A/G6PD*B*	e
	Hemizygous males: Y to identify the male	*G6PD*A/Y* and *G6PD*B/Y*	
	Loci not on the same chromosome are separated by a semicolon	*ENO1*1/ENO*2; PGM2*1/PGM2*2*	
	Loci on the same chromosome with phase known are separated by a space; with phase not known, by a comma	*AMY1*A PGM1*2/ AMY*B PGM1*1 PGM1*1/ PGM1*2,AMY1*A/ AMY1*B*	
DNA segment	Capital letter "D" and following numerals and capital (uppercase) letters indicating chromosome number, and type of sequence (including family designation); see section 20·21 for details	D1F24S3 D21F24S4	
Phenotype	Same characters as for genes and alleles but not underlined or italicized; all characters on the same line; no asterisks; a space separates gene and allele characters; a comma separates alleles; phenotypes of enzymes and proteins are treated as other phenotypes	ADA 1 ADA 1,2 PGM1 1,2 HBB A,S	e

[a]Based on "Guidelines for Human Gene Nomenclature: An International System for Human Gene Nomenclature (Shows and others 1987).

HUMANS (*Homo sapiens*) AND OTHER PRIMATES

20·32 The present conventions for symbolizing human genes, their alleles, and their phenotypes have as their foundation some of the conventions recommended in 1966 by Demerec and colleagues for bacterial genes (Demerec and others 1966), but they go well beyond them. The authority for symbol formation and style is the Human Gene Nomenclature Committee (Shows and others 1987), which functions through the Human Gene Mapping Workshops. Its main guidelines for the presentation of symbols are summarized in Table 20·7. The Committee's guidelines document provides full details on the proper selection of alphanumeric elements for the formation of gene, allele, and phenotype symbols, with special attention to enzymes, proteins, hemoglobin, blood groups, cell surface antigens, DNA segments, protooncogenes, virus-associated markers, fragile sites, mitochondrial genes, homologous genes of different species, and inherited clinical syndromes.

For approved human gene symbols, consult the Genome Data Base, hosted at The Johns Hopkins University School of Medicine; telephone 410 955-9705, fax 410 614-0434.

The conventions for human genes are also applicable to the genes of other primates.

TRANSGENIC ANIMALS

20·33 A committee of the Institute of Laboratory Animal Resources (ILAR) has published recommendations (ILAR 1992) for symbolic representation of transgenic animal lines. Each animal line is assigned a symbol of 3 parts with continuous alphanumeric characters, all in a roman typeface.

Notes to Table 20·7 (*cont.*)

[b]Italic letters need not be used in gene catalogues, or in computer printouts and similar presentations for which an italic typeface may not be available. In such formats a gene symbol can be indicated by following it with an asterisk, and an allele symbol by preceding it with an asterisk. Names and symbols for genes coding for enzymes should be based on the names recommended by the International Union of Biochemistry (see section 16·27).

[c]For details on converting abbreviations to gene symbols, see Shows and others (1987).

[d]This convention is also applicable to gene families with members mapped to different chromosomes (McAlpine and others 1988).

[e]The horizontal-line convention is like that of a numerical fraction in which the numerator and denominator are separated by a horizontal line; the on-line convention illustrated here is more economical of space for text. Additional conventions are available for known and unknown linear order on the same chromosome.

TgX(YYYYYY)#####Zzz	[the basic formula for the symbols]
TgX	[mode of generating the line; for example, TgN for "nonhomologous recombination", TgR for "retroviral infection", TgH for "homologous recombination"]
(YYYYYY)	[insert designation; the main features of the transgene]
#####	[laboratory-assigned number, up to 5 digits]
Zzz	[ILAR-assigned code for the laboratory]

These symbols are used for the 2nd field, "Designation of Transgenic Line", in records in TBASE, the computerized database on transgenic animals and targeted mutations maintained at the Oak Ridge National Laboratory and available through the Johns Hopkins University Computational Biology Gopher Server. Abbreviated forms can be formed by omitting the insert portion, but the full symbol should be given at the 1st mention of the transgene.

TgN(GPDHIm)1Bir	[human glycerol phosphate dehydrogenase gene (*GPDH*) which caused an insertional mutation (Im); 1st transgenic mouse named by Edward H Birkenmeier (Bir)]
TgN1Bir	[abbreviated form]

A specific animal line carrying an insertion should be designated by a stock designator (see section 24·5) preceding, with a hyphen, the transgene symbol.

C57BL/6J-TgN(CD8Ge)23Jwg

[CB57BL/6 mouse from the Jackson Laboratory carrying the human CD8 genomic clone; 23rd mouse screened in a series of microinjections in the laboratory of Jon W Gordon (Jwg)]

See Table 20·8 for further details on selection of characters for symbolization.

Note that the symbol for an animal line should not be used to identify the insertional mutation or phenotype due to the transgene. The locus should be identified and the allele symbolized by the abbreviated symbol applied by the convention for the species involved.

$ho^{TgN447Jwg}$ [insertion of a transgene into the "hotfoot locus" (*ho*)]

Targeted mutations within specific genes are not regarded as transgenic lines; they should be named and symbolized by the conventions for mouse genes (see section 20·30).

Table 20·8 Symbols for transgenes[a]

Feature	Convention
Mode of insertion	"Tg" for "transgene" and letter for mode ("N" for "nonhomologous", "R" for insertion via a "retroviral" vector, "H" for "homologous recombination")
Insert designation; gene feature	Within parentheses, no more than 8 characters. Capital or lowercase letters or an alphanumeric designation. No italics, superscripts, subscripts, internal spaces, and punctuation. Preferred, 6 or fewer characters; the total number of characters plus the laboratory-assigned number may not exceed 11. If a sequence from a named gene is used, its standard symbol should be included, at least its beginning letters; hyphens omitted from an original symbol. No symbols identical with other named genes. Standard abbreviations[b] can be part of the designation.
Laboratory-assigned number and laboratory code	Up to 5 numeric characters, but the total number of characters plus the laboratory-assigned number may not exceed 11. The laboratory code is a unique code; the registry is maintained by the Institute of Laboratory Animal Resources (ILAR)[c]

[a]See section 20·33 for explanation and examples of complete symbols.

[b]Standard abbreviations: An (anonymous sequence), Ge (genomic clone), Im (insertional mutation), Nc (noncoding sequence), Rp (reporter sequence), Sn (synthetic sequence), Et (enhancer trap construct), Pt (promoter trap construct).

[c]For information on the registry and assignments of codes: Dr Dorothy D Greenhouse, ILAR, National Research Council, 2101 Constitution Avenue NW, Washington DC 20418 USA; telephone 202 334-2590, fax 202 334-1687; Bitnet, DGREENHO@NAS).

PLANT-PATHOGENIC FUNGI

20·34 For genetic nomenclature and symbolization for plant-pathogenic fungi, the Genetics Committee of the American Phytopathological Society has prepared recommendations (Yoder and others 1986) based mainly on the conventions for yeast.

A gene locus is designated by a unique 3-letter italic symbol based on the name of the mutant phenotype it represents. The symbol for the locus proper has an initial capital letter; the 2 following letters should be in lowercase type. A dominant allele is represented by 3 capital (uppercase) letters, a recessive allele by 3 lowercase letters. Wild-type alleles can be indicated by a following plus symbol (+) and mutant alleles by a following minus symbol (–). Mutations at different loci yielding similar phenotypes can be identified with identical letter symbols followed by a number unique to that locus.

Table 20·9 Symbols for genes of *Aspergillus nidulans*[a]

Feature	Convention	Examples	Footnote
Genetic locus and mutant	Italicized 3-letter symbol	*arg*	b
Nonallelic locus with the same primary symbol	Suffixed capital italic letter	*argA* *argB*	
Allele	Suffixed italic serial number	*argA1* *argA2*	
Undetermined allelic relationships of a mutant	Hyphenated italicized number replaces the capital letter	*arg-51*	
Wild-type allele	Superscript plus symbol (sign)	$argA^{+}$	
Dominant mutant	Initial italic capital letter	*Acr*	
Phenotype	Nonitalicized gene symbol with initial capital letter	Arg^{-} [for "arginine requirement"]	c
Specific property of mutants	Superscript nonitalicized letter	$areA^{\mathrm{d}}18$	
Mitochondrial gene	Symbols enclosed in square brackets	[*oliA1*]	

[a]Based on Clutterbuck (1973).
[b]Older symbols: 1 to 5 italic letters, for example, *panto* for "pantothenic acid requirement".
[c]A strain not requiring arginine would be phenotypically represented by "Arg^{+}".

Met [for "methionine auxotrophy"] *MET* [dominant allele]
met [recessive allele]
Met1 *Met2* *Met3* [different loci; similar phenotypes]
Met+ [wild type, "methionine-independent growth"]

Similar conventions are applied for phenotypes, but the symbols should be in roman type. A plus symbol indicates the wild-type phenotype, a minus symbol indicates a mutant phenotype.

Met+ [methionine-independent growth]
Met− [methionine auxotrophy]

Cytoplasmically inherited genes can be indicated by enclosing the symbol within brackets.

Further details can be found in Yoder and others (1986).

Aspergillus nidulans

20·35 The symbols for genes of *Aspergillus nidulans* are similar to those for bacteria (Clutterbuck 1973) (see Table 20·9 for a summary of the conventions).

Table 20·10 Symbols for genes of *Neurospora crassa*[a]

Feature	Convention	Examples	Footnote
Gene locus	1 to 4 italic letters	*ad* ["adenine requirement"]	
Recessive genes	All letters lowercase		
Dominant mutant allele	Initial italic capital letter	*Sk* ["Spore killer"]	
Mutant allele	Symbol without superscript		
Wild-type allele	Symbol with superscript plus symbol	ad^{+}	b
Nonallelic locus	Symbol with hyphenated number	*ad-1 ad-2*	

[a]Based on Perkins and others (1982).

[b]Other superscripts (letters) may be used to designate allelic series without a definitive wild type or differing in sensitivity or resistance: $cyh\text{-}1^{R}$ [for "cycloheximide resistance"], $cyh\text{-}1^{S}$ [for "cycloheximide sensitivity"].

Neurospora crassa

20·36 The conventions for *Neurospora* (Perkins and others 1982) are like those for *Drosophila* (see Table 20·10 for a summary).

YEAST (*Saccharomyces cerevisae*)

20·37 Gene symbols for yeast are based essentially on the proposals of Demerec and others (1966) and are set forth in detail by Sherman (1981) (Table 20·11 summarizes the main conventions).

Arabidopsis thaliana

20·38 The guidelines for symbolizing genes of *Arabidopsis thaliana* are particularly concerned with genes identified by mutation (personal communication from DW Meinke). These guidelines are summarized in Table 20·12. Further information can be obtained from the *Arabidopsis* Biological Resource Center, Ohio State University, Columbus OH, USA.

BARLEY (*Hordeum* SPECIES)

20·39 Rules for symbolizing barley genes based on those developed by the International Committee on Genetic Symbols (Tanaka 1957) were published in 1981 in the *Barley Genetics Newsletter;* these rules are now out-of-date. In 1993 new rules were being developed for a system that would be compatible with the concepts of the Plant Molecular Biology Gene Nomenclature Commission applicable to cloned plant genes. The approved rules will be published in the *Barley Genetics Newsletter.*

Table 20·11 Symbols for genes of yeast (*Saccharomyces cerevisae*)[a]

Feature	Convention	Examples	Footnote
Gene symbol	3 italic letters	*ARG arg*	
Gene locus	Italicized number following the symbol	*ARG2*	
Dominant allele	Capitalized italic letters	*ARG2*	
Recessive allele	Lowercase italic letters	*arg*	
Allele designation	Italicized number following the locus number and a hyphen	*arg2-14*	b
Gene cluster	Italicized capital letter following the locus number	*his4A his4B*	c
Wild-type gene	Added plus symbol (sign)	*ARG2*+	d
Gene conferring resistance or susceptibility	Superscript R or S, not italicized	$CUP^{R}1$	e
Phenotype	Same characters as gene symbol but not italicized; superscript + and −	arg^{-} arg^{+}	f

[a]Based on Sherman (1981).

[b]Locus numbers are those of original assignments, but allele numbers may be those of a laboratory.

[c]Also for complementation groups with a gene and domains within a gene having different properties.

[d]Note that the plus symbol is not italicized.

[e]In general, superscripts should be avoided.

[f]In the examples, arg^{-} represents "not requiring arginine" and arg^{+}, "requiring arginine".

COTTON (*Gossypium* SPECIES)

20·40 The recommendations (Kohel 1973) for gene symbols for cotton are based on the 1957 recommendations of the Committee on Genetic Symbols and Nomenclature (CGSN 1957).

Gene symbols are italic letters. Mutant gene names are symbolized by letters representing abbreviations for the adjective, noun, or compound term describing the phenotype. The 1st letter of the symbol is that of the term. Dominant gene symbols begin with a capital letter; symbols for recessive mutants are entirely in lowercase letters. Newly discovered alleles at a given locus are assigned the original locus symbol with a superscript lower-case letter. New mutants with phenotypes similar to previously described mutants are designated with the symbol of the original mutant modified with a numerical subscript.

as_1 as_2 [asynapsis] B_1 [Blight resistance]

ml [mosaic leaf] Sm_3 [Smooth leaf]

Table 20·12 Symbols for genes of *Arabidopsis thaliana* identified by mutations[a]

Feature	Convention	Examples	Footnote
Mutant gene	3 lowercase italic letters; underlined in typescript	*abc*	b
Wild-type allele	Capital (uppercase) italic letters	*ABC*	c
Different genes with the same symbol	Arabic numerals following the letters	*abc1 abc2*	
Different alleles of the same gene	Number following a hyphen	*abc4-1 abc4-2*	d
Phenotype	Gene symbol in roman type; initial capital letter; plus symbol (+) for wild type, minus symbol (−) for mutant	Abc+ [wild type] Abc− [mutant]	
Protein product of gene	All capital letters	ABC	

[a]Based on a personal communication from David W Meinke, Curator of Mutant Gene Symbols for *Arabidopsis,* Department of Botany, Oklahoma State University, Stillwater OK, USA.

[b]Well-known symbols in use before issuance of these guidelines may have only 2 letters.

[c]Underlined in typescript.

[d]If only 1 allele is known, hyphen and number are not needed (for example, *abc3* = *abc3-1*); a D can be added to the allele number to indicate that the allele shows dominance relative to the wild type; for example, *abc5-2D* indicates that allele 2 is dominant to the wild type.

[e]The plus (+) and minus (−) symbols can be on the line or superscript.

MAIZE (CORN, *Zea mays*)

20·41 The definitions and standards for gene nomenclature and symbolization are published periodically in *Maize Genetics Cooperation Newsletter* (NS 1993). The system of symbolization is generally similar to those for other plants and for animals; for a summary, see Table 20·13. A catalogue of symbols for named genes is published annually in the *Maize Genetics Cooperation Newsletter.*

Rules have been specified for naming restriction fragment length polymorphisms (RFLPs): a lowercase 3- or 4-letter code representing the originating university or company, followed, without a space, by a laboratory number. When the probe is a cDNA or gene subclone, the gene symbol or potential functional acronym is added within parentheses after the RFLP locus designation; an example is *umc000(a1).*

OAT (*Avena* SPECIES)

20·42 Recommendations for symbolizing genes of oat were issued in 1978 (Simons 1978). Symbols are derived from the English name of the character

Table 20·13 Symbols for maize (corn) genes, alleles, and phenotypes[a]

Feature	Convention	Examples	Footnote
Gene locus	Lowercase italic characters; 3-letter symbols are recommended; symbols represent gene names and numbers designate different loci with similar phenotypes	*dek12 gdh2*	b
Recessive alleles	Lowercase letters (as in *dek12*)		c
Dominant allele	Same symbol as the italic 3-letter symbol for the recessive allele but with an initial capital letter	*Dek1*	
Codominant allele	First letter capitalized and identified by an allelic specification; symbol italicized	*Pgm1+5*	
Newly identified allele	The symbol plus a laboratory number	*sh2-6801*	d
Gene product (phenotype)	Nonitalic, all capital letters; name of product not capitalized	ADH1; alcohol dehydrogenase	e
Mutation from transposable element insertion	Uses of double colons or apostrophes	*wx-m1::Ds1* *Bz1'+7801*	f

[a]Based on "A Standard for Maize Genetics Nomenclature" (NS 1993).

[b]The numeric suffix is not separated from the gene name by a hyphen; the hyphen is reserved for symbols for mutant alleles. Among these examples, *dek12* represents "defective kernel12" in "defective kernel12, collapsed endosperm lethal, cultured embryos green, narrow-leaved, curled". A hyphen is used to separate the symbol for the allele designation from a suffix specifying the particular allele; a plus symbol (+) is used for nonmutant alleles.

[c]As no "wild type" is recognized for maize, many naturally occurring variants are not properly termed "mutants".

[d]The laboratory number could be a number indicating the date of identification; in the example, "*6801*" would represent "January 1968".

[e]ADH1 is the symbol for the product encoded by *adh1*.

[f]For details see the document cited in footnote a.

(phenotype) or the Latin name of the organism to which the oat reacts; they are in roman type, with an initial capital (uppercase) letter for basic symbols (also representing the dominant allele) or initial lowercase letter for a recessive allele. Nonallelic loci are indicated by an arabic numeral following a hyphen.

sc-1 [for the recessive gene "pinkish straw color"]

Tg-1 [dominant gene for resistance to *Toxoptera graminium*] cda-1 [for "chlorophyll deficiency-albino, Nishiyama (1941)"]

Cda-4 [for "normal chlorophyll production, McGinnis and Taylor (1961)"]

For additional details consult Simons and others (1978).

RICE (*Oryza sativa*)

20·43 The rules for symbolizing the genes of rice have been reported by Kinoshita (1986). Amendments to the rules are announced in the *Rice Genetics Newsletter.*

Gene symbols (2 or 3 italic letters) represent names describing the character modification. Symbols commonly used in the past should be retained even if they do not fit this rule (for example, *C* for "Chromogen"; *wx* for "glutinous endosperm".

The name and symbol for a dominant gene should, if there would be no ambiguity, be represented with a capital (uppercase) letter and for a recessive with a lowercase letter.

Standard or wild-type alleles are designated by a gene symbol with a superscript plus symbol (+), as in al^{+}.

Multiple alleles at the same locus are designated by the appropriate superscript.

$d\text{-}18^{+}$ $d\text{-}18^{k}$ Pgi-1^{1} $Pgi\text{-}1^{2}$

Nonallelic genes that are phenotypically indistinguishable are represented by the same base letter but differentiated by a number or alphabetic subscript. A new gene with an uncertain allelic relationship with previously reported genes with similar effects is denoted by adding a letter t (for "tentative") to its symbol; an example is *d-50*(t).

Inhibitors, suppressors, enhancers, and modifiers are designated, respectively, by the symbols *I, Su, En,* and *M*. The same symbols for recessives are not capitalized. The symbol is followed by a hyphen and the symbol of the affected gene (for example, $Su\text{-}g_1$).

RYE (*Secale cereale*)

20·44 The recommendations for symbolizing genes of rye are based on those for wheat (see section 20.47) and those issued by the International Workshop on Rye Chromosome Nomenclature and Homoeology Relationships (Sybenga 1983). When there might be a misunderstanding with a wheat symbol, a different symbol can be used. Sybenga (1983) summarizes the conventions adopted from wheat-gene symbolization. The style of symbols is governed by 4 rules.

1 Gene symbols (derived from their original names) are italicized.
2 Whenever the usage would be unambiguous, the name and symbol of a dominant allele begins with a capital letter and those of a recessive allele with a lowercase letter.
3 Non-allelic loci (such as mimics and polymeric genes) are designated with additional rules calling for arabic numerals, hyphenation, and genome symbolization.
4 All letters and numerals are on the line.

SOYBEAN (*Glycine max*)

20·45 Rules for gene symbols for the soybean were published by the Soybean Genetics Committee in 1991 (SGC 1991).

1 The symbols are made up of 1, 2, or 3 italicized letters, for example, *Ab*. Subscripts or superscripts may be appended. Allelic genes are designated by the same base symbol.
2 An initial capital letter indicates a dominant or partially dominant allele.
3 Lowercase italic superscripts of 1 or 2 letters are used to represent more than 2 alleles for a locus, for example, *R*, r^m, *r*.
4 Numeric subscripts are added to indicate gene pairs with the same or similar effects.
5 A dash can follow a gene symbol to have it represent any allele at the locus.
6 Cytoplasmic factors are represented with letters preceded by "cyt-".

cyt-G [for "cytoplasmic factor for maternal green cotyledons"]

For additional details, including the use of question marks and plus symbols, consult the *Soybean Genetics Newsletter* (SGC 1978).

TOMATO (*Lycopersicon esculentum*)

20·46 Genes of the tomato are symbolized (Clayberg and others 1970) with 1 or more italic letters; the initial letter should be the same as that of the name of the mutant, which should represent the main feature of the phenotype. The symbols *c, r, s,* and *y* were derived from the normal. For a mutant gene dominant to the normal type, the initial letter is capitalized. A superscript "+" following the gene symbol designates the normal allele of the mutant gene. A subsequently appearing dominant allele is designated with a superscript "D". Additional alleles can be represented by suitable letter or number superscripts.

c $\quad sp^{+}$ $\quad sp^{D}$

For additional information relevant to representing indistinguishable alleles of independent origin and mimics, see Clayberg (1970). It is recommended that these symbols be applied to all species of *Lycopersicon.*

WHEAT (*Triticum* SPECIES)

20·47 Symbols for wheat genes (McIntosh 1988) are composed, in general, of italicized letters representing a gene's original name. The rules recommended by McIntosh in 1983 (McIntosh 1983) allowed "roman letters of distinctive type". Unless an ambiguity would result, the name and symbol for a dominant gene should begin with a capital letter and for a recessive, with a lowercase letter. Superscript and subscript characters should not be used. Two or more genes with similar phenotypes are represented with a common basic symbol followed by an arabic numeral (for sequential polymeric series) or a hyphen with the accepted genome symbol and a homoeologous set number (arabic).

Pc [for "purple culm"] *co* [for "corroded"]
Adh-A1
[for "alcohol dehydrogenase-1, A-genome member of the first *Adh* set"]

Inhibitors, suppressors, and enhancers are represented by the symbols "*I*", "*Su*", and "*En*" (for recessives, "*i*", "*su*", and "*en*") followed by a space and the symbol for the affected allele. The plus symbol (sign) is not to be used. Symbols for extrachromosomal factors are enclosed within brackets and precede the genic formula.

For further details and a catalogue of symbols, consult McIntosh (1988) or more recent proceedings of the International Wheat Genetics Symposium.

TRANSSPECIES GENE FAMILIES: THE P450 SUPERGENE FAMILY

20·48 A system of nomenclature and symbolization for P450 enzymes in eukaryotes and prokaryotes has been proposed by Nebert and others (1991). The system is uniform for all species except for 2 minor exceptions for the mouse; the present version discourages the use of roman numerals that had been used for gene families in the 1987 version (Neb-

Table 20·14 Symbols for the P450 gene superfamily, families, and subfamilies[a]

Feature	Convention	Examples	Footnotes
Gene root symbol for the superfamily	Italic *CYP* [*Cyp* for the mouse]		b
P450 family	Root symbol with an italic arabic numeral for the family	*CYP2*	
Subfamily	Family symbol with an italic capital for the subfamily	*CYP11B*	c
Individual gene	Preceding elements and an arabic numeral for the gene	*CYP2B2*	d
Pseudogene	A closing italic capital letter "*P*"	*CYP2B7P*	
Transcript of gene (mRNA) and its product	The gene symbol in nonitalicized characters (roman typeface)	CYP1A2 [from "*CYP1A2*"]	e

[a]Based on Nebert and others (1991).
[b]"*CYP*" represents "**cy**tochrome **P**450".
[c]Applied when the family has 2 or more subfamilies; lowercase italic letter for the mouse.
[d]For the mouse, the final number is preceded by a hyphen: "*Cyp1a-2*".
[e]For the mouse as well as for other species; "CYP1A2" from "*Cyp1a-2*".

ert and others 1987) of the recommendations. Table 20·14 summarizes the system.

GENES: RELATED FUNCTIONAL ELEMENTS

INITIATION AND ELONGATION FACTORS

20·49 Initiation factors needed to start protein synthesis are designated with the abbreviation "IF" followed by a hyphen and numeric identifier. The symbols for bacterial factors (for example, *Escherichia coli*) have no prefix; those for eukaryotic factors have the prefix "e". A hyphenated number indicates the group assignment; symbols for new initiation factors in a group add the next capital letter available among the group symbols. Subunits of initiation factors are represented by adding greek letters: α, β, γ, and so on.

IF-2   eIF-2   eIF-4B   eIF-3β

Elongation factors have similar representations; the symbols for prokaryotic factors differ in minor ways.

EF-Ts   EF-Tu   eEF-1α   eEF-1βγ

For further details on both groups of factors, consult the recommendations drafted by Safer (IUB 1989).

TRANSPOSONS

20·50 Simple transposons are represented by "IS" (for "insertion sequences") followed by a number specific for a type. Other more complex transposons are represented by "Tn" followed by specific alphabetic and numeric designators; for bacterial transposons, the designators are italicized (but are not for eukaryotic transposons).

IS*2*	IS4	IS10R
Tn*1*	Tn*5 lac*	[bacterial transposons]
Tn9	Tn903	[eukaryotic transposons]

RESTRICTION ENDONUCLEASES

20·51 A 3-letter italicized abbreviation represents the host organism: an initial capital letter that is the 1st letter of the genus name and the first 2 letters of the species epithet. Strain designations consisting only of arabic numerals are spaced; other designations are closed up.

Bce 1229 [for "*Bacillus cereus,* strain IAM 1229"]

*Xma*II [for "*Xanthomonas malvacearum*"]

See Roberts (1977) for a catalogue of symbols.

REFERENCES

Cited References

Ashburner M. 1989. *Drosophila:* a laboratory handbook. Cold Spring Harbor (NY): Cold Spring Harbor Laboratory Pr.

[ASM] American Society for Microbiology. 1985. ASM style manual for journals and books. Washington: ASM.

[ATCC] American Type Culture Collection. 1992. American Type Culture Collection catalogue of cell lines and hybridomas. 7th ed. Rockville (MD): ATCC. Published periodically.

[BGC] Barley Genetics Committee. 1972. Report from the Barley Genetics Committee of the American Barley Research Workers' Conference. Barley Newsl 15:2–6.

[CGSN] Committee on Genetic Symbols and Nomenclature. 1957. Report of the Committee on Genetic Symbols and Nomenclature. In: Union Internationale des Sciences Biologique, Serie B, Nr 10. p 1–6.

Clayberg C, Butler L, Kerr E, Rick C, Robinson R. 1970. Report of Gene List Committee: rules for nomenclature in tomato genetics. Report Tomato Genet Cooperative 20:3–5.

Clutterbuck A. 1973. Gene symbols in *Aspergillus nidulans.* Genet Res 21:291–6.

[CSCP] Committee for Standardization of Chromosomes of *Peromyscus.* 1977. Standardized karyotype of deer mice, *Peromyscus* (Rodentia). Cytogenet Cell Genet 19:38–43.

[CSGNM] Committee on Standardized Genetic Nomenclature for Mice. 1972. Standard karyotype of the mouse, *Mus musculus.* J Heredity 63:69–72.

[CSGNM[Committee on Standardized Genetic Nomenclature for Mice. 1989. Rules for nomenclature of chromosome anomalies. In: Lyon MF, Searle AG. 1989. Genetic variants and strains of the laboratory mouse. New York: Oxford Univ Pr. p 574–5.

[CSKDP] Committee for the Standardized Karyotype of the Domestic Pig. 1988. Standard karyotype of the domestic pig. Hereditas 109:151–7.

[CSKOA] Committee for Standardized Karyotype of the Domestic Sheep. 1985. Standard nomenclature for the G-band karyotype of the domestic sheep (*Ovis aries*). Hereditas 103:165–70.

[CSKVFD] Committee for the Standard Karyotype of *Vulpes fulvus* Desm. 1985. The standard karyotype of the silver fox (*Vulpes fulvus* Desm.). Hereditas 103:171–6.

Demerec M, Adelberg EA, Clark AJ, Hartman PE. 1966. A proposal for a uniform nomenclature in bacterial genetics. Genetics 54:61–76.

De Vries JN. 1990. Onion chromosome nomenclature and homeology relationships: workshop report. Euphytica 49:1–3.

Diberardino D, Hayes H, Fries R, Long S. 1990. International system for cytogenetic nomenclature of domestic animals (1989): The Second International Conference on Standardization of Domestic Animal Karyotypes. Cytogenet Cell Genet 53:65–79.

Dunn L, Grünberg H, Snell G. 1940. Report of the Committee on Mouse Genetics Nomenclature. J Heredit 31:505–6.

Echols H, Murialdo H. 1978. Genetic map of bacteriophage lambda. Microbiol Rev 42:577–91.

Ford C, Pollock D, Gustavsson E. 1980. Proceedings of the First International Conference for the Standardisation of Banded Karyotypes of Domestic Animals. Hereditas 92:145–62.

Gallo R, Wong Staal F, Montagnier L, Haseltine WA, Yoshida M. 1988. HIV/HTLV gene nomenclature [letter]. Nature 333:504.

Gill B. 1987. Chromosome banding methods, standard chromosome band nomenclature, and applications in cytogenetic analysis. In: Heyne E. 1987. Wheat and wheat improvement. Madison (WI): American Soc of Agronomy, Crop Science Soc of America, Soil Science Soc of America. p 243–254.

Horvitz HR, Brenner S, Hodgkin J, Herman RK. 1979. A uniform genetic nomenclature for the nematode *Caenorhabditis elegans.* Mol Gen Genet 175:129–33.

[ILAR] Institute of Laboratory Animal Resources, Committee on Transgenic Nomenclature. 1992. Standard nomenclature for transgenic animals. ILAR News 34:3–10.

[IUB] International Union of Biochemistry, Nomenclature Committee (NC-IUB). 1989. Nomenclature of initiation, elongation, and termination factors for translation in eukaryotes: recommendations 1988. Eur J Biochem 186:1–4.

Kalkman R. 1984. Analysis of the C-banded karyotype of *Allium cepa* L.: standard system of nomenclature and polymorphism. Genetica 65:141–8.

Kidd KK, Bowcock AM, Pearson PL, Schmidtke J, Willard HF, Track RK, Ricciuti F. 1988. Report of the committee on human gene mapping by recombinant DNA techniques. Cytogenet Cell Genet 49:132–218.

Kinoshita T. 1986. Report of the Committee on Gene Symbolization, Nomenclature and Linkage Groups. Rice Genet Newsl (3):4–5.

Klein J, Benoist C, David CS, Demant P, Lindahl KF, Flaherty L, Flavell RA, Hemmerling U, Hood LE, Hunt SW III, and others. 1990. Revised nomenclature of mouse *H-2* genes. Immunogenetics 32:147–9.

Kohel RJ. 1973. Genetic nomenclature in cotton. J Hered 64:291–5.

Lederberg EM. 1986. Plasmid prefix designations registered by the Plasmid Reference Center 1977–1985. Plasmid 15:57–92.

Lindsley DL, Zimm GG. 1992. The genome of *Drosophila melanogaster.* San Diego (CA): Academic Pr.

Lyon MF. 1989. Rules and guidelines for gene nomenclature. In: Lyon MF, Searle AG. 1989. Genetic variants and strains of the laboratory mouse. 2nd ed. Oxford (UK): Oxford Univ Pr. p 1–11.

McAlpine PJ, Boucheix C, Pakstis AJ, Stranc LC, Berent TG, Shows TB. 1988. The 1988 catalog of mapped genes and report of the nomenclature committee. Cytogenet Cell Genet 49:4–38.

McIntosh R. 1983. A catalogue of gene symbols for wheat (1983 edition). In: Proceedings of the 6th International Wheat Genetics Symposium: held at Kyoto, Japan, November 28–December 3, 1983. Kyoto, Japan: Germ-Plasm Inst, Kyoto Univ. p 1197–1254.

McIntosh R. 1988. Catalogue of gene symbols for wheat. In: 1988. Proceedings of the 7th International Wheat Genetics Symposium, Cambridge (UK). p 1225–323.

Nebert DW, Adesnik M, Coon MJ, Estabrook RW, Gonzalez FJ, Guenerich FP, Gunsalus IC, Johnson EF, Kemper B, Levin W, and others. 1987. The P450 gene superfamily: recommended nomenclature. DNA (NY) 6:1–12.

Nebert DW, Nelson DR, Coon MJ, Eastbrook RW, Feyereisen R, Fujii-Kuriyama Y, Gonzalez FJ, Guengerich FP, Gunsalus IC, Johnson E, and others. 1991. The P450 superfamily: update on new sequences, gene mapping, and recommended nomenclature. DNA Cell Biol 10:1–14.

Nesbitt M, Francke U. 1973. A system of nomenclature for band patterns of mouse chromosomes. Chromosoma 41:145–58.

Novick RP, Clowes RC, Cohen SN, Curtiss R, Datta N, Falkow S. 1976. Uniform nomenclature for bacterial plasmids: a proposal. Bacteriol Rev 40:168–89.

[NS] Nomenclature Subcommittee. 1993 Mar 15. A standard for maize genetics nomenclature. Maize Genet Newsl 67:171–3.

Perkins D, Radford A, Newmeyer D, Björkman M. 1982. Chromosomal loci of *Neurospora crassa.* Microbiol Rev 46:426–570.

Rasmussen DI. 1968. Genetics. In: King JA. 1968. Biology of *Peromyscus*: Special Publication 2. American Soc of Mammalogists. p 343–5.

Riddle DA, CGA Advisory Committee. 1988. November 1987 update to *C. elegans* genetic nomenclature. Worm Breeders Gazette [informal publication] 10:1–5.

Roberts RJ. 1977. Appendix D: Restriction endonucleases. In: Bukhari AI, Shapiro JA, Adhya SL, editors. 1977. DNA insertion elements, plasmids, and episomes. Cold Spring Harbor (NY): Cold Spring Harbor Laboratory. p 757–68.

Sandberg A, Hecht BK-M, Hecht F. 1985. Nomenclature: the Philadelphia chromosome or Ph without superscript. Cancer Genet Cytogenet 14:1.

[SCHCN] Standing Committee on Human Cytogenetic Nomenclature. 1978. An international system for human cytogenetic nomenclature (1978) ISCN (1978). Cytogenet Cell Genet 14:313–404.

[SGC] Soybean Genetics Committee. 1982. Report of the Soybean Genetics Committee. Soybean Genet Newsl 9:9–13.

[SGC] Soybean Genetics Committee. 1991. Rules for genetic symbols. Soybean Genet Newsl 18:10–3.

Shaklee JB, Allendorff FW, Morizot DC, Whitt GS. 1990. Gene nomenclature for protein-coding loci in fish. Trans Am Fish Soc 119:2–25.

Sherman F. 1981. Genetic nomenclature. In: Strathern J, Jones E, Broach J, editors. 1981. Molecular biology of the yeast, *Saccharomyces cerevisae.* Cold Spring Harbor (NY): Cold Spring Harbor Laboratory. p 639–40.

Shows TB, McAlpine PJ, Boucheix C, Collins FS, Conneally PM, Frezal J, Gershowitz H, Goodfellow PN, Hall JG, Issitt P, and others. 1987. Guidelines for human gene nomenclature: an international system for human gene nomenclature (ISGN, 1987). Cytogenet Cell Genet 46:11–30.

Simons M, Martens J, McKenzie R, Nishiyama I, Sadanaga K, Sebesta J, Thomas H. 1978. Oats: a standardized system of nomenclature for genes and chromosomes and catalog of genes governing characters. Agriculture Handbook Number 509. Washington: US Department of Agriculture.

Sybenga J. 1983. Rye chromosome nomenclature and homoeology relationships: workshop report. Z Pflanzenzüchtg 80:297–304.

Tanaka Y, chairman. 1957. Report of the International Committee on Genetic Symbols and Nomenclature. Union of International Sci Biol Ser B, Colloquia No. 30.

Yoder O, Valent B, Chumley F. 1986. Genetic nomenclature and practice for plant pathogenic fungi. Phytopathology 76:383–5.

Additional References

King RC, Stansfield WD. A dictionary of genetics. 4th ed. New York: Oxford Univ Pr.

O'Brien SJ, editor. 1993. Genetic maps: locus maps of complex genomes. 6th ed. Plainview (NY): Cold Spring Harbor Laboratory Pr.

Rieger R, Michaelis A, Green MM. 1991. Glossary of genetics: classical and molecular. 5th ed. New York: Springer-Verlag.

Singh RJ. 1993. Plant cytogenetics. Boca Raton (FL): CRC Pr.

Singleton P, Sainsbury D. 1993. Dictionary of microbiology and molecular biology. 2nd ed. New York: J Wiley.

Thompson MW, McInnes RR, Willard HF. 1991. Thompson & Thompson genetics in medicine. 5th ed. Philadelphia: WB Saunders.

21 Viruses

The infection [in tobacco mosaic disease] is not caused by microbes, but by a living liquid virus.

— Martinus Willem Beijerinck, *Über ein Contagium vivum fluidum als Ursache der Fleckenkrankheit der Tabaksblätter,* 1898

The world, said Paul Valery, is equally threatened with two catastrophes: order and disorder. So is virology.

— Lwoff, Horne, and Tournier, 1962. "A system of viruses". Cold Spring Harbor Symposia on Quantitative Biology 27:51–2

21·1 Viruses are neither prokaryotes nor eukaryotes; they are nucleic acid molecules that can enter cells, replicate in them, and code for proteins able to form protective shells around the viral nucleic acid. Because they are not cellular, do not grow, have no observable activity except replication, and function only within living cells, they may not be perceived as living organisms. Viruses appear to be partial or degenerate forms of living systems, and they do not belong to any of Whittaker's 5 kingdoms (Whittaker 1969); see Table 7·6.

Virus particles (virions) are shaped like rods, spheres, polyhedrons, helices, or filaments and are made up of cores of RNA or DNA encased in protein or other kinds of coats. The virion is the infectious form of the virus; it is a parasite that needs host-cell organelles to multiply, is smaller than bacteria, and ranges in size from about 10 to 4000 nm.

Definitions and descriptions of terms used in virology can be found in works by Bojnansky and Fargasova (1991), Calisher and Fauquet

(1992), Diener and Prusiner (1985), Francki and others (1991), Hull and others (1989), Lackie and Dow (1989), Matthews (1991), and Walkey (1991). Among the most frequently used terms are "slow virus", "prion", "phage", "viroid", "virusoid", "satellite virus", and "helper virus".

"Slow virus" refers specifically to a virus in the subfamily *Lentivirinae* that is present in an infected individual for a long time before it becomes active or infectious. It also generally refers to any virus with a very slow onset of clinical manifestations. Degenerative diseases of the nervous system such as scrapie in sheep and kuru or Creutzfeldt–Jakob diseases in humans have been attributed to slow viruses, but prions have also been reported to be causes.

A prion is an infectious agent 100 times smaller than a normal virus but lacking nucleic acid; prions do not have a nomenclature formally established by any official body. Prions have been reported only from studies in humans and other animals (Maramorosch and McKelvey 1985).

A bacteriophage, or "phage", is a virus that infects bacteria. Phages are included in several families, for example, *Inoviridae, Leviviridae, Lipothrixviridae,* and *Microviridae.* A phage consists of a coat protein (capsid) enclosing genetic material, DNA or RNA, and it infects bacteria.

Phage l has been used extensively as a vector in recombinant DNA studies.

Viroids are uncoated RNA species of low molecular weight that never produce recognizable nucleoprotein virus particles and that infect only higher plants (Diener 1987). There are no official proposals for taxonomic grouping of viroids. The best studied viroid is the potato spindle tuber viroid; other examples are citrus exocortis viroid, coconut cadang-cadang viroid, cucumber pale fruit viroid, grapevine viroid 1B, and hop latent viroid. Thirteen viroids have been described.

A virusoid is a viroid-like circular RNA encapsidated in a virus shell that forms an obligatory association with an RNA virus (Diener and Prusiner 1985); the term "virusoid" is, however, no longer favored (Matthews 1991), and virusoids are identified as satellite RNAs or satellite viruses.

A satellite virus is a virus whose small RNA codes for its own protein, but it depends for its replication on another, unrelated helper virus (Diener and Prusiner 1985). Satellite RNAs of plant viruses are not required for virus infection and have little or no homology with either the helper virus or the host plant. Satellite viruses are recognized by helper viruses, are replicated to high levels, and become encapsidated in the helper-virus coat protein (Diener 1987). Components of such a system can be designated by related abbreviations.

STNV [designates] satellite tobacco necrosis virus

STobRV [designates] satellite tobacco ringspot virus

A helper virus may be required for transmission by a vector; for example, Aucuba mosaic virus can be transmitted by aphids only if the source plant is already infected with potato virus A (the helper virus).

TAXONOMY AND NOMENCLATURE

21·2 The International Committee on Taxonomy of Viruses has classified viruses into the following taxonomic ranks: order (1), families, subfamilies, genera, subgenera or groups, and species or individual viruses. For additional discussion of the concepts "taxonomy" and "nomenclature", see sections 23·2 and 24·1.

Names for viruses should be those approved by the International Committee on Taxonomy of Viruses (ICTV) in its 6th report (Murphy and others 1994). There were 72 families and groups of viruses with names approved by the ICTV in its 5th report (Francki and others 1991) (see Table 21·1). A list of virus words and abbreviations approved by the ICTV has been prepared by Calisher and Fauquet (1992); also see section 21·9.

RULES OF VIRUS CLASSIFICATION AND NOMENCLATURE

There are 30 rules on nomenclature of viruses, which were developed by the International Committee on Taxonomy of Viruses (Murphy and others 1994)

General Rules

1 Virus classification and nomenclature shall be international and shall be universally applied to all viruses.
2 The universal virus classification system shall employ the hierarchical levels of *order*, *family*, *subfamily*, *genus*, and *species*. To the extent that species designations are not yet complete, international vernacular names are used for many viruses.
3 The ICTV is not concerned with classification and nomenclature below the species (or virus) level. Delineation of serotypes, genotypes, strains, variants, isolates, and so on is the responsibility of acknowledged international specialist groups.

Table 21·1 Families and groups of viruses with names approved by the International Committee on Taxonomy of Viruses[a], listed by host groups

Host	Names		
Algae	*Phycodnaviridae*		
Bacteria	*Corticoviridae*	*Lipothrixviridae*	*Podoviridae*
	Cystoviridae	*Microviridae*	*Siphoviridae*
	Inoviridae	*Myoviridae*	SSV-1 group
	Leviviridae	*Plasmaviridae*	*Tectiviridae*
Fungi	*Partitiviridae*	*Totiviridae*	
Invertebrates	*Baculoviridae*	*Parvoviridae*	*Reoviridae*
	Birnaviridae	*Picornaviridae*	*Rhabdoviridae*
	Flaviviridae	*Polydnaviridae*	*Tetraviridae*
	Iridoviridae	*Poxviridae*	*Togaviridae*
	Nodaviridae		
Mycoplasmas	*Inoviridae*		
Plants	alfalfa mosaic virus group	*Fabavirus*	pea enation mosaic virus group
	Bromovirus	*Furovirus*	*Potexvirus*
	Capillovirus	*Geminivirus*	*Potyvirus*
	Carlavirus	*Hordeivirus*	*Reoviridae*
	Carmovirus	*Ilarvirus*	*Rhabdoviridae*
	Caulimovirus	*Luteovirus*	*Sobemovirus*
	Closterovirus	maize chlorotic dwarf virus group	*Tenuivirus*
	Commelina yellow rattle virus group	*Marafivirus*	*Tobamovirus*
	Comovirus	*Necrovirus*	*Tobravirus*
	Cryptovirus	*Nepovirus*	*Tombusvirus*
	Cucumovirus	parsnip yellow fleck virus group	*Tymovirus*
	Dianthovirus		
Vertebrates	*Adenoviridae*	*Flaviviridae*	*Parvoviridae*
	Arenaviridae	*Hepadnaviridae*	*Picornaviridae*
	Birnaviridae	*Herpesviridae*	*Poxviridae*
	Bunyaviridae	*Iridoviridae*	*Reoviridae*
	Caliciviridae	*Orthomyxoviridae*	*Retroviridae*
	Coronaviridae	*Papovaviridae*	*Rhabdoviridae*
	Filoviridae	*Paramyxoviridae*	*Togaviridae*

[a]Adapted from Francki and others (1991). Some families are associated with more than 1 host group.

4 Artificially created viruses and laboratory hybrid viruses will not be given taxonomic consideration. Again, delineation of these entities is the responsibility of acknowledged international specialist groups.
5 Taxa will be established only when representative member viruses are sufficiently well characterized and described in the published literature to allow unambiguous identification and discrimination from similar taxa. Likewise, nomenclature will be recognized only when pertaining to viruses that are sufficiently well characterized and described in the published literature to allow unambiguous identification and discrimination from similar viruses.

Rules Pertaining to Naming Taxa and Viruses

21·3

6 Existing names of taxa and viruses shall be retained whenever feasible.
7 The rule of priority in naming taxa and viruses shall not be observed.
8 No person's name shall be used.
9 Names for taxa and viruses should be easy to use and easy to remember. Euphonious names are preferred.
10 Subscripts, superscripts, hyphens, oblique bars, and greek letters may not be used.
11 New names shall not duplicate approved names. New names shall be chosen so as not to be closely similar to names in use currently or in the recent past.
12 Sigla [names formed from compound terms; see section 21·6] may be accepted as names of taxa, provided they are meaningful to virologists in the field and are recommended by acknowledged international specialist groups.
13 Any meaning imparted by a name of a taxon or a virus must 1) not exclude viruses which are legitimate members of the taxon by alluding in the name to characteristics not possessed by all members or potential members, and 2) not apply equally to a different taxon.
14 New names shall be chosen with due regard to national or local sensitivities. When names are universally used by virologists in published work, these or derivatives shall be the preferred basis for creating names, irrespective of national origin. In the event of the advance of more than 1 candidate name, the relevant Study Group or Subcommittee will make a recommendation to the Executive Committee of the ICTV, which will then decide among the candidates.
15 Proposals for new names and name changes shall be submitted to the ICTV in the form of taxonomic proposals.

Rules Pertaining to Species

16 A virus species is defined as a polythetic class of viruses that constitutes a replicating lineage and occupies a particular ecologic niche.
17 A species (or virus) name shall consist of as few words as practicable.
18 A species (or virus) name, usually together with a strain designation, must provide an appropriately unambiguous identification without mention of its genus or family name.
19 Numbers, letters, or combinations thereof may be used as species (or virus) epithets where such numbers and letters already have wide usage. However, newly designated serial numbers, letters, or combinations thereof are not acceptable alone as species (or virus) epithets.
20 Approval by ICTV of newly proposed species, species names, and type species will proceed in 2 stages. In the 1st stage, provisional approval will be given. Provisionally approved proposals will be published in an ICTV Report; then, after a 3-year waiting period, if not withdrawn or modified, proposals will receive final approval.

Rules Pertaining to Genus

21 A genus is a group of species sharing certain common characters.
22 A genus name shall be a single word ending in "*virus*".
23 Approval of a new genus must be linked to approval of a type species.

Rules Pertaining to Subfamily

24 A subfamily is a group of genera sharing certain common characters. It should be used only when needed to solve a complex hierarchical problem.
25 A subfamily name shall be a single word ending in "*virinae*".

Rules Pertaining to Family

26 A family is a group of genera (whether or not these are organized into subfamilies) sharing certain common characters.
27 A family name shall be a single word ending in "*viridae*".
28 Approval of a new family must be linked to approval of a type genus.

Rules Pertaining to Order

29 An order is a group of families sharing certain common characters.
30 An order name shall be a single word ending in "*virales*".

INTERNATIONAL NAMES FOR VIRUS TAXA

21·4 Approved international names for orders, families, subfamilies, and genera are printed in an italic type. The names end in "*virales*" for order, "*viridae*" for family, "*virinae*" for subfamily, and "*virus*" for genus.

VERNACULAR NAMES OF VIRUSES

21·5 Names that have not yet been approved by the ICTV, often designated as "English vernacular names", may be printed in a standard typeface. Because vernacular names are not presented in binary combinations of a genus name and a specific epithet (for the definition of "specific epithet" see section 22·6), they should not be italicized. Note these examples.

1. The English vernacular name for the genus *Phytoreovirus* is "plant reovirus subgroup 1", and the type species (vernacular) is "wound tumor virus"; an international specific epithet has not been designated.
2. Under the international genus name *Orthopoxvirus*, for which there is no corresponding specific epithet, there is the vernacular genus name "vaccinia subgroup" for which the type species is "vaccinia virus".

A virus name derived from a disease name is written as 2 or more words; the 1st letter of a proper noun or proper adjective incorporated into the name of a virus is capitalized. If part of the vernacular name incorporates a scientific (Latin) name, that name is capitalized and italicized. Otherwise a vernacular name of a virus is in a lowercase roman typeface.

maize dwarf mosaic virus alfalfa mosaic virus PM2 phage
avian leukosis virus potato virus X Fiji disease virus
Rous sarcoma virus foot-and-mouth disease virus
Sindbis virus herpes simplex virus type 1
Trichoplusia ni granulosis virus *Lolium* enation virus
yellow fever virus

Genus names of viruses used as common names or virus groups are written as 1 word.

aphthovirus A parvovirus r-1 cytomegalovirus potexvirus
coxsackievirus rhabdovirus ilarvirus rhinovirus luteovirus

ABBREVIATED VIRUS DESIGNATIONS

21·6 Viruses may be designated by acronyms, sigla, or abbreviations. An acronym is a word created by selecting the initial letter of principal words in a compound term.

> ECBO virus
> [an acronym formed from "**e**nteric **c**ytopathic **b**ovine **o**rphan virus"]
>
> ECHO virus, now written as "echovirus"
> [acronyms formed from "**e**nteric **c**ytopathic **h**uman **o**rphan virus"]
>
> CELO virus [formed from "**c**hicken **e**mbryo **l**ethal **o**rphan virus"]
>
> REO virus, now written as "reovirus"
> [formed from "**r**espiratory **e**nteric **o**rphan virus"]

Sigla (plural; singular, "siglum") are names formed from letters or other characters taken from words in a compound term and approved by international study groups (see section 21·3, Rule 12).

arbovirus	[the siglum for "**ar**thropod-**bo**rne virus", formed by the selection of the boldfaced letters combined with "virus"]
cucumovirus	[the siglum for "**cucu**mber **mo**saic virus" with "virus"]
oncornavirus	[from "**onco**genesis" and "**RNA**" with "virus"]
papovavirus	[from "**pa**pilloma", "**po**lyoma" and "**va**cuolating agent" with "virus"]
potyvirus	[the siglum for "**pot**ato virus **Y**"
tombusvirus	[from "**tom**ato **bus**hy stunt virus"]

Some of the viruses named in these examples are virus groups, for example, cucumovirus, potyvirus, and tombusvirus. Potyvirus includes potato virus Y, maize dwarf mosaic virus, and others. "Togavirus", on the other hand, is neither an acronym nor a siglum; "toga" is "a cloak" and refers to the virus particle's being enclosed in an envelope.

Many well-known viruses are known by abbreviations. Some of them are listed in Table 21·2. A substantial compilation can be found in *Stedman's ICTV Virus Words* (Calisher and Fauquet 1992). Abbreviations can be used in a paper for publication if they are used at least 5 times and explained at the 1st mention; otherwise, the names should be written out (ASM 1991).

STRAIN DESIGNATIONS

21·7 Virus strains may be designated within parentheses after the virus names or closed up to the abbreviation for a virus. For example, the KOS strain of herpes simplex virus may be designated "HSV(KOS)" or "HSV

Table 21·2 Some generally accepted abbreviations of names of viruses

Abbreviation	Virus name
AcMNPV	*Autographa californica* nuclear polyhedrosis virus
AcNPV	*Autographa californica* polyhedrosis virus
AMV	avian myeloblastosis virus
ASFV	African swine fever virus
BBV	black beetle virus
CMV	cucumber mosaic virus, cytomegalovirus
CPV	cytoplasmic polyhedrosis virus
CrPV	cricket paralysis virus
DHBV	duck hepatitis B virus
DNV	densovirus
FMDV	foot-and-mouth disease virus
GgV-019/6a	*Gaeumannomyces graminis* virus 019/6a
GSHBV	ground squirrel hepatitis B virus
GV	granulosis virus
HBV	human hepatitis B virus
HDV	human hepatitis D virus
HIV-1, HIV-2	human immunodeficiency virus(es)
HTLV-1, HTLV-2, HTLV-I, HTLV-II	human T-cell leukemia virus(es)
NOV	nonoccludid virus
NPV	nuclear polyhedrosis virus
PLRV	potato leaf roll virus
RKV	rabbit polyoma virus
SIV	simian immunodeficiency virus
SVDV	swine vesicular disease virus
TRV	tobacco rattle virus

(KOS)". The details of the strain designation may be placed within the virus name: "influenza virus A/WS/33". More detailed designations may be needed.

> A/Equine/Prague/1/56 (H7N7)
>
> ["A", the virus type; "Equine", the host of origin (for animal influenza isolates); "Prague", geographic origin; "1", strain number; "56" for "1956", the year of isolation; "(H7N7)", antigenic description]

The abbreviation for "human T-cell leukemia [or "lymphotropic"] virus" when used alone is HTLV. Strains of HTLV can be designated HTLV-I or HTLV-II, but preferably HTLV-1 and HTLV-2. $\mathrm{HIV}_{\mathrm{HTLV\text{-}III}}$ has been used to designate the human immunodeficiency virus strain, instead of HIV (HTLV-III), and HTLV-IIIB or HTLV3b have been equally acceptable to the American Society for Microbiology (ASM 1991) for the strain designation. Note, however, that HTLV-III is preferably not used now for the human immunodeficiency virus; LAV and ARC are also now not acceptable. The proper designation for the HIV viruses is HIV-1, HIV-2, and so on.

VIRUS-MUTANT ABBREVIATIONS

21·8 A virus mutant is designated by an italicized, lowercase abbreviation closed up to an arabic numeral in a roman typeface.

*ts*112 [a temperature sensitive mutant] *hr*4 [a host range mutant]

CATALOGUES OF VIRUS NAMES AND STRAIN DESIGNATIONS

21·9 The American Type Culture Collection publishes a *Catalogue of Plant Viruses & Antisera,* (McDaniel and others 1993) and a *Catalogue of Animal Viruses and Antisera, Chlamydiae & Rickettsiae* (Buck and Paulino 1990), both of which provide names and strain designations. *Stedman's ICTV Virus Words* (Calisher and Fauquet 1992) gives all validly published virus names and valid synonyms, the 1990 rules of nomenclature, and explanations of alphabetization, capitalization, and name formation.

DESCRIPTION OF A VIRUS

21·10 Below is a sample of a typical format, as published in Francki and others (1991), for the formal description of a virus.

GROUP: Commelina yellow mottle virus group.

Compiled by: B.E.L. Lockhart and R. Hill (Francki and others 1991).

TYPE MEMBER: Commelina yellow mottle virus (CoYMV).

PROPERTIES OF THE VIRUS PARTICLE:

Morphology: Bacilliform particles ≈ 130 × 30 nm.

Physicochemical properties: CoYMV has a density in CsCl of 1.37 g/cm^3, cacao swollen shoot virus has a S_{20w} of 218.

Nucleic acid: One molecule of dsDNA: open circular molecules with single-strand discontinuities at specific sites, one in each strand. Mealybug transmitted viruses have genomes ≈ 7.5 kbp (7489 bp in CoYMV), and rice tungro bacilliform virus has a genome of ≈ 8.0 kbp.

Protein: Two protein species $\approx 40 \times 10^3$ and 35×10^3.

Lipid: None determined.

Carbohydrate: None detected.

Antigenic properties: Moderately efficient immunogens, serological relationships among some members.

REPLICATION:

Mechanism not determined but, as the genome has various properties in common with caulimoviruses, it is thought to involve reverse transcription.

BIOLOGICAL ASPECTS:

Host range: Narrow.

Transmission: Most members and possible members not transmissible mechanically; those that are, are only transmitted with difficulty. Members and possible members for which a vector is known are all transmitted by mealybugs in a semi-persistent manner except for rice tungro bacilliform virus which is leafhopper transmitted in association with rice tungro spherical virus, and rubus yellow net which is aphid transmitted.

OTHER MEMBERS:

Banana streak and sugarcane bacilliform viruses

Possible members: aucuba ringspot, cacao swollen shoot (10), canna yellow mottle, *Colocasia* bacilliform, *Dioscorea* bacilliform, *Kalanchoe* top-spotting, *Mimosa* bacilliform, *Rubus* yellow net (188), rice tungro bacilliform, *Schefflera* ringspot, and *Yucca* bacilliform viruses.

REFERENCE:

Lockhart BL. 1990. Evidence for a double-stranded circular DNA genome in a second group of plant viruses. Phytopathology 80: 127–131.

KEYS TO THE IDENTIFICATION OF VIRUS FAMILIES AND GROUPS

21·11 Keys for identifying virus families and groups have structures similar to keys for identifying bacteria, fungi, plants, and animals; see sections 22·13, 23·11–23·12, and Figure 23·1. The key below was excerpted in part from one developed by M A Mayo and C Fauquet (published in the work by Francki and others [1991]) and is a variation of a bracketed key described in Chapter 23 (see section 23·12).

[alternative characteristics]	[next relevant key level or name]
1 Host a prokaryote	2
Host a eukaryote	13
2 Genome of DNA	3
Genome of RNA	12

[alternative characteristics]		[next relevant key level or name]
3	Virion DNA double-stranded	4
	Virion DNA single-stranded	11
4	Virions with lipid-containing envelopes	5
	Virions not enveloped	7
5	Virions rod-shaped	*Lipothrixviridae*
	Virions not rod-shaped	6
6	Virions lemon-shaped, host an archaebacterium	SSV1 group
	Virions not lemon-shaped, host a mycoplasma	*Plasmaviridae*
7	Virions isometric without tails	8
	Virions with tails	9
8	DNA linear, >10 kbp	*Tectiviridae*
	DNA supercoiled, <10 kbp	*Corticoviridae*
9	Tails contractile	*Myoviridae*
	Tails not contractile	10
10	Tails long, DNA about 35 kbp	*Podoviridae*
	Tails short, DNA about 50 kbp	*Siphoviridae*
11	Virions icosahedral	*Microviridae*
	Virions rod-shaped	*Inoviridae*
12	RNA double-stranded	*Cystoviridae*
	RNA single-stranded	*Leviviridae*

REFERENCES

Cited References

[ASM] American Society for Microbiology. 1991. ASM style manual for journals and books. Washington: ASM.

Bojnansky V, Fargasová A. 1991. Dictionary of plant virology: In English, Russian, German, French, and Spanish. New York: Elsevier.

Buck C, Paulino G, editors. 1990. Catalogue of animal viruses & antisera, chlamydiae and rickettsiae. 6th ed. Rockville (MD): American Type Culture Collection.

Calisher CH, Fauquet CM. 1992. Stedman's ICTV virus words. Baltimore: Williams & Wilkins.

Diener TO, editor. 1987. The viroids. New York: Plenum.

Diener TO, Prusiner SB. 1985. The recognition of subviral pathogens. In: Maramorosch K, McKelvey JJ Jr, editors. Subviral pathogens of plants and animals: viroids and prions. New York: Academic Pr. p 3–8.

Francki RIB, Fauquet CM, Knudson DL, Brown F, editors. 1991. Classification and nomenclature of viruses: fifth report of the International Committee on Taxonomy of Viruses. Arch Virol (Suppl 2). New York: Springer-Verlag.

Lackie JM, Dow JAT, editors. 1989. The dictionary of cell biology. New York: Academic Pr.

Maramorosch K, McKelvey JJ Jr, editors. 1985. Subviral pathogens of plants and animals: viroids and prions. New York: Academic Pr.

Matthews REF. 1991. Plant virology. 3rd ed. New York: Academic Pr.

McDaniel LL, Cox RL, Maratos ML, editors. 1993. Catalogue of plant viruses & antisera (includes viroids). 7th ed. Rockville (MD): American Type Culture Collection.

Murphy FA, Fauquet CM, Bishop DHL, Ghabrial SA, Jarvis AW, Martelli GP, Mayo MA, Summers MD, editors. 1994. Classification and nomenclature of viruses: sixth report of the International Committee on Taxonomy of Viruses. New York: Springer-Verlag.

Van Regenmortel MH, Maniloff J, Calisher CH. 1991. The concept of virus species. Arch Virol 120:313-4.

Walkey DGA. 1991. Applied plant virology. 2nd ed. New York: Chapman & Hall.

Whittaker RH. 1969. New concepts of kingdoms of organisms. Science 163:150–60.

Additional References

Coffin J, Haase A, Levy JA, Montagnier L, Oroszlan S, Teich N, Temin H, Toyoshima K, Varmus H, Vogt P, Weiss R. 1986. Human immunodeficiency viruses. Science 232:697–8.

Coffin J, Haase A, Levy JA, Montagnier L, Oroszlan S, Teich N, Temin H, Toyoshima K, Varmus H, and others. 1987. AIDS virus nomenclature. Med Lab Sci 44:113–4.

Fraenkel-Conrat H. 1985. The viruses: catalogue, characterization, and classification. New York: Plenum.

Gallo R, Wong-Staal F, Montagnier L, Haseltine W, Yoshida M. 1988. HIV/HTLV gene nomenclature. Nature 333:504.

Gust ID, Burrell CJ, Coulepsis AG, Robinson WS, Zuckerman AJ. 1986.

Taxonomic classification of human hepatitis B virus. Intervirology 25:14–29.

Prusiner SB. 1991. Molecular biology of prion diseases. Science 252:1515–22.

Singleton P, Sainsbury D. 1987. Dictionary of microbiology and molecular biology. 2nd ed. New York: J Wiley.

22 Bacteria

> In order to prove experimentally that a microscopic organism is truly an agent of disease and contagion, I do not see any other method . . . than to submit the microbe . . . to the method of successive cultures outside the living organism. . . . This is precisely the technique to which M. Joubert and I submitted the anthrax bacterium.
>
> — Louis Pasteur, "La théorie des germes et ses applications à la médicine et à la chirurgie", *Compt rend acad sci* 86:1037–43, 1878

22·1 Bacteria (singular, "bacterium") are, according to Sneath (1986), single-celled organisms in the Kingdom Procaryotae (=Prokaryotae). In the 5-kingdom system (Whittaker 1969), bacteria are placed in the Kingdom Monera (see Table 7·6). They multiply by regular binary fission and separation of daughter cells; some bacteria exchange genetic material through a mechanism that has been compared to a sexual mode (conjugation). Bacteria can be rod-shaped ("bacillus"; plural, "bacilli"), spherical ("coccus"; plural, " cocci"), comma-shaped ("vibrio"; plural, "vibrios"), spiral-shaped ("spirillum"; plural, "spirilli"; known also as "spirochetes"), and other shapes. Bacteria are normally haploid and lack a clearly defined nucleus. DNA passes from 1 cell (the donor) to another (the recipient) by transduction, transformation, and conjugation.

ADDITIONAL SCOPE OF THE TERM "BACTERIA"

22·2 "Bacteria" include rickettsiae (singular, "rickettsia"), chlamydiae (singular, "chlamydia"), mycoplasmas (singular, "mycoplasma"), cyanobacteria (formerly known as blue-green algae, which are prokaryotes and are unicellular or filamentous), and actinomycetes (singular, "actinomycete"). The term "rickettsia" is used for 3 genera: *Rickettsia, Rochalimaea,* and *Coxiella.*

Rickettsia sennetsu Misao and Kobayashi 1956
Rochalimaea vinsonii Weiss and Dasch 1982
Coxiella burnetii (Derrick 1939) Philip 1948

This terminology is based on history and convenience (Weiss and Moulder 1984). Rickettsiae are mainly rod-shaped, sometimes coccoid, and often pleomorphic gram-negative microorganisms with typical bacterial cell walls and no flagella, and they are, with 1 exception, obligate intracellular parasites of eukaryotic hosts. Chlamydiae are a subgroup of rickettsiae.

Chlamydia trachomatis (Busacca 1935) Rake 1957 ATCC VR 571

22·3 Mycoplasmas are prokaryotes in the Class *Mollicutes* and are the smallest free-living cells without cell walls, bounded only by a lipoprotein cell membrane; they evolved from the eubacteria (Maniloff 1983). Cell-wall-less microorganisms associated with plant diseases are of 2 types: helical microbes termed "spiroplasmas", and nonhelical mycoplasma-like organisms ("MLO").

The *Mollicutes* consists of only 1 order, *Mycoplasmatales,* which includes 6 genera: *Mycoplasma, Ureaplasma, Spiroplasma, Acholeplasma, Thermoplasma,* and *Anaeroplasma.* The taxonomy of *Mollicutes* is described by Whitcomb (1984) and that of *Spiroplasma* by Whitcomb and others (1987).

Actinomycetes represent a suprageneric classification of bacteria based on partial sequencing of 16S ribosomal ribonucleic acids and a judicious selection of chemical, morphologic, and physiologic characters. The 7 groups included are actinobacteria, nocardioforms, actinoplanetes, thermomonospora, maduromycetes, streptomycetes, and multilocule sporangia.

Some bacteria are termed "fastidious". These are bacteria that have complex nutritional requirements; such bacteria range from strict anaerobes to archaeobacteria and others such as the rickettsiae, rickettsia-like organisms, mycoplasma-like organisms, spiroplasmas, and both xylem-limited and phloem-limited bacteria.

Table 22·1 Suffixes and modifiers for taxonomic ranks of bacteria[a]

Rank	Suffix or modifier	Example
Order	*ales*	*Pseudomonadales*
Suborder	*ineae*	*Pseudomonadineae*
Family	*aceae*	*Pseudomonadaceae*
Subfamily	*oideae*	*Pseudomonadoideae*
Tribe	*eae*	*Pseudomonadeae*
Subtribe	*inae*	*Pseudomonadinae*
Genus		*Pseudomonas*
Species		*P. syringae*
	sp. (sing.)	*Pseudomonas* sp.
	spp. (plural)	*Pseudomonas* spp.
Pathovar	pv.	*P. syringae* pv. *tabaci*

[a]This table illustrates changes in endings at different taxonomic levels. The correct spelling of names (including the proper endings determined by noun genders; see section 23·8) can be confirmed in standard medical dictionaries and in specialized sources such as culture-collection catalogues and *Stedman's Bergey's Bacteria Words* (Holt and others 1992).

Xylella fastidiosa Wells et al. 1987 is a pathogen of plants and includes all known strains of fastidious gram-negative, xylem-limited bacteria.

"Microbiology" is a term representing the disciplines that study all forms of microscopic life, including bacteria, yeasts and other fungi, 1-celled algae, some nematodes, protozoa, and viruses.

TAXONOMY

22·4 There is no official classification of bacteria by a national or international organization. In nomenclature, however, each taxon can have only 1 name within a classification scheme, but each can have as many names as there are schemes. A name is valid when designated in accordance with the international rules (ICSB 1992); see section 23·5. The systematics of bacteria are similar to those in the botanical sciences. The names of taxa from the rank of order to subtribe, inclusive, are formed by adding a specific suffix to the name of the type genus; all taxonomic epithets (see section 22·6) are italicized, but abbreviations of species ("sp." and "spp.") and pathovar designations are not (Table 22·1).

For additional discussion of the concepts "taxonomy" and "nomenclature", see sections 23·2 and 24·1.

NUMERICAL TAXONOMY

Numerical taxonomy (taxometrics) depends heavily on computer methods, especially to define homogeneous clusters of strains. The data needed for numerical taxonomy are identified by Sneath (1989). The characters used for identification must be coded into a format suitable for use by a computer. Although data from any taxa can be used, most data are derived from characters of strains. These characters are referred to as "operational taxonomic units" (OTUs). The character data must be coded so that the presence of a character is represented by a "+" or a "1", and the absence of a character by "–" or "0"; if the character is an all-or-nothing character, a 2-state coding is applied: for example, "propionate formed from lactate" is represented by + or 1; "propionate not formed from lactate" is represented by – or 0. If characters represent degrees of response, a multistate coding is needed: for example, + for "weakly proteolytic", ++ for "mildly proteolytic", and +++ for "strongly proteolytic". These values are compiled in a table, and the degree of relatedness is determined among strains.

The steps in this process are as follows (Sneath 1989).

1 Collection of data: bacterial strains are selected and examined for the number of relevant properties (taxonomic characters).
2 The data are coded and scaled appropriately.
3 The similarity or resemblance between strains is calculated. This yields a table of similarities (similarity matrix).
4 Similarities are analyzed for taxonomic structure to yield groups or clusters; then strains are arranged into phenons (phenetic groups) that are broadly equated with taxonomic groups.
5 Properties of phenons are tabulated for publication or further study. Appropriate characters are selected (diagnostic characters) to set up identification systems that enable best identification of additional strains.

NOMENCLATURE: SCIENTIFIC NAMES

22·5 The names of bacteria are governed by the International Code of Nomenclature of Bacteria (ICSB 1992). The Code does not cover taxonomic ranks lower than subspecies. Names of taxa have to be both effectively and validly published to be recognized. To be effectively published, the description of the taxon must appear in a printed publication that is sold to the general public or to a bacteriologic institution. To be valid, the name of the taxon must be accompanied by a description published in the *International Journal of Systematic Bacteriology* (IJSB); if the name and description are published elsewhere, a reprint must be sent to this journal for its notation. The notation must include the name, nomenclatural type, author, title, and journal.

Classifications and spellings of scientific names appear in the 4 volumes of *Bergey's Manual of Systematic Bacteriology,* of which J G Holt is editor-in-chief (Krieg 1984, Sneath 1986, Staley 1989, Williams 1989) or in the *Approved Lists of Bacterial Names* (Skerman and others 1980). A list of "bacteria words" taken from *Bergey's Manual* has been prepared by Holt and others (1992). An index of the bacterial and yeast nomenclatural changes from 1 January 1980 to 1 January 1992 in the *International Journal of Systematic Bacteriology* has been published by Moore and Moore (1992). Priority of publication for bacterial names dates from 1 January 1980. Names published before that date are valid only if included in the *Approved List of Bacterial Names* (Skerman and others 1980). Names published before 1980 and not included in this list are invalid. It is the list that is approved, not the names; there can be more than 1 valid name for the same entity, all of which are on the approved list of names. Names often carry a superscript "AL" meaning that the name is on the approved list.

> *Methanococcus vannielii* Stadtman and Barker 1951^{AL} is the type species for the genus.

A nomenclatural type is the constituent element of a taxon to which the name of the taxon is permanently attached. The type of an order, suborder, family, subfamily, tribe, or subtribe is the genus on the name of which the name of the higher taxon is based. The type of a genus is the species and that of the species and subspecies is the strain.

> *Xanthomonas fragariae* Kennedy and King 1962, type strain NCPPB 1469

The type strain may be represented by the original description or by illustrations and specimens when cells cannot be maintained in culture or have been lost. The word "type" does not mean that it is typical; it only means that it is a reference specimen for the name.

22·6 Names of all taxa are italicized. The species name is binomial, consisting of 2 Latin names, a genus name and a specific epithet. The genus name is capitalized and italicized, and the specific epithet is in lowercase letters and italicized, even if the specific epithet is derived from the name of a person.

> *Shigella boydii* Ewing 1949 [named after Sir John Boyd]

Note that "epithet" is used here with this particular biologic meaning and not with the more frequent, common meanings of a substitute for a proper name (such as "The Great Emancipator" for Abraham Lincoln) or a derogatory tag (such as a racial epithet); "specific" is derived from "species".

A genus name should always be followed by a specific epithet or, if the epithet is unknown, by "species," "sp." (singular), or "spp." (plural), for example, *Erwinia* sp. Exceptions are permitted when the entire genus is referred to, for example, "strains or species of *Rhizobium*", "the genus *Erwinia*". If the genus is part of the name of a plant disease, for example "*Erwinia* soft rot of potato" the genus name should be capitalized and italicized to be consistent with convention in all organizations dealing with usage of generic names in disease names. But see section 22·15 on adjectival forms in medicine.

The binomial taxonomic term is always singular.

> *Escherichia coli* (Migula 1895) Castellani and Chambers 1919 was . . .
>
> [but] *E. coli* strains were . . .

A genus name should be spelled out in the title of an article and on 1st mention in the abstract and text. After that, it can be abbreviated to the initial letter followed by a period and the species epithet. When 2 or more genera with the same initial letter are mentioned, genus names should be spelled out in plant and microbiologic publications, but may be abbreviated to 2 or 3 letters elsewhere; also see section 24·2.

> *Thermus aquaticus* Brock and Freeze 1969 [abbreviated] *T. aquaticus* [not "*Th. aquaticus*"]
>
> *Micromonospora echinospora* subsp. *ferruginea* Luedemann and Brodsky 1964 and *M. echinospora* subsp. *pallida* Luedemann and Brodsky 1964 were described . . .
>
> *Campylobacter fetus* (Smith and Taylor 1919) Sebald and Veron 1963 subsp. *venerealis* (Florent 1959) Veron and Chattelain 1973 [on 1st mention] [thereafter] *C. fetus* subsp. *venerealis* [not "subsp. *venerealis*" or "*C. f.* subsp. *venerealis*"]

The taxon "variety" or "var." is not used for bacteria.

22·7 The type strain of a species is a strain on which the original description of the species is based and is the strain designated by the author or a subsequent author as the type strain. This strain should be designated by a strain number of the author and the strain number of at least 1 culture collection where the culture is available. For the *International Journal of Systematic Bacteriology* a superscript T is required to designate the type strain at each occurrence in the text, tables, and figures.

> We obtained strain ATCC 10792^{T} from . . .

Bacterial strains can be designated by letters, numbers, or both and are printed either solid or with a space, depending on the collection.

PAO123 [or] PAO 123

Clostridium argentinense Suen et al. 1988 Type ATCC 27322

Vibrio furnissii Brenner et al. 1984 Type strain CDCB3215

The strain designation should be written with the full binomial.

Pseudomonas aeruginosa ATCC 10145

If the specific epithet is missing, the word "strain" should be inserted.

Pseudomonas strain B13 [or] *Pseudomonas* sp. strain B13.

The American Type Culture Collection (Gherna and Pienta 1992), The Japan Collection of Microorganisms (Nakase 1989), and others have published catalogues of strains.

Diacritical marks are not used in scientific names.

1 ä, ö, ü become ae, oe, and ue, respectively.
2 é, è, and ê become e.
3 ø, æ and å become oe, ae, and aa, respectively.

INFRASUBSPECIFIC TAXA

22·8 Some organisms that cannot be differentiated taxonomically at the level of subspecies are given infrasubspecific designations that are not covered in the Code and are therefore excluded from the *Approved List.* These designations include "pathovars" (pv.), "biovars" (bv.), "serovars" (sv.), "phagovars", "chemovars", and "morphovars".

The term "pathovar" is derived from *pathovarietas.* The pathovar designation, but not its abbreviation (pv.), is written in italics. Pathovars (pathotypes) are distinguished primarily but not solely by their pathogenic characteristics. Authorities are not required for pathotype designations. The suffix "var" is preferred to "type" to avoid confusion with nomenclatural "types".

Corynebacterium michiganense (Smith) Jensen 1977 pv. *tritici* causes spike blight in wheat.

Because the nomenclature of pathovars is not governed by the Code, a committee of the International Society for Plant Pathology drew up a set of international standards for naming pathovars of phytopathogenic bacteria as well as a list of pathovars and their representative pathotype cultures (Dye and others 1980). A newly described pathovar is cited as "*pathovarietas nova* (pv. nov.)".

Xanthomonas campestris (Pammel 1895) Dowson 1939 pv. *caladiae* pv. nov.

The citation of the name of a pathovar reported previously should bear the name of the author(s) of the publication in which the pathovar epithet was proposed formally, followed by the date of publication. Full citation of the publication should include the number of the page in the body of the text (not in the summary or abstract) where the name was proposed.

Xanthomonas campestris (Pammel 1895) Dowson 1939 pv. *cannabina* Severin 1978, 13

[indicates that the pathovar was proposed by Severin in 1978, and the proposed name can be found on page 13 of the publication cited]

22·9 Another infrasubspecific rank is the biovar (biotype) with special biochemical or physiologic properties.

Agrobacterium tumefaciens (Smith and Townsend 1907) Conn 1942 bv. 3

Serovars (serotypes) have distinctive antigenic properties.

Erwinia chrysanthemi Burkholder et al. 1953 sv. IV

Phagovars (phagotypes) are designated for bacteria susceptible to viruses that can lyse bacteria. Phage designations can be numbers or letters, but they are not italicized unless they are written in Latin or they designate genes.

Azotobacter vinelandii Lipman 1903 phagovar A41 ATCC 12518-B10

Bacillus cereus Frankland and Frankland 1887 phagovar *Phagus pertinax* ATCC 12826-B1

Escherichia coli (Migula 1895) Castellani and Chalmers 1919 phagovar Alpha 3 ATCC 13706-B2

Mycobacterium tuberculosis (Zopf 1883) Lehmann and Neumann 1896 phagovar D26A ATCC 25618-B2

Other examples of phage designations are “Charon”, “lambda”, “mu”, “P1*vir*”. Morphovars (morphotypes) have special morphologic features, and chemovars (chemotypes) have special chemical features.

DESCRIPTIONS OF TAXA

22·10 Descriptions of taxa should include the following information.

1 A list of strains and their sources.
2 Descriptions or citations to descriptions of methods used to characterize strains.

3 Characters required for inclusion in the next higher taxon.
4 The diagnostic characters that distinguish the taxon from closely related taxa.
5 Reaction of the type strain for all characters that vary among strains within the species.

To describe a bacterium, certain key characteristics should be listed (Johnston and Booth 1983).

1 Morphologic characteristics of cells and special structures.
2 Staining reactions.
3 Cultural characteristics.
4 Oxygen requirements.
5 Physiologic characters.
6 Biochemical characters.
7 Phage relations.
8 Serologic relations.
9 Nucleic acid data.
10 Sensitivity to antibiotics and bacteriocins.
11 Host range and symptoms, if pathogenic.

If a species is described from examination of more than 1 strain, additional information must be provided.

12 The frequency of each character given as the number of strains with the character compared with the total number of strains studied.
13 A separate description of the type (or neotype) strain.

In describing strains, photomicrographs or electron micrographs that illustrate morphologic or anatomic characteristics pertaining to classification or identification should be included. Methods for presenting such data appear in the "Instructions for Authors" section in the January issues of the *International Journal of Systematic Bacteriology.*

22·11 In describing or referring to a species, the author should follow the specific epithet with a date but without punctuation; "and" is used in place of "et" or the ampersand (&) for 2 authors, contrary to the recommendation in the International Code of Botanical Nomenclature.

> *Erwinia quercina* Hildebrand and Schroth 1967

If there are more than 2 authors for a taxon, "et al." can be used instead of the complete list of authors.

> *Legionella pittsburghensis* Pasculle, Feeley, Gibson, Cordes, Myerowitz, Patton, Gorman, Carmack, Ezzel, and Dowling 1980

[or] *Legionella pittsburghensis* Pasculle et al. 1980.

If the species has been reclassified using the same specific epithet, the 1st author's name appears within parentheses after the specific epithet, followed by the name(s) of the person(s) who reclassified it.

Bacterium herbicola Löhnis 1911 [was reclassified by Dye, so the name becomes] *Erwinia herbicola* (Löhnis 1911) Dye 1964

Descriptions of bacterial species are written in English, not Latin; this is a requirement of the International Code of Nomenclature of Bacteria (ICSB 1992).

Below is an example of the species description of *Spirochaeta thermophila* sp. nov. Aksenova et al. 1992 (Aksenova and others1992). Note that the author citation is not included in the species description.

Spirochaeta thermophila sp. nov. *Spirochaeta thermophila* (ther.mó.phi.la. Gr. n. *therme,* heat; Gr. adj. *phila,* loving; M. L. adj. *thermophila,* heat-loving). Single helical cells are 0.2 to 0.25 by 16 to 50 μm. Negative Gram stain reaction. No lysis of the cells occurs in 3% KOH. An outer sheath encloses a protoplasmic cylinder; two periplasmic flagella are present in a 1-2-1 arrangement and are subterminally anchored by an insertion disc. Strictly anaerobic chemoorganotroph. The temperature range for the type strain is 40 to 73 °C (optimum, 66 to 68 °C). The pH range for the type strain is 5.9 to 7.7 (optimum, 7.5). The NaCl concentration range for the type strain is 0.5 to 4.5% (optimum, 1.5%) (doubling time, 70 min). The temperature, pH, and salinity parameters vary for different strains, reflecting the environmental conditions prevailing at the sites of isolation. No reduction of fumarate, nitrate, oxygen, sulfate, or sulfur occurs. Utilizes various mono-, di-, and polysaccharides but not sugar alcohols, organic acids, or amino acids. Inhibited by penicillin, neomycin, erythromycin, tetracycline, polymyxin B, and novobiocin but resistant to rifampin and streptomycin. The fermentation end products from glucose are acetate, carbon dioxide, hydrogen, and lactate. Glucose is fermented via the Embden–Meyerhof–Parnas pathway, involving a pyrophosphate-dependent phosphofructokinase. Ethanol and succinate are not produced. Indole is not formed; urea is not hydrolyzed. Sulfide is not produced from cysteine, and esculin is hydrolyzed. The guanine-plus-cytosine content of the DNA is 52 mol% (as determined by the thermal denaturation method). Type strain Z-1203 (=DSM 6578) was isolated from a marine hot spring near the beach on Shiashkotan Island, Soviet Far East, USSR.

If an author designated a type strain of a species or subspecies, that strain becomes the holotype (see "Type strain Z-1203 [=DSM 6578]" in the description above). If an author described only a single strain and did not designate a holotype, that strain is referred to as a monotype. If no holotype is designated, 1 of the strains on which the original author based the description may be designated the type strain and referred to as a lectotype. If the strain on which the original description is based is not found, a neotype may be proposed; it must be described and published in the *International Journal of Systematic Bacteriology* in a manner similar to a species description. For additional detail on publication style for bacteriologic literature, consult the *ASM Manual for Journals and Books* (ASM 1991).

CODING GENERIC NAMES FOR COMPUTER STORAGE

22·12 It is convenient to use computers for storing and retrieving information readily. Thus, Krichevsky and others (1980) devised a scheme for coding 458 generic names of bacteria based on the following principles.

1 The 1st letter of the abbreviation is the 1st letter of the generic name.
2 All letters in the abbreviation are found in the generic name, except F for PH.
3 Code letters consist of 2–4 letters.
4 Abbreviations differ in at least 2 letters, where possible.

Note these examples.

Acetobacter (ACBT), *Acetobacterium* (ACBM), *Acetomonas* (ACMN), *Bacillus* (BC), *Bacterium* (BT), *Bactoderma* (BTDA), *Cytophaga* (CTFG), *Enterobacter* (ENBT), *Erwinia* (ERWN), *Rickettsia* (RKTS), *Streptococcus* (STCO), *Xanthomonas* (XNMN).

KEYS TO SPECIES OF BACTERIA

22·13 Bacteria can be identified through dichotomous keys. A key may be either an "indented" key or a "bracketed" key. An indented key is illustrated in Figure 23·1 and discussed in section 23·11, where the rationale and principles used in making keys to organisms for identification are given. A key to the species of *Fusobacterium* Knorr 1922 appears in *Bergey's Manual* (Moore and others 1984) and is revised here as a bracketed key, using the principles described in section 23·11.

1 Esculin hydrolyzed 2
1 Esculin not hydrolyzed 4
  2 Propionate formed from threonine 3
  2 Propionate not formed from threonine . . . . *F. prausnitzii*

3 Lactose acid . . . . *F. mortiferum*
3 Lactose not acid . . . . *F. necrogenes*
4 Propionate formed from threonine 5
4 Propionate not formed from threonine 10
5 Indole produced 6
5 Indole not produced 9
6 Propionate formed from lactate . . . . *F. necrophorum*
6 Propionate not formed from lactate 7
7 Mannose produces weak acid . . . . *F. varium*
7 Mannose does not produce acid 8
8 Hydrogen produced abundantly, cells pleomorphic . . . . *F. gonidiaformans*
8 Hydrogen not produced, cells thin with tapered ends . *F. nucleatum*
9 Mannose produces weak acid . . . . *F. varium*
9 Mannose does not produce acid, cells coccoid . . . . *F. perfoetens*
10 Indole produced . . . . *F. naviforme*
10 Indole not produced . . . . *F. russii*

VERNACULAR NAMES AND ADJECTIVAL FORMS

22·14 Vernacular names of bacteria are always set in roman lowercase letters.

brucella   bacillus   vibrio   pseudomonad

Plural forms may take the Latin or English form.

brucellae   bacilli   vibrios   erwinias   salmonellae

One should avoid using "bacillus" or "bacilli" if there might be confusion as to whether the genus *Bacillus* or a nonspecific rod-shaped bacterium is meant. (The word *Bacilli,* in italics and capitalized, is never correct.) Organisms in the genus *Treponema* can be designated by "treponema" (singular) and "treponemas" or "treponemata" (plural). Note some additional examples of vernacular names (singular and plural).

klebsiella, klebsiellae   streptomycete, streptomycetes
mycobacterium, mycobacteria   citrobacter, citrobacters

Additional vernacular names and plural forms can be found in some general dictionaries and in medical dictionaries.

22·15 Although international rules do not cover adjectives derived from scientific names, some guidelines have been widely followed. Thus

Table 22·2 Names (full and abbreviated) and addresses of major culture collections

Abbreviation	Name and location
ATCC	American Type Culture Collection, 12301 Parklawn Drive, Rockville MD 20852, USA
CBS	Centraalbureau voor Schimmelcultures, Oosterstraat 1, Baarn, The Netherlands
CCM	Czechoslovak Collection of Microorganisms, J.E. Purkyne University, Tr. Obr. Miru 10, Brno, Czech Republic
CIP	Collection of the Institut Pasteur, rue du Dr. Roux, 75015 Paris, France
CNCTC	Czechoslovak National Collection of Type Cultures, Institute of Epidemiology and Microbiology, Srobarova 48, Prague 10, Czech Republic
DSM	Deutsche Sammlung von Mikroorganismen, Mascheroder Weg 1b, D-3300 Braunschweig, Germany
IAM	Institute of Applied Microbiology, University of Tokyo, Bunkyo-ku, Tokyo, Japan
ICMP	International Collection of Microorganisms from Plants, Plant Diseases Division, Department of Scientific and Industrial Research, Auckland, New Zealand
IFO	Institute for Fermentation, 4-54 Jusonishinocho, Osaka, Japan
IMET	Institut für Mikrobiologie und Experimentelle Therapie, Deutsche Akademie der Wissenschaften zu Berlin, Beuthenbergstrasse 11, Jena 69, Germany
JCM	Japan Collection of Microorganisms, Riken, Wako, Saitama 351-01, Japan
KCC	Kaken Chemical Company Ltd, 6-42 Jujodai-1-Chome, Tokyo 114, Japan
LIA	Museum of Cultures, St. Petersburg Research Institute of Antibiotics, 23 Ogorodnikov Prospect, St. Petersburg L-20, Russia
LMD	Laboratorium voor Microbiologie, Technische Hogeschool, Julianalaan 67a, 2623 BC Delft, The Netherlands
NCFB	National Collection of Food Bacteria, AFRC Institute of Food Research, Shinfield, Reading RG2 9AT, England, United Kingdom
NCIMB	National Collection of Industrial and Marine Bacteria, 23 St. Machar Drive, Aberdeen AB2 1RY, Scotland, United Kingdom
NCPPB	National Collection of Plant Pathogenic Bacteria, Plant Pathology Laboratory, Hatching Green, Harpenden, Herts. AL5 2BD, England, United Kingdom
NCTC	National Collection of Type Cultures, Central Public Health Laboratory, 61 Colindale Avenue, London NW9 5HT, England, United Kingdom
NRC	National Research Council, Sussex Drive, Ottawa K1E 0R6, Canada.
NRL	Neisseria Reference Laboratory, US Public Health Service Hospital, Seattle WA 98114, USA
NRRL	Northern Regional Research Center, Agricultural Research Service, US Department of Agriculture, Peoria IL 61604, USA
UQM	Culture Collection, Department of Microbiology, University of Queensland, Herston, Brisbane 4006, Australia
VKM	Institute of Microbiology, Russian Academy of Sciences, Moscow, Russia

adjectival forms usually end in "al", but noun forms can also serve as adjectives. A formal genus name can be used as a vernacular adjective if it is in roman lowercase letters.

streptococcal infections   streptococcus infection
corynebacterium growth   corynebacterial growth
He died of pneumocystis pneumonia.

A formal genus name should not be used as a modifier unless it is meant to refer to all species of the genus. A binomial can be used as a modifier (Huth 1989) when only that species is referred to.

Pseudomonas aeruginosa bacteremia [but not "*Pseudomonas* bacteremia"]
He studied all *Pseudomonas* species.
Pneumocystis carinii pneumonia [but not "*Pneumocystis* pneumonia"]

Adjectival forms must not be derived from specific epithets; for example, do not use "coli" for *Escherichia coli,* or "amylovora" for *Erwinia amylovora.*

CULTURE COLLECTIONS

22·16 There are hundreds of culture collections around the world; some are small, specialized collections maintained by 1 person. *World Directory of Collections of Cultures of Microorganisms* (Staines and others 1986) is a guide to large collections. Smaller but frequently cited collections of bacteria are listed in *Bergey's Manual* (Krieg 1984). Some of the managers of these collections are willing to receive cultures and make them available to others. Sometimes the identification of the organism can be confirmed by the organization housing the collection. For a list of some major collections and the abbreviations of their names, see Table 22·2.

REFERENCES

Cited References

Aksenova HY, Rainey FA, Janssen PH, Zavarzin GA, Morgan HW. 1992. *Spirochaeta thermophila* sp. nov., an obligately anaerobic, polysaccharolytic, extremely thermophilic bacterium. Int J Syst Bacteriol 42:175–7.

[ASM] American Society for Microbiology. 1991. ASM style manual for journals and books. Washington: ASM.

Dye DW, Bradbury JF, Goto M, Hayward AC, Lelliott RA, Schroth MN. 1980. International standards for naming pathovars of phytopathogenic bacteria and a list of pathovar names and pathotype strains. Rev Plant Pathol 59:153–68.

Gherna R, Pienta P, editors. 1992. Catalogue of bacteria and phages. 18th ed. Rockville (MD): American Type Culture Collection.

Holt JG, Bruns MA, Caldwell BJ, Pease CD. 1992. Stedman's Bergey's bacteria words. Baltimore: Williams & Wilkins.

Huth EJ. 1989. Style notes: taxonomic names in microbiology and their adjectival derivatives. Ann Intern Med 110:419–20.

[ICSB] International Committee on Systematic Bacteriology. 1992. International code of nomenclature of bacteria: bacteriological code (1990 revision). Washington: American Soc for Microbiology.

Johnston A, Booth C, editors. 1983. Plant pathologist's pocketbook. 2nd ed. Farnham Royal (UK): Commonwealth Agricultural Bureaux.

Krichevsky MI, Walczak CA, Rogosa M, Johnson R. 1980. Interchange of abbreviations and full generic names in computers. Int J Syst Bacteriol 30:585–93.

Krieg NR, editor. 1984. Bergey's manual of systematic bacteriology. Volume 1. Baltimore: Williams & Wilkins.

Maniloff J. 1983. Evolution of wall-less prokaryotes. Annu Rev Microbiol 37:477–99.

Moore WEC, Holdeman LV, Kelley RW. 1984. Genus II. *Fusobacterium* Knorr 1922,4^{AL}. In: Krieg NR, editor. 1984. Bergey's manual of systematic bacteriology. Volume 1. Baltimore: Williams & Wilkins. p 631–7.

Moore WEC, Moore LVH. 1992. Index of the bacterial and yeast nomenclatural changes: published in the International Journal of Systematic Bacteriology since the 1980 Approved Lists of Bacterial Names (1 January 1980 to 1 January 1992). Washington: American Soc for Microbiology.

Nakase T, editor. 1989. JCM catalogue of strains. 4th ed. Saitama: Japan Collection of Microorganisms, Inst of Physical and Chemical Research.

Skerman VBD, McGowan V, Sneath PHA. 1980. Approved lists of bacterial names. Int J Syst Bacteriol 30:225–420. Reprinted with corrections by American Soc for Microbiology, 1989.

Sneath PHA, editor. 1986. Bergey's manual of systematic bacteriology. Volume 2. Baltimore: Williams & Wilkins.

Sneath PHA. 1989. Numerical taxonomy. In: Williams ST, editor. Bergey's manual of systematic bacteriology. Volume 4. Baltimore: Williams & Wilkins. p 2303–5.

Staines JE, McGowan VF, Skerman VBD, editors. 1986. World directory of collections of cultures of microorganisms. 3rd ed. Brisbane, Australia: World Data Center for Microorganisms, Univ of Queensland.

Staley JT, editor. 1989. Bergey's manual of systematic bacteriology. Volume 3. Baltimore: Williams & Wilkins.

Weiss E, Moulder JW. 1984. Genus I. *Rickettsia* da Rocha-Lima 1916, 567^{AL}. In: Krieg NR, editor. Bergey's manual of systematic bacteriology. Volume 1. Baltimore: Williams & Wilkins. p 688–98.

Whitcomb RF. 1984. International Committee on Systematic Bacteriology Subcommittee on the Taxonomy of *Mollicutes*. Int J Syst Bacteriol 34:361–5.

Whitcomb RF, Bové JM, Chen TA, Tully JG, Williamson DL. 1987. Proposed criteria for an interim serogroup classification for members of the genus *Spiroplasma* (Class *Mollicutes*). Int J Syst Bacteriol 37: 82–4.

Whittaker RH. 1969. New concepts of kingdoms of organisms. Science 163:150–60.

Williams ST, editor. 1989. Bergey's manual of systematic bacteriology. Volume 4. Baltimore: Williams & Wilkins.

Additional References

Cowan ST. 1978. A dictionary of microbial taxonomy. Cambridge (UK): Cambridge Univ Pr.

Jeffrey C. 1989. Biological nomenclature. 3rd ed. London: Edward Arnold.

Lentner C. 1993. Geigy scientific tables. Volume 6, bacteria, fungi, protozoa, helminths. 8th ed. West Caldwell (NJ): Ciba-Geigy.

MacAdoo TO. 1990. Proposed revision of appendix 9, orthography, of the International Code of Nomenclature of Bacteria. Int J Syst Bacteriol 40:103–4.

Singleton P, Sainsbury D. 1993. Dictionary of microbiology and molecular biology. 2nd ed. New York: J Wiley.

23 Plants, Fungi, Lichens, and Algae

> Then God said, "Let the earth bring forth vegetation: plants yielding seed, and the fruit trees yielding fruit of every kind on earth that bear fruit with the seed in it". And it was so.
>
> —*The Holy Bible* (New Revised Standard Version), *Genesis* 1:11

23·1 Early works on taxonomic classifications recognized plants as 1 of 2 kingdoms: Plants and Animals. The 5-kingdom classification developed by Whittaker (1969) has become widely cited but not universally adopted: Monera (bacteria and blue-green algae), Protista (protozoa and simple algae), Fungi, Plantae, and Animalia. The Kingdom Monera comprises prokaryotes and unicellular forms, Protista include eukaryotic and unicellular forms, and the other 3 kingdoms are eukaryotic and multicellular. Margulis and others (1990) use the term "Protoctista" instead of "Protista" for the kingdom name.

PLANTS

TAXONOMY AND NOMENCLATURE

23·2 Taxonomy is the study and description of variation in organisms: its causes and consequences. The data are arranged in a system of classification (Stace 1989). Nomenclature, on the other hand, is the naming of organisms and the establishment, interpretation, and application of regulations governing the system. A classification is a systematic arrangement of categories, each containing any number of organisms, which allows easy reference to its components. Taxonomy and nomenclature are parts of systematics (Woodland 1991).

A rank is a level in a hierarchical classification; a plant taxon is a named group of plants at a rank in a classification. To describe a taxon is to fully state its characters. A diagnosis is a listing of characters that distinguish related taxa from each other. Information on reproductive and vegetative characters is essential, but chromosomal information is often useful.

> In *Krigia oppositifolia* Raf. $2n = 8$, but in *K. biflora* (Walter) S. F. Blake $2n = 10$, where n equals the number of sets of chromosomes.
>
> In *Taraxacum officinale* Wiggers $2n = 16$ to 48, often 24, whereas in *T. spectabile* Dahlst. $2n = 40$. Species are confluent through polyploidy and apomixis, making taxonomy confusing (Gleason and Cronquist 1991).

Chemotaxonomic information is also useful. For example, species of *Daucus* can be distinguished from species of *Torilis* by analysis for specific terpenoids.

> *Daucus* (11 species) produces the terpenoids α-pinene, β-pinene, limonene, geranyl acetate, and caryophyllene, whereas *Torilis* (5 species) does not. *Torilis,* however, produces biphenyl, but *Daucus* does not (Harborne and Turner 1984).

INTERNATIONAL CODE OF BOTANICAL NOMENCLATURE

23·3 The International Code of Botanical Nomenclature establishes the principles and recommendations for naming plants (ICBN).

The 6 principles of the Code (IBC 1988) have been carried forward from earlier versions of the Code.

1 Botanic nomenclature is independent of zoologic nomenclature.
2 Application of names of taxonomic groups is determined by means of nomenclatural types.
3 Nomenclature of taxonomic groups is based on priority of publication.
4 Each taxonomic group with a particular circumscription, position, and rank can bear only 1 correct name, the earliest that is in accordance with the rules, with certain specified exceptions.
5 Scientific names of taxonomic groups are treated as Latin, regardless of derivation.
6 Rules of nomenclature are retroactive unless expressly limited.

These principles form the basis of the system of botanical nomenclature. The rules and recommendations deal with subsidiary points to bring about greater uniformity and clarity to future nomenclature. Rules and recommendations apply to names of all organisms treated as plants (including fungi, algae, and lichen-forming fungi, but not bacteria), whether fossil or nonfossil (see Table 23·1).

The ICBN was last revised in 1988 (IBC 1988) and is published in English, with French and German texts available. The ultimate authority for the Code rests with the International Botanical Congresses held every 6 years. If any differences are perceived among the 3 language versions, the English version is official. The 1988 edition supersedes all previous editions of the Code.

The principle of priority applies only to names of families and taxa below them (excluding *formae speciales*) and starts with names published by Linnaeus in his *Species Plantarum,* 1 May 1753 (Linné 1753), except for names of mosses (other than *Sphagnaceae*), which date from 1 January 1801 (Hedwig 1801), and of some algae; names of fossil plants are dated from 31 December 1820 (Sternberg 1820). The starting point for names of all fungal groups other than fossil fungi is also 1 May 1753.

SCIENTIFIC NAMES

23·4 In formal taxonomy, a binomial method is used for the scientific name of a plant: *Quercus alba* (white oak), written in Latin and italicized, consists of a genus name (*Quercus*) and a specific epithet (*alba*). The genus name can be abbreviated to a capital letter followed by a period once it has been spelled out. For explanations of "epithet" and "specific" see

Table 23·1 Endings and abbreviations for taxa in the plant kingdom, including plants, fungi, and algae

	Plantae		Fungi		Algae	
Rank	Ending or abbreviation	Example	Ending or abbreviation	Example	Ending or abbreviation	Example
Division	*phyta*	Magnoliophyta	*mycota*	Eumycota	*phyta*	Chlorophyta
Subdivision	*phytina*		*mycotina*	Basidiomycotina		
Class	*opsida*	Magnoliopsida	*mycetes*	Urediniomycetes	*phyceae*	Chlorophyceae
Subclass	*idae*	Rosidae	*mycetidae*		*phycidae*	
Order	*ales*	Rosales	*ales*	Uredinales	*ales*	Volvocales
Suborder	*ineae*		*ineae*		*ineae*	Chlorodendrineae
Family	*aceae*	Rosaceae	*aceae*	Pucciniaceae	*aceae*	Volvocaceae
Subfamily	*oideae*	Rosoideae	*oideae*		*oideae*	
Tribe	*eae*	Roseae	*eae*	Puccinieae	*eae*	
Subtribe	*inae*		*inae*		*inae*	
Genus	*us, a, um*	*Rosa*	*us, a, um*	Puccinia	*us, a, um*	*Gonium*
Subgenus	subg.	*Rosa* subg.	subg.	*Puccinia* subg.	subg.	*Gonium* subg.
Section	sect.	*Rosa* sect. *canina*	sect.	*Puccinia* sect.	sect.	

Table 23·1 (*cont.*)

	Plantae		Fungi		Algae	
Rank	Abbreviation	Example	Abbreviation	Example	Abbreviation	Example
Subsection	subsect.					
Series	ser.					
Subseries	subser.					
Species	sp. [singular]	*Rosa* sp.	sp.	*Puccinia* sp.	sp.	*Gonium* sp.
	spp. [plural]	*Rosa* spp.	spp.	*Puccinia* spp.	spp.	*Gonium* spp.
		R. acicularis		*P. graminis*		*G. pectorale*
Subspecies	subsp.	*R. acicularis* subsp. *sayi*	subsp.		subsp.	
Variety	var.	*R. acicularis* var. *rotunda*	var.	*P. graminis* var. *bromi*	var.	
Subvariety	subvar.					
Form	f.		f.			
Special form	f.sp. [singular]		f. sp. [singular]	*P. graminis* f. sp. *tritici*		
	ff. sp. [plural]		ff. sp. [plural]			
Physiologic race			race	*P. graminis* f. sp. *tritici* race 15B		

Table 23·2 Abbreviations of ranks used in taxonomy of plants, bacteria, and fungi[a]

Abbreviation[b]	Rank	Meaning
bv.	biovar	biological variety
comb. nov.	*combinatio nova*	new combination
corrig.	*corrigendum*	to be corrected
cv.	cultivar	cultivated variety
emend.	*emendavit*	he or she corrected
fam. nov.	*familia nova*	new family
f.[c]	*forma*[c]	form
f. sp.	*forma specialis*	special form
gen. nov.	*genus novum*	new genus
nom. approb.	*nomen approbatum*	approved name
nom. cons.	*nomen conservandum*	name to be conserved
nom. nov.	*nomen novum*	new name
nom. nud.	*nomen nudum*	bare name
nom. rej.	*nomen rejiciendum*	name to be rejected
nom. rev.	*nomen revictum*	revived name
pv.	pathovar	pathogenic variety
sp.	species [singular]	species
sp. nov.	*species novum*	new species
spp.	species [plural]	species
subsp.	subspecies	subspecies
var.[c]	*varietas*[c]	variety
var. nov.[c]	*varietas novum*[c]	new variety

[a]Modified from *ASM Style Manual for Journals and Books,* American Society for Microbiology (ASM 1991).

[b]Latin terms italicized when applied with Latin descriptions of new taxa are not italicized in running text. Latin abbreviations are italicized in descriptions of new species.

[c]Rank not used for bacteria.

section 22·6. The name (abbreviated or given in full) of the person credited with naming the species is included; "L." stands for "Linnaeus".

Quercus alba L. [white oak] *Q. coccinea* Muench. [scarlet oak]

The names of genera and higher ranks may stand by themselves as monomials, but categories below the rank of genus must be a combination of a generic name and an epithet. The epithets of combinations other than species must be preceded by the word, often abbreviated, indicating rank. The word or abbreviation indicating rank, however, is not italicized (see Table 23·2).

Desmodium ser. *stipulata* Schub.

["ser." is the abbreviation for "series"; "*stipulata*" is the series name]

Rorippa palustris (L.) Besser subsp. *glabra* (Schulz) Stuckey var. *fernaldiana* (Butters & Abbe) Stuckey

Any combination below the rank of species can be shortened to a trinomial at the lowest rank. Thus, the example just given can be shortened to this form.

> *R. palustris* var. *fernaldiana* (Butters & Abbe) Stuckey

Moreover, the authors need be cited only for the ultimate combination; note their absence in the binomial *R. palustris* just given.

In an "autonym" (a name in which the infraspecific epithet repeats that of the name of the species name), authors are cited only after the species name, not the autonym epithet.

> *R. palustris* (L.) Besser var. *palustris*

The ICBN, unlike the Zoological Code (ICZN 1985), does not allow "undesignated trinomials" such as *Rorippa palustris glabra.*

Names of genera do not have plural forms. If reference is made to a group of species, the genus name followed by the abbreviation (spp.) for plural "species" is recommended.

> *Rhododendron* spp.
>
> *Iris* spp.
>
> *Aster* spp.

Publication of Names

23·5 Persons publishing scientific names are authors, and all scientific names have authors. Citation of authors' names is optional according to the Code and is not part of the scientific name. When a scientific name is followed by the name(s) of its author(s), the latter must include the author(s) who validly published that name under the Code.

> *Tetraneuris herbacea* Greene was reclassified by Cronquist as *Hymenoxys herbacea* (Greene) Cronq.
>
> *Festuca ovina* L. subsp. *saximontana* (Rydb.) St.-Yves var. *rydbergii* St.-Yves was reclassified by Cronquist as *Festuca brachyphylla* J.A. Schultes var. *rydbergii* (St.-Yves) Cronq.

Generic names found to be incorrect according to ICBN rules can be "conserved" when the substitution of "correct" names would cause much inconvenience. Such names are categorized as *nomina conservanda* (nom. cons.) and are listed in an appendix to ICBN.

Maclura Nutt., nom. cons. [the genus for osage-orange]

When an epithet has been published by 2 authors, the names of both authors should be cited and linked by the Latin word "et" or by an ampersand (&).

Panicum ravenelii Scrib. et Merr. [or "Scrib. & Merr."]

If an author publishes a name and validating description or diagnosis in a work by another author, the name(s) of the latter may follow after the word "in".

Solanum sarrachoides Sendt. in Mart.

If an author validly publishes a name (not merely an epithet) coined by another author who did not validly publish it, for example in a manuscript or without a description, and ascribes it to that earlier person, the name of the validating author may be cited before the Latin word "ex" preceding the name(s) of the validating author. If one wants to shorten such citations, the names of the validating authors (that is, the ones before "in" or after "ex") are always the ones to be retained. (See section 23·26 on fungi for the use of "ex" compared with the use of a colon).

Cypripedium candidum Muhl. ex Willd. [for "white lady's-slipper"]

[or]

Cypripedium candidum Willd.

Author citations for scientific names should be omitted from the title of an article unless the article refers to the use of different authors for the same taxon; however, this is not an ICBN rule. For abbreviations of authors' names, consult *Authors of Plant Names* (Brummitt and Powell 1992) or *Taxonomic Literature* (Stafleu and Cowan 1976–1988; Stafleu and Mennega 1992).

Valid Publication

23·6 For the name of a binomial or any other taxonomic rank to be legitimate, it must be validly published with a Latin description, rank (since 1953), and holotype (since 1958); it is recommended in the Code that the author's own name also be included in the scientific description. Validly published generic names are listed in *Index Nominum Genericorum* (Farr and others 1979).

For a name to be validly published, it must meet certain requirements.

1 Effective publication in printed matter distributed at least to accessible libraries.
2 A Latin description or diagnosis (or reference to one); names published before 1 January 1935 as well as names for fossils and for algae prior to 1958 do not require Latin.
3 For names of algae since 1950 and for species of fossils since 1912, a figure or illustration.
4 Clear indication of the rank of the name (as of 1 January 1953).
5 Indication, as of 1 January 1958, of a holotype must be indicated, that is, a single specimen (1 collection from 1 locality on 1 date) that will fix forever the application of that name.
6 Compliance with special requirements, such as the form of names at different ranks.

It is conventional in many journals when a new taxon is being described to place a comma after the author's name followed by the abbreviation for the taxon name: "*species novum*" (*sp. nov.*) for a new species, "*genus nova*" (*gen. nov.*) for a new genus, and "*familia nova*" (*fam. nov.*) for a new family.

Gilbertellaceae Benny, *fam. nov.*

Dictyocoprotus Krug & Khan, *gen. nov.*

Piper allardii Yuncker, *sp. nov.*

Strotheria gypsophila B.L. Turner, *gen. et sp. nov.*

×*Dryostichum singulare* W.H. Wagner, *nothogen. et nothosp. nov.*

A new combination requires, since 1 January 1953, explicit citation of the name on which it is based and the bibliographic reference (author and place of valid publication) but does not require a description or diagnosis.

Cycadeoidea wyomingensis (Ward) Wieland, *comb. nov.*

Authors and Latin descriptions should be used also for infraspecific ranks such as subspecies, varieties, and forms (but not *formae speciales* or autonyms).

Medicago sativa L. subsp. *falcata* (L.) Arcangeli

Fraxinus caroliniana Mill. var. *cubensis* (Griseb.) Lingelsh. f. *lasiophylla* Fern. & Schub.

Salix candida Flügge f. *denudata* (Anderss.) Rowlee

A voucher specimen (nonliving) of a taxon must be deposited in a recognized herbarium (Holmgren and Holmgren 1992). The scientific or cultivar name of the specimen and the source or collection site should be included in the publication. When several specimens are used in the study, they should be listed in a table with the scientific name, locality, collector, collector's identification number, and the place of deposit for each specimen (see description of the newly described species that follows).

In describing a new taxon, citations of all specimens of the taxon known to the author are usually appropriate. If many specimens of the taxon have been examined, citation of representative collections may suffice, in which case the choice of specimens should preferentially include 5 kinds.

1 Specimens of the author, indicating the author's familiarity with the plant in the field.
2 Specimens identified by a collector's serial number.
3 Specimens represented by duplicates in many herbaria and more accessible for examination.
4 Specimens documenting the geographic extremities of the range.
5 Specimens of possible historic value.

Specimen citation may range from reference by collector's name and number to a listing of all field data. Data in the specimen citation are preferably arranged in the following order: country, state or province, locality, elevation, date, phenology, collector's name and number, and herbaria of deposit, cited according to the abbreviation system of *Index Herbariorum* (Holmgren and others 1990, Holmgren and Holmgren 1992).

Taxonomic Description and Diagnosis

23·7 Below is an example of a Latin description of a newly described species *Burlemarxia pungens* Menezes & Semir, sp. nov. (Menezes and Semir 1991).

> *Burlemarxia pungens* Menezes & Semir sp. nov.—Type: Brasil, Minas Gerais, road from Diamantina to Conselheiro Mata, km 185, 16 Dec. 1985, *N.L. Menezes, V.C. Souza & R. Simão CFCR 8833* (Holotype SPF, Isotype UEC,K,US). —Fig. 22-32.
>
> Caulis haud conspicuus. Planta pulvinata, in rupium rimis crescens. Folia utrinque pilosa vel tomentosa, apice pungentia, senectute haud convoluta. Flores singuli vel glumini. Hypanthium validum. Filamenta geniculis inconspicuis, basi complanata atque basi coronae adnata.
>
> Plants growing inside cracks in the rock in such a way that the stem is not visible; stem simple or branching, 1.0–2.0 cm in diameter including the sheaths. Leaves 4.0–7.0 cm long, 0.4–0.6 cm broad, subulate, erect to spreading, those at the base reflexed but not twisted; tips acuminate, pungent; margins long-ciliate, strigose to hispid, the hairs usually

ascending, pale brownish, sometimes breaking off so that their broad bases give the margin a slightly serrate appearance; adaxial and abaxial surfaces densely hirsute, sparsely strigose to hispid, silvery gray. Flowers normally single, rarely paired, the bracts linear-lanceolate, somewhat hirsute, up to 3.0 cm long, 0.3 cm broad. Peduncle puberulent or tomentose to sparsely glandular-hairy, 6.0–7.5 cm long. Ovary and hypanthium puberulent or tomentose to sparsely pilose, both together 1.8–2.0 cm long, 0.7–0.8 cm broad. Perianth segments puberulent on the back at the base and over the entire central part, 4.0–5.5 cm long, 0.4–0.5 cm broad. Corona oval-lanceolate with a bifid tip, 2.0–2.2 cm long, 0.35–0.4 cm broad. The three antipetalous staminal filaments very slightly geniculate at the base, or none of the filaments geniculate, all slightly joined to the corona base, the free and erect portion about 4.0 cm long; anthers 1.0–1.2 cm long. Style 5.5–6.0 cm long. Fruit puberulent to slightly tomentose, somewhat ribbed, about 1.5 cm long and 1.3 cm broad.

Material examined: Brasil, Minas Gerais, Diamantina, dirt road to São João da Chapada, 14 km from Diamantina, 16 Apr. 1987, *N.L. Menezes, V.L. Scatena, D.C. Zappi, F.S. Pires & J. Prado* CFCR 10574 (SPF); Diamantina, road to Conselheiro Mata, km 183, 10 Sep. 1986, *N.L. Menezes, T.B. Cavalcanti & J.C.C. Gonçalves CFCR 10277a* (SPF); Diamantina, road to São João da Chapada, 13 Mar. 1989, *N.L. Menezes CFCR 12154* (SPF).

A single type specimen designated by the original author of a taxon is a holotype, and duplicates of this specimen are isotypes. If no holotype was indicated for a previously published name at the time of publication or if it is missing, a lectotype can be designated, which is a specimen, or collection, selected from the original cited material to serve as the nomenclatural type. A specimen other than the holotype on which the first account of a species is based is a paratype. A syntype is any 1 of 2 or more specimens cited by the author when no holotype is designated, and a neotype is a specimen selected to serve as a nomenclatural type as long as material on which the taxon is based is missing.

Another example of a diagnosis, which is shorter than a description and which emphasizes distinguishing characters, follows.

Viola grandisepala W. Becker; ex affinitate *V. smithianae* W. Becker et specierum affinium sepalis late ovatis conspicuis distinguenda (W. Becker, 1928). From the alliance of *V. smithiana* and related species to be distinguished by its conspicuous broadly ovate sepals (Stearn 1983).

If the name of a species is published without a Latin description or a diagnosis (after 1 January 1935), it should be cited as a "*nomen nudum*" (*nom. nud.*), this designation often being preceded by a comma.

Carex bebbii Olney, *nom. nud.* *Leucographia* Nyl., *nom. nud.*

Nomina nuda and other names not validly published have no standing whatsoever under the Code. They are as if never published, and do not, for example, prevent later publication and use of the same names for the same taxa or for totally different taxa. They are not really "names" as defined in Article 6 of the ICBN.

Note that specimens examined by the author can be indicated by an exclamation mark.

Lectotype here designated: P!; isolectotypes: K!, NY!.

Botanical Latin

23·8 Botanical Latin is more formalized in style and vocabulary than classical Latin. As in classical Latin, all adjectives must agree in gender, number, and case with the nouns they modify. Latin and Greek words adopted as generic names generally retain their classical gender unless an exception is specified in the Code (Article 76).

1 Latin names ending in "*us*" are masculine, unless they are trees, which are feminine; thus *Helianthus annuus* L. (sunflower) is masculine (*Helianthus*), but *Pinus resinosa* Aiton. (red pine) is feminine with a masculine ending (*Pinus*).
2 Names ending in "*a*" are feminine (*Cicuta maculata* L. for "water hemlock"), and names ending in "*um*" are neuter (*Viburnum dentatum* L. for "arrow-wood").
3 Names ending in "*on*" also are masculine, for example, *Rhododendron canadense* (L.) Torr., and names ending in "*ma*" are neuter, for example, *Alisma subcordatum* Ref.
4 Names ending in "*is*" are usually feminine or masculine but treated as feminine, for example, *Orchis rotundifolia* Pursh.
5 Names ending in "*e*" also are neuter, for example, *Daphne mezereum* L.
6 Some other feminine endings are "*ago*", as in *Plantago cordata* Lam.; "*es*" as in *Prenanthes aspera* Michx.; "*ix*", as in *Larix laricina* (Duroi) K. Koch; "*odes*", as in *Erythrodes querceticola* (Lindl.) Ames; and "*oides*", as in *Typhoides arundinacea* Moench.

Latin grammar and gender are covered in detail by Gledhill (1989), IBC (1988), Manara (1991), and Stearn (1992).

23·9 Although names of plant families end in "aceae", there are exceptions for 8 families: Compositae for Asteraceae, Cruciferae for Brassicaceae, Gramineae for Poaceae, Guttiferae for Clusiaceae, Labiatae for Lamiaceae, Leguminosae for Fabaceae, Palmae for Arecaceae, and Umbelliferae for Apiaceae. Authors should be consistent in a given work in the choice of endings to family names.

Names at the rank of family and above are plural in form and therefore require plural verbs and pronouns.

> The Rosales are estimated to comprise 6600 species.
>
> The Liliaceae are very diverse, and they have been separated into numerous smaller families by some botanists.

Synonyms and Homonyms

23·10 Synonyms are any of 2 or more names for a taxon, including names not in current use. Synonyms also include names at different ranks and positions. If 2 or more names are based on the same type, they are considered as nomenclatural (obligate) synonyms and are designated with the following symbol: ≡.

> *Ostrya* J. Hill (1757) ≡ *Carpinus* Linnaeus (1753)
>
> *Lycopersicon esculentum* P. Miller (1762) ≡ *Lycopersicon lycopersicum* (L.) H. Karsten (1882)

If names treated as synonyms are based on different types, they are taxonomic (facultative) synonyms and are designated by an equals symbol (=).

> *Zinnia* Linnaeus (1759) ≡ *Crassina* Scepin (1758)
> [a nomenclatural synonym]
> = *Lepia* J. Hill (1759) [a taxonomic synonym]
>
> *Polygonum lapathifolium* L. = *P. nodosum* Pers. [a taxonomic synonym]

The name of any plant cannot duplicate a name used previously for a different plant. Such duplicate names are homonyms. All homonyms except the earliest name are illegitimate.

> *Agrostis arachnoides* Ell. (1816) and *A. arachnoides* Poiret (1810) are homonyms because they are the same name, but Elliott applied it to a different species. Schultes renamed Elliott's plant as *A. elliottiana* Schult., and this is the correct name for it.

KEYS TO IDENTIFICATION OF PLANTS

23·11 Dichotomous keys consist of pairs (couplets) of contrasting statements (leads) that offer the user a choice depending on the material being identified. Each lead of a couplet leads to a further couplet or, if it is an ultimate lead, to a name of a taxon. Strictly dichotomous keys are the rule in modern-day systematics, and an occasional diversion from them, for example, a trichotomy, could lead to confusion.

Consecutive numbering of the couplets aids in clarity and is prefera-

1 Leaflets 7-13, entire .... *T. vernix*
1 Leaflets 3, entire to often toothed or lobed 2
  2 Stems simple or sparingly branched, suberect shrubs, without aerial roots, not climbing 3
    3 Petiole and young fruits pubescent; leaflets and their lobes or teeth mostly blunt; southern ... *T. pubescens*
    3 Petiole and fruit glabrous; leaflets and their teeth ± pointed; northern or western.................. *T. rydbergii*
  2 Stems climbing or straggling (vines) with aerial roots; leaflets pointed; widespread .... *T. radicans*

Figure 23·1 An indented key to species of poison ivy (*Toxicodendron* Miller), modified from Gleason and Cronquist (1991)

ble to alphabetic lettering. It is not necessary to distinguish the 1st lead from the 2nd in each couplet, as some authors do with the letters "a" and "b" or the prime sign (′) on the 2nd lead. Those groups are introduced with variable character states in a key more than once when necessary. Both leads of a couplet start with the same word, usually the plant part, and follow with contrasting descriptive phrases. This facilitates orientation of leads of any 1 couplet.

1 Leaves petiolate

2 Leaves sessile

["Leaves opposite", for example, is not a contrasting character with either choice.]

Where possible, the leads should be phrased to read as positive statements, especially the initial lead of a couplet. When more than 1 character is used as a couplet, priority should go to the character most differentiating or easiest to access. Geographic distribution and habitat data when used in a key should always be placed last in a couplet.

Structurally, there are 2 basic types of keys, indented (Figure 23·1) and bracketed (see section 22·13 for an example), both of which incorporate the same kinds of data and differ only in organization. Most botanists prefer the indented type, and most zoologists and typesetters prefer the bracketed type.

Indented Keys

23·12 A well-organized indented key can be scanned rapidly and followed backwards when one goes astray. It has the advantage of making the relationships of the groups more apparent to the eye and can be read easily in both directions. A disadvantage arises only in long keys where successive shifts to the right result in wasted space, and members of a couplet are separated by more than 1 page. Such problems can be avoided if printers are instructed to use minimal indentation, or if long keys are broken into groups, the 1st key being a key to groups.

In constructing an indented key, the leads of the 1st couplet begin at the left margin and each successive subordinate couplet is indented beyond that preceding (leading to) it. When the lead runs more than 1 line, the "run-over" lines should, for the sake of brevity, be indented slightly farther than the next subordinate lead. Often 1 section of an indented key contains fewer leads than the other. In a key intended for identification, it is better to have the smaller division precede the larger so that the leads of the dividing couplet are closer together. One should try to make a balanced key, a key in which each lead of each couplet accounts for approximately half the remaining number of taxa.

Bracketed Keys and Other Types

The bracketed key takes less space on a page, and some readers prefer to see the opposing leads side-by-side (bracketed). It does not show relationships, and it is difficult to scan backwards and forwards. Typing is easy because all couplets or alternate couplets begin flush left.

Keys have also been developed in which characters can be accessed in any order or at any place of entry (Duncan and Meacham 1986).

INTERNATIONAL CODE OF NOMENCLATURE FOR CULTIVATED PLANTS

23·13 The International Code of Nomenclature for Cultivated Plants (ICNCP 1980) was developed by the International Commission for the Nomenclature of Cultivated Plants of the International Botanical Congress and the International Commission for Horticultural Nomenclature and Registration at the International Horticultural Congress held in London in 1952. The nomenclature for cultivated plants supplements that for wild plants and was published in 1953 but has been revised several times (ICNCP 1980). The ICNCP of the International Union of Biological Sciences has responsibility for new editions of the Code, and the Code can be modified only by this Commission.

Plants brought from the wild into cultivation retain the names that were applied to the same taxa growing in nature. Additional, independent designations for plants used in agriculture, forestry, and horticulture are dealt with in the International Code of Nomenclature for Cultivated Plants. However, nothing precludes using for cultivated plants names published in accordance with the requirements of the ICBN (Article 28 of ICBN).

Because this Code has no legal status, the commercial interests of plant breeders are protected by the Council of the International Union for the Protection of New Varieties of Plants. To ensure that a cultivar has only 1 correct name, the ICNCP requires priority acts, which are achieved by publication and registration that are dated and distributed to the public. There are societies that maintain statutory registers of names. The aim of the Code is to promote uniformity, accuracy, and stability in the naming of agricultural, horticultural, and silvicultural cultivated varieties (cultivars).

Cultivated plants are named at 3 main levels: genus, species, and cultivar. There is only 1 rank below the species—the cultivar, which comprises all varieties or derivatives of wild plants that have been grown under cultivation, that can be distinguished by any characters (morphologic, physiologic, cytologic, chemical, or others), and that when reproduced, sexually or asexually, retain the distinguishing characters (Pringle 1975). A cultivar name cannot be registered as a trademark. Common names of genera and species are not regulated by the Code, for example, "rye" for "*Secale*", or "potato" for "*Solanum tuberosum* L.".

23·14 The term "cultivar" includes clones that are derived vegetatively from a single parent; lines of selfed (self-pollinated) or inbred individuals; series of crossbred individuals; and collections of individuals that are resynthesized only by crossbreeding, for example, F_1 hybrids. Cultivar names can include a forestry provenance, or a particular growth-habit form that can be retained by appropriate methods of propagation. This means that parity exists between names, not taxonomic entities (Gledhill 1989).

Cultivars have names, sometimes referred to as "fancy" names, which are indicated either by placing them within single quotation marks or by preceding them with the abbreviation "cv.". Cultivar names are written with an initial capital letter and are not italicized.

Rubus arctius L. var. *grandiflorus* Ledeb. [a botanical variety]

Rubus flagellaris Willd. 'American Dewberry'
[a cultivated variety, or cultivar]

Triticum aestivum L. 'Era' [or] *T. aestivum* L. cv. Era

Hordeum vulgare L. cv. Proctor [or] *H. vulgare* L. 'Proctor'

Juniperus communis L. var. *depressa* Pursh. 'Plumosa'

Single quotation marks are not necessary when cultivar names are written alone unless their absence would be confusing.

A widely grown cultivar of wheat is named Era.

Hedera helix L. 'Chicago' is a popular ivy because of its fast growing habit and good keeping qualities. It resembles *H. helix* 'Pittsburgh' except that plants of 'Chicago' grow more slowly than those of 'Pittsburgh'.

HYBRIDS

23·15 Rules for naming hybrids (nothotaxa) are in Appendix I to the ICBN and apply to naturally occurring hybrids as well as to agronomic and horticultural ones. There are 2 ways of designating hybrids: by a formula or by a name. In a formula, the names of the known or putative parents are separated by a multiplication symbol (sign) (×) equidistant between them; the symbol is not italicized. A formula consists of existing names and requires none of the formalities for new names.

Digitalis lutea L. × *D. purpurea* L.

When a name, published under essentially the same rules as any other name of the same rank, is applied to a hybrid, the multiplication symbol (×) is placed without any space between it and the epithet or name.

Mentha ×*piperita*

×*Stiporyzopsis* [= *Stipa* × *Oryzopsis*]

The names are listed in alphabetical order according to the Code; however, an alternative such as listing the seed (female) parent 1st can be used if explained.

If the letter "x" is used instead of the multiplication symbol (×), because the symbol is not available, a space is needed between "x" and the adjoining taxon name to avoid ambiguity, and it is set in lowercase type.

Digitalis lutea L. × *D. purpurea* L.
[or] *Digitalis lutea* L. x *D. purpurea* L.

A naturally occurring nothospecies is *Castanea* ×*neglecta* Dode [or "*Castanea* x *neglecta*"]— a natural hybrid with *C. dentata* (Marshall) Borkh.

The multiplication symbol (×) is not used for infraspecific ranks, but the term denoting their rank carries the prefix "notho".

Polypodium vulgare L. nothosubsp. *mantoniae* (Rothm.) Schidlay

[the epithet is part of the name of the nothosubspecies and is termed a collective epithet]

The multiplication symbol (×) separates generic names (with spaces) for intergeneric hybrid formulae.

Fatsia × *Hedera* [for the hybrid ×*Fatshedera*]

A multiplication symbol (×) precedes a nothogeneric name.

×*Mahoberberis* sp. is a hybrid between species of the genus *Mahonia* and species of the genus *Berberis.*

23·16 In horticulture, names are sometimes used incorrectly or have no botanical standing. Such names are shown in double quotation marks (") or are followed by the abbreviation "hort.," for "hortulanorum" ("of the garden"), or both, and they are not capitalized.

Caladium hort. "Ace of Spades"

Dracaena baptisii hort.

[the name of this taxon has no botanical standing and there is no author citation]

Atriplex "*nova*" Winterl

["*nova*" refers to 4 different species of *Atriplex,* and *A.* "*nova*" is an illegitimate name for this species, hence the double quotation marks]

A chimera (plural, "chimeras") is an individual plant or organ consisting of tissues of different genetic constitution in a plant resulting from a graft union (graft chimera). Because they are not true hybrids, they are not represented by a multiplication symbol (×). Instead, a plus symbol (+) appears before the genus name.

+*Laburnocytisus* C.K. Schneid. (*Cytisus* + *Laburnum*)

+*Laburnocytisus adamii* (Poit.) C.K. Schneid.

+*Crataegomespilus dardarii* 'Bronvaux' and 'Jules d'Asnieres'

Syringa +*correlata*

If an interspecific or intergeneric somatic cross is made by a parasexual process, such as by protoplast fusion, the species are connected with a multiplication symbol (×) within parentheses.

Nicotiana glauca (×) *Nicotiana langsdorffii*

ORCHID NOMENCLATURE

23·17 A *Handbook on Orchid Nomenclature and Registration,* published by the International Orchid Commission (IOC 1993), is not part of the ICNCP. There are 20 000 species of orchids and 70 000 hybrid swarms or grexes with complex ancestral histories. According to the orchid code, the Latin term "grex" (plural, "grexes") is used to designate a population of cultivated hybrid individuals with the same parentage but not necessarily containing cultivars. Orchid fanciers use the expression "grex epithet" as a shorter name for "collective epithet" for artificial hybrids. They retain the term "collective epithet" for natural hybrids. The name of the collective epithet for a natural hybrid must be in Latin and be published with a Latin description.

> *Cymbidium* Sw. ×*ballianum*

An artificial hybrid is designated by an epithet that is a fancy name and must be formulated by the rules of the International Code of Nomenclature of Cultivated Plants and must, after 1 January 1967, be registered by the International Registration Authority for Orchid Hybrids.

> *Odontoglossum* Orpheon
>
> [the grex name for an artificial interspecific orchid hybrid for which *Odontoglossum* is the generic name and "Orpheon" is the name of the grex epithet]

VERNACULAR NAMES FOR PLANTS

23·18 Unlike scientific names, which are treated as Latin, vernacular names are by definition in local languages, and the discussion here applies only to English common names. When common, or vernacular, names are used, the scientific names should, however, be also given at least once in the abstract and once in the text of the manuscript. Although there is no generally accepted work on standarized common names, Rickett (1965) and Kartesz and Thieret (1991) have suggested guidelines on developing common names and whether or not to hyphenate them.

Common names for plants consist of 2 parts: the 1st part is the modifier and the 2nd is the group name. The modifier provides the unique character more or less at the species level, whereas the group name represents characters of taxa in ranks above species, such as family, genera, or subgenera. Group names need not have a modifier. Group names may include single words describing a particular family, genus, subgenus, tribe, or section.

23·19 Guidelines for style in group names have been suggested by Kartesz and Thieret (1991).

1 Simple group names consist of a single word.

> ash aster clover fern grass lily mallow
> mustard orchid stopper tulip willow

2 Singly compound group names are composed of 2 root words or elements connected as 1 name. Such names consist of a pair of single syllable words, or both a single- and a double-syllable word. These words should be joined as a single word.

bloodleaf chickenthief goldenrod hawkweed hawthorn
lousewort mousetail nipplewort quillwort rockcress
sneezeweed waternymph

3 Group names that have the stem words "Indian", "false", "true", "mock", or "wild" are written as separate, nonhyphenated words.

Indian bean false spikenard true watercress
mock pennyroyal wild rice

4 Common names are not capitalized unless they are derived from a proper noun.

English ivy Dutchman's-pipe Jerusalem artichoke
Good King Henry Aunt Lucy Blue Ridge gayfeather

5 A generic name used as a vernacular name is neither capitalized nor italicized and forms a plural as in English.

Aster, aster, asters *Camellia,* camellia, camellias *Iris,* iris
Rhododendron, rhododendron, rhododendrons
crocuses [not "crocci"]

23·20 Guidelines for using hyphens in group names have been suggested by Kartesz and Thieret (1991).

1 Words of 1 or 2 syllables referring to plants in general or to some part of a plant are joined to a preceding word without a hyphen, unless the resulting word is long and unwieldy, or it brings together a collection of consonants not readily grasped by a reader.

cat-tail carpenter-weed desert-thorn five-eyes
partridge-berry, unicorn-plant Kenilworth-ivy
monkey-flower morning-glory popcorn-flower
pygmy-melon roving-sailor treasure-flower
trumpet-creeper water-horehound yellow-saucers
butterfly-weed burr-cucumber pincushion-plant
rattlesnake-root strawberry-tree scorpion-tail
unicorn-plant vegetable-sponge

2 If a 2nd word is not taxonomically correct, it is hyphenated to the preceding word.

Douglas-fir [not a fir]
poison-oak [not an oak of the genus *Quercus*]
skunk-cabbage [not in cabbage family] star-grass [not a grass]
water-lily [not a lily]

3 If a 2nd word is itself taxonomically correct, it is given as 2 words without a hyphen, or joined as a single word [often because of long usage].

sugar beet [because it is a beet]   wood lily [because it is a lily]
moth mullein   water cress   poison ivy   crested wheatgrass
bluegrass

4 If a group name consists of words that have 2 or more syllables, or if either element of a pair has 3 or more syllables, the name is hyphenated.

evening-primrose   morning-glory   mountain-laurel
sea-purslane   pincushion-plant   butterfly-weed

If the name consists of 3 or more words, it is hyphenated.

pale alpine-forget-me-not
["alpine" is part of the group name, not a modifier]

arctic sweet-colt's-foot ["sweet" is part of the group name]

[In these 2 examples, the words "alpine" and "sweet" precede taxonomically incorrect names and are therefore set off by hyphens.]

fringed yellow star-grass
["yellow" is part of the group name "yellow star-grass"]

Sonoran false prairie-clover ["false" is part of the group name]

5 If a word or element of a group name includes an apostrophe, the words are hyphenated.

adder's-mouth orchid   bishop's-cap   Jacob's-ladder
lady's-slipper   mare's-tail   Solomon's-seal   St. John's-wort

6 All genera of the following plant families (or major plant groups) represent true types; thus, their group names, or any names referencing them, should not be hyphenated: Arecaceae [palm], Cactaceae [cactus], Cucurbitaceae [gourd], Cyperaceae [sedge], Orchidaceae [orchid], Poaceae [grass], and pteridophytes [fern and "fern-allies"].

Some words are of indeterminate application and do not represent true groups; thus these names can be used in various group names or fanciful phrases: "balm", "balsam", "bay", "briar", "creeper", "cress", "daisy", "flag", "haw", "hedge", "ivy", "mampoo", "mangrove", "osier", "rocket", and "rodwood".

General guidelines for selecting group names and modifiers, as well as their spelling and capitalization, are provided by Kartesz and Thieret (1991); generic names with their "true group" names are listed in Table 23·3.

Table 23·3 List of generic plant names with their "true group" names. Only these genus names are valid in texts using these common names[a]

Genus	Common name(s)	Genus	Common name(s)
Abies	fir	*Convolvulus*	bindweed
Abutilon	velvetweed	*Corallorrhiza*	coralroot
Achillea	yarrow	*Corchorus*	jute
Achyranthes	chaff-flower	*Corylus*	hazel
Aesculus	buckeye	*Croton*	croton
Ageratina	snakeroot	*Cucumis*	cucumber, melon
Agropyron	wheat grass	*Cucurbita*	pumpkin, squash
Alisma	water-plantain	*Cupressus*	cypress
Allium	garlic, leek, onion	*Cydista*	with
Alnus	alder	*Cydonia*	quince
Alocasia	taro	*Cynara*	artichoke
Aloe	aloe	*Cytisus*	broom
Amaranthus	pigweed,	*Dianthus*	pink
	tumbleweed	*Digitalis*	foxglove
Anchusa	bugloss	*Diodia*	buttonweed
Andropogon	bluestem, broom	*Dioscorea*	yam
	grass	*Dodecahema*	spinyherb
Antirrhinum	snapdragon	*Dracocephalum*	dragonhead
Apocynum	dogbane	*Drypetes*	rosewood
Arachis	peanut	*Elymus*	wild rye
Arctostaphylos	manzanita	*Epilobium*	fireweed,
Aristolochia	birthwort,		willowherb
	Dutchman's-pipe	*Erica*	heath
Aster	aster	*Eucalyptus*	gum
Bambuseae	bamboo	*Eugenia*	stopper
Brandegea	starvine	*Euphorbia*	spurge
Brassica	mustard, cabbage,	*Fagopyrum*	buckwheat
	rape	*Fendlera*	Fendlerbush
Brickellia	brickelbush	*Foeniculum*	fennel
bryophyte	moss	*Fragaria*	strawberry
Buxus	box	*Fraxinus*	ash
Calluna	heather	*Gaylussacia*	huckleberry
Camassia	camas	*Gentiana*	gentian
Campanula	bellflower	*Geum*	avens
Capparis	caper	*Gnaphalium*	cudweed
Capsicum	pepper	*Gossypium*	cotton
Carex	sedge	*Helianthus*	sunflower
Carum	caraway	*Helleborus*	hellebore
Castanea	chestnut	*Hemizonia*	tarweed
Castanopsis	chinquapin	*Houstonia*	bluet
Cedrus	cedar	*Humulus*	hop
Cichorium	chicory	*Hyacynthus*	hyacinth
Cimicifuga	bugbane	*Hyssopus*	hyssop
Cinnamomum	cinnamon	*Ilex*	holly
Cirsium	thistle	*Indigofera*	indigo
Cissus	treebine	*Ipomoea*	morning-glory
Citrus	orange, lemon, lime	*Isoetes*	quillwort

Table 23·3 (*cont.*)

Genus	Common name(s)	Genus	Common name(s)
Jasminum	jasmine	*Pimenta*	allspice
Juglans	walnut	*Pinguicula*	butterwort
Juncus	rush	*Pinus*	pine
Lactuca	lettuce	*Plantago*	plantain
Lagerstroemia	crape-myrtle	*Polygala*	milkwort
Laurus	laurel	*Pontederia*	pickerelweed
Lavendula	lavender	*Portulaca*	purslane
Levisticum	lovage	*Potamogeton*	pondweed
Ligustrum	privet	*Primula*	primrose
Lilium	lily	*Proboscidea*	unicorn-plant
Linaria	toadflax	*Prunus*	plum, cherry, almond, peach
Linum	flax		
Liriodendron	tuliptree	*Psidium*	guava
Lithospermum	gromwell	*Pyrola*	wintergreen
Loeseliastrum	calico	*Pyrus*	pear
Lomatium	desert-parsley	*Quercus*	oak
Lonicera	honeysuckle	*Ranunculus*	buttercup
Lychnis	campion	*Raphanus*	radish
Lythrum	loosestrife	*Rhamnus*	buckthorn
Malus	apple	*Rheum*	rhubarb
Malva	mallow	*Rhus*	sumac
Marrubium	horehound	*Ribes*	currant, gooseberry
Matthiola	stock	*Robinia*	locust
Mentha	mint	*Rosa*	rose
Mercurialis	mercury	*Rosmarinus*	rosemary
Mesembryanthemum	iceplant	*Rubia*	madder
Mimulus	monkey-flower	*Rudbeckia*	coneflower
Mirabilis	four-o'clock	*Rumex*	sorrel
Morus	mulberry	*Ruta*	rue
Musa	banana	*Sabal*	palmetto
Myosotis	forget-me-not	*Salix*	willow
Myrrhis	anise	*Salvia*	sage
Myrtus	myrtle	*Sambucus*	elder
Nelumbo	lotus	*Santalum*	sandalwood
Nicotiana	tobacco	*Sarcodes*	snowplant
Obolaria	pennywort	*Satureja*	savory
Ocimum	basil	*Saxifraga*	saxifrage
Olea	olive	*Scirpus*	bulrush
Oryza	rice	*Scrophularia*	figwort
Paeonia	peony	*Scutellaria*	skullcap
Panicum	millet, panic grass	*Sedum*	stonecrop
Papaver	poppy	*Selinocarpus*	moonpod
Pastinaca	parsnip	*Sequoia*	redwood
Penstemon	beardtongue	*Sideritis*	ironwort
Petroselinum	parsley	*Solanum*	nightshade
Phaseolus	bean	*Solidago*	goldenrod
Phoradendron	mistletoe	*Spinacia*	spinach
Phragmites	reed	*Sullivantia*	coolwort

Table 23·3 (*cont.*)

Genus	Common name(s)	Genus	Common name(s)
Swertia	felwort	*Tillandsia*	airplant
Swietenia	mahogany	*Tragopogon*	salsify
Symphoricarpus	snowberry	*Trichostema*	bluecurls
Symphytum	comfrey	*Trifolium*	clover
Symplocarpus	skunk-cabbage	*Tsuga*	hemlock
Tagetes	marigold	*Tussilago*	colt's-foot
Talinum	fameflower	*Ulmus*	elm
Tamarindus	tamarind	*Urtica*	nettle
Tanacetum	tansy	*Vallisneria*	eel-grass
Taraxacum	dandelion	*Verbascum*	mullein
Teucrium	germander	*Verbena*	vervain
Thalictrum	meadow-rue	*Wolffia*	watermeal
Thuja	arborvitae	*Zea*	corn, maize
Thymus	thyme	*Zingiber*	ginger

[a]Source: Kartesz and Thieret (1991).

23·21 Some names of weeds adopted by the Weed Science Society of America (WSSA 1984) differ from names used in other sources (DPL 1990; Kartesz and Thieret 1991; Rickett 1965) in being solid or without hyphens or spaces.

VEGETATIVE CHARACTERS USED IN TAXONOMY

23·22 Architectural features of leaves of dicotyledonous plants are useful in plant identification. Hickey (1973) has defined and illustrated leaf architecture using criteria of leaf orientation, shape, margin, texture, gland position, petiole characters, and venation. Plant hair (trichome) morphology is useful in identification because trichomes are almost universally present and are diverse in shape and size. Payne (1978) has defined 490 terms used to describe individual trichomes and their appearance in mass, and 47 trichomes are illustrated. *Plant Surfaces* (Juniper and Jeffree 1983) provides terms useful in identifying plants.

PLANT PHYSIOLOGY

23·23 Journals of plant physiology generally use SI (Système International d'Unités) units for measurement. Journals such as *Plant Physiology* and *Physiologia Plantarum* list some of these units in the standing matter as "Instructions to Contributors". In addition, guidelines for measuring and reporting environmental conditions in controlled environment studies have been published by Krizek (1982) and updated by Krizek and

McFarlane (1983). These guidelines were developed by the North Central Region Technical Committee on Growth Chamber Use (NCR 101), a committee appointed by the Cooperative State Research Service of the US Department of Agriculture, in recognition of the widely used plant growth chambers and other controlled chamber facilities that provide a reproducible environment for growing plants, plant tissues, or plant cells. Typically used units of measurement appear in Table 23·4.

COMMON NAMES OF PLANT DISEASES

23·24 Holliday (1989) has listed 2 principles for naming plant diseases.

1. The name should be readily usable by the grower because the grower is concerned with the disease, not the pathogen.
2. The name should include the host and the most conspicuous abnormality instead of the name of the pathogen.

cotton wilt [for "*Verticillium* wilt"]

eucalyptus canker [for "*Cytospora* canker"]

soybean root rot [for "*Phytophthora* root rot"]

tomato leaf spot [for "*Septoria* leaf spot"]

wheat downy mildew

Dutch elm wilt [instead of "Dutch elm disease"; wilt is the primary symptom, and the word "disease" should not appear in the name of the disease]

If a pathogen name were used in the name of a disease and if subsequently nomenclatural changes were made in naming the pathogen, the name of the disease would also have to be changed. For example, *Fomes* root rot of pine was reported to be caused by *Fomes annosus*; the genus name *Fomes* was replaced by the genus name *Heterobasidion*. Thus the disease name became inappropriate, and the disease had to be renamed.

Disease names are listed also in *Thesaurus of Agricultural Organisms, Pests, Weeds, and Diseases* (DPL 1990).

FUNGI

TAXONOMY AND NOMENCLATURE

23·25 The International Code of Botanical Nomenclature (IBC 1988) applies to fungi (singular, "fungus") as well as to plants in the usual sense.

Table 23·4 Measurement of environmental conditions in controlled environments for growing plants, plant tissues, plant cells

Observation[a]	Unit
Radiation	
Photosynthetically active radiation (PAR)	
Photosynthetic photon flux density (PPFD) (400–700 nm with cosine correction)	μmol s^{-1} m^{-2}
[or]	
Photosynthetic irradiance (PI) (400–700 nm with cosine correction)	W m^{-2}
Total irradiance (with cosine correction; indicate bandwidth)	W m^{-2}
Spectral distribution	
Spectral photon flux density (λ_1–λ_2 nm in <20 nm bandwiths with cosine correction)	μmol s^{-1} m^{-2} nm^{-1} (λ_1–λ_2 nm) (quanta)
[or]	
Spectral irradiance	
Spectral energy flux density (λ_1–λ_2 in <20 nm bandwidths with cosine correction)	Wm^{-2} nm^{-1} (λ_1–λ_2 nm)
Illuminance (380–780 nm with cosine correction)	klx
Temperature	
Air; shielded and aspirated ($\geq$3 m s^{-1}) device	°C
Soil or liquid	°C
Atmosphere moisture	
Shielded and aspirated ($\geq$3 m s^{-1}) psychrometer, dew point or infrared analyzer	%RH [or] dewpoint temperature in °C [or] g m^{-3}
Air velocity	m s^{-1}
Carbon dioxide	mmol m^{-3}
Watering	L, 1
Nutrition	
Solid media	mol m^{-3} [or] mol kg^{-1}
Liquid culture	μ [or] mmol L^{-1}
pH	pH units
Electrical conductivity[b]	dS m^{-1}

[a]Selected from Krizek and McFarlane (1983).
[b]Given as decisiemens per meter: 1 dS m^{-1} = 1 mmho cm^{-1} (millimho per centimeter).

Austrogautieria costata Stewart & Trappe, *sp. nov.*

Fusarium acuminatum subsp. *armeniacum* Forbes, Windels et Burgess, *subsp. nov.*

Fungi were included in the plant kingdom in earlier taxonomic works based on a 2-kingdom classification. In the 5-kingdom classification, fungi are placed in their own kingdom (Fungi): they are nongreen eukaryotic organisms whose cell walls differ chemically from those of green plants, and they have different modes of nutrition.

The starting point for names of fungi, other than fossil fungi, is 1 May 1753, the same date as for plants. Fossil names are dated from 31 December 1820 (Sternberg 1820). Fungus names used in previous starting dates of 1821 for Fries and 1801 for Persoon are sanctioned, that is, they are not affected by and take priority over homonymous and synonymous names published earlier. Current usage of fungus names appears in works such as those by Farr and others (1989), Hawksworth and others (1983), and Jong and Edwards (1991).

SCIENTIFIC NAMES

23·26 As with plants, the binomial method applies to the scientific (Latin) name of a fungus. *Amanita phalloides* Fries (death cap) is written in Latin and italicized. The genus name is *Amanita,* and the specific epithet is *phalloides.* Once spelled out, the genus name can be abbreviated to a single capital letter followed by a period.

Amanita phalloides Fr. is a poisonous mushroom, but *A. caesarea* (Fr.) Schw. is generally regarded as nonpoisonous.

The Code permits different states of fungi with pleomorphic life cycles to be given separate names for the anamorph and teleomorph. The correct name of the holomorph is that of its teleomorph.

Fusarium graminearum Schwabe [anamorph]

Gibberella zeae (Schwein.) Petch [teleomorph]

[These are names for the same fungus, the holomorph.]

The infraspecific rank of form (f.) (see Table 23·1) is usually based on morphologic characteristics, and names at the rank of form are the same as for any other rank in their requirements for publication.

Pyrenophora teres Drechs. f. *teres* Smedeg. causes net blotch of barley.

Other infraspecific ranks are subspecies and variety.

Another infraspecific category used with fungi is *forma specialis* ("f. sp.", singular), or *formae speciales* ("ff. sp.", plural), which is characterized by physiologic criteria (host adaptation and not morphologic characteristics), and is different from "form" (f.); see Tables 23·1 and 23·2. Nomenclature for this rank is not governed by the Code, and no Latin description or author citation is required.

Fusarium oxysporum Schlechtend.:Fr. f. sp. *lini* [The name should not be written as a trinomial, "*Fusarium oxysporum lini*" or "*F.o. lini*".]

Fungus taxa adopted by Persoon or Fries in certain of their works are "sanctioned" as privileged names with special priority and typification status (Korf 1982). When a name was taken up again by Fries or Persoon after the starting point, a colon (:) is placed between the originating and the sanctioning author to protect names designated by Fries ("Fr.") and Persoon ("Pers.").

Valsa coronata (Hoffm.:Fr.) Fr.

Uromyces polygoni-aviculare (Pers.: Pers.) P. Karst.

This use of the colon has virtually replaced the use of the Latin word "ex" in fungus nomenclature and has limited the use of "ex" for the person who 1st validly publishes a name ascribed to another person. Such names will continue to use the "ex" formulation. (See section 23·5.)

Ramichloridium Stahel ex De Hoog 1977
[Stahel did not provide a Latin diagnosis.]

23·27 Type specimens are as described for plants. For newly described names of fungus taxa, a living culture cannot serve as the holotype (Article 9.5 of the Code). Instead, a dried culture must be deposited in a recognized herbarium (Holmgren and others 1990). It is recommended in the Code, however, that a live culture also be deposited in a recognized culture collection.

23·28 Fossil fungi may be named with new form-generic names the same way as other fungi, and, in fact, as any fossil plants are named.

Grilletia Renault & Betrand   *Plectosclerotia* Stach & Pickh.

[both are examples of form-genera for fossil fungi]

Fossil fungi may also be given currently used names that have been modified by adding a suffix to that name, usually "ites".

Clasterosporites Pia
[fossil, from "*Clasterosporium* Schwein", a living fungus]

Graphiolites Fritel [fossil, from "*Graphiola* Poit.", a living fungus]

Pleuricellaesporites v.d. Hammen   *Fungites* Hallier

[both are fossil fungi, but their names are not derived from current names]

VERNACULAR NAMES OF GENERA

Yeast

23·29 A yeast is a unicellular fungus that reproduces by budding or fission. In some genera, pseudomycelium or mycelium may be present. Yeasts that produce spores have teleomorphs in the Ascomycotina or Basidiomycotina; those without spores are in the Hyphomycetes. The nomenclature and typification of yeasts are as described for fungi in the International Code of Botanical Nomenclature (IBC 1988).

> *Saccharomyces cerevisiae* Meyen ex E. Hansen var. *ellipsoideus* (Hansen) Dekker is the common example of a yeast.

See section 23·4 on shortening names.

The term "yeast" applies also to genera such as *Candida*, *Cryptococcus*, *Torulopsis*, and *Schizosaccharomyces*, but the term has been applied erroneously to fungus genera such as *Aspergillus* and *Neurospora*, probably because in these 2 genera there is a yeast-like phase with budding cells at some stage in the life cycle. Baker's or brewer's yeast is *Saccharomyces cerevisiae* Meyen ex E. Hansen, but "black yeast" is jargon and refers to yeast-like states of *Aureobasidium*, *Cladosporium*, *Moniliella*, and other non-yeast fungi. Cells that settle to the bottom of fermented liquids are "bottom yeasts (wort)", and those that come to the top are "top yeasts (wort)"; these may or may not be yeasts. Industrial yeasts such as those used, for example, in distilling, brewing, and wine-making are often not identified by genus or species.

If the word "yeast" is used, the organism should be identified by its scientific name at least once in the abstract and once in text. Although the word "yeast" is a noun, it may be used as an adjective if only 1 kind of yeast is described in the paper.

> yeast cells   yeast DNA   yeast culture   yeast colony

Slime Molds

23·30 The term "mold" usually refers to any fungal growth. Historically, mycologists have considered slime molds to be fungi and have applied nomenclature consistent with the International Code of Botanical Nomenclature (Alexopoulos 1973). Mycologists classify slime molds as myxomycetes in the Myxomycota, and they are characterized by a mobile, ameba-like swarm cell in the life history.

> *Fuligo septica* (L.) A.Wigg, a myxomycete, grows on lawns in wet weather.

Physarum nutans Pers. is a cosmopolitan myxomycete that often grows on bark of oaks.

Dictyostelium discoideum Raper is a cellular slime mold parasitic on bacteria.

Olive (1975), however, considered their status as fungi to be uncertain, and he classified them in the Kingdom Protista (which includes protozoa and unicellular algae) and gave names consistent with usage in the International Code of Zoological Nomenclature (ICZN 1985). Margulis and others (1990) classified these organisms in the Kingdom Protoctista (these biologists accepted the 5-kingdom classification but substituted the term "Protoctista" for "Protista").

GENUS NAMES

23·31 As is true for all plants, names of genera of fungi do not have plural forms. If reference is made to a group of species in a genus, the genus name followed by the abbreviation of plural "species" is recommended.

Pythium spp. [not "pythia"] *Fusarium* spp. [not "fusaria"]
Helminthosporium spp. [not "helminthosporia"]

The genus name has been used as part of the name of a plant disease (Hansen 1985). In such cases, the genus name should be italicized to be consistent with conventions used in other disciplines, and as used by Johnston and Booth (1983). (See also section 23·24.)

Cytospora canker *Phytophthora* root rot
Septoria leaf spot *Verticillium* wilt

Adjectival endings for names of fungus genera are not recommended.

Fusarium head blight [not "Fusarial head blight" or "fusarial head blight"]

For adjectival forms in medical terms, see section 22·15.

LICHENS

23·32 Lichens represent a biological, not a systematic, group, and each lichen is basically a stable, self-supporting association (consortium) of a fungus (mycobiont) and an alga or cyanobacterium (photobiont). For nomenclatural purposes, names given to lichens are regarded as applying to their fungal components, and lichens are classified mainly in the Lecanorales of the Ascomycotina. Algae in lichens are usually green algae or occasionally blue-green algae and have separate names.

Cladonia cristatella Tuck. is a lichen in the Lecanorales of the fungi, and its algal component is *Trebouxia* sp., a green alga.

Peltula polysora (Tuck.) Wet. is a lichen (mycobiont) and its algal component (photobiont) is *Anacystis montana* (Lightf.) Dr. et Daily, a blue-green alga.

ALGAE

23·33 Algae (singular, "alga") is a nontaxonomic word for a group of taxa that have been classified traditionally as plants, but today blue-green algae are usually included in the Kingdom Monera and simple algae in the Kingdom Protista. An alga is any plantlike organism that carries on photosynthesis and differs structurally from ordinary land plants such as mosses, ferns, and seed plants. The rules for nomenclatural style and format given for plants apply also to the algae, and names of algae are governed by the International Code of Botanical Nomenclature. Silva (1980) has compiled a list of names of classes and families of living algae.

REFERENCES

Cited References

Alexopoulos CJ. 1973. Myxomycetes. In: Ainsworth GC, Sparrow FK , Sussman AS, editors. The fungi: an advanced treatise. Volume 4B. New York: Academic Pr. p 39–60.

[ASM] American Society for Microbiology. 1991. ASM style manual for journals and books. Washington: ASM.

Brummitt RK, Powell CE, editors. 1992. Authors of plant names. Kew, London (UK): Royal Botanic Gardens.

[DPL] Derwent Publications Ltd. 1990. Thesaurus of agricultural organisms, pests, weeds, and diseases. 2 volumes. New York: Chapman & Hall.

Duncan T, Meacham CA. 1986. Multiple-entry keys for the identification of angiosperm families using a microcomputer. Taxon 35:492–4.

Farr DF, Bills GF, Chamuris GP, Rossman AY. 1989. Fungi on plants and plant products in the United States. St Paul (MN): American Phytopathological Soc.

Farr ER, Leussink JA, Stafleu FA, editors. 1979. Index nominum genericorum (plantarum). Regnum vegetabile 100, 101, 102. Utrecht (Netherlands): Bohn, Scheltema & Holkema.

Gleason HA, Cronquist A. 1991. Manual of vascular plants of northeastern United States and adjacent Canada. 2nd ed. New York: New York Botanical Garden.

Gledhill D. 1989. The names of plants. 2nd ed. Cambridge (UK): Cambridge Univ Pr.

Hansen JD. 1985. Common names for plant diseases. Plant Dis 69:649–76.

Harborne JB, Turner BL. 1984. Plant chemosystematics. New York: Academic Pr.

Hawksworth DL, Sutton BC, Ainsworth GC. 1983. Ainsworth & Bisby's dictionary of the fungi. 7th ed. Kew, London (UK): Commonwealth Mycological Inst.

Hedwig J. 1801. Species muscorum. Paris (France): Koenig.

Hickey LJ. 1973. Classification of the architecture of dicotyledonous leaves. Am J Bot 60:17–33.

Holliday P. 1989. A dictionary of plant pathology. Cambridge (UK): Cambridge Univ Pr. Reprinted with corrections, 1990.

Holmgren PK, Holmgren NH. 1992. Plant specialists index. Regnum vegetabile 124. Königstein (Germany): Koeltz Scientific Books.

Holmgren PK, Holmgren NH, Barnett LC. 1990. Index herbariorum. Part I, The herbaria of the world. 8th ed. Regnum vegetabile 120. Bronx (NY): New York Botanical Garden.

[IBC] International Botanical Congress. 1988. International code of botanical nomenclature, W Greuter, committee chairman. Regnum vegetabile 118. Königstein (Germany): Koeltz Scientific Books.

[ICNCP] International Commission for the Nomenclature of Cultivated Plants. 1980. International code of nomenclature for cultivated plants, CD Brickell, commission chairman. Regnum vegetabile 104. Utrecht (Netherlands): International Assoc of Plant Taxonomists.

[ICZN] International Commission on Zoological Nomenclature. 1985. International code of zoological nomenclature. 3rd ed. London: International Trust for Zoological Nomenclature.

[IOC] International Orchid Commission; Greatwood J, Cribb PJ, Stewart J. 1993. The handbook on orchid nomenclature and registration. 4th ed. London: IOC.

Johnston A, Booth C, editors. 1983. Plant pathologist's pocketbook. 2nd ed. Farnham Royal (UK); Commonwealth Mycological Inst.

Jong SC, Edwards MJ, editors. 1991. American Type Culture Collection catalogue of filamentous fungi. 18th ed. Rockville (MD): American Type Culture Collection.

Juniper BE, Jeffree CE. 1983. Plant surfaces. London: Edward Arnold.

Kartesz JT, Thieret JW. 1991. Common names for vascular plants: guidelines for use and application. Sida Contrib Bot 14:421–34.

Korf RP. 1982. Citation of authors' names and the typification of names of fungal taxa published between 1753 and 1832 under the changes in the code of nomenclature enacted in 1981. Mycologia 74:250–5.

Krizek DT. 1982. Guidelines for measuring and reporting environmental conditions in controlled-environment studies in plant growth chambers. Physiol Plant 56:231–5.

Krizek, DT, McFarlane JC. 1983. Controlled-environment guidelines. Hort Science 18: 662–4.

Linné C [Linnaeus]. 1753. Species plantarum. Stockholm: Holmiae, impensis L. Salvii.

Manara B. 1991. Some guidelines on the use of gender in generic names and species epithets. Taxon 40:301–8.

Margulis L, Corliss JO, Melkonian M, Chapman DJ, editors. 1990. Handbook of Protoctista: the structure, cultivation, habitats and life histories of the eukaryotic microorganisms and their descendants exclusive of animals, plants and fungi. Boston: Jones and Bartlett.

Menezes NL de, Semir J. 1991. *Burlemarxia,* a new genus of Velloziaceae. Taxon 40: 413–26.

Olive LS. 1975. The mycetozoans. New York: Academic Pr.

Payne WW. 1978. A glossary of plant hair terminology. Brittonia 30:239–55.

Pringle JS. 1975. The concept of the cultivar. J Arboric 1:30–4.

Rickett HW. 1965. The English names of plants. Bull Torrey Bot Club 92:137–9.

Silva PC. 1980. Names of classes and families of living algae. Utrecht (Netherlands): Bohn, Scheltema & Holkema.

Stace CA. 1989. Plant taxonomy and biosystematics. 2nd ed. London: Edward Arnold.

Stafleu FA, Cowan RS. 1976–1988. Taxonomic literature. Volumes 1–7. Regnum vegetabile 94, 98, 105, 110, 112, 115, 116. Utrecht (Netherlands): Bohn, Scheltema & Holkema.

Stafleu FA, Mennega EA. 1992. Taxonomic literature, Suppl 1. Regnum vegetabile 125. Utrecht (Netherlands): Bohn, Scheltema & Holkema.

Stearn WT. 1992 Botanical Latin. 4th ed. Newton Abbot (UK): David & Charles.

Sternberg KM. 1820. Flora der Vorwelt. Versuch 1:1–24.

Whittaker RH. 1969. New concepts of kingdoms of organisms. Science 163:150–60.

Woodland DW. 1991. Contemporary plant systematics. Boston: Allyn and Bacon.

[WSSA] Weed Science Society of America. 1984. Composite list of weeds, D Patterson, committee chairman. Weed Sc 32 (suppl 2):1–137.

Additional References

Bailey LH, Bailey EZ. 1976. Hortus third. New York: Macmillan.

Brummitt RK. 1992. Vascular plant families and genera. Kew, London (UK): Royal Botanic Gardens.

Cronquist A. 1988. The evolution and classification of flowering plants. 2nd ed. New York: New York Botanical Garden.

Flora of North America Editorial Committee, editors. 1993. Flora of North America north of Mexico. New York: Oxford Univ Pr. 14 volumes to cover all vascular plants and mosses and liverworts.

Hawksworth DL. 1974. Mycologist's handbook. Kew, London (UK): Commonwealth Mycological Inst.

Hitchcock AS, Chase A. 1951. Manual of the grasses of the United States. Washington: US Department of Agriculture. Miscellaneous Publication 200.

Jong SC, Birmingham JM, Ma G. 1993. Stedman's ATCC fungus names. Baltimore: Williams & Wilkins.

Jong SC, Edwards MJ, editors. 1990. American Type Culture Collection catalogue of yeasts. 18th ed. Rockville (MD): American Type Culture Collection.

Kartesz JT. 1994. A synonymized checklist of the vascular flora of the United States, Canada, and Greenland. 2nd ed. 2 volumes. Portland (OR): Timber Pr.

Kreger-van Rij NJW, editor. 1984. The yeasts: a taxonomic study. 3rd ed. Amsterdam: Elsevier.

Lentner C. 1993. Geigy scientific tables. Volume 6, bacteria, fungi, protozoa, helminths. 8th ed. West Caldwell (NJ): Ciba-Geigy.

Little EL Jr. 1979. Checklist of United States trees (native and naturalized). Washington: US Department of Agriculture. Agriculture Handbook 541.

Mabberley DJ. 1987. The plant-book. Cambridge (UK): Cambridge Univ Pr. Reprinted with corrections, 1989.

McVaugh R, Ross R, Stafleu FA. 1968. An annotated glossary of botanical nomenclature. Regnum vegetabile 56. Utrecht (Netherlands): International Assoc for Plant Taxonomy.

Miller RB, Ilic J. 1992. A comprehensive database of common and scientific names of world woods. IAWA [International Association of Wood Anatomists] Bull 13:246.

Miyachi S, Nakayama O, Yokohama Y, Hara Y, Ohmori M, Komagata K, Sugawara H, Ugawa Y, editors. 1989. World catalogue of algae. 2nd ed. Tokyo: Japan Scientific Soc Pr.

Morin NR, editor. 1993. Flora of North America. Volume 1, introduction; Volume 2, ferns and gymnosperms. New York: Oxford Univ Pr.

Nerad TA. 1991. American Type Culture Collection catalogue of Protista. 17th ed. Rockville (MD): American Type Culture Collection.

Reed PB Jr. 1988. National list of plant species that occur in wetlands: national summary. Washington: US Fish and Wildlife Service. Biological Report 88(24).

Soil Conservation Service. 1982. National list of scientific plant names. Volume 1, list of plant names. Washington: US Department of Agriculture. SCS-TP-159.

Terrill EE, Hill SR, Wiersema JH, Rice WE. 1986. A checklist of names for 3,000 vascular plants of economic importance. Rev ed. Washington: US Government Printing Office. Agriculture Handbook 505.

Tutin TG, editor. 1993. Flora Europaea. Volume 1, *Psilotaceae* to *Platanaceae*. 2nd ed. New York: Cambridge Univ Pr. This edition will appear in 5 volumes.

US Fish and Wildlife Service. 1992. Endangered and threatened wildlife and plants. 50 Code of Federal Regulations 17.11 & 17.12. Washington: [available from US Fish and Wildlife Service, Washington DC 20240, USA].

24 Human and Animal Life

God said, 'Let the waters teem with countless living creatures. . . .' God said, 'Let the earth bring forth living creatures according to their kind.'

— *The New English Bible, Genesis* 1:20, 24

The conventions in nomenclature and symbolization for the structure and functions of human beings and other animals have developed mainly within zoology and the human medical sciences. Some of these conventions are formally documented, but many have become established simply through usage. Where conventions for the animal sciences are lacking, those developed for the human sciences may be applicable.

TAXONOMY AND NOMENCLATURE

24·1 The rules and recommendations for giving an animal (or a taxonomic group of animals) a scientific name are found in the *International Code of Zoological Nomenclature* (ICZN 1985) and its amendments. The essential features of zoologic nomenclature and taxonomy are explained in *Principles of Systematic Zoology* (Mayr and Ashlock 1991). Current developments are reported in *Systematic Biology* (SSB [quarterly]). Applications to the International Commission on Zoological Nomenclature for approval of scientific names (and comments thereon) are published in the *Bulletin of Zoological Nomenclature* (BZN [quarterly]) which also includes official decisions of the Commission regarding names and works. *Official Lists* of approved and *Official Indexes* of rejected names and works (publications) are produced by the Commission (Melville and Smith 1987).

TAXONOMIC CATEGORIES

The 7 basic systematic categories ("taxa"; singular, "taxon") are, in descending order, kingdom, phylum (or division), class, order, family, genus, and species. Terms below "kingdom" may be prefixed with "sub" or "super" to provide additional taxonomic levels; the use of subkingdom is sometimes seen. Other supplementary taxa (such as brigade, cohort, and tribe) are sometimes used to further extend the structure of the hierarchy. The names of taxa from kingdom through family are set in roman type. The names of taxa at the level of genus and below (including subgenus, species, and subspecies) are set in italics. The names of all taxa down through genus are written with an initial capital letter, and any may stand alone. The 2nd part (the specific epithet) of the binomial names of all species and the names of lower taxa are written in lowercase letters and are usually not meaningful alone. Table 24·1 illustrates the complete hierarchy of taxa for human beings and the conventions for capitalization and type style. See sections 23·1 and 23·2 for additional comment on taxonomic concepts.

When the term for the rank precedes the name in text, the rank term

Table 24·1 The hierarchy of taxa for the classification of human beings[a]

Taxon	Name
Kingdom	Animalia
Phylum	Chordata
Class	Mammalia
Order	Primates
Family	Hominidae
Genus	*Homo*
Species	*Homo sapiens*[b]

[a]The name column illustrates the conventions for capitalization and type style for names at the successive taxonomic levels.

[b]The specific epithet (2nd part of the species name) is never capitalized, even in titles; see section 24·2.

is in lowercase letters; for example, "the phylum Protozoa", "the family Terebratulidae".

The Zoological Code (ICZN 1985) requires the omission of all diacritical marks, apostrophes, diaereses, and hyphens (except when hyphens are used to set off individual letters in a compound species-group name, as in "*c-album*"). For example, in names derived from German words, the umlaut is dropped and an "e" is added after the umlauted vowel to replace the umlaut; thus a "ü" becomes "ue".

SPECIES NAMES

24·2 The scientific name of a species is binomial and italicized; it consists of the genus name followed by the specific epithet; the genus name always takes an initial capital, and the species epithet never does. These conventions should be applied wherever a species name is used, including titles, tables, indexes, and dictionary entry-terms.

Homo sapiens *Dama dama* *Aphis gossypii*
Mus musculus *Drosophila melanogaster*

The Metabolism of *Drosophila melanogaster* [a title]

Musca domestica 25–37 [index entry]

A species name may be in a title that itself must be italicized. An example is a title italicized in running text to distinguish it from the rest of the text. In such a title the species name may be in roman type to maintain the distinction between the title and the species name.

His major monograph was *The Metabolism of* Drosophila melanogaster: *A New Treatise.*

Note, however, that this convention can confuse the reader as to the full extent of the title in many text settings.

When a species name is first used in a document, both the genus and specific epithet should be spelled out in full; subsequently, the genus name may be abbreviated to its initial capital letter. If confusion is possible among genus names having the same initial letter, use more than 1 letter in the abbreviations to distinguish the genera unambiguously. For example, when used in the same manuscript, *Salmo clarki* and *Salmo salar* become *S. clarki* and *S. salar.* However, *Mesocricetus auratus* and *Meriones unquiculatus* would become *M. auratus* and *Mer. unquiculatus* to avoid possible misreading of the genus abbreviation (the 1st used would take the single letter abbreviation of the genus). If *Salvelinus fontinalis* and *Salmo clarki* were discussed together, the former should be *S. fontinalis* and the latter should be written out in full, there being little saving in length with the abbreviation "Salm.". The genus name should not be abbreviated if no specific epithet is stated.

Leishmania sp. [not "*L.* sp."]

AUTHOR NAMES

24·3 Appendix E of the Zoological Code (ICZN 1985) recommends that each use of a name of a genus or taxon of lower rank include the name of the author of the taxon and the year of naming at least once in the document: "*Aphis gossypii* Clover, 1877". In a taxonomic work, a full reference to the original naming publication should be included in the document's list of references; the form of citation in the text for this example would be "(Clover 1877)". The author's name and the year need appear only once in an article, preferably on 1st mention of the taxon; it usually should not appear as part of the title. In nontaxonomic articles, when a full reference is included in the list of references, an author-year citation should be provided if the name-year system is being used (as in the 1st example below); a citation number is used in the citation-sequence system (as in the 2nd example).

Genus albus Smith, 1900 (Smith 1900)

Genus albus Smith, 1900[10]

When a species is transferred from its originally designated genus to another on the basis of new information or a new interpretation of the species' characters, the name of the original author is included within parentheses.

Limax ater Linnaeus, 1758 [becomes] *Arion ater* (Linnaeus, 1758)

Unlike the practice in microbiology and botany (see section 23·6), the author of the new combination is not added in the zoologic sciences. The ending of the specific epithet will change if the new genus name is applied for the organism and it differs in gender from that of the previous genus name: *Taeniothrips albus* (masculine) became *Frankliniella alba* (feminine).

COMMON NAMES

24·4 A genus name used as a common or vernacular name is neither italicized nor capitalized: *Gorilla,* gorilla; *Octopus,* octopus; *Python,* python. A common (vernacular) name may be formed from the scientific name of a family by making the initial letter lowercase and dropping the terminal "ae": for example, "sciurid" from "Sciuridae" and "chironomid" from "Chironomidae".

Many societies in the zoologic sciences have published standards for, and checklists of, preferred common names. The Entomological Society of America publishes a list of common names for insects (ESA 1989); note that in common names having 2 parts, the 2nd part is kept separate when it is a systematically correct name, but that it is combined with, and preceded by, a modifier when it is not.

bed bug [a true bug] house fly [a true fly] butterfly [not a true fly]

In common names for insect larvae, the suffix "worm" is always combined with the modifier; larvae are not true worms (annelids).

silkworm [not "silk worm"] striped cutworm [not "striped cut worm"]

Lists of proper common names and scientific names for species in a number of phyla have been published: amphibians and reptiles (Collins 1990), aquatic invertebrates (AFS 1991a), birds (AOU 1983) (also see Sibley and Monroe [1990]), decapod crustaceans (CS 1989), fishes (AFS 1991b, AFS 1991c), mammals (Hall 1965), and mollusks (CSM 1988).

LABORATORY ANIMALS

24·5 Nomenclature for outbred laboratory animals should conform to that recommended by the Committee on Nomenclature, Institute of Laboratory Animal Resources (ILAR 1970). A compact alphabetic-numeric symbolic system for designation of outbred strains has been published (ICLA 1972): the components consist, in the following sequence and unspaced, of a capital letter and 2 or more lowercase letters to represent the organization that bred or supplied the organism; a colon (:); 1 or more capital letters and numerals for current and basic stock designation; capital letters within parentheses for the original stock; lowercase letters to indicate rearing by means other than the natural mother (fostered, f; ova transplant, e; ovary transplant, o; hand-reared, h); and capital letters designating environmental factors (gnotobiote, GN; defined flora, DF; conventional rearing, CV; barrier rearing, BR).

Hqf:A5(SD)hGN
["Hqf" = "High Quality Farm"; "A5(SD)" = "A5 rats", "Sprague-Dawley descent"; "h" = "hand-reared"; "GN" = "gnotobiote"]

Information on inbred animal strains can be found in the *International Index of Laboratory Animals* (Festing 1987). A symbolic system for designating inbred strains of mice has been established by the Committee on Standardized Genetic Nomenclature for Mice (CSGNM) (Lyon 1989).

1 Inbred strains: designated by 1 or more capital letters; numerals are less preferred and the symbol should begin with a letter. Examples: DBA, NZB.
2 Inbreeding: an "F" followed by number of inbred generations. Example: F87.
3 Substrains: name of the parent, slant line, substrain symbol (numbers, or abbreviation with capital letter and, sometimes, lowercase letters for the name of person or laboratory originating the strain or substrain, or a combined name abbreviation and number). Examples: DBA/1, CBA/J, C57BL/6J.
4 Sublines developed by manipulative procedures: indicated by adding an appropriate lowercase letter. Example: C3Hf [foster-reared]

For further details on codes for inbred strains and sublines and on designating recombinant inbred, coisogenic, congenic, and segregating strains and strains preserved by freezing, consult the CSGNM document (Lyon 1989) and the related chapters of the same work, "Inbred strains of mice" and "Subline codes for holders and producers"; note that some of these symbolizations may include italic characters. Y consomic strain symbols include a superscript, all-capital-letter designation (following "-Y") for the strain or wild population of origin.

C57BL/6-Y^{AKR} C57BL/6J-$\text{Y}^{\text{CZECH II}}$

Similar conventions are recommended (Hedrich 1990) for inbred strains and substrains of laboratory rats.

Guidelines have been published for describing the diets used for ex-

Table 24·2 Symbols for the designation of generations

Symbol	Meaning	Example
$F_1, F_2, \ldots$	Filial generations	$P_1 \times P_1 \rightarrow F_1$ $F_1 \times F_1 \rightarrow F_2$ [and so on]
$P_1, P_2, \ldots$	Parental generations	P_1 = parents of F_1 P_2 = grandparents of F_1 [and so on]
$B_1, B_2, \ldots$	Backcross generations	$F_1 \times P_1 \rightarrow B_1$ $B_1 \times P_1 \rightarrow B_2$ [and so on]
$S_1, S_2, \ldots$	Self-fertilized generations; only for plants	Parental self-fertilization $\rightarrow S_1$ S_1 self-fertilization $\rightarrow S_2$ [and so on]
$I_1, I_2, \ldots$	Inbred generations	
$E_1, E_2, \ldots$	Generations after experimental manipulation	
X_2	Offspring of F_1 testcross	

perimental animals (AIN 1987). Terms and their definitions for such diets have also been published (AIN 1977).

Information expected by readers for assurance that animals in experimental studies have had care and treatment in accordance with any of the numerous standards of granting agencies and other institutions should be included in the methods section of research reports. If practices in the laboratory adhered to a published standard, that document should be cited and referenced.

INHERITANCE

For nomenclature and symbols for chromosomes and genes, see Chapter 20.

PEDIGREES

24·6 The symbols commonly used in genetics to designate generations are illustrated in Table 24·2. Each consists of a single roman capital letter and a subscript to indicate the number of the generation.

The most common symbols in use for pedigree diagrams are illus-

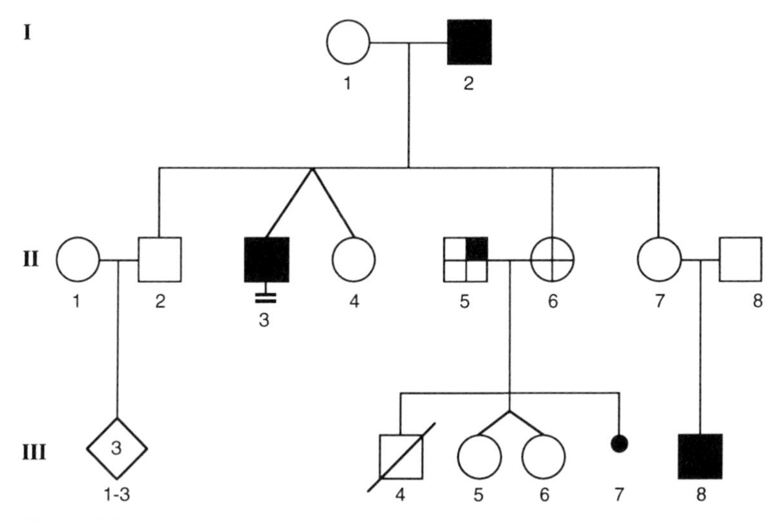

Figure 24·1 Common symbols for pedigree diagrams

trated in Figure 24·1. Squares represent males, circles females, and diamonds individuals of unknown sex. The symbols are left open except for those representing individuals carrying the trait of concern, which are shaded. Generations are assigned roman numerals in consecutive order, beginning with "I" for the uppermost (1st) generation in the diagram. Individuals on the same horizontal generation line are referred to by arabic numbers assigned from left to right.

Parents are connected by a horizontal line. Offspring are arrayed below on a parallel line in birth order from left to right. Identical (monozygotic) twins are connected to the horizontal line to their siblings by a single line. Fraternal (dizygotic) twins have separate lines that join the horizontal sibling line at the same point. Individuals with no offspring are shown by a vertical line exiting the symbol downward and terminating in a double horizontal line. Stillborn or aborted fetuses are shown by a solid dot. Traditionally, a cross fills the symbols of individuals who died before their manifestation of the particular trait of concern could be determined. More recently, a diagonal line slanting down from right to left has been used (Figure 24·1, individual III, 4). This allows the interior of the symbol to be subdivided into quadrants, each when filled representing a symptom or trait manifested (individuals II, 5 and II, 6). When not all members of a generation are shown, a numeral representing the number of individuals not shown is placed in the appropriate symbol. Additional symbols can be found in Thompson and others (1991).

GENETIC SYNDROMES

24·7 Human phenotypes such as congenital anomalies, inherited disorders and diseases, and other conditions determined by genetic mechanisms may have more than 1 name, descriptive or eponymic; in general, a descriptive term should be preferred to an eponymic term when both are available for the same phenotype. Many of the names of genetic syndromes and their synonyms can be found in standard medical dictionaries, but some may be found only in detailed catalogues, of which the 2 probably most comprehensive examples in print form are *Mendelian Inheritance in Man* (McKusick 1992) and *Birth Defects Encyclopedia* (Buyse 1990). Both include phenotypic symbols (see section 20·27 and Table 20·7).

STRUCTURE AND FUNCTION

ANATOMIC DESCRIPTION

24·8 The official list of human anatomic terms is published by the International Anatomical Nomenclature Committee. The 6th edition of *Nomina Anatomica* (IANC 1989) also contains the lists of human histologic and embryologic terms, *Nomina histologica* and *Nomina embryologica*. The current edition of *Nomina Anatomica Veterinaria* is the 3rd (WAVA 1983). A special compilation, *Nomina Anatomica Avium* (Baumel and others 1979) covers birds.

Anatomic terms derived from Latin are regarded as common nouns and thus should not be capitalized or italicized. The Latin form of any term is the official term approved by the International Committee, but English equivalents that either translate the Latin terms directly or provide idiomatic substitutes may be used in most scientific texts. For example, "brachial plexus" can be used instead of "plexus brachialis", and "stomach" instead of "ventriculus". Some terms retain the Latin form because formal equivalents are lacking in English (for example, "cisterna chyli").

BLOOD, INCLUDING COAGULATION FACTORS

Blood Groups

24·9 Blood groups have been designated within various classification systems (for example, the ABO, Rh, Lewis, and MNS systems). The antigens and factors are usually represented by capital and lowercase letters; some of these designations are arbitrary (the A, B, and O blood groups of the

ABO system), and others stand for the name of the person ("Le" for "Lewis", "Fy" for "Duffy") who served as the source of the initially identified component. In systems defined since 1945, antigens, phenotypes, and genes for blood groups are all represented by the same letter or letters for a particular group, with the gene designations set in italics. However, the specifics of how each system represents subclasses differ.

[antigen]	[phenotype]	[gene]
K	K+	*K*
A_1	A_1	A^1
Fy^a	Fy(a+)	Fy^a
Sc1	Sc:1	Sc^1

Some systems, including Kell, Lutheran, and Duffy, have both letter and numeric designations, for example, K and K1 for the same antigen, K- and K:-1 for the same phenotype, and *K* and K^1 for the same gene. One of the most complex symbolizations is that for the Rh system ("Rh" for "rhesus monkey"); 3 systems have been advocated and applied: the Rh-Hr notation of Wiener and Race, the CDE notation of Fisher, and the numeric system of Rosenfeld.

[Rh-Hr notation]	[CDE notation]	[numeric notation]
Rh_0	D	1
rh″	E	3
hr″	e	5
rh^G	G	12
Hr^B	Bas	34

These examples illustrate some of the uses of capital and lowercase letters and superscripts. Further details can be found in Chapter 5, "The Rh Blood Group System", of *Blood Transfusion in Clinical Medicine* (Mollison and others 1993); this monograph includes descriptions and examples of symbolic systems for other groups; also see Issitt and Crookston (1984).

The ISBT Working Party on Terminology for Red Cell Surface Antigens has published a system of 6-digit numeric codes for the established, serologically determined blood groups, the first 3 numbers representing the system and the last 3 the antigenic specificity (Lewis and others 1990). Existing symbols have not been altered, but all have been restricted to capital letters without superscripts or subscripts. For example, in the ABO system, the A antigen becomes ABO1, and in the Lutheran system, Lu^{ab} becomes LU3. For text use, leading zeros in the 3 digits designating the specificity are dropped. Phenotype designations are shown by the system symbol, a colon, and the appropriate numbers of the specificities; genotypes are shown as the italicized system symbol, a space, and italic numbers separated by a slash.

[Colton]	Co(a+b–)	[becomes]	CO:1,–2
	Co^a/Co	[becomes]	*CO 1/0*
[Kell]	K– k+ Kp(a–b+) Js(a–b+)	[becomes]	KEL:–1,2,–3,4,–6,7
	k,Kp^b,Js^b/K^o	[becomes]	*KEL 2,4,7/0*

Within this system, other antigens and specificities are grouped under "collections" of specificities that have serologic, biochemical, or genetic connections (for example, "Gerbich", "Cromer", "Auberger"). These collections are given 6-digit designations beginning with "2xx" followed by 3 digits for each member of the collection. For example, the Gerbich collection is number 201, and the code for the Ge2 specificity is 201002 (which may be displayed in text as 201.2). Antigens not assigned to systems or collections are put into 1 of 2 "series", the 700 series for low-incidence antigens and the 901 series for high-incidence antigens.

Hemoglobin

24·10 Human hemoglobins have been conventionally designated by the symbol "Hb" followed by a space and then a letter (A, C through Q, or S); new additions to the list of hemoglobin variants are now given a word indicating the location or laboratory of discovery instead of a letter.

Hb A   Hb S   Hb Providence   Hb Hopkins-2

The normal components of human hemoglobin are Hb A, Hb A_2, and Hb F. Each molecule of normal hemoglobin comprises 2 pairs of the 4 globin chains, which are designated by the lowercase greek letters alpha (α), beta (β), gamma (γ), and delta (δ); the chains designated epsilon (ε) and zeta (ζ) are found in embryonal hemoglobins.

The composition of the normal major fraction of human hemoglobin, Hb A, is 2 alpha and 2 beta chains, $\alpha_2\beta_2$; Hb A_2 contains 2 alpha and 2 delta chains, $\alpha_2\delta_2$. Some forms of hemoglobin contain only non-alpha chains.

Hb H: β_4 Hb Portland: $\gamma_2\zeta_2$

New variants of hemoglobin differing in a single chain may be identified by the sole change in the chain.

Hb Gα Philadelphia

The protein structure of the globin chains is conventionally designated by capital letters (note that the symbols for the terminal segments do not correspond to the standard protein nomenclature):

NA	[amino-terminal segment]
A through H	[helical segments]
AB through GH	[interhelical segments]
HC	[carboxy-terminal segment]

Mutations in the primary sequence are referred to in terms of position within these segments. For example, the substitution of tyrosine for histidine at position 87 of the chain is represented by

Hb M Iwate, α87(F8)His→Tyr

Other details of the nomenclature of hemoglobins can be found in the literature (TICH 1964, ISH 1965).

Platelets

24·11 A nomenclature for platelet antigens has been put forward by the ICSH/ISBT Working Party on Platelet Serology (von dem Borne and Decary 1990). The systems are designated as HPA (for human platelet antigen) and numbered in order of date of publication (HPA-1, HPA-2, and so on). Allelic antigens are to be designated alphabetically in order of their frequency of occurrence in the population, from high to low (for example, HPA-1a occurs more frequently than HPA-1b).

Coagulation Factors

Human clotting factors (ICNBCF 1962) are represented by the term "factor" with a roman numeral (from I to XII; for example, "factor III", "factor XI", but note, there is no factor VI). In common with general genetic rules, roman characters are used for symbols of phenotypes and italic characters are used to represent the genotype (ICTH 1973). Factor VIII and the von Willebrand factor have special terminology (Marder and others 1985).

BONE

24·12 A system of nomenclature, symbols, and units for bone histomorphometry has been published (Parfitt 1988). A set of 123 abbreviations are used in combination to produce standard representations of the variables reported in histomorphometry; these symbols are to replace the terminology and symbols of the International Society of Stereology. Symbols are given for the primary measurements in both 2-dimensional and 3-dimensional systems.

With 1 exception, the abbreviations are 1 or 2 letters. Both capital and lowercase letters are used for 1-letter abbreviations; 2-letter abbreviations are an initial capital letter followed by a lowercase letter. The sole 3-letter abbreviation is all capitals (BMU). When used in combinations, single-letter abbreviations are run together and multiletter abbreviations are separated by periods from any adjacent abbreviation.

BS	[bone surface]	Ob.S	[osteoblast surface]
B.Pm	[bone perimeter]	Ct.Th	[cortical thickness]

CIRCULATION

Hemodynamics

24·13 Symbols for hemodynamic quantities should, as far as possible, follow the international conventions for symbolizing quantities (see section 3·5 and 9·3). Single-letter symbols for quantities should be italicized; subscript or superscript symbols modifying the main symbol are in roman typeface (unless the modifier is itself a symbol for a quantity). See Table 24·3 for symbols, modifiers, and applications for specific measurements.

Multiterm abbreviations have been widely used to represent quantities (such as "LVEDP" for "left ventricular end-diastolic pressure"), but these should be replaced as much as possible by symbols developed in accordance with the international convention.

Authors must note that, because of the lack of an international and rationalized standard specifically for symbolic representation of cardiovascular structures and functions, many journals differ in their publication styles for symbols. Hence, authors must take care to follow a journal's particular requirements. A good guide is Volume 5, *Heart and Circulation* (Lentner 1990) of the *Geigy Scientific Tables* series.

Clinical Notations

ELECTROCARDIOGRAPHIC RECORDINGS

24·14 Leads (recording electrodes and connecting wires) for recording electrocardiographic tracings are designated by alphabetic and numeric characters. The standard leads are designated with roman numerals. In the unipolar lead system, the central terminal is designated by "V" for the

Table 24·3 Hemodynamic and related variables: symbols, modifying symbols, and units[a]

MAIN SYMBOLS

Variable	Symbol	SI unit[b]
area	A	mm^2
quantity (volume)	Q	mL, L
flow rate (volume/unit time)	$\dot{Q}$ (V/t)	$mL \cdot s^{-1}$, $L \cdot min^{-1}$
flow velocity	v	$cm \cdot s^{-1}$
flow acceleration	v/t	$cm \cdot s^{-2}$
tissue or organ mass	M	g, kg
perfusion (flow rate/mass perfused)	$\dot{Q}/M$	$mL \cdot min^{-1} \cdot g^{-1}$, $mL \cdot min^{-1} \cdot (100\ g)^{-1}$
pressure	P	kPa [for clinical use, "mm Hg", "cm H_2O"]
resistance	R	$kPa \cdot L^{-1}$
velocity	v	$mm \cdot s^{-1}$, $cm \cdot s^{-1}$
volume	V	mL, L

MODIFYING SYMBOLS[b]

Anatomic structure or time	Symbol
arterial	a
venous	ven
capillary	cap
atrium	A
right atrium	RA
left atrium	LA
ventricle	V
right ventricle	RV
left ventricle	LV
systemic	S
aortic	aor
mitral	mit
valve	val
aortic valve	aor val
mitral valve	mit val
pulmonary	P
blood	B, b
systolic	syst
diastolic	diast
end-diastolic	ED
end-systolic	ES

Table 24·3 (*cont.*)

EXAMPLES OF MODIFIED SYMBOLS			
Variable	Symbol	SI unit	Older metric unit[c]
pressure, arterial	P_a	kPa	mm Hg
systolic	$P_{a, syst}$		
diastolic	$P_{a, diast}$		
mean	$\overline{P}_a$		
aortic	P_{aor}		
pulmonary artery	P_{Pa}		
pulmonary artery, systolic	$P_{Pa, syst}$		
mean pulmonary artery	$\overline{P}_{Pa}$		
pressure, venous	P_{ven}	kPa	cm H_2O
pressure, right atrial	P_{RA}	kPa	cm H_2O
pressure, left ventricular	P_{LV}	kPa	mm Hg
pressure, left ventricular end-diastolic	$P_{LV, ED}$	kPa	mm Hg
volume, blood	BV, V_b	ml, L	ml, L
cardiac output	CO, $\dot{Q}_{LV}$	$L \cdot min^{-1}$	
cardiac index[d]	CI	$L \cdot min^{-1} \cdot kg^{-1}$	
stroke volume	SV	mL	
vascular resistance	R	$kPa \cdot L^{-1}$	mm $Hg \cdot L^{-1}$
vascular resistance, systemic	R_S		
vascular resistance, pulmonary	R_P		

[a]Symbols shown here are illustrative (see section 24·13 for explanation); a comprehensive list can be found in the appendix to *Geigy Scientific Tables,* Volume 5, Heart and Circulation (Lentner 1990).

[b]Single-letter symbols representing quantities (variables) are italicized, but multiletter symbols that are abbreviations of multiword terms for quantities (variables), for example "CO" for cardiac output, are not italicized. Modifying symbols that are themselves quantities (variables) should be italicized (see sections 3·5 and 9·3); others should be in a roman (upright) typeface. Most abbreviations of multiword terms are in capital letters, of single words in lowercase letters; note, however, that "A" is reserved for "atrium", "V" for "ventricle", and "P" for "pulmonary", and that when multiword terms might use any of these abbreviations for another term the lowercase version is applied.

[c]Where given, the unit is the currently preferred unit for clinical applications in the US.

[d]In this context, "kg" represents body mass in kilograms: thus, "CI" represents liters of blood per minute ("cardiac output") per kilogram of body mass.

central terminal and a letter for "right arm", "left arm", or "left foot". In the augmented lead system, the central terminal is "aV".

lead I   lead II   lead III   VR   VL   VF   aVR   aVL   aVF

Chest leads are designated with "V" for the central terminal and an arabic numeral for the chest electrode ("R" on the right chest and roman numerals for intercostal-space locations above the standard locations).

Esophageal leads have the added designation "E", and the number is the distance in centimeters from the nares to the esophageal electrode.

lead V1 V2 lead VR1 lead V3III lead VE28

Electrocardiographic waves (deflections from the tracing baseline) are designated by the capital letters P, Q, R, S, T, and U. Wave complexes are represented by unspaced letters, a minor wave in a complex by a lowercase letter. Prime signs indicate waves located after their usual position. An interval between 2 waves is represented by an en-dash (preferred) or hyphen between the wave symbols.

P wave QRS complex rS rR′ P–R interval

GRADING OF CLINICAL CARDIOVASCULAR MANIFESTATIONS

The severity of manifestations of cardiac and vascular disease is commonly represented by roman or arabic numerals.

grade IV hypertensive retinopathy

mitral stenosis grade 2/6 ["2" on a scale of "6"]

The New York Heart Association system (NYHA 1979) uses a 2-number system for representing the severity of disease ("status") and the prognosis, both with scales of 1 to 4.

3.2 [status, "3"; prognosis, "2"]

CLINICAL CHEMISTRY

24·15 SI units are used in most countries for reporting clinical chemistry findings, but the use of SI units in the United States has lagged, and authors may be required by many US journals to report with older metric units or to provide both metric units and SI units. The measurement of clinical laboratory data in SI units is addressed by Siggaard-Anderson and others (1987). Young (1987) has compiled tables of conversion factors from the older metric units to SI units, along with recommendations on significant digits and minimum increments for clinical hematology and clinical chemistry measurements.

Information on the clinical chemistry of laboratory animals has been compiled and published (Loeb and Quimby 1989); this work covers the mouse, rat, guinea pig, hamster, rabbit, and dog, and nonhuman primates. Over 90 pages of reference values for various analyses are provided, arranged by substance analyzed and then by species, with references to the literature from which the values were taken.

IMMUNOLOGIC SYSTEMS

The HLA System

24·16 Human histocompatibility leukocyte (HLA) antigens determine a person's tissue type and are important in tissue transplantation. The gene loci for the class I antigens, found on most nucleated cells and on platelets, are lettered A, B, C, E, F, G, and H and are symbolized as follows.

HLA-A *HLA-C* *HLA-E* *HLA-H*

The D region, which contains the loci for class II antigens, is divided; another capital letter distinguishes subregions (N, O, P, Q, or R). Immediately following is a 3rd capital letter indicating an alpha (A) or a beta (B) chain and, if needed, a number to distinguish among several chains.

HLA-DRA *HLA-DOB* *HLA-DNA* *HLA-DRB4* *HLA-DPA2*

The class III region of the major histocompatibility complex carries various genes, including those for some of the complement factors.

A "w" in an HLA designation (for the International Workshop on HLA, at which the status decisions are made) indicates that the specificity assigned is provisional. When the designation is made official, the "w" is dropped.

HLA-DRw10 [would become] HLA-DR10

Alleles of the HLA genes are designated by the letter(s) of the locus, followed by an asterisk and then 4 numerals (5 if needed to distinguish silent nucleotide substitutions).

*A*1101* *Cw*0301* *DQB1*0602* *DRB1*08022*

Current nomenclature for the HLA system is provided in a continuing series of reports from the WHO Nomenclature Committee for Factors of the HLA System. The 1987 report (WHO NC 1988) provided guidelines for the naming of genes and alleles that are identified by nucleotide sequencing. Updating reports are issued from time to time. An interim report (WHO NC 1990) covered the naming of genes and alleles identified by sequencing. The 1990 report (Bodmer and others 1991) contains complete lists of all officially recognized HLA class I and class II genes and alleles.

The major histocompatibility complexes have been described for other species (*RT1* in the rat, *DLA* in the dog, *RhLA* in the rhesus monkey, and so on). A revised nomenclature (Klein, Benoist, and others 1990) covers the mouse *H-2* genes. Klein, Bontrop, and others (1990) have put forward recommendations for a uniform system of symbols for these complexes across species lines (except for the human and mouse

HLA and *H-2* designations). A monograph (Warner and others 1988) covers the major histocompatibility complex in domestic animals.

Immunoglobulins

24·17 The 5 classes of human immunoglobulins are represented by the symbol "Ig" followed by an unspaced capital letter.

IgA IgD IgE IgG IgM

Subclasses are indicated by an arabic numeral after the class letter, for example, IgG3. A potential new subclass is signified by adding distinguishing letters within parentheses, for example, IgG(Pr) for a potential subclass identified in Prague.

Immunoglobulin molecules are composed of heavy (H) and light (L) polypeptide chains. The heavy chains of each immunoglobulin class are different and are symbolized by the greek letter that corresponds to the roman capital letter representing the class (A, α; D, δ; E, ε; G, γ; and M, μ); the heavy chains of IgG1 and IgA2 would be $\gamma 1$ and $\alpha 2$. The light chains of all classes are either kappa (κ) or lambda (λ) chains.

The basic structure of immunoglobulin molecules is a 4-chain structure, 2 heavy chains (specific to the class) and 2 light chains (kappa and lambda chains are never mixed on a single structure). Thus, an IgG molecule would contain 2 γ chains and either 2 κ or 2 λ chains: $\gamma_2\kappa_2$ or $\gamma_2\lambda_2$. IgA may exist as a dimer, for example, $(\alpha_2\lambda_2)_2$, and IgM is a pentamer of the 4-chain structure, for example, $(\mu_2\lambda_2)_5$. Because only 1 light chain type is found in a molecule, another shorthand representation of the immunoglobulins is IgG-κ, IgM-λ, and so on. The immunoglobulin molecule can be fragmented into portions that are designated with a capital F and unspaced lowercase letters.

Fab $F(ab')_2$ Fd Fd′ Fv Fc

Each light and heavy chain contains variable (V) and constant (C) regions that are designated by capital letters, with subscripts showing the type of chain. The heavy-chain constant region has subgroups (numbered 1 through 4) that are designated by numerals attached to the symbol. Shown with roman numerals attached to the symbol are the 4 subgroups of the heavy chain variable region, the 4 subgroups of the kappa chain variable region, and the 6 subgroups of the lambda chain variable region.

[Variable region]	[Generic]	[Specific]
Light chain	V_L	$V_\kappa I$
Heavy chain	V_H	$V_\alpha III$

[Constant region]	[Generic]	[Specific]
Light chain	C_L	C_λ
Heavy chain	C_H	$C_\gamma 1$

To designate hypervariable regions, the letters C and V are switched with H and L, and the number of the region (1, 2, or 3) is added.

L_V1 H_V3

An overview of immunoglobulins (Hasemann and Capra 1989) can be found in *Fundamental Immunology.* A nomenclature for synthetic peptides that represent portions of immunoglobulin sequences was published in 1990 (WHO–IUIS 1990).

24·18 The surface membrane molecules that specifically bind immunoglobulins via the Fc portion are called Fc receptors and symbolized by FcR (IUIS 1989). Receptors specific for a specific class of immunoglobulin are designated by a greek letter subscripted between the c and R. Subtypes are shown by unspaced roman numerals; species abbreviations are prefixes (such as "hu" for "human"; "mo" for "mouse").

$Fc_\alpha R$ $moFc_\gamma RII$

Component chains of the receptors may be shown as greek letters following the roman numeral, with different transcripts shown as subscripts to the greek letters.

$Fc_\gamma RII\alpha_a$ $Fc_\gamma RII\alpha_{b2}$

Soluble molecules having an affinity for immunoglobulins are known as binding factors and are designated as "BF".

IgG-BF IgM-BF

Although there is no standard system of terminology for antibodies, a general nomenclature for antibodies and their derivatives has been proposed by Baumgarten and Kürzinger (1989). The abbreviated designations for the monoclonal antibodies to T-lymphocyte surface antigens may be devised by commercial suppliers or institutional laboratories; for a large collection of designations, see the "Hybridomas–Monoclonal Antibody Index" in *American Type Culture Collection Catalogue of Cell Lines and Hybridomas* (ATCC 1992).

Complement

24·19 Complement is a system of proteins involved in antigen-antibody reactions in all vertebrates. The components of the classical complement pathway are designated by the capital letter C, to which are attached arabic numerals (WHO 1968); the numerals do not follow the order of the reaction sequence of the classical pathway, which is C1, C4, C2, C3, C5, C6, C7, C8, C9.

Component C1 has subcomponents C1q, C1r, and C1s. Other complement factors, which take part in the alternate complement pathway, are represented by the letters B, D, H, I, and P (WHO 1981).

Fragments of complement are designated by added lowercase letters, generally with the larger fragment designated by the letter "b" and the smaller by "a", for example, C3a and C3b. Inactive fragments are designated by the letter "i" as a suffix: C4bi. The loss of activity from a protein through peptide hydrolysis without fragmentation is indicated by a prefixed letter "i", for example, iC3b. A bar over the numbers and suffixes of a complement component designates biological activity.

$\mathrm{C}\overline{4}$ $\mathrm{C}\overline{3\mathrm{bBb}}$

An overview of the complement pathways can be found in *Fundamental Immunology* (Frank and Fries 1989).

Lymphocytes and Surface Antigens of Immune Cells

24·20 The use of monoclonal antibodies to identify lymphocyte cell-surface antigens has resulted in development of the CD ("cluster of differentiation") system of nomenclature (Erber 1990, Knapp and others 1991). A cluster of antibodies that display the same cellular reactivity are assigned numbers prefixed by "CD".

CD2 CD11a CD11b CD45 RB CDw75

A lowercase "w" indicates a provisional designation ("w" for the International Workshop on Leukocyte Typing, at which the decisions about designations are made). The system of human antigens now includes over 75 members. The CD designation may also be used for the antigen, but this should be made explicit in use, for example, "CD2 antigen" or "CD2 molecule".

No analogous system is available for the cell surface antigens of mice; Morse and others (1987) have put forward a proposed system (the Ly system). The human CD system (Knapp and others 1991) is being used for mouse antigens.

Interleukins and Interferons

24·21 Nomenclature for interleukins is established by the WHO-IUIS Nomenclature Subcommittee on Interleukin Designation (WHO-IUIS 1991).

As of this date, the designations are numerically sequenced up to "interleukin 10".

interleukin 1α interleukin 1β
interleukin 2 [and so on to] interleukin 10

The symbol for interferon is "IFN" (Anonymous 1983, Myers 1984). Prefixes "Hu" and "Mu" indicate, respectively, "human" and "murine". A hyphen-attached greek letter specifies the type of interferon. Nonallelic variants can be indicated by an arabic numeral or capital letter immediately following the greek letter.

IFN-β HuIFN-α3 MuIFN-γB

Allergens

24·22 A formal nomenclature system (Marsh and others 1988) for allergens has been recommended by the International Union of Immunological Societies (IUIS). Highly purified allergens are designated by the 1st 3 letters of the genus, in italics; a space; the 1st letter of the species name, in italics; a space; and a roman numeral. For example, pollen from perennial rye grass, *Lolium perenne,* would be designated as "*Lol p* I". An allergen from the honey bee, *Apis mellifera,* would be "*Api m* IV". If this system produces identical designations, 1 or more letters (as necessary) are added to the species designation. Fungal allergens that have strain differences will have the applicable numeric designation (for example, the ATCC number) is added to the roman numeral as a subscript: "*Alt a* I_{6663}".

HORMONES

24·23 The trivial or common names of many hormones are frequently represented by abbreviations, many of which are standard for journals in medicine and endocrinology; see Table 24·4. These abbreviations may not be acceptable in journals of biochemistry; see section 16·40. In most biological and medical contexts, steroid hormones can be represented by their common names; also see section 16·47.

aldosterone [for "11β,18-epoxy-18χ-hydroxypregn-4-ene-3,20-dione"]
progesterone [for "preg-4-ene-3,20-dione"]

Tests for Thyroid Hormones and Related Proteins in Serum

24·24 An extensive set of abbreviations (ATA 1987) is relevant to reporting tests for thyroid hormones in serum; see Table 24·5.

Table 24·4 Abbreviations for common (trivial) names of hormones[a]

Hormone name	Abbreviation
ACTH-releasing factor	CRH
adrenocorticotropin, adrenocorticotrophin	ACTH
arginine vasopressin	AVP
atrial natriuretic factor	ANF
calcitonin	CT
chorionic gonadotropin, human	hCG[b]
follicle-stimulating hormone	FSH
FSH- and LH-releasing hormone	GnRH
GH-inhibiting hormone (somatostatin)	SRIH
GH-releasing hormone	GHRH
growth hormone (somatotropin)	GH[b]
bovine GH	bGH
human GH	hGH
mouse GH	mGH
ovine GH	oGH
porcine GH	pGH
rat GH	rGH
insulin-like growth factor I, factor II	IGF-I, IGF-II
luteinizing hormone	LH
lysine vasopressin	LVP
melanocyte-stimulating hormone	MSH
oxytocin	OT
parathyroid hormone	PTH
proopiomelanocortin	POMC
prolactin	PRL
thyroid-stimulating autoantibodies	TSab
thyrotropin, thyrotrophin (thyroid-stimulating hormone)	TSH
thyroxine[c]	T4
3,5,3′-triiodothyronine[c]	T3
3,3′,5′-triiodothyronine[c]	rT3
TSH-releasing hormone	TRH

[a]In biochemical contexts the trivial name or the abbreviation may not be acceptable; see section 16·40.

[b]Species prefixes need not be used in contexts in which their absence would not produce ambiguity.

[c]The numeral in the abbreviation has sometimes been placed subscript; the online position is preferred because it helps reserve subscript numbers for standard uses such as those in chemical formulas (and reduces work at the keyboard).

The Renin-Angiotensin System

24·25 Systematic names and abbreviations for the renin-angiotensin system were established in 1979 by a nomenclature committee of the International Society of Hypertension (ISH 1979). The conventions include abbreviations for components of the system and superscript numbers indicating amino acid substitutions and their position.

Table 24·5 Abbreviations relevant to tests for thyroid hormones in serum[a]

Test	Abbreviation
Iodine concentration	
protein-bound iodine	PBI
butanol-extractable iodine	BEI
Total thyroid hormone concentration	
thyroxine	T4
triiodothyronine (3,5,3′-triiodothyronine)	T3
Free thyroid hormone concentrations	
free thyroxine	FT4
free triiodothyronine	FT3
Estimates of thyroid-hormone binding in serum	
thyroid-hormone binding ratio	THBR
Other hormones and thyroid-related serum proteins	
thyroglobulin	Tg
reverse T3 (3,3′,5′-triiodothyronine)	rT3
T4-binding globulin	TBG
T4-binding prealbumin (transthyretin)	TBPA (TTR)
Tests for autoimmune thyroid disease	
thyroid microsomal antibody titer	TmAb
thyroglobulin antibody titer	TgAb
TSH receptor antibodies	TRAb
Indirect estimates of free thyroid-hormone concentration	
free-T4 index	FT4I
free-T3 index	FT3I

[a]This table is based in large part on recommendations of a committee of the American Thyroid Association (ATA 1987). Note, however, that the numeral in abbreviations such as "T3" and "T4" is placed on-line in this table. The recommendations of the American Thyroid Association and the Endocrine Society are for subscript numerals (as in "T_3"). The on-line position is to be preferred because it helps reserve subscript numbers for standard uses elsewhere in science such as those in chemical formulas indicating the number of atoms of an element in a molecule; the on-line position also reduces work at the keyboard for authors, typists, and compositors.

ANG II [and] ANG-(1–8) [for "angiotensin-(1–8)octapeptide"]
[Sar^1, Val^5, Ala^8]ANG II
[for "[Sar^1,Val^5,Ala^8]angiotensin-(1–8)octapeptide"]

A detailed source for these conventions is "Renin-angiotensin system, Table 1" in *Geigy Scientific Tables: Volume 5, Heart and Circulation* (Lentner 1990).

KIDNEY: STRUCTURE AND FUNCTION

24·26 With the increasingly detailed research in the biochemistry and physiology of renal function has come a need for a more detailed, standardized nomenclature for all components of renal structure which has led to the development of such a nomenclature (IUPS 1988). Terms are presented in that document in an anatomically logical sequence; a table summarizes the sequence and relationships and gives abbreviations for the names of segments of the renal tubule and cell types.

OMCD [for "outer medullary collecting duct"]

DST cells [for "distal straight tubule cells"]

Symbols representing renal functions have not been well standardized, but many in use are generally accepted. Note that single-letter symbols representing quantities should be italicized (see sections 3·5 and 9·3); modifiers are not italicized.

C_{cr} [for "creatinine clearance"]　　U_{osm} [for "urine osmolality"]

$T\mathrm{m}_{\mathrm{PAH}}$
[for "tubular maximal secretory capacity for paraaminohippuric acid"]

RESPIRATION

24·27 The symbols and abbreviations used in respiratory physiology developed to describe human pulmonary function are also applied to reporting studies of other animals. The major symbols in respiratory physiology, representing variable quantities (see section 3·5, Table 3·5, and section 9·3) should be set in italic capital letters (ECSC 1993), but note that this convention is ignored by some journals and book publishers. Also note that multiletter symbols representing abbreviations of terms are not italicized. Modifiers (specifications) of the main symbols are set in roman small capital or lowercase letters on the line with the main symbol or are subscripted. Multiple subscripts are separated by commas. The specification of main symbols by modifiers should follow a specific order: anatomic location (where), time (when), and condition or quality (what, how). Note that the same letter may be used for different meanings (for example, "*C*" for both "concentration" and "compliance") and as a small capital and in lowercase may carry a different meaning ("A" for "alveolar" and "a" for arterial). More detailed descriptions of nomenclature and symbols are available (ACCP–ATS 1975, Macklem 1986, Fishman 1988); an especially detailed account has been published by a Working Party of the European Community for Steel and Coal (ECSC 1993).

$V_{\mathrm{CO_2}}$ [for "CO_2 production per unit time"]

$\mathrm{FEF}_{200\text{–}1200}$ [for "forced expiratory flow between 200 and 1200 ml of the forced vital capacity"]

*P*aw [for "pressure at any point along the airways"]

See Table 24·6 for the main symbols and examples of modified symbols.

Recent discussions of standardization of the symbols used in this field have been published (Clausen 1990, Zander and Mertzlufft 1990). The proposal of Zander and Mertzlufft argues for standardizing symbols to remove ambiguity and bringing the symbols used in physiology into better accord with those in chemistry.

THERMAL REGULATION

24·28 The symbols for thermal physiology (Gagge and others 1969) represent physical quantities and may include modifying subscripts defining physical and physiologic specificities. A subsequent publication (Bligh and Johnson 1973) included a glossary to improve the precision of meaning and the uniformity of usage of technical terms in the field and a list of symbols, abbreviations, and SI units; note that some of the symbols are given more than 1 meaning. Also note that these 2 papers do not specify italicization of symbols for quantities as is required by the relevant ISO standard; see section 3·5, Table 3·5, and section 9·3. This convention has been applied in Table 24·7, which illustrates many of the symbols and illustrative modifiers and gives the proper SI units.

DISEASES

24·29 Many standard nomenclatures, national and international, for diseases and histologic classifications of tumors are available, but many authors do not follow these standards. Generally accepted names and synonyms can be found in standard medical, dental, and veterinary dictionaries; authors unfamiliar with English terms can often verify the correct English form in a multilingual thesaurus (see section 6·24). A truly comprehensive compilation relevant to usage in the United States is *SNOMED International: The Systemized Nomenclature of Human and Veterinary Medicine* (CAP 1993), available in print, computer-diskette, CD-ROM, and magnetic-tape media. This nomenclatural system covers anatomy; diseases, disorders, syndromes, and their manifestations; clinical procedures and devices; animals and plants; chemicals; pharmaceutical products; and occupations.

Standard nomenclatures for narrower fields of importance have been published, and some of these may be specified by journals covering those fields; examples include those for animal fungal and parasitic diseases (Kassai and others 1988, Odds and others 1992).

Table 24·6 Examples of symbols for respiratory functions, including modifiers[a]

MAIN SYMBOLS

Variable	Symbol	Character[b]
partial pressure in blood or gas	P	italic capital
arterial CO_2 pressure ("tension")	Pa_{CO_2}	on-line "a", subscript "CO_2"
mean pressure	$\overline{P}$	italic capital, overbar
volume of gas	V	italic capital
inspired volume	$V\text{I}$	roman small-capital "I"
expired volume	$V\text{E}$	roman small-capital "E"
alveolar volume	$V\text{A}$	roman small-capital "A"
residual volume	RV	roman capitals
closing volume	CV	roman capitals
inspiratory capacity	IC	roman capitals
total lung capacity	TLC	roman capitals
forced vital capacity	FVC	roman capitals
flow of gas (time derivative, rate)	$\dot{V}$	italic capital, overdot
CO_2 per unit time	$\dot{V}_{CO_2}$	italic capital, subscript "CO_2"
expired volume per minute	$\dot{V}\text{E}$	italic capital, overdot, on-line roman small-capital "E"
fractional concentration of a gas	F	italic capital
volume of blood	Q	italic capital
blood flow (time derivative, rate)	$\dot{Q}$	italic capital, overdot
blood flow via shunts	$\dot{Q}\text{s}$	italic capital, overdot, on-line roman "s"
concentration in the blood phase	C	italic capital
concentration of O_2 in arterial blood	Ca_{O_2}	italic capital, "a" on-line, subscript "O_2"
compliance (volume-pressure relationships)	C	italic capital
dynamic compliance	$C\text{dyn}$	italic capital, on-line lowercase "dyn"
diffusing capacity	D	italic capital
standard conditions of temperature, pressure, and "dry" (0 mm Hg water vapor)	STPD	roman small capitals
ambient temperature, dry	ATPD	roman small capitals

[a]This table illustrates only a few of the possible symbols derived from the main symbols and available modifiers. For a complete compilation, with symbols arranged by gas phase, blood phase, ventilation and lung mechanics, diffusing capacity, and pulmonary shunt representations, see "Appendix D" in *Pulmonary Diseases and Disorders* (Fishman 1988); also see the extensive tabulation of symbols published by a committee of the European Community for Steel and Coal (ECSC 1993).

[b]"Capital" and "capitals" are used as shorter forms for "capital letter" and "capital letters".

Table 24·7 Symbols for thermal physiology: examples with subscript modifiers; SI units[a]

Symbol	Quantity	SI unit
A_b	area, total body	m^2
BMR	metabolic rate, basal	W, $W \cdot m^{-2}$, $W \cdot kg^{-1}$
ε	emissivity	[no dimension]
H	metabolic heat production	$W \cdot m^{-2}$
H_r	radiant flux, effective	$W \cdot m^{-2}$
I_λ	radiant intensity, spectral	$W \cdot sr^{-1} \cdot nm^{-1}$
I_{cl}	thermal insulation, clothing	$m^2 \cdot {}^\circ C \cdot W^{-1}$
γ	humidity, absolute	$kg \cdot m^{-3}$
$\dot{m}$	mass transfer rate	$kg \cdot s^{-1}$
$P_{s,T}$	pressure, vapor (saturated) at temperature T	Pa
P_w	pressure, water vapor	Pa
T_a	temperature, ambient	°C

[a]This table is based on Bligh and Johnson (1973). Note, as commented on in section 24·28, that single-letter symbols for quantities should be italicized. Because some identical capital letters are used for different quantities, an editor may have to accommodate nonitalicized symbols for some quantities; for example, the table of symbols in Bligh and Johnson (1973) gives "E" for "irradiance" and "E" for "evaporative heat transfer". See Bligh and Johnson for a complete tabulation of symbols and modifiers.

EPONYMIC NAMES FOR DISEASES

24·30 In general, descriptive names should be preferred to eponymic names. Medical dictionaries provide synonyms for eponymic names, usually under the headings "diseases" and "syndromes".

Nonpossessive forms of eponymic names should be preferred to possessive forms; see section 5·38. Adjectival and derivative forms of proper names used in names for diseases should not be capitalized.

Crohn disease [not "Crohn's disease"]
the Carvallo sign [not "Carvallo's sign"]
the Korotkoff test [not "Korotkoff's test"]
the Weber law [not "Weber's law"]
addisonian anemia [from "Thomas Addison"]
parkinsonism [from "James Parkinson"]

Names ending with terms like "law", "sign", "syndrome", and "test" should be preceded by "the" when they are used in running text, titles, or headings.

He confined his studies to the acquired immunodeficiency syndrome.
[not "to acquired immunodeficiency syndrome"]

The reason for this usage is that a "disease" may have widely differing clinical manifestations in different patients and is usually defined definitively mainly by pathologic manifestations. A "syndrome", in contrast, is usually defined by a specified group of clinical manifestations; that group is "the syndrome". Likewise, "laws", "signs", and "tests" are sharply defined and are not variable phenomena.

NEOPLASIA

24·31 Malignant diseases (cancers) characteristically progress from being a small tumor of limited extent to a tumor spreading into adjacent tissues or disseminating to distant tissues. The successive steps in the progession are called "stages". Stages are frequently designated by capital roman numerals, for example, "stage I", "stage IV". Suffixes (either letters or numerals) may be added to these symbols ("stage IVA") and sometimes subscripted ("stage III_{E+S}"). The term "stage 0" is used to indicate the most limited extent or stage, carcinoma in situ.

The TNM ("**t**umor, **n**ode, **m**etastasis") system (ACS 1992) is a commonly used, but not universally applicable, system for describing cancers. There is no common equivalence for different cancers between the TNM designation and the stages described in the previous paragraph. The designations of the system are presented as T with suffixes, N with suffixes, and M with suffixes, all run together with no spaces. In some specialized applications of the system the symbols are not run together.

cT2,N0,M1LYM   aT3,N2,M1BRA   T2 N1 M0

A lowercase letter (c, p, s, r, or a) prefixed to the T indicates the chronology of the classification for the case: "c" for "clinical-diagnostic", "p" for "postsurgical treatment-pathologic", "s" for "surgical-evaluative", "r" for "retreatment", "a" for "autopsy". The suffixes used with the T component of the system may include numerals or letters.

X	[tumor size or involvement cannot be assessed]
0	[no evidence of a primary tumor]
is	[**in s**itu tumor]
1,2,3,4	[size or involvement of the tumor]

The numbers used do not carry the same meaning for different types of tumors. These designations may be followed by additional suffixes (for example, "T3b", "T1a2")

The letter N may be followed by the letter "X" or by numerals.

X	[lymph node involvement cannot be assessed]
0	[no involvement of lymph nodes]
1,2,3	[degree of lymph node involvement]

The numbers carry different meanings for different types of cancer and may carry additional suffixes.

The letter M may be followed by letters, numerals, or both.

X [extent of metastasis cannot be determined]
0 [no evidence of metastasis]
1 [distant metastasis present]

The numeral "1" may be followed by 3-letter codes indicating the locations of metastasis, for example, "BRA" for "brain", "MAR" for "bone marrow", "PLE" for "pleural", or "OTH" for "other".

Other indicators of cancer staging and examples of their symbols include grade (for example, "GX", "G1"), host performance ("H0", "H4"), lymphatic invasion ("LX", "L0", "L2"), residual tumor ("R0", "R2"), and venous invasion ("VX", "V1").

These elements of notation combine to provide a concise statement of the nature and stage of the malignant disease in a particular patient.

aT3,N2,M1BRA

[autopsy basis for the description, the tumor size is designated as 3 in a scale of 4, the degree of lymph node involvement is 2, there is distant metastasis (spread) to the brain]

PROTEIN-STRUCTURE ABNORMALITIES: AMYLOIDOSIS AND RELATED DISORDERS

24·32 An expert committee of the World Health Organization (WHO-IUIS 1993) has recommended standardized abbreviations for designations of amyloid fibril proteins.

1 The 1st letter of all designations is capital A; the abbreviation designating the specific protein precursor follows the A without a space.
2 If this specifying element is a single letter, it should be a capital roman letter or greek lowercase letter.
3 If the specifying element is a multiletter abbreviation, the letters should all be roman capital letters. When, however, the abbreviation is formed of initial letters of a single word, only the 1st letter of the abbreviation is capitalized; the rest should be lowercase.

A particular amyloid protein may be characteristic of more than 1 clinical disorder or syndrome.

AA [protein precursor, apoSAA; characteristic of secondary amyloidosis]
Aβ [β protein precursor; characteristic of Alzheimer disease]

ATTR	[protein precursor, transthyretin; characteristic of systemic senile amyloidosis]
AGel	[protein precursor, gelsolin; characteristic of Finnish familial amyloidosis]
AIns	[protein precursor, insulin; characteristic of islet amyloid in the degu, a rodent]

PSYCHIATRIC SYNDROMES

24·33 Psychiatric diagnoses may be specified with classification numbers and terms from the system described in *Diagnostic and Statistical Manual of Mental Disorders* (APA 1994), widely known as "DSM–IV".

292.11 cannabis delusional disorder

and the case met the criteria for antisocial personality disorder (DSM-IV 301.70) in both evaluations during the patient's hospitalization.

VIRAL INFECTIONS

Hepatitis

24·34 Various combinations of capital and lowercase letters and arabic numerals are used to represent hepatitis caused by a particular virus and to represent specific hepatitis viruses, antigens, and antibodies.

HB	[for "hepatitis B"]
HBV	[for "hepatitis B virus"]
HDV	[for "hepatitis D virus"]
HAAg	[for "antigen of hepatitis A virus"]
anti-HBe	[for "antibody to HBeAg"]
HBsAg	[for "surface antigen of hepatitis B virus"]
HBeAg/3	[component of hepatitis B e antigen]

Human Immunodeficiency Virus Infection and Related Diseases

24·35 Various alphabetic arabic-numeric and roman-numeric abbreviations have been used to represent retroviruses causing or associated with the acquired immunodeficiency syndrome (AIDS) or a related disease; also see section 21·7.

HIV-1	[for "human immunodeficiency virus, strain 1"]
HTLV-2	[for "human T-cell lymphotropic virus, strain 2"]

HTLV-II [for "human T-cell lymphotropic virus, strain 2", but now obsolete]

LAV [for "lymphadenopathy-associated virus", but now obsolete]

The clinical manifestations of AIDS have been classified into staging systems that use various alphanumeric symbols.

group II group IVE category 4B WR3CNS

REFERENCES

Cited References

[ACCP–ATS] American College of Chest Physicians–American Thoracic Society Joint Committee on Pulmonary Nomenclature. 1975. Pulmonary terms and symbols. Chest 67:583–93.

[ACS] American Cancer Society, National Cancer Institute; American Joint Committee on Cancer. 1992. Manual for staging of cancer. 4th ed. Philadelphia: Lippincott.

[AFS] American Fisheries Society, Committee on Common Names of Cnidaria and Ctenophora. 1991a. Common and scientific names of aquatic invertebrates from the United States and Canada: Cnidaria and Ctenophora. Bethesda (MD): AFS. Special Publication 22.

[AFS] American Fisheries Society, Committee on Names of Fishes. 1991b. Common and scientific names of fishes from the United States and Canada. 5th ed. Bethesda (MD): AFS. Special Publication 20.

[AFS] American Fisheries Society, Committee on Names of Fishes. 1991c. World fishes important to North Americans: exclusive of species from the continental waters of the United States and Canada. Bethesda (MD): AFS. Special Publication 21.

[AIN] American Institute of Nutrition, Ad Hoc Committee on Standards for Nutrition. 1977. Report of the American Institute of Nutrition Ad Hoc Committee on Standards for Nutritional Studies. J Nutrition 107:1340–8.

[AIN] American Institute of Nutrition, Experimental Animal Nutrition Committee. 1987. Guidelines for describing diets for experimental animals. J Nutrition 117:16–7.

[Anonymous]. 1983. Interferon nomenclature. Arch Virol 77:283–5.

[AOU] American Ornithologists' Union, Committee on Classification and Nomenclature. 1983. Check-list of North American birds: the species of birds of North America from the Arctic through Panama, including the West Indies and Hawaiian Islands. 6th ed. Washington: AOU. Should be ordered from Max C Thompson, Assistant to the Treasurer, AOU, Biology Department, Southwestern College, 100 College Street, Winfield KS 67156, USA; supplements have been published since 1983 in *The Auk*; the 7th edition is scheduled for publication in 1994.

[APA] American Psychiatric Association. 1994. Diagnostic and statistical manual of mental disorders. 4th ed rev. Washington: APA.

[ATA] American Thyroid Association, Committee on Nomenclature. 1987. Revised nomenclature for tests of thyroid hormones and thyroid-related proteins in serum. J Clin Endocrinol Metab 64:1089–94.

[ATCC] American Type Culture Collection. 1992. American Type Culture Collection Catalogue of Cell Lines and Hybridomas. 7th ed. Rockville (MD): ATCC.

Baumel JJ, King AS, Lucas AM, Breazile JE, Evans HE, editors. 1979. Nomina anatomica avium: an annotated anatomical dictionary of birds; prepared by the International Committee on Avian Anatomical Nomenclature, World Association of Veterinary Anatomists. New York: Academic Pr.

Baumgarten H, Kürzinger K. 1989. Designation of antibodies and their derivatives: suggestions for a general nomenclature—a discussion document. J Immunol Methods 122:1–5.

Bligh J, Johnson KG. 1973. Glossary of terms for thermal physiology. J Appl Physiol 35:941–61.

Bodmer JG, March SGE, Albert ED, Bodmer WF, Dupont B, Erlich HA, Mach B, Mayr WR, Parham P, Sasazuki T, and others. 1991. Nomenclature for factors of the HLA system, 1990. Tissue Antigens 37:97–104.

Buyse ML. 1990. Birth defects encyclopedia: the comprehensive, systematic, illustrated reference source for the diagnosis, delineation, etiology, occurrence, prevention, and treatment of human anomalies of clinical relevance. Dover (MA): Center Birth Defects Information Service.

[BZN] Bulletin of Zoological Nomenclature. [quarterly]. London (UK): International Trust for Zoological Nomenclature, British Museum (Natural History), Cromwell Road, London SW7 5BD, UK.

[CAP] College of American Pathologists. 1993. SNOMED international: the systemized nomenclature of human and veterinary medicine. Northfield (IL): CAP.

Clausen JL. 1990. Blood gas terminology: efforts to standardize terminology and calculations. Scand J Clin Lab Invest 50(Suppl 203):169–75.

Collins JT. 1990. Standard common and current scientific names for North American amphibians and reptiles. 3rd ed. Lawrence (KS): Society for the Study of Amphibians and Reptiles. Herpetological Circular No. 19; inquiries should be directed to Douglas H Taylor, Department of Zoology, Miami University, Oxford OH 45056, USA.

[CS] Crustacean Society, Committee on the Names of Decapod Crustaceans. 1989. Common and scientific names of aquatic invertebrates from the United States and Canada: decapod crustaceans. Bethesda (MD): American Fisheries Soc. Special Publication 17.

[CSM] Council of Systematic Malacologists, Committee on Scientific and Vernacular Names of Mollusks. 1988. Common and scientific names of aquatic invertebrates from the United States and Canada: mollusks. Bethesda (MD): American Fisheries Soc. Special Publication 16.

[ECSC] European Community for Steel and Coal, Working Party, Standardization of Lung Function Tests. 1993. Symbols, abbreviations and units. Eur Respir J 6(Suppl 16):85–100.

Erber WN. 1990. Human leucocyte differentiation antigens: review of the CD nomenclature. Pathology 22:61–9.

[ESA] Entomological Society of America, Committee on Common Names of Insects. 1989. Common names of insects and related organisms approved by the Entomological Society of America. College Park (MD): ESA. A companion spelling-checker for IBM-compatible computers and Macintosh computers is available.

Festing MFW. 1987. International index of laboratory animals giving sources of animals used in laboratories throughout the world. 5th ed. Newbury, England (UK): Laboratory Animals Ltd.

Fishman AP, editor. 1986. Handbook of physiology. Bethesda (MD): American Physiological Soc.

Fishman AP. 1988. Pulmonary diseases and disorders. 2nd ed. New York: McGraw-Hill. Appendix D, p D-1–D-6.

Frank MM, Fries LF. 1989. Complement. In: Paul WE, editor., Fundamental immunology. 2nd ed. New York: Raven. p 679–701.

Gagge AP, Hardy JD, Rapp GM. 1969. Proposed standard system of symbols for thermal physiology. J Appl Physiol 27:439–46.

Hall ER. 1965. Names of species of North American mammals north of Mexico. Lawrence (KS): Univ Kansas Museum Nat Hist. Miscellaneous Publication 42.

Hasemann CA, Capra JD. 1989. Immunoglobulins: structure and function. In: Paul WE, editor. Fundamental immunology. 2nd ed. New York: Raven. p 209–33.

Hedrich HJ, editor. 1990. Genetic monitoring of inbred strains of rats: a manual on colony management, basic monitoring techniques, and genetic variants of the laboratory rat. New York: Gustav Fischer.

[IANC] International Anatomical Nomenclature Committee. 1989. Nomina anatomica: authorized by the 12th International Congress of Anatomists in London, 1985. 6th ed. Edinburgh: Churchill Livingstone; 1989. Includes: Nomina histologica, 3rd ed. and Nomina embroyologica, 3rd ed.

[ICLA] International Committee on Laboratory Animals. 1972. International standardized nomenclature for outbred stocks of laboratory animals. ICLA Bull Nr 30.

[ICNBCF] International Committee for the Nomenclature of Blood Clotting Factors. 1962. The nomenclature of blood clotting factors. JAMA 180:733–5.

[ICTH] International Committee of Haemostasis and Thrombosis 1973. Genetic nomenclature for human blood coagulation. Thromb Diath Haemorrh 30:2–11.

[ICZN] International Commission on Zoological Nomenclature. 1985. International code of zoological nomenclature. 3rd ed. London: International Trust for Zoological Nomenclature.

[ILAR] Institute of Laboratory Animal Resources, Committee on Nomenclature, NAS-NRC. 1970. A nomenclatural system for outbred animals. Lab Anim Care 20:903–6.

[ISH] International Society of Haematology. 1965. Nomenclature of abnormal haemoglobins. Br J Haematol 11:121–2.

[ISH] International Society of Hypertension, Nomenclature Committee. 1979. Nomenclature of the renin-angiotensin system. Hypertension 1:654–6.

Issitt PD, Crookston MC. 1984. Blood group terminology: current conventions. Transfusion 24:2–7.

[IUIS] International Union of Immunological Societies, Subcommittee on Nomenclature. 1989. Nomenclature of the Fc receptors. Bull World Health Organ 1989;67:449–50.

[IUPS] International Union of Physiological Sciences, The Renal Commission. 1988. A standard nomenclature for structures of the kidney. Kidney Int 33:1–7.

Kassai T, Cordero del Campillo M, Euzeby J, Gaafar S, Hieppe T, Himonas CA. 1988. Standardized nomenclature of animal parasitic diseases. Vet Parasitol 29:299–326.

Klein J, Benoist C, David CS, Demant P, Lindahl KF, Flaherty L, Flavell RA, Hämmerling U, Hood LE, Hunt SE III, and others. 1990. Revised nomenclature of mouse *H-2* genes. Immunogenetics 32:147–9.

Klein J, Bontrop RE, Dawkins RL, Erlich HA, Gyllensten UB, Heise ER, Jones PP, Parham P, Wakeland EK, Watkins DI. 1990. Nomenclature for the major histocompatibility complexes of different species: a proposal. Immunogenetics 31:217–9.

Knapp W, Stockinger H, Majdic O, Shevach EM. 1991. The CD system of leukocyte surface molecules. In: Coligan JE, Kruisbeck AM, Margulies DH, Shevach EM, Strober W, editors. Current protocols in immunology. New York: J Wiley. Appendix 4: A.4.1–A.4.28.

Lentner C. 1990. Geigy scientific tables. Volume 5, heart and circulation. 8th ed. West Caldwell (NJ): Ciba-Geigy.

Lewis M, Anstee DJ, Bird GWG, Brodheim E, Cartron J-P, Contreras M, Crookston MC, Dahr W, Daniels GL, Engelfriet CP, and others. 1990. Blood group terminology 1990. Vox Sang 58:152–69.

Loeb WF, Quimby FW. 1989. The clinical chemistry of laboratory animals. New York: Pergamon.

Lyon MF. Rules for nomenclature of inbred strains. 1989. In: Lyon MF, Searle AG, editors. Genetic variants and strains of the laboratory mouse. 2nd ed. Oxford: Oxford Univ Pr. p 632–5.

Macklem PT. 1986. Symbols and abbreviations. In: Fishman AP, editor. Handbook of physiology. Bethesda (MD): American Physiological Soc; 1986. Section 3, Volume 1: p ix.

Marder VJ, Mannucci PM, Firkin BG, Hoyer LW, Meyer D. 1985. Standard nomenclature for factor VIII and von Willebrand factor: a recommendation by the International Committee on Thrombosis and Haemostasis. Thromb Haemost 54:871–2.

Marsh DG, Goodfriend L, King TP, Lowenstein H, Platts-Mills TAE. 1986. Allergen nomenclature. Bull World Health Organ 64:767–74.

Mayr E, Ashlock PD. 1991. Principles of systematic zoology. 2nd ed. New York: McGraw-Hill.

McKusick VA. 1992. Mendelian inheritance in man: catalogues of autosomal, dominant, autosomal recessive, and X-linked phenotypes. 10th ed. Baltimore: Johns Hopkins Univ Pr.

Melville RV, Smith JDD, editors. 1987. Official lists and indexes of names and works in zoology. London (UK): International Trust for Zoological Nomenclature.

Mollison PL, Engelfriet CP, Contreras M. 1993. Blood transfusion in clinical medicine. Boston: Blackwell Scientific.

Morse HC III, Shen F-W, Hämmerling V. 1987. Genetic nomenclature for loci controlling mouse lymphocyte antigens. Immunogenetics 25:71–8.

Myers MW. 1984. Interferon nomenclature. ASM News 50(2):68–9.

[NYHA] New York Heart Association, Criteria Committee. 1979. Nomenclature and criteria for diagnosis of diseases of the heart and great vessels. Boston: Little, Brown.

Odds FC, Arai T, Disalvo AF, Evans EG, Hay RJ, Randhawa HS, Rinaldi MG, Walsh TJ. 1992. Nomenclature of fungal diseases: a report and recommendations from a subcommittee of the International Society for Human and Animal Mycology (ISHAM). J Med Vet Mycol 301:1–10.

Parfitt AM. 1988. Bone histomorphometry: standardization of nomenclature, symbols and units (summary of proposed system). Bone 9:67–9.

Sibley CG, Monroe BL Jr. 1990. Distribution and taxonomy of birds of the world. New Haven (CT): Yale Univ Pr.

Siggaard-Anderson O, Durst RA, Maas AHJ. 1987. Approved recommendation (1984) on physico-chemical quantities and units in clinical chemistry. J Clin Chem Clinical Biochem 25:369–91.

[SSB] Society of Systematic Biologists. Systematic biology. Published quarterly; for information, inquire of the National Museum of Natural History, NHB163, Washington DC 20560, USA.

Thompson MW, McInnes RR, Willard HF. 1991. Thompson & Thompson genetics in medicine. 5th ed. Philadelphia: WB Saunders.

[TICH] Tenth International Congress of Haematology. 1964. Nomenclature of hemoglobins. Br Med J 2:1258.

von dem Borne AEG, Decary F. 1990. ICSH/ISBT Working Party on Platelet Serology: nomenclature of platelet-specific antigens. Vox Sang 58:176.

Warner CM, Lamont SJ, Rothschild MF, editors. 1988. The molecular biology of the major histocompatibility complex of domestic animal species. Ames (IA): Iowa State Univ Pr.

[WAVA] World Association of Veterinary Anatomists. 1983. Nomina anatomica veterinaria, 3rd ed. Ithaca (NY): NY State Coll Vet Med.

[WHO] World Health Organization. 1968. Nomenclature of complement. Bull World Health Organ 39:935–8.

[WHO] World Health Organization. 1981. Nomenclature of the alternative activating pathway of complement. Bull World Health Organ 59:489–91.

[WHO–IUIS] WHO-IUIS Nomenclature Sub-Committee. 1990. Nomenclature for synthetic peptides representative of immunoglobulin chain sequences. Bull World Health Organ 68:109–11.

[WHO-IUIS] WHO-IUIS Nomenclature Subcommittee on Interleukin Designation. 1991. Nomenclature for secreted regulatory proteins of the immune system (interleukins). Bull World Health Organ 69:483–4.

[WHO-IUIS] WHO-IUIS Nomenclature Subcommittee. 1993. Nomenclature of amyloid and amyloidosis. Bull World Health Organ 71:105–8.

[WHO NC] [WHO Nomenclature Committee on Leukocyte Antigens]. 1988. Nomenclature for factors of the HLA system, 1987. Vox Sang 55:119–26.

[WHO NC] [WHO Nomenclature Committee]. 1990. Nomenclature for factors of the HLA system, 1989. Vox Sang 58:140–9.

Young DS. 1987. Implementation of SI units for clinical laboratory data: style specifications and conversion tables. Ann Intern Med 106:114–29. Reprinted in J Nutrition 1990;120:20–35.

Zander R, Mertzlufft F. 1990. Oxygen parameters of blood: definitions and symbols. Scand J Clin Lab Invest 50(Suppl 203): 177–85.

Additional References

Allaby M. 1991. The concise Oxford dictionary of zoology. Oxford: Oxford Univ Pr.

[ASIH] American Society of Icthyologists and Herpetologists, Herpetological Catalogue Committee. 1963–70. Catalogue of American amphibians and reptiles. New York: ASIH.

American Thoracic Society. 1991. Lung function testing: selection of reference values and interpretative strategies. Am Rev Resp Dis 144:1202–18.

Anderson S, Jones JK. 1984. Orders and families of recent mammals of the world. New York: J Wiley.

Banks RC, McDiarmid RM, Gardner AL. 1987. Checklist of vertebrates of the United States, the U. S. Trust Territories, and Canada. Washington: US Fish and Wildlife Service. US Fish and Wildlife Service Resource Publication 166.

Barclay AN, Birkeland ML, Brown, MH, Beyers AD, Davis SJ, Somoza, C, Williams AF. 1993. The leucocyte antigens factsbook. San Diego (CA): Academic Pr.

Barnard PC. 1993. Zoological nomenclature. In: Enckell PH, editor. Science editors' handbook. London (UK): European Assoc of Science Editors. p D4:1–3.

Blood DC, Studdert VP. 1988. Baillière's comprehensive veterinary dictionary. London (UK): Baillière Tindall.

Cameron JN. 1989. The respiratory physiology of animals. New York: Oxford Univ Pr.

Corbet GB, Hill JE. 1991. A world list of mammalian species. 3rd ed. Oxford: Oxford Univ Pr.

Critchley Macdonald I. 1980. Butterworth's medical dictionary. 2nd ed. Stoneham (MA): Butterworth-Heinemann.

Férard G. 1992. International Union of Pure and Applied Chemistry, Commission on Quantities and Units in Clinical Chemistry, and International Federation of Clinical Chemistry, Committee on Quantities and Units. Quantities and units for metabolic processes as a function of time, (recommendation 1992). Eur J Clin Chem Clin Biochem 30:901–5.

Frost DR. 1985. Amphibian species of the world: a taxonomic and geographical reference. Lawrence (KS): Allen Pr and Assoc Systematics Collections.

Gibson GJ. 1993. Standardised lung function testing [editorial]. Eur Respir J 6:155–7.

Gilbert P, Hamilton CJ. 1990. Entomology: a guide to information sources. 2nd ed. London (UK): Mansell Publishing.

Greiff M. 1985. Spanish–English–Spanish lexicon of entomological and related terms. Kew, London (UK): Commonwealth Agric Bur. Also available through Univ Arizona Pr, Tucson AZ.

Hauschka PV, Mann KG, Price PA, Termine JD. 1986. Nomenclature recommendations: bone proteins and growth factors. Collagen Relat Res 6:453–4.

Hay ID, Bayer MF, Kaplan MM, Klee GG, Larsen PR, Spencer CA. 1991. American Thyroid Association assessment of current free thyroid hormone and thyrotropin measurements and guidelines for future clinical assays. Clin Chem 37:2002–8.

Henderson IF. 1989. Henderson's dictionary of biological terms. 10th ed. New York: J Wiley.

Hensyl WR, editor. 1990. Stedman's medical dictionary. 25th ed. Baltimore: Williams & Wilkins.

Honacki JH, Kinman KE, Koeppl JW. 1982. Mammal species of the world: a taxonomic and geographic reference. Lawrence (KS): Allen Pr and Assoc Systematics Collections.

Jablonski S. 1991. Jablonski's dictionary of syndromes and eponymic diseases. 2nd ed. Melbourne (FL): Krieger.

Jablonski S. 1992. Dictionary of medical acronyms and abbreviations. 2nd ed. Philadelphia : Hanley & Belfus.

Jablonski S. 1992. Jablonski's dictionary of dentistry. Reprinted with new material. Melbourne (FL): Krieger.

Jeffrey C. 1989. Biological nomenclature. 3rd ed. London: Edward Arnold.

Jobling JA. 1991. A dictionary of scientific bird names. New York: Oxford Univ Pr.

King RC, Stansfield WD. 1990. A dictionary of genetics. 4th ed. New York: Oxford Univ Pr.

Knapp W, Dorken B, Rieber P, Schmidt RE, Stein H, Von dem Borne AEGK. 1989. CD antigens 1989—summary of nomenclature system for human leucocyte surface antigens. Am J Pathol 135: 420–1.

Kriz W, Bankir L; The Renal Commission of the International Union of Physiological Sciences. 1988. A standard nomenclature for structures of the kidney. Kidney Int 33:1–7.

Landau SI, editor. 1986. International dictionary of medicine and biology. New York: J Wiley.

Lee JJ, Hutner SH, Bovee EC. 1985. An illustrated guide to the protozoa. Lawrence (KS): Soc of Parasitologists.

Lentner C. 1981. Geigy scientific tables. Volume 1, units of measurement, body fluids, composition of the body, nutrition. 8th ed. West Caldwell (NJ): Ciba-Geigy.

Lentner C. 1982. Geigy scientific tables. Volume 2, introduction to statistics, statistical tables, mathematical formulae. 8th ed. West Caldwell (NJ): Ciba-Geigy.

Lentner C. 1984. Geigy scientific tables. Volume 3, physical chemistry, composition of blood, hematology, somatometric data. 8th ed. West Caldwell (NJ): Ciba-Geigy.

Lentner C. 1986. Geigy scientific tables. Volume 4, biochemistry, metabolism of xenobiotics, inborn errors of metabolism, pharmacogenetics, and ecogenetics. 8th ed. West Caldwell (NJ): Ciba-Geigy.

Lentner C. 1993. Geigy scientific tables. Volume 6, bacteria, fungi, protozoa, helminths. 8th ed. West Caldwell (NJ): Ciba-Geigy.

Miller JY, editor. 1992. The common names of North American butterflies. Washington: Smithsonian Inst Pr.

Morton LT, Godbolt S. 1992. Information sources in the medical sciences. 4th ed. London: Bowker-Saur.

Parker SP. 1982. Synopsis and classification of living organisms. New York: McGraw-Hill.

Peck S, Peck L. 1993. A time for change of tooth numbering systems. J Dent Educ 57:643–7.

Pennak RW. 1978. Fresh-water invertebrates of the United States. New York: J Wiley.

Pittaway AR. 1992. Arthropods of medical and veterinary importance. Kew, London (UK): Commonwealth Agric Bur. Also available through Univ Arizona Pr, Tucson AZ.

Putnam Paul A. 1991. Handbook of animal science. San Diego (CA): Academic Pr.

Rosen FS. 1989. Macmillan dictionary of immunology. New York: Macmillan.

Simpson GG. 1945. The principles of classification and a classification of the mammals. Bull Am Mus Nat Hist 85:i–xvi, 1–350.

Simpson GG. 1990. Principles of animal taxonomy. New York: Columbia Univ Pr.

Verwilghen RL. 1991. Standardized format for abbreviations and units to be used when expressing blood count results, International Committee for Standardization in Haematology. Haematologica 76:165.

West GP. 1988. Black's veterinary dictionary. 16th ed. Totowa (NJ): Barnes and Noble.

Wilson DE, Reeder DM. 1993. Mammal species of the world: a taxonomic and geographic reference. 2nd ed. Washington: Smithsonian Inst Pr.

Wyatt HV. 1987. Information sources in the health sciences. 3rd ed. London: KG Saur.

25 Human History and Society

> History is the witness of the times, the torch of truth, the life of memory, the teacher of life, the messenger of antiquity.
>
> —Cicero, *Pro Publio Sestio,* 53 BC

25·1 Anthropology comprises 4 fields: archaeology, cultural anthropology, anthropologic linguistics, and physical anthropology. All have been characterized by individualism that extends to report-writing and beyond. Their subject matter is dynamic, or was in life, making hard-and-fast rules sometimes difficult to formulate and apply. Often a "fact" is true only in relation to certain peoples at a particular period. Tidy classification is often difficult or impossible. But some generalizations for style are possible, and various publications have laid ground rules. In the United States the standard for anthropologic archaeology is the journal *American Antiquity* (AA 1983); for cultural anthropology, *The American Anthropologist;* for classical archaeology, *The American Journal of Archaeology* (AJA 1991); and for physical anthropology, the *American Journal of Physical Anthropology.* In Great Britain, the standard for prehistoric archaeology is *Antiquity;* and for social and cultural anthropology, *Man.* Also note the guidelines of the *Journal of Field Archaeology* (JFA 1988).

Some American publications (including government publications and *The American Anthropologist*) prefer the spelling "archeology".

NOMENCLATURE AND CAPITALIZATION

NON-ENGLISH WORDS

25·2 Sometimes only a non-English word expresses a concept accurately. Such terms should be used sparingly. When they are helpful, they should be italicized or surrounded by quotation marks on 1st use and defined parenthetically. In subsequent uses in text they should be set in roman type as though they were English terms. Non-English words in common usage and defined in a recent edition of a standard dictionary do not appear in italics (for example, "in situ", "a priori", "barranca", "barrio").

> In the 17th century, *hyoryumin* (drifters after a shipwreck) were Japan's accidental ambassadors.
>
> The *mindalaes* (Ecuadorian exchange specialists who supplied northern chiefs with valued goods from distant sources) also brought information and customs from afar.
>
> Among the Northwest Coast Skokomish, authority within the household was vested in *sčël'áqs* (a senior family head), and within the village in *sčátšëd* (the socially most prominent household head).
>
> Among the Nez Perce, power may be viewed as a composite of specific attributes, including systemic relationships with tutelary spirits called *naqsní·x hi ˀmtá·ˀ lam.*
>
> [but]
>
> The skeleton appeared in situ to be that of a man.
>
> ["in situ" is not italicized]
>
> The day after the failed coup d'état, an unidentified helicopter fired on the palace.
>
> ["coup d'état" is not italicized but retains the accent]
>
> We found the 1st site on a windswept llano in Chile.
>
> ["llano" is not italicized]

A word without its diacritical marks is misspelled and may not convey the intended meaning. In the absence of a typewriter or computer that can make the appropriate mark, it should be inserted by hand rather than omitting it or substituting another character (for example, a comma for a cedilla).

PEOPLES AND OTHER HUMAN GROUPS

25·3 Styles and preferences for designating peoples change over time and differ from region to region (Table 25·1).

"Band", "tribe", and "chiefdom" are not synonymous; nor are "culture", "nation", "state", "society", and "civilization" (Table 25·1). Note

Table 25·1 Peoples: social systems and groups[a]

A **band** is a small, highly mobile egalitarian society based on hunting and gathering and characterized by a lack of a formal governmental institution and economic specialization. Political dominance is gained through achieved leadership rather than ascribed leadership; as many leadership positions exist as circumstances call for and there are qualified persons to fill them.

A **chiefdom** is the kind of political system characteristic of most ranked societies, politically dominated by chiefs.

A **civilization** is a complex sociopolitical form defined by the institutions of the state and the existence of a distinctive "great tradition" (sets of elite values and behaviors that emerge from folk traditions and that are expressed in distinctive rituals, art, writing, or other symbolic forms).

A **ranked society** is a society in which persons are ranked in relation to each other in terms of kinship status and social prestige, which are largely ascribed. Fewer positions of authority or high status exist than there are persons capable of filling them, and access to these positions is generally ascribed.

Sodalities are organizations such as age grades, warrior societies, curing societies, and religious societies that take membership from across several local groups and kinship segments.

A **state** is a **stratified society** that has developed the institutions for effectively upholding an order of stratification. States are strongly territorial, with complex, well-defined political leadership, hierarchies of settlement, and often elaborate and highly specialized bureaucracies. Populations can range from the thousands to the millions.

A **tribal society** is an egalitarian society larger and more complex than a band and often sedentary, that practices either hunting and gathering or food production and has politically autonomous communities. Political dominance is gained through achieved leadership rather than ascribed leadership, and sodalities are important in integrating the social system.

[a]Sources: W D Lipe, correspondence 1993; Webster and others (1993).

the useful distinction between "peoples" and "persons"; see "people" in section 6·4.

"Ethnicity" distinguishes persons or a people by cultural characteristics. "Race" distinguishes by similar visible physical characteristics that may not be defined by scientific criteria. Because of its inherent vagueness, the term "race" should be used with discretion.

In references to Americans of black African descent (African Americans) "colored" is no longer used in a context of present-day US culture; "black" as both a noun and an adjective is used in preference to "Negro". In a South African context "Bushman" is a pejorative term and should not be used. "Asian" is used as both a noun and an adjective and is preferred to "Oriental"; "Asiatic" is never used to refer to a person. "Caucasian" (or "Caucasoid"), "Mongoloid", and "Negroid" are terms based on an outmoded theory of racial distinction and are no longer used.

25·4 Names of ethnic groups are capitalized. When the name refers to a tribe or group, only the singular form is used generically. In the generic sense, add "s" (with an apostrophe) only to form the possessive. But add "s" when referring to a specific number of persons. (An exception is "Inuk", singular, but "Inuit", plural and adjectival.) The adjectival form of a tribal name does not differ from the name itself. Adding an "n" to the noun usually denotes the language group, for instance, the Mayan language group or the Piman language.

The Maya calendar begins at the year equivalent to 3114 BC.

Awaiting Cortés were 25 Mayas.

The Maya inhabited Tikal in AD 800.

The Chinook are 1 of the tribes that make up the Quinault Indian Nation, now centered on the Olympic Peninsula, Washington.

Chinook canoes were ocean-going.

Speakers of Chinookan lived along rivers.

Smallpox killed hundreds of Chinooks in 1770.

The Chippewa (also known as Ojibwa) inhabit the region around Lake Superior and westward.

Chippewa beadwork is a prized art form.

The reward of several of the Chippewas who went to Washington, DC, to further their cause was burial in Congressional Cemetery.

25·5 Names of groupings of humans are capitalized if they are derived from proper names of geographic entities, if they are names for ethnically or culturally homogeneous groups living within a specific region, or are names for adherents to organized bodies.

African American   Kiowa   Asian

Eagle clan   Hispanic   Latino, Latina   Italian

Frenchman   New Yorker   Nordic   Pygmy   Communist

Odd Fellow   Boy Scout   Liberal   Republican   Allied Powers

Democrat [member of the political party, but "democrat", person who espouses democracy]

Designations based only on color, size, or local usage as well as names representing scientifically ill-defined groups of wide distribution are set in lowercase type.

aborigine   black   white   highlander

Do not hyphenate "American" in a compound proper noun.

an African American the Mexican American
10 Irish Americans

But do hyphenate "American" as a compound proper adjective when the term reflects a geographic place-name rather than a general cultural entity and when it is combined with part of a word.

Afro-Caribbean	African-American students
Afro-American music	Anglo-American exchanges
Polish-American influence	[but "Latin American businessman"]
Japanese-American art	[but "North American politics"]

See section 6·9 for a discussion of biased usage in references to ethnicity, race, nationality, citizenship, or religion.

ARTIFACT TAXONOMY

25·6 If a formal taxonomic system for artifacts has been established (consult authorities in the field if necessary), the names of the types and other categories are treated as proper nouns. In the absence of a formal system, the names of archaeologic classes that are proper nouns are capitalized when standing alone, but generic terms are not.

Folsom points Oldowan chopper Levallois flake
Maya stela Acheulian hand-axe Venus figurines
[but "end scraper", "utilized flake", "red-ware sherds"]

CERAMIC NOMENCLATURE

25·7 In the American Southwest, the following system applies. Terms for specific pottery types, wares, and varieties are capitalized. All components of the term are capitalized, but not the 2nd and succeeding words in hyphenated descriptive terms. A type is separated from its variety by a colon and 1 space if both are given. The term "ware" is not joined to pottery type-names.

Alma Plain [not "Plainware"]
Alma Fingernail Incised
Grasshopper Obliterated Corrugated
Mesa Verde Black-on-white [not "Black-on-White"]
Tusayan Corrugated: Tusayan Variety
Mimbres Bold Face Black-on-white
Urita Gouged-incised: Urita Variety

Table 25·2 Leyden-system bracketing conventions for recording inscriptions[a]

[ ]	Letters that once existed on the stone, but have disappeared and are supplied by the editor.
( )	*Scriptio plena* of abbreviations, contractions, or symbols supplied by the editor.
{ }	Letters erroneously engraved on the stone, which are suppressed by the editor.
< >	Additions or corrections of the editor, supplying letters accidentally omitted by the stonecutter, or correcting the stonecutter's errors of engraving (the literal reading of the stone being given in the commentary).
[[ ]]	Letters erased on the stone.
[[αβ]]	Letters erased on the stone, traces of which are still visible.
[[[αβ]]]	Letters erased on the stone that are supplied by the editor.
αβ	Letters of doubtful reading.

[a]Sources: Dow (1969), Gifford and Heathington (1989).

Ixcanrio Orange-polychrome: Ixcanrio Variety

[but "Hohokam buff wares", "Salado polychromes", "Mogollon brown wares"]

SITE NAMES AND ARCHAEOLOGIC FEATURES

25·8 Both the name of a site or feature and the generic term that follows, such as "village", "pueblo", "ruin", or "site", are capitalized.

Naco Site Promontory Village Grasshopper Pueblo

Hodges Ruin Ventana Cave Lascaux Cave

Sudden Shelter Red Bow Cliff Dwelling Avenue of the Dead

Pyramid of the Sun Monument 1 Stela 27

INSCRIPTIONS

Inscriptions should be bracketed by the Leyden system (Table 25·2; see also Dow 1969).

The inscription "INΓ BAPF" was interpreted as an abbreviation for "INΓ[ΛΙΝΟγ] E [[ ]], "English Varangians", but a much simpler interpretation would be the Norse name "Ingvar E . . . ", the rest having been either lost or cut off.

"MLQRT 'BD" ["slave of Melkart"]

"MLQRT '[BD]" ("[slave] of Melkart")

"MLQRT <'> BD" ["slave <of> Melkart"]

Table 25·3 The 3-age chronologic system for describing prehistory[a]

Stone Age:

Paleolithic or Old Stone Age: only chipped stone tools.

Upper Paleolithic.

Middle Paleolithic (Mousterian).

Lower Paleolithic.

Mesolithic or Middle Stone Age: chipped stone tools, microliths, beginnings of pottery and ground stone in some areas.

Neolithic or New Stone Age: ground and polished stone tools, pottery, agriculture.

Chalcolithic, Eneolithic, Copper Age (terms used interchangeably): period between the Neolithic and the Bronze Age when unalloyed copper was used rather than true tin bronze.

Bronze Age: emergence of an urban way of life.

Iron Age: tools and weapons of iron.

[a]Some cultures remained in "prehistory" in historical periods of the main civilizations, for example, the aboriginal culture of Australia.

TIME

SPANS OF TIME

25·9 The so-called 3-age system (published in 1836 by Christian Jürgensen Thomsen [Spjeldnaes 1976]), divides prehistory into 3 successive periods (subsequently subdivided) defined by the main material in the period for making tools: stone, bronze, iron (Table 25·3). These periods are defined by cultural developments; the actual dates differ among regions. Because these developments are evinced by artifacts, "upper", as in "Upper Paleolithic", refers to strata closer to the surface and hence newer, the older artifacts being buried under the newer ones; hence "upper" = "late", "lower" = "early". The 3-age system is less useful in Africa, where bronze was not used south of the Sahara, and in the Americas, where bronze was never important and iron was not used until Europeans introduced it (Table 25·4).

25·10 Names of cultural periods, eras, epochs (spans of time associated with dates), and names of archaeologic features are capitalized. "Age" and "early", "middle", and "late" in this context (but not "period") are also capitalized.

Late Bronze Age Middle Kingdom Iron Age

Late Hellenic Chalcolithic Age Late Stone Age

[but 'Early Woodland period", "final Jomon period", "Viking period"]

Table 25·4 Culture types[a,b]

Lithic or **Paleoindian** (New World) or **Upper Paleolithic** (Old World): The stage of adaptation by immigrant societies to the late glacial and early post-glacial climatic and physiographic conditions in the New World. The Lithic is preeminently a hunting stage, although other economic patterns were certainly present whose local dominance, under certain conditions, is not included.

Archaic (New World) or **Mesolithic** (Old World): The stage of migratory hunting and gathering cultures continuing into environmental conditions approximating those of the present. There is now a dependence on smaller and perhaps more varied fauna. There is also an apparent increase in gathering. Sites begin to yield large numbers of stone implements and utensils that are assumed to be connected with the preparation of wild vegetable foods.

Formative (New World) or **Neolithic** (Old World): The stage defined by the presence of agriculture, or any other subsistence economy of similar effectiveness, and by the successful integration of such an economy into well-established, sedentary village life. Pottery-making, weaving, stone-carving, and a specialized ceremonial architecture are usually associated with these American Formative cultures.

Classic (New World) or **Early Bronze Age** (Old World): The American Classic stage is characterized by urbanism and by superlative performance in many lines of cultural endeavor. The Classic is the stage of great artistic achievements and of monumental and ambitious architecture. Fine, specialized craft products were turned out in profusion; there is evidence of strong social class distinctions. With the perfection of writing and astronomy, intellectual interests as well as the arts flourished. There was active trade between the regional centers in ceremonial and luxury goods.

Postclassic (New World) or **Late Bronze Age** (Old World): Marked by the breakdown of the old regional styles of the Classic stage, by a continuing or increased emphasis upon urban living, and inferentially, by tendencies toward militarism and secularism.

[a]Sources: W D Lipe, correspondence 1992; Willey and Phillips (1958).

[b]These culture types occurred at different times in different places, and general dates for the Americas as a whole cannot be applied. All stages from Lithic through Postclassic occurred in the Valley of Mexico, but the Interior Northwest was essentially Archaic on 1st European contact.

No general chronology is applicable to all cultures. Mankind and groups of mankind are dynamic entities: people move around, marry into other groups, die out over a span of time, and incorporate other groups and innovations often over a span of not necessarily continuous years. Theories about the relative ages of cultures change. Hence a chronology of cultural groups is not useful for more than the specific purpose for which it was constructed and for a specific time: for example, the Paleolithic Age lasted until about 8000 BC in the Middle East and until the 19th century in Australia.

CHRONOMETRIC DATING

25·11 Absolute, or chronometric, dating is the measurement of age with reference to a specific time scale that dates from or to a fixed reference point. In the predominantly Christian world, the convention is to use the birth of Christ, set in the year AD 1 (= 1 CE), as that point. Years are counted back before Christ (BC) and forward after Christ (AD for *Anno Domini,* Latin for "in the year of our Lord", or AC for "after Christ", or CE for "common era"); there is no year 0. In the ancient Greek world, the calendric starting point was the holding of the 1st Olympic Games (776 BC in the Christian calendar); for Muslims it is the date of the Hegira, the Prophet's departure from Mecca (AD 622); for the Maya it is equivalent to 3114 BC.

An international system, without reference to a particular calendar system, counts years before the present (BP). AD 1950 is usually used for "present," so a date of 400 BP is not about 400 years ago (AD 1590), but about 444 years ago (AD 1550). Archaeologists differ in their usage, but most confine the use of BC and AD terms to the most recent 5000 to 10 000 years.

AD, AC, CE, BC, and BP are frequently set in small capitals, but this manual recommends capitals, to keep their capitalization consistent with the usage recommended for abbreviations in general; see section 10·2. AD and AC are placed before the date, BC, CE, and BP after. In ranges, the earliest date appears 1st.

> Río Azul may have been settled as early as 500 BC, but its period of greatness was approximately AD 390 to 540.
>
> The Greek colony at Metaponto, Italy, flourished between 600 and 250 BC.
>
> Radiocarbon tests indicate a date for Hallan Çemi from 10 500 BP.

RADIOMETRIC DATING

25·12 Radiometric dating refers to all methods of age measurement that rely on the nuclear decay of naturally occurring radioactive isotopes, whether they are short-life radioactive elements (for example, ^{14}C) or long-life radioactive elements and their decay products (for example, ^{40}K–^{40}Ar; the uranium series, the decay of ^{238}U and ^{235}U to thorium [^{230}Th] and protactinium [^{231}Pa]).

Radiocarbon is still the most frequently useful tool for dating organic samples as old as perhaps 50 000 years, as is potassium–argon dating (the ratio of potassium to argon) for rock samples 100 000 years old or older. For strata or artifact assemblages between those ages, dating is far less reliable.

In early practice, archaeomagnetic dates were reported in the same form as ^{14}C dates, that is, as a mean value plus or minus a certain number of years (AD 1015 ± 35). It is now considered more accurate to report dates as ranges, for example, AD 980 to 1050. Both conventions for reporting archaeomagnetic dates continue to be used. When such dates are cited, follow the original exactly. Note that ± designations should be accompanied by a symbol indicating what the numeric value following the ± represents, such as "*s*" for "standard deviation", "2*s*" for "2 standard deviations", or "$S_{\bar{x}}$" for "standard error of the mean", for example, "AD 1015 ± 35(*s*)"; see section 11·24.

25·13 The conversion of radiocarbon ages to calendric dates must take into account temporal variations in the natural radiocarbon content of atmospheric carbon dioxide. Consult Klein and others (1982) for calibration tables (95% confidence level), or the journal *Radiocarbon* for the most up-to-date calibrations.

Uncalibrated dates (dates not checked against those derived from another system) are not linked to any system of reckoning in calendar years. If the date has not been calibrated with another dating method (for example, dendrochronology), some archaeologists, but not all, express the era in lowercase letters (ad, bc, bp). A more cumbersome method, espoused by radiocarbon laboratories, gives the calibrated date as "Cal BC" or "Cal AD".

[uncalibrated]	ad 1500	3000 bc	400 bp
[calibrated]	AD 1400	3700 BC	Cal AD 1550

For BP dates, the author must indicate the date the laboratory used to represent "present" (generally 1950). The international symbol BP indicates radiocarbon years ago (dates not calibrated); BC and AD indicate calendric dates derived by subtracting 1950 from uncalibrated radiocarbon ages. The laboratory number follows the age and date if age is announced for the 1st time. Each laboratory has its own letter code, for example, P = "Philadelphia"; Q = "Cambridge, England". If a published reference exists, the citation follows the age and date. There should be 1 space before and 1 after the ± symbol; also note the recommendation in section 25·12 for an indication of the specific meaning of the ± value.

950 ± 100 BP (P 1234) 11 950 ± 100 years: 10 000 BC (P 16779)

950 ± 100 radiocarbon years: AD 1000 (Smith 1992)

See also Chapter 12, "Time and Dates", section 26·1 ("Geologic Time"), and sections 27·17–27·23 ("Celestial Time").

REFERENCES

Cited References

[AA] American Antiquity. 1983. Editorial policy and style guide for American Antiquity. Am Antiquity 48:429–42.

[AJA] American Journal of Archaeology. 1991. Editorial policy, notes for contributors, and abbreviations [of journals, books, collections, organizations]. Am J Archaeol 95:1–16.

Dow S. 1969. Conventions in editing. Greek, Roman and Byzantine studies. Scholarly Aids [irregular nr] 2.

Gifford CA, Heathington CA. 1989. Arizona State Museum style guide. 2nd ed. Tucson (AZ): Arizona State Museum, Univ Arizona. Archaeological Series 180.

[JFA] Journal of Field Archaeology. 1988. Guidelines for contributors—1989. J Field Archaeol 15:485–9.

Klein J, Lerman JC, Damon PE, Ralph EK. 1982. Calibration of radiocarbon dates: tables based on the consensus data of the Workshop on Calibrating the Radiocarbon Time Scale. Radiocarbon 24:103–50.

Spjeldnaes N. 1976. Christian Jürgensen Thomsen. In: Dictionary of scientific biography: Volume XIII:357–8. New York: Charles Scribner's Sons.

Webster DL, Evans ST, Sanders WT. 1993. Out of the past: an introduction to archaeology. Chicago: Univ Chicago Pr.

Willey GR, Phillips P. 1958. Method and theory in America archaeology. Mountain View (CA): Mayfield Publishing.

Additional References

Association of American University Presses, Task Force on Bias-Free Language. [in preparation]. Guidelines for bias-free usage.

Beals RL, Hoijer H, Beals AR. 1977. An introduction to anthropology. 5th ed. New York: Macmillan.

Crown CH. 1986. The growth of ethnobiological nomenclature. Curr Anthropol 27:1–19.

Fagan BM. 1986. People of the earth. 5th ed. Boston: Little, Brown.

Renfrew C, Bahn P. 1991. Archaeology: theories, methods, and practice. New York: Thames & Hudson.

[UCP] University of Chicago Press. The Chicago Manual of Style. 1993. 14th ed. Chicago: UCP.

Whitehouse RD. 1983. The Macmillan dictionary of archaeology. London: Macmillan.

26 The Earth

> And God said, "Let the waters under the heaven be gathered together under one place, and let the dry land appear": and it was so.
>
> — *The Holy Bible* (The Authorized Version), *Genesis* 1

The atmosphere, soils, rocks, and oceans have been the subjects of intensive scientific study in the disciplines of geology, soil science, and related fields. This chapter covers the conventions used to report and describe research on these subjects.

GEOLOGIC TIME

26·1 Geologic time is divided into eons, which are subdivided into eras; eras are subdivided into periods, periods into epochs, and epochs into ages ("stages" in the recommendations by Harland and colleagues [1990]). Time terms should not be abbreviated in text. Formal time terms have initial-letter capitalization (Harland and others 1990, Hansen 1991); in informal terms, the modifier (such as "early") is not capitalized. For terms for North American and European geologic divisions of time see Tables 26·1 and 26·2. Hansen (1991) and Harland (1990) are authoritative sources for formal terms.

Phanerozoic Eon Mesozoic Era
Cretaceous Period Maastrichtian Stage
early Mesozoic Era

Geochronologic tables delineate the age of the earth's strata; chronostratigraphic tables delineate the relative positions of the earth's strata. The 2 are conventionally combined and listed in downward order of increasing geologic age; see Tables 26·1 and 26·2. Geologic units (rocks) are divided into eonothems, which are subdivided into erathems; erathems are subdivided into systems (Tertiary and Carboniferous systems are further divided into subsystems); systems (or subsystems) are divided into series, and series into stages. Each generic term (for example, "eonothem", "stage") is capitalized when used with its specific modifier.

Phanerozoic Eonothem Mesozoic Erathem
Cretaceous System Upper Cretaceous Series Maastrichtian Stage

The US Geological Survey does not accept formal names of subdivisions except within the Cenozoic. Series (rocks) are often termed "Upper" and "Lower" for the same epoch (time) that is termed "Late" or "Early". The paired terms are not interchangeable.

[position] [age]
Upper Triassic strata were deposited in Late Triassic time.
Lower Cambrian rocks contain Early Cambrian fossils.

Table 26·1 North American divisions of geologic time[a]

Division in time EON ERA Period Subperiod Epoch	Estimated age of boundary Millions of years ago, Ma	Characteristic life forms, 1st known appearance
PHANEROZOIC EON		
CENOZOIC ERA		
Quaternary Period		
Holocene Epoch		
	0.010	
Pleistocene Epoch		*Homo sapiens,* Neanderthal
	1.6 (1.6–1.0)	
Tertiary Period		*Homo erectus*
Neogene Subperiod		
Pliocene Epoch		*Australopithecus africanus*
	5 (4.9–5.3)	
Miocene Epoch		hominids
	24 (23–26)	
Paleogene Subperiod		
Oligocene Epoch		
	38 (34–38)	
Eocene Epoch		LATE EOCENE EXTINCTION
	55 (54–56)	
Paleocene Epoch		anthropoids, rodents, horses, grasses
	66 (63–66)	
MESOZOIC ERA		primates
Cretaceous Period		mammals diversify
Late Cretaceous Epoch		TERMINAL CRETACEOUS EXTINCTION
	96 (95–97)	
Early Cretaceous Epoch		diatoms, angiosperms, placental mammals
	138 (135–141)	
Jurassic Period		
Late Jurassic Epoch		
Middle Jurassic Epoch		
Early Jurassic Epoch		bony fishes
	205 (200–215)	
Triassic Period		
Late Triassic Epoch		early mammals
Middle Triassic Epoch		TERMINAL TRIASSIC EXTINCTION
Early Triassic Epoch		
	~240	
PALEOZOIC ERA		
Permian Period		
Late Permian Epoch		
Early Permian Epoch		dinosaurs, corals

Table 26·1 (*cont.*)

Division in time EON ERA Period Subperiod Epoch	Estimated age of boundary Millions of years ago, Ma	Characteristic life forms, 1st known appearance
	290 (290–305)	
Carboniferous Period		
Pennsylvanian Subperiod		
Late Pennsylvanian Epoch		
Middle Pennsylvanian Epoch		
Early Pennsylvanian Epoch		conifers, winged insects
	~330	
Mississippian Subperiod		
Late Mississippian Epoch		
Early Mississippian Epoch		
	360 (360–365)	
Devonian Period		
Late Devonian Epoch		
Middle Devonian Epoch		gymnosperms, sharks
Early Devonian Epoch		land plants
	410 (405–415)	
Silurian Period		
Late Silurian Epoch		
Middle Silurian Epoch		
Early Silurian Epoch		
	433 (435–440)	
Ordovician Period		
Late Ordovician Epoch		
Middle Ordovician Epoch		
Early Ordovician Epoch		
	500 (495–510)	
Cambrian Period		
Late Cambrian Epoch		
Middle Cambrian Epoch		
Early Cambrian Epoch		
	~570[b]	
PROTEROZOIC EON		
LATE PROTEROZOIC ERA[c]		
	900	
MIDDLE PROTEROZOIC ERA		
	1600	
EARLY PROTEROZOIC ERA		
	2500	
ARCHEAN EON		
LATE ARCHEAN ERA		

(*continued*)

Table 26·1 (*cont.*)

Division in time EON ERA Period Subperiod Epoch	Estimated age of boundary Millions of years ago, Ma	Characteristic life forms, 1st known appearance
	3000	
MIDDLE ARCHEAN ERA		
	3400	
EARLY ARCHEAN ERA		
	3800 ?	
pre-Archean		

[a] Source: Hansen (1991). The terms "eon", "era", "period", "stage" refer to geologic time; "eonothem", "erathem", "system", "series" are equivalent generic terms that refer to stratigraphic positions dated for a particular time. The specific name in the corresponding chronologic and stratigraphic terms is the same: "Jurassic Period" (chronologic term), "Jurassic System" (stratigraphic name representing rock formations assigned to the Jurassic Period. Authors frequently use only the specific name of a term and imply the generic element ["Most of his work was concerned with the Jurassic."].

In the North American nomenclature, when "early", "middle", and "late" are part of the formal names of epochs (time), "upper", "middle", and "lower" are part of the formal names of series (position); both elements of the names are capitalized ["Upper Cretaceous", "Late Cretaceous", "Middle Devonian"]. These modifiers are considered descriptive only and are not capitalized when used with any division of time or position other than Cretaceous, Jurassic, Triassic, Permian, Pennsylvanian, Mississippian, Devonian, Silurian, Ordovician, and Cambrian.

[b] Rocks older than 570 Ma are also called Precambrian, a time term without specific rank.

[c] Informal term without specific rank.

Rocks older than 570 Ma (mega-annum or millions of years) are divided into geochronometric units, their boundaries usually internationally agreed upon. No type localities have been designated.

Archean Eon   Late Archean Era

26·2 Throughout geologic time, the Earth's polarity has repeatedly reversed. The following polarity–chronostratigraphic units distinguish the primary magnetic-polarity record imposed when the rock was deposited, in order of increasing age: Brunhes, Matuyama, Gauss, and Gilbert.

ICE AGES

26·3 Except for "Glacial Epoch" (Pleistocene Epoch), the generic terms in names of ice ages are set in lowercase type.

Illinoian glaciation   Riss glaciation

"Neoglacial" refers to the small-scale glacial advance that occurred in the Holocene. The most recent advance was the Little Ice Age, AD 1550

Table 26·2 European divisions of geologic time[a]

Division in time EON ERA Period Subperiod Epoch	Estimated age of boundary Millions of years ago, Ma	Characteristic life forms, 1st known appearance
PHANEROZOIC EON		
CENOZOIC ERA		
Quaternary Period		
Holocene Epoch		
Versilian		
	0.01	
Pleistocene Epoch		
Tyrrhenian		
Milazzian		*Homo sapiens*
Sicilian		Neanderthal man
Calabrian		*Homo erectus*
	1.64	
Tertiary Period		
Neogene Subperiod		
Pliocene Epoch		
Piacenzian		*Australopithecus africanus*
Zanclean		
	5.2	
Miocene Epoch		
Messinian		
Tortonian		
Serravallian		
Langhian		hominids
Burdigalian		
Aquitanian		
	23.3	
Paleogene Subperiod		
Oligocene Epoch		
Chattian		
Rupelian		
	35.4	
Eocene Epoch		
Priabonian		LATE EOCENE EXTINCTION
Bartonian		anthropoids
Lutetian		rodents
Ypresian		
	56.5	
Paleocene Epoch		
Thanetian		horses
Danian		grasses
	65.0	

(*continued*)

Table 26·2 (*cont.*)

Division in time EON ERA Period Subperiod Epoch	Estimated age of boundary Millions of years ago, Ma	Characteristic life forms, 1st known appearance
MESOZOIC ERA		primates
Cretaceous Period		mammals diversify
Gulf Epoch		TERMINAL CRETACEOUS EXTINCTION
Senonian Subepoch		
Maastrichtian		
Campanian		
Santonian		
Coniacian		
	88.5	
Gallic Subepoch		
Turonian		
Cenomanian		diatoms
Albian		angiosperms
Aptian		
Barremian		
	131.8	
Neocomian Subepoch		
Hauterivian		placental mammals
Valanginian		
Berriasian		
	145.6	
Jurassic Period		
Malm Epoch		
Tithonian		
Kimmeridgian		birds
Oxfordian		bony fishes
	157.1	
Dogger Epoch		
Callovian		
Bathonian		
Bajocian		
Aalenian		
	178.0	
Lias Epoch		
Toarcian		
Pliensbachian		
Sinemurian		
Hettangian		
	208.0	

Table 26·2 (*cont.*)

Division in time EON ERA Period Subperiod Epoch	Estimated age of boundary Millions of years ago, Ma	Characteristic life forms, 1st known appearance
Triassic Period		
Triassic 3 Epoch		
Rhaetian		early mammals
Norian		TERMINAL TRIASSIC
Carnian		EXTINCTION
	235.0	
Triassic 2 Epoch		oysters
Ladinian		
Anisian		
	241.1	
Scythian Epoch		
Spathian		
Nammalian		
Griesbachian		
PALEOZOIC ERA		
Permian Period		
Zechstein Epoch		dinosaurs, corals
Lopingian Subepoch		
Changxingian		
Longtanian		
Guadalupian Subepoch		
Capitanian		
Wordian		
Ufimian		
	256.1	
Rotliegendes Epoch		
Kungurian		
Artinskian		
Sakmarian		
Asselian		
	290.0	
Carboniferous Period		
Pennsylvanian Subperiod		
Gzelian Epoch		conifers, winged insects
Noginskian		
Klazminskian		
	295.1	
Kasimovian Epoch		
Dorogomilovskian		
Chamovnicheskian		
Krevyakinskian		
	303.0	

(*continued*)

Table 26·2 (*cont.*)

Division in time EON ERA Period Subperiod Epoch	Estimated age of boundary Millions of years ago, Ma	Characteristic life forms, 1st known appearance
Moscovian Epoch		
Myachkovskian		
Podolskian		
Kashirskian		
Vereiskian		
	311.3	
Bashkirian Epoch		
Melekeskian		
Cheremshanskian		
Yeadonian		
Marsdenian		
Kinderscoutian		
	322.8	
Carboniferous 1 Subperiod [Mississippian]		
Serpukhovian Epoch		
Alportian		
Chokierian		
Arnsbergian		
Pendleian		
	332.9	
Visean Epoch		
Brigantian		
Asbian		
Holkerian		
Arundian		
Chadian		
	349.5	
Tournaisian Epoch		
Ivorian		
Hastarian		
	362.5	
Devonian Period		
Devonian 3 Epoch		
Famennian		
Frasnian		
	377.4	
Devonian 2 Epoch		gymnosperms, sharks
Givetian		
Eifelian		
	386.0	

Table 26·2 (*cont.*)

Division in time EON ERA Period Subperiod Epoch	Estimated age of boundary Millions of years ago, Ma	Characteristic life forms, 1st known appearance
Devonian 1 Epoch		
Emsian		
Pragian		
Lochkovian		land plants
	408.5	
Silurian Period		
Pridoli Epoch		
	410.7	
Ludlow Epoch		
Ludfordian		
Gorstian		
	424.0	
Wenlock Epoch		
Gleedonian		
Whitwellian		
Sheinwoodian		
	430.4	
Llandovery Epoch		
Telychian		
Aeronian		
Rhuddanian		
	439.0	
Ordovician Period		
Bala Subperiod		
Ashgill Epoch		
Hirnantian		
Rawtheyan		
Cautleyan		
Pusgillian		
	433.1	
Caradoc Epoch		
Onnian		
Actonian		
Marshbrookian		
Longvillian		
Soudleyan		
Harnagian		
Costonian		
	463.9	

(*continued*)

Table 26·2 (*cont.*)

Division in time EON ERA Period Subperiod Epoch	Estimated age of boundary Millions of years ago, Ma	Characteristic life forms, 1st known appearance
	463.9	
Dyfed Subperiod		
Llandeilo Epoch		
Late		
Mid		
Early		
	468.6	
Llanvirn Epoch		
Late		
Early		
	476.1	
Canadian Subperiod		
Arenig Epoch		
Tremadoc Epoch		
	510.0	
Cambrian Period		
Merioneth Epoch		
Dolgellian		
Maentwrogian		
	517.2	
St David's Epoch		
Menevian		
Solvan		
	536.0	
Caerfai Epoch		
Lenian		
Atdabanian		
Tommotian		
	570	
PROTEROZOIC EON		
SINIAN ERA		
Vendian Period		
Ediacara Epoch		
Poundian		
Wonokan		
	590	
Varanger Epoch		
Mortensnes		
Smalfjord		
	610	

Table 26·2 (*cont.*)

Division in time EON ERA Period Subperiod Epoch	Estimated age of boundary Millions of years ago, Ma	Characteristic life forms, 1st known appearance
Sturtian Period	0.80 Ga[b]	
RIPHIAN ERA		
Karatau Period		
Yurmatin Period		
Burzyan Period		
	1.65 Ga	
ANIMIKEAN ERA		
Gunflint Period		
	2.2 Ga	
HURONIAN ERA		
Cobalt Period		
Ourke Lake Period		
Hough Lake Period		
Elliot Lake Period		
	2.4–2.5 Ga	
ARCHEAN EON		
RANDIAN ERA		
Ventersdorp Period		
Central Rand Period [4 divisions]		
Dominion Period		
	2.8 Ga	
SWAZIAN ERA		
Pongola Period		
Moodies Period		
Figtree Period		
Onverwacht Period		
	3.5 Ga	
ISUAN ERA		
	3.8 Ga	
HADEAN ERA		
Early Imbrian Period [2 epochs]		
Nectarian Period [2 epochs]		
	3.95 Ga	
PRISCOAN EON		
Pre-Nectarian Period		
Basin Groups [1–9 epochs]		
Procellarum Epoch		
	4.15 Ga	

(*continued*)

Table 26·2 (*cont.*)

Division in time EON ERA Period Subperiod Epoch	Estimated age of boundary Millions of years ago, Ma	Characteristic life forms, 1st known appearance
Cryptic Division [many dated events]		
	4.55 to 4.5 Ga	
PRE-HADEAN ERA: ORIGIN OF EARTH AND MOON		

[a]Source: Harland and others (1990).
[b]Giga annum.

to 1850. "Neoglaciation" is an informal term used to designate glacial expansions later than the Holocene climatic optimum.

RECENT AND PRESENT TIME

The term "Holocene" refers to the epoch of the last 10 000 years; it replaces "Recent". Now "recent" is always set in lowercase type and connotes any recent time of unspecified duration. Adding the terms "ago" and "before the present" produces a redundancy when they follow a geochronologic date correctly expressed, as, for instance, 1990 BP or 1990 Ma BP. Abbreviations for the number of years without reference to the present differ widely (for example, y, yr [years]; Ma [mega-annum], my, m. y., m.yr. [millions of years]). The boundaries of the Early Cretaceous Epoch are calibrated at 96 Ma to 138 Ma, but the interval of time this epoch represents is 42 my.

GEOLOGIC NAMES

26·4 A geologic name is the name of a defined unit of rock, or the local name applied to a mapped rock unit. The map unit is recognized by its lithologic content and its boundaries. It is assigned a place within the geologic age sequence, has a stratigraphic rank, and is mappable.

The Geologic Names Committee of the US Geological Survey considers all names of geologic formations and other divisions of rock classifications used by members of the Survey to ensure that the names comply with nomenclature used in previously published US Geological Survey reports and to recommend policy on stratigraphic nomenclature.

Most geologists in the United States submit a proposed name to the Geologic Names Committee before introducing it into the literature.

Formal, full terms should always be used instead of shortened terms; tampering with formal names for the sake of brevity endangers clarity.

Permian and Pennsylvanian	[not "Permo-Penn"]
Cambrian and Ordovician	[not "Cambro-Ordovician"]
Middle Cambrian	[not "Mid-Cambrian"]
Westwater Canyon Member	[not "Westwater Member"]

Map symbols, such as those for names of formations, should not be used in text.

The US Geological Survey has capitalized formal geologic names since 1961. Names in material quoted before that time should follow the original author's usage.

Degrees of Doubt

A question mark immediately after a term to show doubt of accuracy is often more specific than words can be. For example, does "probably Late Devonian" mean "Late(?) Devonian" or "Late Devonian(?)"?

STRATIGRAPHIC NAMES

26·5 Rocks are usually discussed chronologically, oldest (bottommost) 1st, youngest (topmost) last.

On 1st use, informal terms, including local or commercial names (as in mining or the oil industry), should be defined and described. For all informal stratigraphic units, only the place-name is capitalized; simply descriptive geologic terms are not capitalized.

tuff of Stony Point   formation of Madeira Canyon

Formal names of formations often include the name of the rock type; that name is also capitalized as part of the proper name.

Stony Point Tuff   Baltimore Gneiss   St Louis Limestone

For formal nomenclature, refer to the *North American Stratigraphic Code* (NACSN 1983).

When accompanied by the name of its unit, a petrologic term becomes part of the proper name and should be capitalized. The terms may be abbreviated in illustrations or tables, but not in text.

Nevada Formation [but "the formation"]

Uinta Mountain Group [but "group in the West"]

The terms "system", "series", "period", and "epoch" are not used in map explanations.

MAP UNITS

26·6 The description of a map unit is an abbreviated account of, from left to right, the rock type, color, and thickness described in order of increasing age. To the right of the stratigraphic column is the name of the stratigraphic unit, followed by its position (usually the series term) or age within parentheses.

Kr Raritan Formation (Upper Cretaceous)

Letter symbols for map units are unique to each geologic map and are not used in text.

CORRELATION CHARTS; STRATIGRAPHIC TABLES AND COLUMNAR SECTIONS

26·7 Correlation charts show the author's interpretation of rock units and their ages as they relate to units other workers have recognized elsewhere. The relative or radiometric time is usually designated at the left. Rock units should be identified by name, whether formal or informal, without abbreviations if space permits.

Stratigraphic columnar sections are graphic illustrations that describe and show in a vertical column the sequence and relationships of rock or soil in a defined area. Color or symbols used to distinguish units are unique to the particular illustration. Color terms used to describe rocks should be as specific as possible. Refer to *Rock-Color Chart* (GSA 1991). Standard patterns indicate rock types (see Compton 1985, p 376–7).

For details on the construction of these charts and illustrations, consult Hansen (1991).

PHYSICAL DIVISIONS OF THE UNITED STATES

26·8 The physiographic divisions, provinces, and sections of the United States are specific physiographic entities and are therefore capitalized as proper nouns (see Table 26·3).

Table 26·3 Physical divisions of the United States[a]

Major division	Province	Section
Laurentian Upland	Superior Upland	
Atlantic Plain	Continental Shelf	
	Coastal Plain	Embayed Section
		Sea Island Section
		Floridian Section
		East Gulf Coastal Plain
		Mississippi Alluvial Plain
		West Gulf Coastal Plain
Appalachian Highlands	Piedmont Province	Piedmont Upland
		Piedmont Lowland
	Blue Ridge Province	Northern Section
		Southern Section
	Valley and Ridge Province	Tennessee Section
		Middle Section
		Hudson Valley
	St Lawrence Valley	Champlain Section
		Northern Section
	Appalachian Plateaus	Mohawk Section
		Catskill Section
		Southern New York Section
		Allegheny Mountain Section
		Kanawha Section
		Cumberland Plateau
		Cumberland Mountain Section
	New England Province	Seaboard Lowland
		New England Upland
		White Mountain Section
		Green Mountain Section
		Taconic Section
	Adirondack Province	
Interior Plains	Interior Low Plateaus	Highland Rim
		Lexington Plain
		Nashville Basin
	Central Lowland	Eastern Lake Section
		Western Lake Section
		Wisconsin Driftless Section
		Till Plains
		Dissected Till Plains
		Osage Plains
	Great Plains	Missouri Plateau, glaciated
		Missouri Plateau, unglaciated
		Black Hills
		High Plains
		Plains Border
		Colorado Piedmont
		Raton Section
		Pecos Valley
		Edwards Plateau
		Central Texas Section

Table 26·3 (*cont.*)

Major division	Province	Section
Interior Highlands	Ozark Plateaus	Springfield–Salem Plateaus
		Boston "Mountains"
	Ouachita Province	Arkansas Valley
		Ouachita Mountains
Rocky Mountain System	Southern Rocky Mountains	
	Wyoming Basin	
	Middle Rocky Mountains	
	Northern Rocky Mountains	
Intermontane Plateaus	Columbia Plateaus	Walla Walla Plateau
		Blue Mountain Section
		Payette Section
		Snake River Plain
		Harney Section
	Colorado Plateaus	High Plateaus of Utah
		Uinta Basin
		Canyon Lands
		Navajo Section
		Grand Canyon Section
		Datil Section
	Basin and Range Province	Great Basin
		Sonoran Desert
		Salton Trough
		Mexican Highland
		Sacramento Section
Pacific Mountain System	Sierra–Cascade Mountains	Northern Cascade Mountains
		Middle Cascade Mountains
		Southern Cascade Mountains
		Sierra Nevada
	Pacific Border Province	Puget Trough
		Olympic Mountains
		Oregon Coast Range
		Klamath Mountains
		California Trough
		California Coast Ranges
		Los Angeles Ranges
	Lower California Province	

[a]Source: USGPO (1984). p 228.

ROCKS AND MINERALS

26·9 In general, nomenclature for sedimentary and metamorphic rocks has not been standardized. Detailed comments on nomenclature for igneous rocks and on application of modifiers to existing rock names can be found in Hansen (1991).

Fleischer and Mandarino (1992) list the correct spellings and chemical formulas for more than 3400 mineral species; this list is periodically updated and corrected. Authors should avoid colloquial, outdated, nonspecific, and varietal (unless the parent mineral species is referred to) names. Names of minerals are not abbreviated in text. Nickel and Mandarino (1987) summarize the procedure and criteria for proposing designation of a "new" mineral and coining a term for it. *American Mineralogist* periodically publishes new mineral names with characterizations of the newly designated minerals.

Hansen (1991) recommends 4 guidelines for the use of mineral-phase and mineral-component abbreviations to be used as symbols or subscripts or superscripts.

1 Abbreviations should consist of 2 or 3 letters, the 1st capitalized, the rest lowercased.
2 The 1st letter should be the 1st letter of the mineral name, the other letters come from the rest of the name but preferably from the consonants.
3 Symbols should not be identical with those of the chemical elements (see Table 15·1).
4 A symbol should not form a word likely to be used in scientific writing in any language.

Mineral-phase and mineral-component symbols must differ from each other: mineral-phase symbols begin with a capital letter; mineral-component symbols should be entirely in lowercase letters.

Di [the mineral phase "diopside"]   di [the diopside component]

Below are examples of abbreviations formed in accordance with these recommendations. Compilations can be found in Table 3 of Hansen (1991) and in Kretz (1983).

aegirine-augite Agt   åkermanite Ak   ferrotschermakite Fts
halite Hl   pyrite Py   rhodochrosite Rds   ulvöspinel Usp

CRYSTALS

26·10 Crystals are distinguished by class, family, and system. A crystal class designates 1 of 32 categories determined by operations (such as inversion, rotation; see Table 26·4) or combinations that leave a crystal invariant. "Family" refers to 1 of 17 plane groups and 230 space groups. "System" refers to 1 of 7 categories classified by the unit-cell shape of its Bravais lattice or by the main symmetry elements of its class (see Table 26·5).

Table 26·4 Symbols for basic symmetry operations of a space group in crystallography[a]

Symbol	Description
$\{E/0\}$	identity operation; no rotation, no translation
$\{C_n/0\}$	n-fold rotation
$\{\sigma/0\}$	reflection
$\{I/0\}$	inversion
$\{S_n/0\}$	n-fold rotation followed by a fractional translation in a direction parallel to the plane
$\{E/\tau\}$	translation
$\{\sigma/\tau/m\}$	reflection followed by a fractional translation in a direction parallel to the plane
$\{C_n/\tau/\mathrm{m}\}$	rotation followed by a fractional translation parallel to rotation axis

[a]Source: Bennett and others (1965).

SYMMETRY, PLANES, AND AXES

A crystal is a periodic repetition of atoms or a group of atoms at equal intervals throughout the volume of a specimen. An arrangement of points in space, where each point has identical surroundings of the same orientation, is called a lattice. There are 14 types of 3-dimensional (or space) lattices (Bravais lattices).

The Miller indices of a plane are numbers derived from the reciprocals of the intercepts of a plane with the coordinate axes multiplied by the smallest number that will cause all numbers in the set of reciprocals to be integers. There are 3 principal axes. Principal axes are chosen parallel to translation directions, such that the axes are aligned with symmetry elements, and are characteristically labeled *a, b,* and *c.* If the intercepts are all equal, the reciprocals of the intercepts are in the ratio 1:1:1. The Miller indices of this plane are conventionally written within parentheses with no spaces between the numbers: (111). When each of the 3 indices is less than 10, no commas separate the numbers; when even 1 is greater than 10, commas are included to prevent confusion, for example, (11,2,1). A negative intercept is indicated by a bar above the number (for example, $11\bar{1}$, which is read "1, 1, bar 1").

To refer to a group of planes, the parentheses are replaced by curly brackets; "+" means included within the set.

$$\{abc\} = (abc) + (acb) + (bac) + (bca) + (cab) + (cba)$$
$$\{100\} = (001) + (010) + (100)$$

The Miller indices for direction are denoted by square brackets: [*abc*]. The set of directions that is equivalent by symmetry is indicated by angle brackets: <*abc*>.

Table 26·5 Symbols in crystallography

Symbol	Description
′	prime applied to the symbol for any movement: the movement is accompanied by change of color
′	prime applied to the prefix: change of color with translation
1	not a diad
$\bar{1}$	pure inversion
2	diad
$\bar{2}$	reflexion in a plane perpendicular to the axis of rotation
A	centered in yz plane
a	glide reflection in the x direction
B	centered in zx plane
b	glide reflection in the y direction
C	centered in xy plane
c	centered net; glide rotation in the z direction
d	diamond glide rotation
F	centered in all 3 planes (xy, yz, zx)
g	glide reflection; for plane groups (layer patterns) glide reflection in both x and y directions
I	body-centered
i [or] I	inversion
l'	reflection on line l
l''	reflection line of l'
m	reflection lines; in parallel mirror lines, reflections and glide reflections
n	glide reflection in a diagonal direction for plane groups (layer patterns)
$\bar{n}$	rotation through 360°/n ($2\pi/n$) combined with inversion
n_r	rotation of $2\pi/n$ combined with an axial movement of r/n units
(n)	number in parentheses as a superscript: number of colors in a polychromatic crystal
O	center of inversion
P	primitive plane
p	primitive net
r	row
σ	reflection
T	translation
∞	infinity

The symmetry group of an object is the totality of all operations that leave the object invariant. A group that must remain fixed in space during every symmetry operation is designated a "point group". Symmetry considerations can give information only about whether a specific phenomenon is or is not allowed. They cannot give information about the magnitude of the effect, except to say that when it is forbidden, it is 0.

A crystal is said to be right-handed or left-handed depending on the direction in which it rotates polarized light. Some crystals are twinned. ("Japan", "Dauphiné", and "Brazil" are 3 kinds of twinning.) References to the faces of 1 twin are in lowercase italic letters; references to the faces of the other twin are in the same lowercase italic letters with a bar underneath.

a, $\underline{a}$ z, $\underline{z}$

ROTATION, REFLECTION, INVERSION, AND TRANSLATION

26·11 Rotation leaves the coordinate system (either right-hand or left-hand) unchanged, whereas inversion changes a right-hand coordinate system to a left-hand one, and vice versa. An integer n subscript denotes the number of types of rotation about a fixed point 0: C_3, C_4, C_2, and so on; but more often the numbers are used alone: 3, 4, 2, and so on. For example, D_5 indicates a dihedral crystal with pentad rotation and 5 mirror lines. In international notation, a number indicating rotation is followed by the letter "*m*" for the mirror lines. The mirror (reflection) lines pass through the rotation center. For any even-numbered types the lines are divided into 2 sets, those along the arms of the cross and those bisecting the angles so formed, whereas for odd-numbered types the 2 sets are the same. Therefore, the mirror lines for even-numbered types are denoted by "*mm*".

D_1	D_2	D_3	D_4	D_5	D_6	...
1*m*	2*mm*	3*m*	4*mm*	5*m*	6*mm*	...

Translation means moving from 1 point in an object to some other point in the same object so that the environment about the 2 points is exactly the same. (For example, a point on the lower right side of a square in a cyclone fence is identical to all the other such points on the lower right sides of all the other squares in a cyclone fence.)

INTERNATIONAL (HERMANN–MAUGUIN) SYMBOLS

26·12 The "Hermann–Mauguin" symbol (Hahn 1987), both the short and the full version, consists of 2 parts: a letter indicating the centering type of the conventional cell, and a set of characters indicating symmetry elements of the space group (modified point-group symbol). Italic lowercase letters are used for 2 dimensions (nets), and italic capital letters for 3 dimensions (lattices).

[full]	[short]
$C1m1$ or $C11m$	Cm
$P2_1/n2_1/m2_1/a$	$Pnma$
$P6_3/m\ 2/m\ 2/c$	$P6_3/mmc$

PATTERSON SYMMETRY

Patterson symmetry gives the space group of the Patterson function $P(x,y,z)$. The Patterson function represents the convolution of a structure with its inverse or the pair-correlation function of a structure.

LAUE CLASS AND CELL

The space-group determination starts with the assignment of the Laue class to the weighted reciprocal lattice and the determination of the cell geometry. The Laue class determines the crystal system.

The axial system should be taken right-handed. The symmetry directions and the convention that the cell should be taken as small as possible determine the axes uniquely for crystal systems with symmetry higher than orthorhombic. For orthorhombic crystals 3 directions are fixed by symmetry, but any of the 3 may be called a, b, or c. The convention is for $c<a<b$. For monoclinic crystals there is 1 unique direction. If there are no special reasons to decide otherwise, the standard choice "b" is preferred. For triclinic crystals usually the reduced cell is taken, but the labeling of the axes remains a matter of choice.

FOSSILS

26·13 Fossils must be described in detail and illustrated (photographs, line drawings, charts, as appropriate) in the 1st published description. On 1st publication, a species description should include at least the following information.

1 A brief diagnosis.
2 Morphologic description.
3 Accurate information as to the locality from which the fossils came, including stratigraphic and geographic detail.
4 Current location of the types and other specimens.

An author should also be encouraged to include this additional information.

5 Comparison with other similar species.
6 Remarks regarding variability of features.
7 Discussion of significant results regarding phylogeny, ontogeny, functional morphology, paleoecology, and biostratigraphy.

Countless abbreviations are used on illustrations of specimens; they may differ among authors and specimens and should be explained in the illustration caption or other appropriate place.

SYNONYMY

26·14 If a species has been described previously, the author should give a synonymy to show the history of usage of names applied to the taxon. If a species is newly described, the author may or may not give the etymology of the name, plus the type locality. Formerly the complete bibliographic reference was given in the synonymy, but now the reader is usually referred to the bibliography. Below is an example of a standard format.

> *Incadelphys* n gen
>
> Etymology: Inca-, for the Inca civilization and its descendants who inhabit Tiupampa; and -delphys, in reference to didelphid marsupials.
>
> Type species: *Incadelphys antiquus* n sp
>
> Diagnosis: *Incadelphys antiquus* [another acceptable form: "Diagnosis: Same as for type and only known species"]
>
> Subfamily Caroloameghiniianae
>
> *Roberthoffstetteria nationalgeographica* Marshall, de Muizon, & Sigé 1983

NAMES OF FOSSILS

26·15 Names of fossils consist of a genus name and a specific epithet and should conform to the appropriate international code, for example, *International Code of Zoological Nomenclature, International Code of Botanical Nomenclature*. See sections 23·3 and 24·1.

The name of the 1st describer of a taxon should be included in all references to that taxon (but can be omitted from tables if the earlier reference to the describer is clear). If the species has been transferred out of the genus in which it was originally placed, the name of the author is placed within parentheses. When dealing with fossil plants, the name of the author(s) who made the transfer must follow the original authors' name(s) within parentheses.

The description of taxa may be in telegraphic style or in complete sentences.

CAPITALIZATION

26·16 Informal names and adjectives based on fossil names (for example, "bryozoans", "foraminifers") are neither italicized nor capitalized.

All generic and suprageneric names are capitalized; specific epithets are never capitalized. In older literature some patronyms were capitalized, and when quoting from 1 of these documents, the original capitalization should be retained.

Roberthoffstetteria Nationalgeographica *Paraconularia Ulrichana*
Paraconularia Africana *Conularia Kozlowskii*

Generic names may be abbreviated to the 1st letter if the full name of the species has been given in the paper and there is no confusion with other generic names having the same initial letter.

The skull of *Rhabdodon priscus* is known only from isolated specimens . . .

The squamosal bone of *R. priscus* is well preserved . . .

The dentition of *R. priscus* is distinctive.

Among the best-known taxa from Transylvania is *Telmatosaurus transsylvanicus* . . .

Because embryonic material clearly pertains to Hadrosauridae, the most likely candidate to have laid the eggs is *T. transsylvanicus.*

Suprageneric names in a classification are capitalized but not italicized.

LATIN AND ENGLISH TERMS

26·17 Formal generic and specific fossil names are in Latin and italicized. But English, rather than Latin, is preferred (but not required) for some taxonomic terms.

n gen [new genus] n sp [new species]
indet [indeterminate] undet [undetermined]

Note that this usage is not consistent with that in bacteriology and botany; see sections 22·6 and 23·6.

A name used on a label or published without a description ("nomen nudum" or "nude name") is invalid and should be used in a report only if a description of the species will be published before the report is published.

Table 26·6 Excerpt from the US Soil Classification System[a]

Order	Suborder	Great group
Alfisols	Aqualfs	Albaqualfs
		Duraqualfs
		Fragiaqualfs
		Glossaqualfs
		Kandiaqualfs
		Natraqualfs
		Ochraqualfs
		Plinthaqualfs
		Umbraqualfs
	Boralfs	Cryoboralfs
		Eutroboralfs
		Fragiboralfs
		Glossoboralfs
		Natriboralfs
		Paleboralfs

[a]Source: Soil Survey Staff (SSS 1975).

SOIL

SOIL TAXONOMY

26·18 Revisions of the US system of soil taxonomy (SSS 1975) have been issued in the *National Soil Taxonomy Handbook* (SCS 1982–1986) and in *Keys to Soil Taxonomy* (SMSS 1985). The soil horizon designations of the US Department of Agriculture Soil Classification System (Guthrie and Witty 1982) should be used for nomenclature of soil horizons.

In soil science, the following 24 proper names may be found in older literature; the US Soil Classification System (Table 26·6) supersedes this list. See section 26·21 on the nomenclatural structure of the classification.

Alpine Meadow	Laterite	Sierozem (Gray)
Bog	Pedalfer	Solonchak
Brown	Pedocal	Solonetz
Chernozem (Black)	Podzol	Soloth
Chestnut	Prairie	Terra Rossa
Desert	Ramann's Brown	Tundra
Gray-Brown Podzolic	Red	Wiesenboden
Half Bog	Rendzina	Yellow

Table 26·7 The downward sequence of soil horizons by master horizons (capital letters) and subordinate distinctions (lowercase letters)

Oi	
Oe	
Oa	
A	
E	
AB; EB	transition to B, more like A or E than B
BA; BE	transition to A or E, more like B than A or E
B	
BC	transition to C, more like B than C
C	
R	bedrock

UNIFIED CLASSIFICATION

26·19 The Unified Soil Classification System is a system for classifying mineral and organomineral soils for engineering purposes that is based on particle-size characteristics and Atterberg limits. It is based on soil with particles smaller than 3 inches in diameter. Atterberg limits are the results of arbitrarily defined tests that show properties of soils that have cohesion in that they represent changes in state or water content from solid to plastic to liquid.

The plastic limit (PL) is the minimum moisture content at which soil can be rolled into a thread 3 mm in diameter without breaking. The liquid limit (LL) is the minimum moisture content at which soil can flow under its own weight. The shrinkage limit (SL) is the moisture content at which further loss of moisture does not further decrease the volume of the sample. The plasticity index is (PI) = LL–PL. The liquidity index is (LI) = (100 m–PL) (LL–PL) where m is the natural moisture content of the soil. PI is often plotted against LL.

Other soil classification systems are used in the former Soviet Union, France, Canada, Australia, and some other nations (Buol and others 1980, CDA 1974).

SOIL PROFILE

26·20 Each soil is characterized by a given sequence of horizons. A vertical exposure of a sequence is termed a soil profile. The layers resulting from soil-forming processes are grouped under 5 headings: O, A, E, B, and C. The master horizons are designated with capital letters and the subordinate distinctions with lowercase letters (Table 26·7).

Table 26·8 Diagnostic horizons in minerals soils

Surface horizons (epipedons)	Subsurface horizons
Mollic (A)	Argillic (Bt)
Umbric (A)	Natric (Btn)
Ochric (A)	Spodic (Bhs)
Histic (O)	Cambic (B)
Anthropic (A)	Agric (A or B)
Plaggen (A)	Oxic (Bo)
	Duripan (m)
	Fragipan (x)
	Albic (E)
	Calcic (k)
	Gypsic (y)
	Salic (z)

Diagnostic surface horizons (Table 26·8) are called "epipedons". The epipedon includes the upper part of the soil darkened by organic matter, the upper eluvial horizons, or both. It may include part of the B horizon if the latter is significantly darkened by organic matter. Six epipedons are recognized, but only 4 are important in soils in the United States. Anthropic and Plaggen are the result of intensive use of the soils and are found in parts of Europe and, probably, Asia.

NOMENCLATURE

26·21 There are 6 categories in soil taxonomy: order, suborder, great group, subgroup, family, and series.

The names of orders are derived from French, Greek, or Latin terms; see Table 26·9. The names of suborders identify by their stems the order of which they are a part (Table 26·6). For example, soils of the suborder "Aquoll" are the wetter soils (Latin *aqua* = water) of the Mollisol order. Likewise, the name of the great group identifies the suborder and order of which it is a part. "Argiaquolls" are "Aquolls" with clay or argillic (Latin *argilla* = white clay) horizons. The relation of the nomenclature to the different categories in the classification system can be illustrated thus.

Mollisol [order]
 Aquoll [suborder]
  Argiaquoll [great group]
   Typic Argiaquoll [subgroup]

Table 26·9 Names of US soil orders and their derivations[a]

Name	Derivation
Alfisols	"Alfi" (an arbitrary root[b]) + "sol"
Aridisols	Latin *aridus* ("dry") + "sol"
Entisols	English "entire" + "sol"
Histosols	Greek *histos* ("tissue") + "sol"
Inceptisols	Latin *inceptum* ("beginning") + "sol"
Mollisols	Latin *mollis* ("soft") + "sol"
Oxisols	French *oxide, oxyde* ("oxide"; Greek *oxys,* "acid") + "sol"
Spodosols	Greek *spodos* ("wood ash") + "sol"
Ultisols	Latin *ultimus* ("last") + "sol"
Vertisols	Latin *verto* ("turn") + "sol"

[a]Source: *Glossary of Soil Science Terms* (SSSA 1987).

[b]*The New Shorter Oxford English Dictionary* (Brown 1993); the root may be *alpha,* the 1st letter of the Greek alphabet, in a variant spelling.

Note that the stem of 3 letters "oll" identifies each of the lower categories as being in the Mollisol order. Likewise, the suborder name "Aquoll" is included as part of the great group and subgroup name.

Family names, in general, identify groups of soil series similar in texture, mineral composition, and soil temperature at a depth of 50 cm.

All names of orders end in "sol" (from the Latin *solum* = soil).

The ending of the suborder name identifies the order in which the soils are found.

Arqu**ent**s and Ar**ent**s	[suborders of "**Ent**isols"]
And**ept**s and Ochr**ept**s	[suborders of "Inc**ept**isols"]
Alb**oll**s and Bor**oll**s	[suborders of "**Moll**isols"]
Torr**ert**s and Xer**ert**s	[suborders of "**Vert**isols"]

Great groups, subdivisions of suborders, are defined largely on the presence or absence of diagnostic horizons and the arrangement of those horizons.

Argi**abolls**	[a great group of "**Abolls**"]
Argi**xerolls**	[a great group of "**Xerolls**"]
Argi**udalfs**	[a great group of "**Udalfs**"]

The family category is based on properties important to the growth of plant roots. Series (the lowest category of the system) are distinguished by soil properties and in the United States are usually named for a city, village, river, or county.

For an example of the classification down to great groups, see Table 26·6.

CLASSIFICATION OF LAND CAPABILITY

26·22 The US Soil Conservation Service recognizes 8 classes of land capability, numbered I to VIII (SSS 1975). Soils with the greatest capabilities for response to management and least limitations in how they can be used (land that could be cropped intensively; used for pasture, range, or wildlife preserves; well-drained, level, fertile) are in Class I. Those with the least capabilities and greatest limitations (land that should not be used for production of commercial plants, but restricted to recreation, wildlife, water-supply, or aesthetic uses) are in Class VIII.

IDENTIFICATION OF SOILS

26·23 Soils should be identified to the lowest level possible, in accordance with the US soil taxonomic system, the 1st time each soil is mentioned (Table 26.6) (ASA and others 1988). For soil science terms see SSSA (1987).

A soil series (for example, "the Holdrege series") is the lowest category of the national soil classification system and the most common reference term used to name soil units on maps. The series must be described as to soil content, slope, temperature, and precipitation. Each layer within the series is assigned a 2-letter designation, which is described in telegraphic style as to depth, color, soil type, granularity, and range. Color is described using the Munsell color notation (for example, "10YR 5/4") either with or without the descriptive words (for example, "yellowish brown"), but the descriptive words cannot be used without the Munsell color notation: "10YR 5/4 (yellowish brown)".

> Ap—0 to 15 cm; dark grayish brown (10YR 3/2) silt loam, very dark grayish brown (10YR 4/2) dry; moderate fine and very fine granular structure; soft, very friable; medium acid; abrupt smooth boundary. (10 to 23 cm thick)
>
> BA—15 to 30 cm; dark grayish brown (10YR 3/2) silty clay loam, very dark grayish brown (10YR 4/2) dry; weak medium and fine granular structure; slightly hard, friable; slightly acid; clear smooth boundary. (0 to 20 cm thick)

The singular form is used for a single pedon or polypedon or a single class. The plural form is used to refer to several or all the soils (polypedons) of a class. Any unit less than a pedon, such as a single horizon, is only part of a soil. Bodies of soil less than a full pedon are considered as samples of soil or as soil material. For field experiments, the soil in the plots or fields studied should be identified, preferably as phases of soil series, so that surface texture and slope are known in addition to profile properties. Any dissimilar inclusions should be named and their extent indicated. It also may be appropriate to name and briefly describe the common soils of the area surrounding the study site. The present tense is used if the soil exists or is likely to exist.

> The 5-ha study area is mapped as Yolo silt loam, 0 to 2% slopes. The Yolo soils are fine-silty, mixed, nonacid, thermic Typic Xerorthents. Small areas of Cortina very gravely sandy loam soils (loamy-skeletal, mixed, nonacid, thermic Typic Xerofluvents) occupy about 10% of the study area.
>
> The soil material used in this study was collected from A horizon of a Brookston pedon (a fine-loamy, mixed, mesic Typic Argiaquoll).
>
> A Cisne soil, fine, montmorillonitic, mesic Mollic Albaqualf, was described and sampled at this site.
>
> Criteria for the Typic Hapludult subgroup were examined.
>
> Ontario soils, in the fine-loamy, mixed, mesic Glossoboric Hapludalf family, were studied in greater detail.

NON-SI UNITS

26·24 Some terms convenient for crop and soil scientists are not converted to SI units, for example, area when it refers to the 160-acre parcels of land the US and Canadian governments surveyed in the 19th century; lumber is still measured in boardfeet.

AQUIFERS

26·25 An aquifer may encompass thousands of square hectares. The term is difficult to define and map; Bates and Jackson (1987) define it as "a body of rock that is sufficiently permeable to conduct ground water and to yield economically significant quantities of water to wells and springs". The following terms are not capitalized even when named: aquifer, aquifer system, zone, confining unit. Terms such as "sand and gravel aquifer" and "limestone aquifer" are neither capitalized nor hyphenated.

Adjectival modifiers and relative-position terms are not capitalized unless they are part of the formal geographic name.

> Mississippi River alluvial aquifer   Upper Klamath aquifer
> Upper Canada aquifer

Quotation marks are not used for aquifer names unless the term is a misnomer. Although the term "aquifer" may be imprecise, it is widely accepted and used; coining new terms will only add to the confusion. Terms intended to be synonymous with "aquifer" or "aquifer system" should not be used. "Aquigroup" should not be used in place of "aquifer system". Hydrologic and geologic terms should be distinguished.

water from the Madison aquifer [not "Madison water"]

wells completed in Madison Limestone [or "aquifer", not "Madison wells"]

TRACTS OF LAND

26·26 The location of tracts of US public land is designated by rows of townships (each 6 miles square) and columns of ranges that make up the grid system used to survey all land west of the Ohio River, except for Texas, in relation to various east–west baselines and named north–south meridians. Note the following formats.

SE1/4 NW1/4 sec 4, T 12 S, R 15 E, Boise Meridian

[designates the southeast quarter of the northwest quarter of section 4, which is 12 townships [T] south and 15 ranges east of the Boise Meridian]

N1/2 sec 20, T 7 N, R 2 W, 6th Principal Meridian

[designates the northern half of section 20, which is 7 townships north and 2 ranges west of the Sixth Principal Meridian]

The plural of T (for "township") is Tps; of R (for "range") is Rs.

Tps 9, 10, 11, and 12 S, Rs 12 and 13 W

If fractions are spelled out in land descriptions, "half" and "quarter" are used, not "one-half" or "one-quarter".

south half of T 47 N, R 64 E

If breaking the line of a land-description symbol group is unavoidable, it should be broken after a fraction and without a hyphen.

NE1/4 SE1/4
sec 4

REFERENCES

Cited References

[ASA] American Society of Agronomy, Crop Science Society of America [CSSA], and Soil Science Society of America [SSA]. 1988. Publications handbook and style manual. Madison (WI): ASA, CSSA, SSSA.

Bates RL, Jackson JA. 1987. Glossary of geology. 3rd ed. Alexandria (VA): American Geological Inst.

Bennett A, Hamilton D, Maradudin A, Miller R Murphy J. 1965. Crystals perfect and imperfect. New York: Walker.

Brown L, editor. 1993. The new shorter Oxford English dictionary. Oxford: Oxford Univ Pr, Clarendon Pr.

Buol SW, Hole FD, McCracken RJ. 1980. Soil genesis and classification. 2nd ed. Ames: Iowa State Univ Pr.

[CDA] Canada Department of Agriculture. 1974. The system of soil classification for Canada. Ottawa: Information Canada. Publication 1455.

Compton RR. 1985. Geology in the field. New York: J Wiley.

Fleischer M, Mandarino J. 1992. Additions and corrections to the glossary of mineral species, 1991. Mineralog Rec 22:263–6.

[GSA] Geological Society of America. 1991. Rock-color chart. Boulder (CO): GSA.

Guthrie RL, Witty JE. 1982. New designations for soil horizons and layers and the new soil survey manual. Soil Sci Soc Am 46:433–44.

Hahn T, editor. 1987. International tables for crystallography. Volume A, Space-group symmetry. 2nd ed. Boston: D Reidel. Published for the International Union of Crystallography.

Hansen WR. 1991. Suggestions to authors of the reports of the United States Geological Survey, 7th ed. Washington: US Government Printing Office.

Harland WB, Armstrong RL, Cox AV, Craig LE, Smith AG, Smith DG. 1990. A geologic time scale 1989. New York: Cambridge Univ Pr.

Kretz R. 1983. Symbols for rock-forming minerals. Am Mineralog 68:277–9.

[NACSN] North American Commission on Stratigraphic Nomenclature. 1983. North American stratigraphic code. Am Assoc Petroleum Geol Bull 67:841–75.

Nickel EH, Mandarino JA. 1987. Procedures involving the IMA Commission on New Minerals and Mineral Names and guidelines on mineral nomenclature. Am Mineralog 72:1031–42.

[SCS] Soil Conservation Service. 1982–1986. National soil taxonomy handbook, Issues 1–9. Washington: US Department of Agriculture.

[SMSS] Soil Management Support Services. 1985. Keys to soil taxonomy. Ithaca (NY): Cornell Univ, Department of Agronomy. Technical monograph 6.

[SSS] Soil Survey Staff. 1975. Soil taxonomy: a basic system of soil classification for making and interpreting soil surveys. Washington: US Government Printing Office. USDA–SCS Agriculture Handbook 436.

[SSSA] Soil Science Society of America. 1987. Glossary of soil science terms. Madison (WI): SSSA.

[USGPO] United States Government Printing Office. 1984. Government Printing Office style manual 1984. Washington: USGPO.

Additional References

Ashworth W. 1992. The encyclopedia of environmental studies. New York: Facts on File.

Bates RL. 1988. Writing in earth science. Alexandria (VA): American Geological Inst.

Benton MJ, Whyte MA. 1993. The fossil record. 2nd ed. New York: Chapman & Hall.

Brady NC. 1990. Soil formation, classification, and survey. Chapter 3 in: The nature and properties of soils. 10th ed. New York: Macmillan. p 47–90.

[CSPG] Canadian Society of Petroleum Geologists. Lexicons of Canadian stratigraphy. Calgary: CSPG.

Clark A. 1992. The mineral index. 3rd ed. New York: Chapman & Hall.

Fanning DS, Fanning MC. 1989. Soil morphology, genesis, and classification. New York: J Wiley.

[GSC] Geological Survey of Canada. 1979. Guide to authors. Ottawa: GSC. Miscellaneous Report Nr 29. A new edition is in preparation.

Goudie A, Atkinson BW, Gregory KJ, Simmons IG, Stoddart DR, Sugden D, editors. 1985. The encyclopaedic dictionary of physical geography. Oxford: Blackwell.

Hardy L, Wood DN, Harvey AP. 1989. Information sources in the earth sciences. 2nd ed. New York: KG Saur.

Klein C, Hurlbut CS. 1993. Manual of mineralogy; after James D Dana. 20th ed. New York: J Wiley.

Lozet JMC. 1991. Dictionary of soil science. 2nd ed. Brookfield (VT): Ashgate.

Mayr H; Dineley D, Windsors G, translators. 1992. A guide to fossils. Princeton (NJ): Princeton Univ Pr.

Porteus A. 1992. Dictionary of environmental science and technology. New York: J Wiley.

Shimer HW, Shrock RH. 1944. Index fossils of North America. Cambridge (MA): MIT Pr.

Wilson AJC. 1992. International tables for crystallography. Volume C, Mathematical, physical, and chemical tables. Boston: Kluwer Academic. Published for the International Union of Crystallography.

Witty JE, Arnold RW. 1987. Soil taxonomy: an overview. Outlook Agric 16:8–13.

27 Astronomical Objects and Time Systems

> What a masterpiece is the clear vault of the sky!
> How glorious is the spectacle of the heavens!
> . . . The brilliant stars are the beauty of the sky,
> a glittering array . . .
>
> —*The New English Bible: The Apocrypha, Ecclesiasticus* 43:1, 9

Through the past half-century, new instruments and methods for seeing what could not be seen as well, or at all, through optical telescopes have rapidly expanded our knowledge of what lies beyond the earth's surface and atmosphere. This change has led to increasing needs to accurately represent in print a still-growing array of astronomical objects and phe-

Table 27·1 Abbreviations for the titles of astronomical catalogues[a]

Abbreviation	Catalogue
AGK *n*	*Astronomischer Gesellschaft Katalog Nummer n*
BD	*Bonner Durchmusterung*
BS [BSC]	*Catalogue of Bright Stars* (Yale)
CD	*Cordoba Durchmusterung*
CPD	*Cape Photographic Durchmusterung*
FK *n*	*Fundamental Katalog Nummer n*
GC	*General Catalog* (Washington, 1936)
GCVS	*General Catalogue of Variable Stars*
HD	*Henry Draper Catalogue*
HDE	*Henry Draper (Catalogue) Extension*
HR	*Harvard Revised Photometry Catalogue*
IC	*Index Catalogue* (Dreyer)
IDS	*Index Catalogue of Visual Double Stars*
NGC	*New General Catalogue*
SAO	*Smithsonian Astrophysical Observatory Catalogue*

[a]For additional titles and their abbreviations see Ridpath (1989) and Zombeck (1990).

nomena. The leading authority on this specialized and rapidly growing nomenclature is the International Astronomical Union (IAU); its style manual (Wilkins 1989) is the source for most of the recommendations in this chapter.

NAMES AND ABBREVIATIONS

27·1 An object may be designated by a formal or proper name, a catalogue number, or a composite name that indicates the type of object and its position; the 3rd system should normally be used for new designations.

Formal names may be traditional names established long before telescopic observation or names applied to objects of a class not too numerous for ready naming. Some examples are the names of constellations, stars prominent to the naked eye, the planets, and minor planets.

Catalogue numbers are the numbers assigned by the compiler of a catalogue, and they are preceded by an abbreviation of the catalogue name; see section 27·11 for examples. For a list of catalogue names and their abbreviations, see Table 27·1; for more comprehensive lists, see *Handbook of Space Astronomy and Astrophysics* (Zombeck 1990) and *Norton's 2000.0 Star Atlas and Reference Handbook* (Ridpath 1989).

Some objects may have 2 or more names of equal standing.

Ring Nebula = M 57 [Messier catalogue number] = NGC 6720 [New General Catalogue number]

the Pleiades = 7 Sisters = M 45

Additionally, a temporary name may be given to an object at the time of 1st observation, with a new permanent name given when the object's properties have been established. Proper names and catalogue numbers may be inadequate designations.

A satisfactory composite name gives the essential characteristics of the object (by alphanumeric code-name) and its position. Abbreviations should not be introduced unnecessarily; when necessary they should be unambiguous and have at least 2 letters for each. For examples, see sections 27·8 and 27·13.

CAPITALIZATION

27·2 The names of constellations (Table 27·2); the names of planets and their satellites, asteroids (minor planets), and stars; and the names of other unique celestial objects should be capitalized. Generic words forming part of the name are not capitalized; such terms (for example, "comet", "nebula") are set in lowercase type whether they precede or follow the proper name. The names of space programs are capitalized. A generic term is capitalized when it is the only noun in the proper name, as in "Large Magellanic Cloud".

the Crab nebula the Coalsack Project Apollo Solar System

Halley's comet comet Biela Milky Way

the North Star Orion Large Magellanic Cloud

Designations by catalogue number have a capitalized catalogue-name abbreviation or full name.

NGC 6165 Bond 619 M 81

The words "earth", "sun", and "moon" are capitalized in contexts with the names of other bodies of the Solar System, but are often set in lowercase type for more general usage.

the 4 corners of the earth the salt of the earth

The sun provides energy for photosynthesis.

The moon is made of green cheese.

[but]

The Soviets planted flags on the Moon and Venus.

The Sun is a run-of-the-mill star.

Table 27·2 Constellation names and their abbreviations

Name	Abbreviation	Name	Abbreviation
Andromeda	And	Indus	Ind
Antlia	Ant	Lacerta	Lac
Apus	Aps	Leo	Leo
Aquarius	Aqr	Leo Minor	LMi
Aquila	Aql	Lepus	Lep
Ara	Ara	Libra	Lib
Argo[a]	Arg	Lupus	Lup
Aries	Ari	Lynx	Lyn
Auriga	Aur	Lyra	Lyr
Boötes	Boo	Mensa	Men
Caelum	Cae	Monoceros	Mon
Camelopardalis	Cam	Musca	Mus
Cancer	Cnc	Norma	Nor
Canes Venatici	CVn	Octans	Oct
Canis Major	CMa	Ophiuchus	Oph
Canis Minor	CMi	Orion	Ori
Capricornus	Cap	Pavo	Pav
Carina	Car	Pegasus	Peg
Cassiopeia	Cas	Perseus	Per
Centaurus	Cen	Phoenix	Phe
Cepheus	Cep	Pictor	Pic
Cetus	Cet	Pisces	Psc
Chamaeleon	Cha	Piscis Austrinus[b]	PsA
Circinus	Cir	Puppis	Pup
Columba	Col	Pyxis	Pyx
Coma Berenices	Com	Reticulum	Ret
Corona Austrina[b]	CrA	Sagitta	Sge
Corona Borealis	CrB	Sagittarius	Sgr
Corvus	Crv	Scorpius	Sco
Crater	Crt	Sculptor	Scl
Crux	Cru	Scutum	Sct
Cygnus	Cyg	Serpens[c]	Ser
Delphinus	Del	Sextans	Sex
Dorado	Dor	Taurus	Tau
Draco	Dra	Telescopium	Tel
Equuleus	Equ	Triangulum	Tri
Eridanus	Eri	Triangulum Australe	TrA
Fornax	For	Tucana	Tuc
Gemini	Gem	Ursa Major	UMa
Grus	Gru	Ursa Minor	UMi
Hercules	Her	Vela	Vel
Horologium	Hor	Virgo	Vir
Hydra	Hya	Volans	Vol
Hydrus	Hyi	Vulpecula	Vul

[a]In modern usage Argo is divided into Carina, Puppis, and Vela.

[b]"Australis" is sometimes used as the 2nd term.

[c]Serpens may be divided into Serpens Caput and Serpens Cauda.

Impact scars indicate that the Moon's early history was exceedingly violent.
Much of Earth's surface structure is concealed by liquid water.

Lowercase type is used for 3 main classes of terms.

1 Descriptive terms applied to celestial phenomena.

the gegenschein   the rings of Saturn   the Cassini division
the laplacian plane of Saturn's rings   Kirkwood gaps

2 Terms applied to meteorologic and other atmospheric phenomena.

aurora borealis   sun dog   meteor shower   Tunguska fireball

3 Adjectival forms of usually capitalized nouns.

Jupiter [but] jovian
Moon [but] lunar
Sun [but] solar
Mercury [but] mercurial

Names of spacecraft and artificial satellites are capitalized and italicized. Note that after a spacecraft is launched into orbit, it becomes a satellite (Schmidt 1987).

Luna   *Columbia*   *Vega*   *Landsat*   *SPOT*

CELESTIAL COORDINATES

27·3 Right ascension (α) is given in hours, minutes, and seconds of sidereal time, with no spaces between numbers and units. In less precise designations, the general convention for units may be acceptable; see section 11·10.

$14^h6^m7^s$ [in less precise designations, "14 h 40 min"]

Declination (δ) is given in degrees, minutes, and seconds of arc north (marked + or unmarked) or south (marked −) of the celestial equator.

$49°8'11''$ $-87°41'08''$

If decimal fractions of the basic units are used, the decimal point is placed between the unit symbol and the decimal value.

$26^h6^m7^s.2$ $+34°.26$

The mean equator and equinox serve to define a coordinate system by ignoring small, short-period variations in the motion of the celestial equator, but this system is affected by precession (the slow continuous westward motion of the equinoxes around the ecliptic that results from precession of the earth's axis). The positions of the mean equator and equinox at particular epochs are used to define standard reference systems, such as those for B1950.0 and J2000.0 (see section 27·16 on the besselian and julian epochs).

OBJECTS IN THE SOLAR SYSTEM

PLANETS, SATELLITES, AND RINGS

27·4 The International Astronomical Union Working Group for Planetary System Nomenclature is responsible for the adoption of names for the surface features of planets and satellites and for newly discovered members of the planetary system other than minor planets and comets. Satellites are also designated by numbers assigned in the chronologic sequence of discovery. The decisions of the working group are reported in the *Transactions of the International Astronomical Union.* A tabulation of natural satellite numbers and names can be found in Zombeck (1990). When listed, satellites of planets are given in order of their mean distances from the planet.

The symbols for planets common in astrologic and other informal literature should not be used in scientific literature.

MINOR PLANETS (ASTEROIDS) AND COMETS

27·5 Minor planets (asteroids) are given numbers serially and proper names when reliable orbital elements have been defined.

878 Mildred 719 Albert 1627 Ivar 288 Glauke
216 Kleopatra 4 Vesta 3200 Phaethon 1862 Apollo

Newly discovered comets are assigned temporary designations, each consisting of the year of discovery followed by a letter (lowercase) that indicates the chronologic sequence within that year of the discoveries. Previously discovered comets are usually not given a date. The word "comet" preceding the name is set in lowercase type.

comet West (1975n) comet Humason (1961e)
comet Bradfield (1979x) comet Bradfield (1987s)
[but for names in use before the formal system of temporary naming was established]
comet Halley comet Biela comet 1858 Donati

When the discovery has been confirmed, the name(s) of the discoverer(s) also usually forms part of the comet's name. Subsequently, permanent numbers (in capital roman numerals) are assigned; they indicate for each year the chronologic sequence of the passages of all of the observed comets through perihelion (the point in its path at which an astronomical body is closest to the Sun).

comet Arend–Roland 1957 III
comet Tago–Sato–Kosaka 1969 IX
comet Kohoutek 1973 XII

Periodic comets are distinguished by the code "P/".

comet P/Encke 1977 XI   comet P/Halley
comet P/Oterma   comet P/d'Arrest 1976 XI

METEOR SHOWERS, METEORS, AND METEORITES

27·6 Meteor showers are usually named for the constellations in which their radiant points appear or according to the name of the comet with which they are associated.

the Arietids   the Lyrids   the Eta Aquarid   the Quadrantid

A large meteorite is usually identified by the name of the place near where it was found; a meteor crater, by its necessarily fixed geographic location.

Seymour meteorite (Indiana)   Brent crater (Canada)

A fireball is usually identified by the date on which it was seen.

For a comprehensive catalogue of meteorite names and their synonyms, see Graham and others (1985).

OBJECTS OUTSIDE THE SOLAR SYSTEM

BRIGHT STARS

27·7 The Bayer designation of a bright star consists of a greek letter followed by the name of (or abbreviation for) the constellation; the greek letters are usually assigned in the order of brightness (α, β, χ, δ, and so on). They should be printed in an upright (roman) typeface with the lowercase greek letter followed by the standard 3-letter abbreviation for the name of the constellation (Table 27·2). If greek type is not available, the English name of the greek letter is spelled out.

α Cen A [or] alpha Centauri A

Table 27·3 Examples of alternative designations for bright stars

Traditional name	Bayer system	Flamsteed system
Betelgeuse	α Ori	58 Ori
Sirius	α CMa	9 CMa
Ras-Alhague	α Oph	55 Oph
Polaris	α UMi	14 UMi

In references to 1 of the some 1000 stars that have proper names derived from early Arabic names or peculiar characteristics, the star should be identified more precisely in addition, especially if the star's name differs in other languages. The Yale *Catalogue of Bright Stars* (BS) (Hofleit 1964) lists about 900 such names and gives information on more than 9000 stars brighter than magnitude 6.5. It also gives names based on the Bayer system (greek letter followed by the star's constellation name) or on the Flamsteed catalogue (number and the star's constellation) (Table 27·3).

About 1500 bright stars are listed in the *Astronomical Almanac* (published annually by the US Government Printing Office); the tabulation gives the corresponding Bayer, Flamsteed, and BS designations, as well as the mean position and other information about each star.

Table 27·1 gives the only acceptable abbreviations for catalogues of stars; these abbreviations should not be used for catalogues of other kinds of entities (except that "ZC" is already in use for the Zwicky catalogue of clusters of galaxies).

Stars may be classified by spectral class into 1 of 7 groups according to spectral temperature: O (hottest stars), B, A, F, G, K, M (coolest, very red). In turn, these groups can be subdivided into divisions designated by numerals 0 to 9 appended to the alphabetic group designation: for example, A4, F8, G0.

Stars may also be classified by luminosity class with roman numerals and alphabetic divisions; see Table 27·4.

The spectral and luminosity classifications of stars and other similar symbols and abbreviations should be printed in roman type.

HD 190406 Algol (β Per) ξ UMa B λ And
70 Oph A F8 AU Mic ε Eri ξ Boo A HZ 43
Feige 24 K2 III

Note, however, that symbols for quantities should be italicized, in accordance with the relevant ISO standard; see section 3·5 and Table 3·5.

Table 27·5 lists abbreviations and symbols frequently used in astronomical literature.

Table 27·4 Luminosity classes of stars

Class	Luminosity type
Ia	Supergiants
Ib	Supergiants
II	Bright giants
III	Giants
IV	Subgiants
V	Main sequence (dwarfs)
VI	Subdwarfs
VII	White dwarfs

FAINT STARS

27·8 For stars not listed in the catalogues and for which designations have not been given, designations should be assigned in the form of an acronym and position in the standard form. (See "Radiation Sources", section 27·14.)

DOUBLE OR BINARY STARS

The standard form for the name of a component of a double or multiple star gives its position by equatorial coordinates for the equinox of 1900.0, indicates northern (N) (+) or southern (S) (–) declinations, and specifies a component by a capital letter. The *Index Catalogue of Visual Double Stars* (IDS) is updated at the US Naval Observatory and then known as VDS. The brighter double stars have alternative designations in bright-star catalogues, as well as in special lists of double or variable stars.

V444 Cyg PSR 1913 + 16 PSR 2303 + 46
PSR 1855 + 09 BD +56° 1450 BD +4° 3561

VARIABLE STARS

27·9 Names of variable stars consist of the name of the constellation preceded by a 1- or 2-letter code (capitalized), or the letter V followed by a number. When a classical name exists, it should be used in titles; consult the *General Catalogue of Variable Stars* or the CDS database SIMBAD.

M 3 W Vir RV Tau BL Her R Cor Bor U Gem
VZ Can BY Dra V356 Sgr V444 Cyg V827 Her 1987

Table 27·5 Abbreviations and symbols frequently used in astronomical literature

Abbreviation or symbol	Meaning
α	right ascension; the brightest star in a constellation, sometimes indicates a star's position in a group
AU	astronomical unit (mean Earth–Sun distance); UA in French
B	besselian
β	2nd brightest star in a constellation; sometimes indicates a star's position in a group
c	speed of light
Δ T	the increment to be added to universal time to give terrestrial dynamical time
δ	declination; 4th brightest star in a constellation; sometimes indicates a star's position in a group
E	color excess
ET	ephemeris time, a measure of time for which a constant rate was defined; used 1958 to 1983
ϕ	phase; luminosity
G	gravitation constant
γ	3rd brightest star in a constellation; sometimes denotes position in a group
GST	Greenwich sidereal time
H π region	a volume of hydrogen photoionized (into protons and electrons) by the ultraviolet radiation from a central, hot object
HA	hour angle (GHA and LHA)
HR	Hertzsprung–Russell
IAT	international atomic time
Jy	Jansky
k	curvature index of space; Gaussian gravitational constant
kpc	kiloparsec
L	luminosity, stellar
λ	celestial longitude
LAST	local apparent sidereal time
LHA	local hour angle
LMC	Large Magellanic Cloud
LMST	local mean sidereal time
LSR	local standard of rest
LST	local sidereal time
M	absolute magnitude
M	mega (= 1 million)
m	mass; apparent magnitude
MBR	microwave background radiation
MERIT	monitor Earth-rotation and intercompare the techniques of observation and analysis (1980–1987)
MHD	magnetohydrodynamics (also known as hydromagnetics)
mJy	milli-Jansky
Mpc	a million parsecs or 1 000 000 pc

Table 27·5 (*cont.*)

Abbreviation or symbol	Meaning
μG	microgauss
pc	parsec
PM	postmodulation
PZT	photographic zenith telescope [out of date]
Q	aphelion (the point in solar orbit farthest from the Sun)
QSO	quasistellar object, quasar
R	cosmic scale factor (a measure of the size of the Universe as a function of time)
ρ	mean density of matter
R	radius, stellar
RA	right ascension
R_E	Earth radius
rv	radial velocity
SIMBAD	CDS database (set of identifications, measurements, and bibliography for astronomical data)
SMC	Small Magellanic Cloud
SRS	southern reference system
t	hour angle
TAI	international atomic time
TDB	barycentric dynamical time
TDT	terrestrial dynamic time
UT	universal time
V	visual magnitude
ν	frequency
VLBI	very-long-baseline radio interferometry
z	redshift parameter
ZAMS	zero-age main sequence
ZHR	zenithal hourly rate

NOVAE OR CATACLYSMIC VARIABLES

27·10 The format for the designation of a nova is "Nova", constellation name or its abbreviation, year. The year is followed by a letter (lowercase) if more than 1 nova has been discovered in a constellation in the same year. Subsequently, a nova may be assigned a standard designation as a variable star.

Nova Sag 1987   Nova Sag 1987a
Nova Per 1901   Nova Sge 1783

SUPERNOVAE

The designation for a supernova is "SN" followed by the year and an alphabetic designation for order of discovery.

SN1985A SN1985B SN1985aa SN1985ab SN1985az SN1985ba

The California Institute of Technology maintains an archival list. Newly discovered supernova remnants are designated SNR, followed by the position with respect to the system of J2000.0 (see section 27·16).

NEBULAE, GALAXIES, CLUSTERS, PULSARS, AND QUASARS

27·11 Two catalogues of nebulae, galaxies, and clusters are in wide use. The one compiled by Charles Messier between 1771 and 1784 now comprises 109 entries. The objects in the catalogue (the brightest nebulae, galaxies, and star clusters visible from France) are referred to by Messier's initial and his serial number, separated by a space.

M 1 = Crab Nebula M 45 = Pleiades (the Seven Sisters)

The 2nd catalogue, compiled by Johan Dreyer in 1888, is called the *New General Catalogue* (NGC). It lists 7840 objects, which may be designated by the abbreviation of the catalogue name, a space, and the catalogue number for the object. An additional 5086 objects were added to the NGC in 2 supplements called the *Index Catalogue* (IC).

NGC 1952 = Crab Nebula h Perseus = NGC 869

χ Perseus = NGC 884

For the many catalogues since compiled, the objects are usually referred to by an abbreviation for the name of the catalogue or of its compiler or by its compiler's initial and a serial number. Abbreviations for the names of catalogues (Table 27·1), spelled out at 1st use, are set in roman type.

FK5 SAO HD M

27·12 Another means of designating objects is by reference to the constellation, or large-scale star pattern, in which they reside (Table 27·2). The word "constellation" now refers to the whole of a specific region of the sky in which the pattern formed by the bright stars is seen; the ancient constellation patterns are still used, however, as a convenient guide to parts of the sky. Thus Centaurus A was the 1st radio source to be discovered in the Southern Hemisphere constellation of Centaurus. Similarly, many especially famous galaxies, nebulae, and clusters are still known by individual names, often derived from their appearance and their more or less fanciful resemblance to terrestrial objects.

The Coalsack   the Whirlpool galaxy   the Eskimo

The zodiacal symbols for constellations common in nonscientific literature should never be used in scientific literature.

Planetary Nebulae

27·13 Some planetary nebulae have proper names, but the designation usually consists of the letters PN followed by its galactic coordinates in degrees (without a decimal point), and a serial number. The appropriate catalogue designation should also be given when available.

Owl Nebula = M 97 = NGC 3587 = PN $11^{h}14^{m}$ +55°01′

Galaxies

Some galaxies have proper names; if so, they should also be identified by their Messier, NGC, or IC numbers.

Radio Sources

The IAU-recommended designation (as of 1985) is a catalogue acronym followed by the right ascension and declination (see "Celestial Coordinates", section 27·3) with respect to the reference system of J2000.0 (see "Besselian and Julian Epochs", section 27·16).

Star Clusters

Some clusters have proper names; many are identified by their numbers in the Messier catalogue or by their numbers in the *New General Catalogue* (NGC) and supplementary *Index Catalogues* (IC).

NGC 4755 (Jewel Box)   NGC 6656 (M 22)

NGC 7078 (M 15)   NGC 104 (47 Tuc)

NGC 5139 (ω Cen)   Terzan 2   Terzan 5   Grindlay 1

Pulsars

The position, in condensed notation, of the pulsar follows the abbreviation PSR. Older names, such as CP1919 (for the 1st pulsar discovered), are now obsolete.

PSR 1937 + 21 [the numeric designation "1937 + 21" = the right ascension of 19^h37^m and declination of 21°.4]

Quasistellar Objects, QSO, or Quasars

No standard form has been assigned to the designation of quasars, but catalogue numbers are used ("3C" refers to the *Third Cambridge Catalogue*).

3C 273 3C 48 3C 175

RADIATION SOURCES

27·14 Designations of X-ray sources are now taken from the catalogues of surveys made by particular satellites. When existing designations are used in listings, they should never be altered. Object listings should contain a 2nd designation or positional information or both for objects with unfamiliar names. The designation of an astronomical source shall consist of the following parts: originΔsequenceΔ(specifier), where the Δ denotes a blank space. The designation of origin and sequence are necessary; the specifier is optional and must be cited within parentheses. The number of blanks may be increased to accommodate machine-readable files to right justify numeric or tabular data.

NGCΔ205 PKSΔ1817–43 COΔJ0326.0+3041.0

H2OΔG123.4+57.6Δ(VSLR=−185)

The origin element specifies the catalogue or collection of objects. It may be constructed from catalogue names (for example, NGC, BD), the names of authors (RCW), types of objects (PSR, PN), types of sources (13CO, HCN), instruments or observatories used (1E, IRAS), and so on. It should consist of at least 2 letters or numerals and not special characters. The origin element should never be further abbreviated.

The sequence element is normally a numeric field that uniquely places the object within a catalogue or collection. It may be a sequence number within a catalogue (for example, HDΔ224801), or it may be based on coordinates. Coordinates are preceded by a code for the reference frame: G for galactic coordinates, B for besselian 1950, J for julian 2000 equatorial coordinates. Coordinates used in designations are considered names; therefore, they are not changed even if the positions become more accurately known (for example, at a different epoch: "BD –25° 765" remains the designation, even though the source's declination is now –26°). Subcomponents or multiples of objects are designated with letters or numerals added to sequence with a colon: for example, "NGC 1818:B12".

The optional specifier allows indication of association with larger radiating sources (for example, MΔ31, WΔ3) or other object parameters. Because they are not required, they are enclosed within parentheses. Note these examples of complete designations.

[Designation]	[Position]	
[OriginΔSequenceΔ(Specifier)]	[RA (2000)]	[Dec (2000)]
BD –3° 5750	00 02 02.4	–02 45 59
H20 B0446.6+7253.7	04 46 37.3	+72 53 47
AC 211 (=1E 2127+119; in M 15) 21	21 30 15.54	+11 43 39.0
PN G001.2–00.3	17 49 36.9	–28 03 59
R 136:a3	05 38 42.4	–69 06 03

[not]

BD 4° [the 2nd part of the number is missing]

N221 [unclear source: NGC, or N in LMC?]

IRAS 5404–220 [leading zero missing; poor position]

P 43578 [1-letter origin is ambiguous]

DATES AND TIME

DATES

Definitions of astronomical dates include a specification of universal time (UT). For discussion of UT and coordinated universal time (UTC), see sections 27·20 and 27·22.

Julian Date

27·15 The date on which an event occurred is usually best represented by giving the julian date (JD) with an appropriate conventional form. The julian date corresponding to any instant is the interval in mean solar days since 4713 BC January 1 at Greenwich mean noon (1200 UT). (Midnight, 1 January 1961 = 0000 UT 1 January 1961 = JD 2 437 500.5.) In an astronomical context, the calendar date is expressed as year–month–day whether the month is spelled out or represented by a number. Months and days should be represented by 2-digit numbers from 01 to 12 for

January to December and from 01 to 31 for the days of the month; arabic, not roman, numerals should be used. The elements of a numeric date are separated by hyphens.

1988 December 31   1988-12-31

The basis of the date and time system used must be clearly stated, because no system is free from ambiguity. The calendar year 1 BC is followed by the year AD 1; for astronomical purposes it is convenient to denote the year 1 BC as year 0 and the year n BC as year –(n–1).

In converting from calendar date and time (see sections 12·2 and 12·3) to julian date, remember that the julian day begins at 12^hUT (noon on the Greenwich meridian), whereas current calendar days begin at 0^hUT. Before 1925 Greenwich mean time (GMT) often, but not always, referred to the day that began at noon (rather than at midnight) on the Greenwich meridian. There is still some ambiguity about the precise meaning of GMT, and so UTC (see section 27·20) should be used instead of GMT for reporting of astronomical observations.

The JD system may be used in conjunction with other time systems such as ephemeris time and international atomic time. The name "julian ephemeris day" (JED) was introduced for the former, but it is now more appropriate to use the abbreviation JD together with the abbreviation for the time scale. For example, a column heading in a table could be JD (TAI) while for individual values the time scale may be indicated after the numeric value.

The modified julian date (MJD) is in widespread use for current dates to provide a shorter number for which the decimal part is zero at 0^hUTC (or TAI or UT, as specified); MJD = JD–240 0000.5.

Besselian and Julian Epochs

27·16 One besselian year is the period of a complete circuit of the mean Sun in right ascension beginning at the instant when its right ascension is 18^h40^m. Such an instant, or epoch, occurs near the start of the calendar year and is denoted by adding ".0" to the number of the year. For example, the epoch 1950.0 is the instant 1949 December 31 at 22^h09^m UT. This instant may be denoted by 1950 January 0.923. The use of the besselian epoch has been superseded by the use of the julian epoch, which is based on a julian year of 365.25 days and for which 2000.0 is the instant of mean noon at Greenwich on 2000 January 1.

Certain epochs (or instants of time) have special significance in relation to the definition and use of celestial reference systems and are indicated by adding the suffix .0 or .5 to the number for the year. They may be defined either in terms of the besselian or tropical year or in terms of the julian year, and so the appropriate letter B or J should precede the numerical value that specifies the epoch, which may be either the beginning or middle of a besselian or julian year.

J1900.0 = 1900 Jan 0.5 = JD 241 5020.0
B1950.0 = 1950 Jan 0.923 = JD 243 3282.423
J1986.5 = 1986 July 2.625 = JD 244 6614.125
J2000.0 = 2000 Jan 1.5 = JD 245 1545.0

These dates refer to instants of UT unless otherwise stated. When a precision of better than 0d.001 is needed, the time scale should be specified. Such dates may be understood to refer to besselian epochs for years before 1984 if no prefix letter is given. The letters B and J often indicate that positions are referred to the reference systems for B1950.0 and J2000.0, respectively, but they have a much wider use.

TIME SYSTEMS

27·17 Four time systems are used in astronomy: international atomic time, universal time, sidereal time, and dynamical time (which, since 1983, has replaced ephemeris time). The precision with which time should be specified varies with circumstances. Current observations are usually reported in coordinated universal time (UTC; see section 27·22), but the difference from universal time is less than 1 second and may often be ignored. Observations that are timed very precisely and that span several years may be better reported in international atomic time, since this is free from step-adjustments. Current papers that deal with observations made before 1972 should specify what assumptions have been made in reducing the time scale used in the original record to that used in the new paper.

Solar Time: Mean Sun and Mean Solar Time

27·18 Mean sun is an abstract reference point introduced to define mean solar time. It has a constant rate of motion and is used in timekeeping in preference to the real Sun, whose observed motion is nonuniform because of Earth's elliptical orbit and the inclination of the plane of the orbit to the plane of the equator.

Mean solar time is time measured with reference to the uniform motion of the mean sun; it is the local hour angle, plus 12 hours, of a fictitious mean sun, which moves along the equator at a constant rate equal to the average annual rate of the Sun. Mean solar time at 0° longitude is called universal time (formerly Greenwich mean time). The mean solar day is the interval of 24 hours between 2 successive passages of the mean sun across the meridian. The mean solar second is 1/86 400 of the mean solar day.

Apparent solar time is the local hour angle of the true Sun, expressed in hours, plus 12 hours. The difference between mean solar time and apparent solar time on any particular day is the equation of time.

Since the mean sun is an abstract, unobservable point, mean solar time is defined in terms of sidereal time. One mean solar day equals 24 hours, 3 minutes, and 56.555 seconds of mean sidereal time.

Mean solar time was originally devised to measure time uniformly, based on the assumption, now known to be incorrect, that Earth's rate of rotation is constant. (The difference between the mean solar day and a day of 86 400 SI seconds is, however, only a few milliseconds.)

The interval between 2 successive passages of the Sun through the vernal equinox equals the civil or mean solar day, which is 1/365.422 of a tropical year.

Sidereal Time

27·19 Sidereal (meaning "relating to stars or constellations") time is measured in sidereal days. The sidereal day is the interval between 2 successive passages of a equinox across a given meridian, or the interval between 2 successive upper culminations or transits of the equinox. It is divided into 24 sidereal hours. Because of Earth's revolution around the Sun, the sidereal day is 3 minutes and 56 seconds shorter than the 24-hour day.

Sidereal time is given by the sidereal hour angle of an equinox and ranges from 0 to 24 hours during 1 day. The day starts at sidereal noon (the instant at which the equinox crosses the local meridian). The hour angle at a particular location gives the local sidereal time, the hour angle at Greenwich being Greenwich sidereal time. Apparent sidereal time is measured by the hour angle of the true equinox and thus suffers from periodic inequalities.

Mean sidereal time relates to the motion of the mean equinox, which is affected only by long-term inequalities from precession (the slow change of direction of Earth's axis). Apparent minus mean sidereal time equals the equation of the equinoxes. Sidereal time is directly related to universal time and mean solar time and is used in their determination.

The mean sidereal day is the interval between 2 successive upper culminations or transits of the mean equinox.

Universal Time

27·20 The definitions and significance of universal time (UT) have changed in recent years as the methods and precision of timekeeping and of observation have changed and improved. Universal time is now a precise measure of the rotation of Earth that has been defined in terms of mean sidereal time in such a way as to preserve continuity with the earlier use of Greenwich mean (solar) time. It is now determined indirectly from observations of artificial satellites of Earth and of very distant quasars, rather than from observations of stars. These observations are also analyzed to determine the variations in the direction of the axis of rotation with respect to Earth and with respect to a standard reference system. It is no longer necessary to recognize the approximations to universal time that were denoted by UT0, UT1, and UT2.

For ordinary purposes the variability of UT with respect to atomic-time standards, which are even more precise, may be ignored, but a current knowledge of universal time to an accuracy of 1 second or better is needed for some purposes, such as navigation. This accuracy can be obtained from the UTC time signals that are broadcast for general use (see section 27·22).

Atomic Time

27·21 International atomic time (TAI) is the most precisely determined time scale now available. Established by the Bureau Internationale de l'Heure in Paris, it was adopted 1972 January 1. Its fundamental unit is the SI second. The SI second is the duration of 9 192 631 770 periods of the radiation corresponding to the transition between 2 hyperfine energy levels of the ground state of the cesium-133 atom. The astronomical unit of time is often considered to be the interval of 1 day (86 400 seconds).

Coordinated Universal Time and Standard Time

27·22 Coordinated universal time (UTC) is based on international atomic time. It was introduced so that the broadcast time signals could be based on atomic time (and on the SI second) and yet would give universal time with an accuracy that was sufficient for most purposes, including navigation at sea and in the air. Since 1972 January 1 the time broadcast by most time services has been UTC.

UTC is not convenient for general use in all countries, so the world is divided into time zones that are separated by 15° in longitude, that is by 1 hour in local solar time. The exact boundaries of the time zones may be adjusted to match boundaries between states. The time kept in any given time zone differs from UTC by an exact number of hours, which may be changed to give daylight saving during the summer (see also sections 12·4 and 12·5).

4:15 eastern standard time = 2:15 mountain standard time

4:15 eastern standard time = 3:15 eastern daylight time

UTC differs from international atomic time by an integral number of seconds. UTC is kept within 0.90 seconds of UT by the insertion or deletion of exactly 1 second when necessary, usually at the end of December or June; these adjustments are called leap seconds.

An approximation (DUT) to the value of UT minus UTC is transmitted in code on broadcast time signals.

US standard and daylight times are calculated from UTC by subtracting 5 hours for eastern standard time, 4 for eastern daylight time, 6 for central standard time, 5 for central daylight time, 7 for mountain standard time, 6 for mountain daylight time, 8 for Pacific standard time, 7 for Pacific daylight time.

2012 UTC = 1:12 PM PDT = 4:12 PM EDT

Dynamical Time

27·23 Dynamical time scales were introduced when it became necessary to take into account relativistic effects in the computation of precise orbits in the Solar System, and it was therefore no longer possible to use a unique time scale for these purposes. It was, however, possible to define the unit and the epoch of 1 such time scale in terms of international atomic time and the SI second.

Dynamical time is a family of time scales introduced in 1984 to replace ephemeris time (ET) as the independent argument of dynamical theories and ephemerides. (Ephemeris time is based on the time interval of a tropical year; 1 ephemeris second is 1/31 556 925.9747 of the tropical year 1900.)

TERRESTRIAL DYNAMICAL TIME

Terrestrial dynamical time (TDT) is a unique time scale independent of any theory. It is used for apparent geocentric ephemerides: TDT = TAI + 32.184 seconds.

Continuity with pre-1984 practices has been achieved by setting the difference between TDT and international atomic time to the 1984 estimates of the difference between ephemeris time and International Atomic Time. The increment to be applied to universal time (UT1) to give TDT is called ΔT (T).

Table 27·6 Abbreviations for the names of astronomical and related organizations

Abbreviation	Organization
AAS	American Astronomical Society
AG	Astronomische Gesellschaft
AGU	American Geophysical Union
AIAA	American Institute of Aeronautics and Astronautics
AIG	Association Internationale de Geodesie; IAG in English (see below)
ASA	Astronomical Society of Australia
ASP	Astronomical Society of the Pacific
ASSA	Astronomical Society of South Africa
BAA	British Astronomical Association
BIH	Bureau International de l'Heure
BIPM	Bureau International des Poids et Mésures
CCIR	Comité Consultatif International des Radiocommunications
CDS	Centre de Données Stellaires
CETEX	Committee on Contamination by Extraterrestrial Exploration
CNRS	Centre National de la Recherche Scientifique
CODATA	Committee on Data for Science and Technology (ICSU)
COSPAR	Committee on Space Research (ICSU)
CSIRO	Commonwealth Scientific and Industrial Research Organization
EPS	European Physical Society
ESA	European Space Agency
ESO	European Southern Observatory
FAGS	Federation of Astronomical and Geophysical Data Analysis Services
GSFC	Goddard Space Flight Center
IAF	International Astronautical Federation
IAG	International Association of Geodesy
IAGA	International Association of Geomagnetism and Aeronomy
IAU	International Astronomical Union
ICSTI	International Council for Scientific & Technical Information
ICSU	International Council of Scientific Unions
ICSU-AB	Abstracting Board of ICSU; replaced by ICSTI
IEEE	Institute of Electrical and Electronics Engineers
IERS	International Earth Rotation Service
INSPEC	Information Services for the Physical and Engineering Communities
IPMS	International Polar Motion Service; now replaced by IERS
ISO	International Standards Organization
ITU	International Telecommunication Union
IUCAF	Inter-Union Committee on Frequency Allocation for Radio and Space Science
IUCI	Inter-Union Committee on the Ionosphere
IUCS	Inter-Union Committee on Spectroscopy
IUCSTP	Inter-Union Commission on Solar-Terrestrial Physics
IUGG	International Union of Geodesy and Geophysics
IUPAC	International Union of Pure and Applied Chemistry
IUPAP	International Union of Pure and Applied Physics
IUTAM	International Union of Theoretical and Applied Mechanics
IUWDS	International Ursigram and World Days Service
JOSO	Joint Organisation for Solar Observations

Table 27·6 (*cont.*)

Abbreviation	Organization
JPL	Jet Propulsion Laboratory
MPI	Max Planck Institut
NASA	National Aeronautics and Space Administration
RAS	Royal Astronomical Society
RASC	Royal Astronomical Society of Canada
RASNZ	Royal Astronomical Society of New Zealand
SAAO	South African Astronomical Observatory
SAF	Société Astronomique de France
SAI	Societa Astronomica Italiana
SAO	Smithsonian Astrophysical Observatory
SERC	Science and Engineering Research Council
SPARMO	Solar Particles and Radiation Monitoring Organization
UAI	Union Astronomique Internationale; IAU in English (see above)
URSI	Union Radio-Scientifique Internationale
WDC	World Data Center
WMO	World Meteorological Organization

The unit of TDT is the day, 86 400 seconds, 24 hours, at mean sea level.

BARYCENTRIC DYNAMICAL TIME

Barycentric dynamical time (TDB) is used for ephemerides and equations of motion referred to the barycenter of the Solar System. TDB differs from TDT only by periodic variation.

ASTRONOMICAL AND RELATED ORGANIZATIONS

27·24 Many organizations in, or important for, astronomy are frequently referred to by capitalized abbreviations; see Table 27·6.

REFERENCES

Cited References

Graham AL, Bevan WR, Hutchison R. 1985. Catalogue of meteorites: with special reference to those represented in the collection of the British Museum (Natural History). Tucson (AZ): Univ Arizona Pr.

Hofleit D. 1964. Catalogue of bright stars. New Haven (CT): Yale Univ Observatory.

Ridpath I, editor. 1989. Norton's 2000.0 star atlas and reference handbook. 18th ed. New York: J Wiley.

Schmidt A. 1987. Editorial standards [for NASA]. Beltsville (MD): Engineering & Economics Research.

Wilkins GA. 1989. IAU style manual. Norwell (MA): Kluwer Academic Publishers.

Zombeck MV. 1990. Handbook of space astronomy and astrophysics. 2nd ed. Cambridge (UK): Cambridge Univ Pr.

Additional References

Beatty JK, Chaikin A, editors. 1990. The new Solar System. 3rd ed. Cambridge (MA): Sky Publishing.

Curtis AR. 1990. Space almanac: facts, figures, names, dates, places, lists, charts, tables, maps covering space from Earth to the edge of the Universe. Woodsboro (MD): Arcsoft Publishers.

de Vaucouleurs G, de Vaucouleurs A, Corwin HG, Buta RJ, Paturel G, Fouqué P. Third reference catalogue of bright galaxies. 1991. New York: Springer-Verlag.

Maran SP, editor. 1992. The astronomy and astrophysics encyclopedia. New York: Van Nostrand Reinhold.

Murdin P, Allen D. 1979. Catalogue of the Universe. New York: Crown Publishers.

Schmadel LD. 1994. Dictionary of minor planet names. 2nd ed. New York: Springer-Verlag.

Seidelmann PK. 1992. Explanatory supplement to the Astronomical Almanac. Mill Valley (CA): Univ Science Books.

PART

4 Journals and Books

28 Journal Style and Format

To study, to finish, to publish. —Benjamin Franklin

While the artist's communication is linked forever with its original form, that of the scientist is modified, amplified, fused with the ideas and results of others.

—Max Delbrück (cited in *The Eighth Day of Creation,* H F Judson, 1979)

28·1 The usefulness and ease of use of a journal are determined not only by its content but also by its style and format. Decisions on these characteristics should aim to serve the needs of readers, librarians, and bibliographers. Priority should go to clarity and adequacy of information; these can be served by carefully attending to the many details of style and format described here. The recommendations in this chapter are derived mainly from the relevant standards of the American National Standards Institute (ANSI 1977, ANSI–NISO 1992) and the National Information Standards Organization (NISO 1985, 1987, 1990, 1992) and from recommendations of the Institute for Scientific Information (ISI 1983). A comprehensive collection of relevant standards can be found in *Documentation and Information: ISO Standards Handbook 1* (ISO 1988); see the description of this handbook in Appendix 3.

Standards for electronic journals have not been established (but see sections 30·62 and 30·63 on references). Many of the recommendations in this chapter may, however, be relevant in principle to elements of the presentation of electronic journals on monitor screens and to printouts from such journals.

Note that librarians and bibliographers often designate journals under the term "serials", a category that includes books related in topic or discipline but not having the characteristics of journals in the usual sense of "journal" in science except for, usually, consistent formats and sequential numbering as members of a series with a series title.

JOURNAL TITLE

FUNCTION AND LENGTH

28·2 A new title should be drafted with great care. The title should represent the best compromise between a title long enough to adequately convey the journal's scope and content and short enough to not annoy persons having to record it in bibliographic records.

Current Research

[Does this journal cover all current research in all fields? Some fields within a broad discipline such as biology? Research only by members of the sponsoring society?]

The International Archives of Botany, Plant Physiology, Zoology, and Related Fields

[The scope appears to be all of biology. Why try to specify some components and not others? Why not devise the shortest title that reasonably well represents the scope, for example, *The International Archives of Biology*?]

A single-word title can be amplified in meaning with a compound title or subtitle that details the journal's scope.

Medusa: A Multidisplinary Journal of Mythology

Note, however, that such a title might be represented bibliographically only by the initial term.

Because many computer monitors are limited to displays with a width of 80 characters, the title could helpfully be held to fewer characters.

Acronymic titles such as "*PABA*" for *Proceedings of the American Bariatrics Association* are undesirable. They may be readily understood by readers in the journal's field but will be cryptic to others and fail to indicate the scope and content of the journal. If they are used along with the full title they represent, users of the journal may not know which version is preferred for references.

If the scope of a journal changes, a title change may be desirable; see section 28·6.

UNIQUENESS

28·3 Care should be taken to ensure that the title does not duplicate another title in use or owned by another publisher. Duplication might represent infringement on the legal right of a publisher to a title and, at the very least, could confuse readers, librarians, and bibliographers.

Existing titles can be identified in many different sources, both print catalogues and database listings of serial titles; examples are the SERLINE database and *List of Serials Indexed for Online Users* of the National Library of Medicine, *Serial Sources for the BIOSIS Previews® Database,* and the *Bowker International Serials Database,* available through the DIALOG online system. These and other sources will be searched by the National Serials Data Program in response to requests for information on the uniqueness of a proposed title or to a request for an International Standard Serial Number (ISSN); see section 28·5. Information on these sources can be obtained from the National Library of Medicine, 8600 Rockville Pike, Bethesda MD 20894, USA; BIOSIS, 2100 Arch Street, Philadelphia PA 19103-1399, USA; or Dialog Information Services Inc., 3460 Hillview Avenue, Palo Alto CA 94304, USA. Information on additional sources will usually be available from librarians in science and other academic libraries.

LOCATIONS AND FORM

28·4 The identical and complete form of the title should be used wherever it appears: cover, spine, table of contents, masthead, officers' page, running heads and feet, and text. If the title must be abbreviated to save space, the abbreviation should have the form specified by *Documentation—Rules for the Abbreviation of Title Words and Titles of Publications* [ISO 4-1984(E)], published by ISO (1986). Examples of title-word abbreviations and abbreviated titles prepared in accordance with this standard can be found in the serials lists cited above in section 28·3; also see section 30·24 and Appendix 1.

The location and typography of the title should be such as to avoid any ambiguity with other text near it. An acronymic designation of a society or publisher's name should not be near a title if it might be thought to be an acronymic form of a title.

INTERNATIONAL STANDARD SERIAL NUMBER AND CODEN

28·5 The International Standard Serial Number (ISSN) (ISO 1986, NISO 1992) is an 8-digit code number that uniquely identifies a journal title.

Journal of Cyclic Nucleotide Research	ISSN 0095-1544
[continued as]	
Journal of Cyclic Nucleotide and Protein Phosphorylation Research	ISSN 0746-3898

Note that the 8th character is a check digit, which may be the letter X (when the check-digit number would be 10) rather than a numeral. In the United States a request for an ISSN for the title of a new journal or a new title for an existing journal should be made to the National Serials Data Program, Library of Congress, Washington DC 20540. The request should be made before publication of the new title so that the ISSN can appear on the 1st issue under this title.

A CODEN is a unique 6-character identifier of a serial.

Journal of Cyclic Nucleotide Research	JCNRDU
Journal of Cyclic Nucleotide and Protein Phosphorylation Research	JCNREV

CODENs are assigned by the International CODEN Service, c/o Chemical Abstracts Service, PO Box 3012, Columbus OH 43210, USA.

CHANGE OF TITLE

28·6 A title should not be changed except for important reasons. These might include a change in scope or organizational sponsorship, or a need for more accurately representing the journal's scope. A change should be made in the issue beginning a new volume, preferably the 1st volume of a year. The uniqueness of the title can be verified with the means used for the title of a new journal; see section 28·3. If a title is changed, readers, librarians, bibliographers, and subscription agencies should be alerted to the change; see sections 28·20 and 28·27.

JOURNAL UNITS: VOLUMES AND ISSUES

28·7 The division of journals into units and subunits usually designated as "volumes" and "issues" is a long-standing convention that brings readers the journal in readily handled dimensions, facilitates its display and storage in libraries, and provides conventions for bibliographic identification. Volumes and issues should be organized and formatted with a view to efficiently serving the needs of readers for many kinds of information.

The rest of this chapter applies to journals published on paper. Electronic journals may need different conventions, but the principle of serving user needs efficiently should be kept in mind when deciding on what conventions will be used for monitor displays of contents and for bibliographic identification; see sections 28·36 and 30·63.

JOURNAL VOLUMES

PERIOD OF PUBLICATION

28·8 Each volume should represent no more than 1 full calendar-year of issues. If a single volume carrying all issues in a calendar year would be too bulky for reliable binding and easy handling, a year's issues can be grouped into 2, 3, or more volumes, but all volumes should include the same number of issues.

Because the recommended formats for bibliographic references to journals and their articles include the year of publication (see section 30·25) as a useful although usually redundant datum, no volume should include issues from more than 1 calendar year. Thus for a journal with monthly issues to be gathered in 2 volumes, the starting date would be January or July; for a semimonthly (twice-a-month) or weekly journal, the appropriate starting date would be, respectively, 1 January (semimonthly) or the date of the 1st Monday in January (a weekly), or the counterparts in July. For semimonthly journals the successive dates are the arbitrary choices of the 1st and 15th day of each month.

NUMBERING

28·9 Volumes should be numbered sequentially with arabic numerals starting with 1, the numeral to follow "Volume". If the journal's title is changed, the sequence should not continue with the new title but begin again with 1. Roman-numeral sequences should not be used.

PAGINATION

28·10 Page numbers of the text pages must begin at 1 and run sequentially through all text pages of the issues making up the volume, rather than restarting sequential numbers with 1 in individual issues, as is usually done in popular magazines. That practice can make finding articles in bound volumes more difficult.

The preliminary pages of a volume (title page, table of contents, information pages; see sections 28·11–28·14) must be numbered so as to distinguish them from text pages. Lowercase roman numerals (i, ii, iii, iv, v, and so on) have been conventionally used in books and journals, but an alternative system of alphanumeric designators (such P1, P2, P3, and so on) is a logical substitute for the now generally archaic system of roman numerals.

Pages not to be bound into the volume, such as advertising pages, can be also be numbered alphanumerically to distinguish them from text pages, for example, as A1, A2, A3, and so on.

The pages of the index to the volume should continue with the numeric sequence of the text pages if the index is to be at the end of the bound volume. If the journal recommends placing the index just before the main text of the bound volume, the pagination should continue that of the preliminary pages opening the volume.

Note that publishers and printers often refer to page numbers as "folios".

ELEMENTS OF VOLUMES

28·11 Each volume should include at least 4 parts.

1 Title page.
2 Table of contents.
3 Text of the volume.
4 Index.

These components are needed because libraries generally bind journals for storage in volume units, and each volume of the journal should be comprehensible by itself. With the widespread indexing of journals in readily available bibliographic data bases, some publishers have considered omitting indexes; they should keep in mind that indexes in bound volumes of journals can sometimes help readers find articles when the page number in a reference is wrong but the volume number is correct.

An additional part, information pages, can be helpful to many readers. These pages can provide the information given on the masthead page of each issue (see section 28·21 and Table 28·5) and on information-for-authors pages (especially if they are published within the advertising pages of 1 or more issues rather than within the text pages; see section 28·23 and Table 28·6).

To facilitate prompt binding of the issues of a volume, the title page, the volume table of contents, the information pages, and the index are usually published as the end pages of the last issue of the volume. If these pages are, instead, routinely sent to subscribers later in a separate form or are sent only on request, this practice should be noted on the masthead of each issue. In separate form these pages must have the same trim size as issues.

Title Page

28·12 The title page for the volume should carry information needed to identify fully the journal, the volume, the issues included, the 1st and last dates of issues, and the publisher's name and location (see Table 28·1); it should be a right-hand (recto) page but need not carry a page number.

If bound volumes would be too thick and heavy for easy handling, a volume can be divided into 2 or more parts. If it is, a title page should be provided for each part, with designations such as "Volume 17: Part 1" and "Volume 17: Part 2". This solution can be avoided by increasing

Table 28·1 Volume title-page: recommended information

Information	Notes
Complete title (sections 28·2, 28·4)	The title should be the dominant element and preferably near the top. If the complete title is a compound title or includes a subtitle, all elements should be included. If an acronymic verison of the title is used, there should be no ambiguity as to the proper formal and complete title; the acronym should be inferior to the formal title both in location and font size.
ISSN and CODEN (section 28·5)	These can be in a subordinate location, such as the foot of the page.
Publisher's name and location (section 28·12)	If the journal has a sponsoring organization (such as a professional society) that is not the publisher, the organization's name and location should also appear.
Editor's name (sections 28·11, 28·21)	This information need not appear here if it is on the masthead page or other information pages included among the preliminary pages.
Volume number and issue numbers	First and last issue numbers of its issues (in arabic numerals) separated by an en-dash (see section 4·22); if volumes are divided into parts (see section 28·12), a part title page should carry only the numbers of the issues included.
Dates of the 1st and last issues	Including the year, as in "January–December 1991".

the number of volumes per year and decreasing the number of issues per volume.

The reverse (verso) of the title page is usually blank but can be used to carry additional information about the journal such as that usually appearing in the masthead of issues (see section 28·21).

Table of Contents

28·13 The volume table of contents should provide all the information carried on issue tables of contents (see section 28·20) and in the same sequence. It is conveniently formed from the tables of contents of individual issues, but maintaining separation of the contents listing into issues (and their dates) rather than providing an uninterrupted sequence for the entire volume will probably be preferred by most readers. If additional preliminary pages follow the table of contents (such as a masthead page or information-for-authors pages), these should be listed before the listing of the contents of the 1st issue.

The table should begin on a right-hand (recto) page, which will usually be "P3" (or "iii" in a roman-numeral sequence; see section 28·10).

Information Pages

28·14 These pages can include information carried in pages of each issue that would not be available in bound volumes, for example, information-for-authors pages published within the advertising pages of issues. For recommendations on the content of a masthead page and of information-for-author pages, see sections 28·21 and 28·23 and Tables 28·5 and 28·6.

Index

28·15 The text pages should be indexed for subjects of articles and other separate units (such as editorials and letters) and their authors. Separate subject and author indexes are easier to consult than single indexes carrying both categories. Index pages should not be interrupted by advertising or news pages, and their pagination should be continuous with that of the preceding text pages.

The index is preferably the last element of the text pages of the last issue in a volume; see section 28·11 for policies on availability of indexes.

JOURNAL ISSUES

TRIM SIZE

28·16 All issues should have the same trim size (height and width). Many scientific journals have the trim size of 8.5 by 11.0 inches (21.5 by 28.0 cm or a similar metric standard). This size is desirable for several reasons: it accommodates more text per weight of paper than smaller trim sizes, which may reduce both paper and postal costs; it facilitates the sale of advertising space; it allows conveniently for 2- or 3-column page formats; and bound volumes will fit on most library shelving. Smaller trim sizes, such as 6.875 by 10.0 inches (17.5 by 25.5 cm), are widely used in the humanities; they make up a bulkier issue for a given amount of text.

FREQUENCY, NUMBERING, AND DATES

28·17 Issues are collections of papers and other documents assembled and published with dimensions and a frequency that are economical for the pub-

Table 28·2 Issue dates

Frequency of issues	Recommended dating
Weekly	The same day of each week.
Quasiweekly (for example, 50 issues per year)	The same day of each week in which an issue is published.
Semimonthly (twice-a-month)	Dates representing approximately equal intervals between issues. Convenient dates are the 1st and 15th days of each month (such as 1 January 1997 and 15 January 1997) even if the mailing date does not coincide with the issue date.
Monthly	The month of issuance; a day of the month need not be specified. The date of mailing should be as constant as possible from month to month.
Bimonthly (2-month intervals)	The pair of months for which issued, even if the mailing date (which should be as constant as possible) is not in the 1st month; for example, January–February 1997.
Quarterly or thirdly	The span of months represented by each issue as with bimonthly issuance, for example, January–March 1997 (quarterly) and January–April 1997 (thirdly). Designators such as "Winter 1997", which could mean mailing in December 1997 or in February 1997, should be avoided.

lisher and convenient for the reader. The choice of frequency (quarterly to weekly for most scholarly journals) is usually determined by the amount of text to be published, the relative merits of getting papers to readers without unnecessary delay and in a conveniently handled format, these factors being judged against the lower costs of printing, binding, and distributing with a lower frequency of publication.

Issues are identified sequentially with arabic numerals beginning with "1" within each volume; for example, the January issue of a monthly journal is "Issue 1".

The needs of publishers, readers, and librarians are best served with a constant frequency and issuance on fixed dates; see Table 28·2 for recommendations on dates of issues. See sections 28·12, 28·18–28·20 and Tables 28·1 and 28·3 on locations for issue dates.

ISSUE ELEMENTS: SEQUENCE AND CONTENTS

Cover

INFORMATION

28·18 The front cover must carry the information needed to identify the journal and the issue unequivocally; see Table 28·3. The covers of particular issues may need additional information; see Table 28·4.

Table 28·3 Issue cover: needed information and location

Element	Recommendation
Complete title	With subtitle if used. See section 28·4. In a dominant position on the upper half of the cover.
Volume number and issue number	In arabic numerals, for example, "Volume 68, Number 9", placed near the title. If issues of a usually monthly journal are combined, perhaps for economy, the numbers of the combined issues and their dates should appear (for example, "Volume 68, Numbers 9–10, September–October 1998"), but this practice should be avoided if possible.
Frequency: Issue date	Placed near the title.
Weekly and semimonthly	A specific date, not a specific day, in ascending or descending sequence of elements, for example, "15 January 1998" or "1998 January 15", not "January 15, 1998" (see section 12·9).
Monthly	Month and year.
Bimonthly	Single issue number, paired months, year, as in "Number 2, March–April 1996".
Quarterly or lower frequency	Issue number, months represented, and year; see "Bimonthly" above. Seasonal designators such as "Winter 1997" should be avoided.
International Standard Serial Number (ISSN) (see section 28·5)	In the form ISSN nnnn-nnnn.
CODEN (see section 28·5)	With volume number, issue number, inclusive page numbers, and year, as in "AIMEAS 115(9)665–752(1991)".
Name of sponsor and of publisher	Not duplicated if they are the same.
Location of the table of contents (page numbers)	Only if the table of contents is not on the front or back cover or within the first 3 pages.
Issue contents	If length of contents information and space available on the cover permit.
Bar code	See section 28·18.

Bar codes (Code 128) identifying a journal's ISSN, its title, and the particular issue are widely used by popular magazines to speed accession of issues in libraries. The recommended location for the code is the lower left-hand corner of the front cover in the horizontal orientation. Information can be obtained from the Serials Industry Systems Advisory Committee, 160 Fifth Avenue, New York NY, USA, telephone 212 929-1393.

Table 28·4 Issue cover: additional information for particular issues

Information	Notes
Volume index	Location given in page numbers.
Supplements, special issues	Identified as part of a regular issue, for example, "Number 9, Part 2", with the usual part identified as "Number 9, Part 1". See sections 28·26, 28·27.
Change of title	Previous title within parentheses or after "Previously" under the new title on the covers of all issues in the 1st year of the volume with the new title. See section 28·27.
Changes in trim size, frequency; irregularities	See section 28·27.

FORMAT

The information recommended in Table 28·3 should appear on the front cover of every issue in the same location, typefaces, and fonts. The title should be the most prominent element on the cover and preferably be placed in the upper half of the cover. If titles are presented in more than 1 language, the primarily known title should be emphasized by its size and location. Formatting that might produce ambiguity or obscurity of this information from, for example, advertising should be avoided. The cover should not be included in the pagination of the text.

Spine

28·19 If the spine can accommodate legible type, it should carry the most important information identifying an issue.

1 The title of the journal or its standard abbreviation (see section 28·4).
2 The volume and issue numbers; the issue date.
3 The page numbers of the issue if the volume is paginated continuously.

If the spine is wide enough, this information should be printed so that it can be read when the issue stands upright on a shelf. On a narrower spine, the information should be positioned so that it can be read from left to right when the issue is lying with the front cover up.

Table of Contents

28·20 The table of contents of each issue must identify fully all of the articles and the sections into which they are grouped. Material of minor and ephemeral value might be identified solely by a section title, for example, "News Notes".

LOCATION

The front cover is the preferred location in scientific journals for the contents listing; the next preference is the right-hand (recto) page facing the inside (verso) of the front cover.

Even if the table of contents is on the front cover, it should also appear within the first 5 pages as a precaution against loss or damage of the front cover. If the front cover carries only an abbreviated version, the location (page number) of the full version should be identified on the cover.

Tables of contents of past or coming issues should be clearly identified and placed so as to eliminate confusion with the table for the present issue.

BIBLIOGRAPHIC IDENTIFICATION

All information needed to identify the journal and the issue should be included at the head of the table-of-contents page.

1 Title: complete title (and subtitle).
2 Abbreviated title in standard form (see sections 28·4 and 30·24; Appendix 1) and the International Standard Serial Number (ISSN), both of which need not be adjacent to the complete title but should be in an obvious place and not buried in text.
3 Volume and issue numbers; date of issue (exact for a weekly, biweekly, or semimonthly (twice-a-month); only month or months for a monthly, bimonthly, or quarterly).
4 Notice of a change in title or frequency.

INFORMATION ON ARTICLES AND SECTIONS

All articles except those of minor and ephemeral value must be fully identified.

Titles must have the same complete form as on the title pages of articles. Complete titles must be provided for articles of more than ephemeral value that appear under section headings, for example, "Editorials".

Author names should have the same form as on the articles if space permits; if it does not, given names should be represented by initials as in bibliographic references (see section 30·22).

Only initial page numbers need be given for articles paged continuously; otherwise, the initial page of each appearance of the article should be given, with comma and space separations.

Section-grouped articles and other content should be completely identified under the headings of the sections if space allows. Sections with many short articles, for example, book reviews, letters-to-the-editor, news notices, may be represented only by the section heading and the initial page of the section.

Special features that do not appear in every issue must be represented by title and initial page number when they do, for example, an information-for-authors section, an erratum or retraction notice, a volume index. If an information-for-authors page is published only at infrequent intervals, the issues in which it does appear should be identified on the contents page of all other issues. A feature appearing in every issue and serving important needs of readers and authors, such as a journal-policy page or an information-for-authors page, should be represented with a page number even if carried in an advertising section.

FORMAT AND TYPOGRAPHY

Bibliographic data, including the journal title, should precede the title "Table of Contents"; see section 28·4.

Section and article titles must appear in the same sequence in the table of contents as in the issue. The most legible arrangement for article titles is probably that of 3 columns: article titles in the 1st; author names in the 2nd; page numbers in the 3rd. If space does not allow this format, article titles and author names should be listed on the left and page numbers in a right-hand column. Different typefaces for titles and author names may help with legibility, for example, a roman face for titles and an italic face for names.

Black ink is preferred to allow for clear reproduction of the page in a publication such as *Current Contents* or in xerographic copying. The type should be 10-point or larger if the text page is more than 7 inches wide.

The Masthead

28·21 Each issue should include a section describing the journal, its availability, its ownership, and any additional information often needed by authors, readers, subscribers, librarians, dealers, indexers, bibliographers, archivists, and advertisers. See Table 28·5 for details on the information to be provided on the masthead page; note that some of the recommenda-

Table 28·5 The masthead

Minimally needed information	Notes
Title, subtitle, International Standard Serial Number (ISSN), CODEN	If the journal is published with titles in more than 1 language or with an acronymic title, this information should be included; the complete (not the acronymic) title most widely used, should be given 1st place.
Publisher	Name and complete address (postal [including postal code], cable, electronic-mail, TELEX, facsimile-transmission ["fax", "telefax"] addresses); telephone numbers (including "800" number), with area (or regional) code.
Volume number per year, issue number per volume, issue frequency	Frequency as "weekly", "biweekly" (not "fortnightly"), "semimonthly" (or "twice-a-month"), "monthly", "bimonthly", "quarterly", "semiannually" (or "twice-a-year").
Postal data required	Information required by the postal service of the country in which the journal is published.
Copyright information	The legally required notice of copyright with the copyright sign (©), the copyright year, and the name of the copyright holder. Also the publisher's policy on fair-use copying and mechanism for copying royalties (Copyright Clearance Center or other means) should be stated.
Subscription rates, single-issue price, ordering information	The address to which orders should be sent even if it is the same as that of the publisher. Ordering information: terms of payment (currency, payment media [check, money order, cabled payment, credit-card payment, including telephone number for ordering]), availability of back issues and other media (such as microforms, online services).
Address-change procedure	Address to which notice of changes should be sent.
Sponsorship	Name and address if the publisher is not the sponsor.
Advertising management	Names, addresses, and telephone numbers of staff persons or agencies able to give information on display advertising, classified advertising, or both kinds.
Logo and name of circulation auditor	
Statement and logo indicating publication on acid-free paper	The statement recommended by the relevant NISO standard (ANSI–NISO 1992), "This paper meets the requirements of ANSI/NISO Z39.48-1992 (Permanence of Paper)". The logo is an infinity symbol within a circle, ♾.

Table 28·5 (*cont.*)

Additional useful information	Notes
Editors and other staff persons	Names, addresses, and telephone numbers of persons receiving manuscripts and other mail intended for the editor.
Publication committee	Representatives of the sponsoring or publishing organization.
Peer-review policy	Details useful to authors.
Indexing, abstracting	Identification of the print and computer services providing index data and abstracts from the journal.
Statement on use of recycled paper	

tions are relevant only to publication in the United States. The content specified by "Additional Useful Information" should be considered for inclusion if it is not in an information-for-authors section.

The masthead should appear within the first 5 pages, preferably on the page with, or immediately adjacent to, the table of contents. Its location should be constant from issue to issue. If it appears on the inside front cover or within pages not likely to be bound in the volume, it should be included with the preliminary pages (title page, table of contents) that are provided with the conclusion of each volume; see section 28·11. If the masthead information is on the table-of-contents page, the layout and typography should be such as to make clear the information most likely to be needed by a majority of the users of the journal.

Pages

FORMAT

28·22 The text can be carried as a solid text block on each page or divided into 2 or 3 columns, separated from each other by white space. Text in columns must begin at the upper left-hand corner of the text block, proceed down the 1st column, and continue at the top of the next column.

The choice of format from among the single-, double-, and triple-column possibilities will hinge on various considerations. The single-column format used in a journal with a relatively large trim size (such as the 8.5 by 11-inch size widely used in the United States) may have lines of type too long for convenient reading. The double-column format produces shorter and more easily read lines, which, in addition, can have less space (leading) between lines and permit variation in placement of figures and tables, with saving in space needed for a given amount of text. A 3-column format may produce lines that are too short for efficient reading unless the margins around the text block are narrow. The choice among the possible formats should be made in consultation with a typographic designer.

The margins must be wide enough to allow for some trimming in the binding process and for the binding itself. Suggested minimums for the inner margin ("gutter" is the printer's term) are 1 inch (2.5 cm) and for the top, outer, and bottom margins, 0.75 inch (1.9 cm).

Figures and tables should be aligned within the margins of the text block and be readable in the same orientation as the text. If a table is too large for this orientation, it can be rotated counterclockwise 90° so that its bottom is against the inner margin if it is on a left-hand page and against the outer margin on a right-hand page. A large table can be split between 2 facing pages, but the row headings must be repeated on the right-hand half of the table. Rules for the most pleasing placement of figures and tables on pages can be developed by the journal's designer, but, in general, tables and figures are best placed in the upper part of a page and the text in the lower part.

INFORMATION CARRIED ON PAGES

Facing pages of text (adjacent left-hand and right-hand pages) should together carry the information that identifies the journal and issue: journal title, abbreviated if necessary; the volume number, and the issue number (or issue date). If space permits, facing pages should also carry a short version of the article title or the name of the journal section (such as "Book Reviews"). These elements can be in the top margin (as a "running head") or in the bottom margin as a "running foot" or footline. A widely used convention is using the top margin for running heads and the bottom margin for journal title, volume number, and issue number (or date).

All pages must be numbered sequentially: with right-hand pages odd numbered and left-hand pages even numbered. The recommended position for page numbers ("folios") is in the bottom margin below the text block, aligned with the left-hand edge of the text block on left-hand pages and with the right-hand edge on right-hand pages. See section 28·10 on sequential numbering through each volume.

Information for Authors

28·23 Journals can help themselves and potential authors who are preparing and submitting papers by publishing at regular intervals 1 or more pages describing journal policies and manuscript requirements; for recom-

Table 28·6 Information-for-authors page: potential content

Potential content	Specific items
Distribution	Circulation: domestic, foreign. Categories of subscribers.
Other availability	Online sources; CD-ROM formats; microforms.
Indexing and abstracting services	Names of services covering the journal, addresses, and sources of additional information.
Content	Categories of papers considered; specifications on their specific content, expected format, allowed length. Descriptions of appropriate content for sections of the journal.
Submission of manuscripts	Content of submission letter: must include name of author responsible for further communication, address, and communications numbers (telephone, fax, E-mail); information on prior or expected publication of a paper's content (whether identical or not). Number of copies; spacing; kind of paper. Requirements for figures. Mailing specifications. Alternative formats, such as word-processing diskettes and acceptable word-processing programs. Copyright-transfer requirements; other required information, such as financial interests. Submission and page charges.
Ethical considerations	Criteria for authorship; basis for sequence of author names. Conflict-of-interest identification. Definition of prior publication.
Editorial and review process	Procedure followed by the editors and time constraints. Information provided to authors. Policy on peer reviewing.
Manuscript style and format	Details on manuscript format: title page, limit on title length, abstract structure and length, text and its headings. Details on scientific style, including units of measurement, nomenclature. Formats for references. Details on requirements for figures and tables. Charges for color figures. Recommended style manual and dictionary.
Publication	Scheduling; handling of proofs; reprint orders; prepublication release of information.

mended content, see Table 28·6. These pages should appear in the same place in each issue, preferably inside the issue rather than on the cover, with the page listed in the table of contents (see section 28·20). If they are not published in each issue, the issues in which they have appeared should be identified in the table of contents, on the masthead page, or on the cover, with mention of the source from which copies are available.

Corrections and Retractions

28·24 Statements of authors' errors, editor's or publisher's errors, omissions, retractions or other matters that merit being called to the attention of readers should be published in the same place in the journal in any issue carrying such statements. The location should be prominent, such as on a page with an editorial or with letters to the editor, or in a section titled "Corrections". If the statement is not on the table of contents page, its location should be identified on the contents page, with page number. The statement should make clear its function, for example, with a title beginning "Correction" or "Notice of Error" and including a brief term or phrase identifying the subject; the statement should fully identify the corrected item with bibliographic data. The material corrected and the correction should be made clear in the notice.

Retraction notices should be published with attention to the same considerations of format, but they must also identify the person responsible for the retraction.

Both correction and retraction notices must be indexed in volume indexes. In addition to the index entries for the notices themselves, the index entries for the articles that are the subject of a notice should include a parenthetic statement cross-referencing the related notice. In fields with indexing databases that flag indexed articles with notices of retraction or correction (such as MEDLINE of the National Library of Medicine), these notices should be called to the attention of the relevant database producer for possible action.

Articles in Installments or in Series

28·25 If an article is too long for a single issue and must be published in 2 or more installments, the successive sections should carry the article's title with the additional designation of part, such as "Part 1", "Part 2", and so on. The concluding part should, however, include notice of conclusion, for example, "Part 2 (Conclusion)". If possible, such articles should be scheduled so that all parts appear in consecutive issues of a single volume, with notices on their title pages of the titles and pages of the preceding parts. If the subjects of the divisions of such an article and space in the table of contents permit, each part can be helpfully titled to represent its content.

Recent Developments in Geomorphology of the United States: Part 1, The East

Recent Developments in Geomorphology of the United States: Part 2, The Midwest

Recent Developments in Geomorphology of the United States: Part 3, The West (Conclusion)

Individual articles representing a series should also be published within a single volume if possible, with titles relating the articles in the series.

The Economics of Health Services: Canada	[by Jones, Smith, and Canaday]
The Economics of Health Services: France	[by Guilliard, Boissett, and Mendez]
The Economics of Health Services: Great Britain	[by Macmillan, Stuckert, and MacDougal]

Supplements and Special Issues

28·26 To serve special editorial objectives, journals may occasionally publish supplements or special issues. They should have the same trim size, design, and format as regular issues to avoid confusing readers or producing difficulties for library binding. See Table 28·7 for details. The pagination should be determined by whether the supplement is designated as Part 2 of a specific issue and is to be bound immediately after Part 1 of an issue or is to be bound at the end of the usual set of issues (but before the volume index); in either case, the pagination of the supplement should continue from the preceding text (Part 1 of an issue or of the final issue). If separate pagination is desired, the page numbers should include an alphabetic modifier (such as "S" for "supplement"): 1S, 2S, 3S, and so on; this choice should be considered only if the supplement is to follow the last issue of a volume.

The content of the supplement should be indexed in the volume's index whether the supplement has the volume's pagination or its own.

CHANGES AND IRREGULARITIES

28·27 In general, changes should not be made in the characteristics of a journal (title, frequency, trim size, design, and so on) without careful thought as to the possible consequences and the expectation that the changes will apply for an indefinitely long period. Frequent and hastily adopted

Table 28·7 Supplements and special issues

Elements	Notes
Supplement	
Trim size	Same as for regular issues.
Cover	Separate cover; design, format, and identification same as for regular issues.
Title page, table of contents	Specific for the supplement.
Bibliographic identification	The supplement: "Part 2 of 2"; the regular issue it supplements, "Part 1 of 2". In a bibliographic reference the identifier follows the issue number, as in "(6 Pt 2)".
Title	Supplement title; if the supplement carries a single article (such as a very long review article or a dissertation) the supplement title is the article title. A multiarticle supplement (such as a conference proceedings) must carry individual article titles and preferably also carry a supplement title.
Pagination	Preferably continuous with regular issues.
Indexing	Indexed with issues and other supplements in the volume.
Special issue	Same trim size, design, format, and pagination as regular issues; cover and table of contents may carry a title identifying the theme of the special issue.

changes can injure a journal's reputation among authors, readers, and librarians.

TITLE CHANGE

A change in the title of a journal should be made only for carefully considered reasons, such as a change in topical scope or sponsorship or the need for a clearer, more specific title. The change should be made at the beginning of a volume, preferably the 1st volume of a calendar year. The precautions that should be taken in selecting a new title are the same as for drafting the title of a new journal; see sections 28·2, 28·3. The title will have to be changed in all of its locations; see section 28·4. In these locations, the former title should be in positions and a size subordinate to the new title. The former title (preceded by "Formerly") should appear in these locations for 1 year after the change.

When 2 journals are combined and 1 of the previous titles is continued, the recommendations in the paragraph above apply; the volume-number sequence of the retained title should continue. When a new title is applied to 2 combined journals, apply the recommendations for a new journal; see sections 28·2, 28·3. For 1 year the new title should be accompanied in the stipulated locations by a clarifying statement, "[New Title] represents the combined [Previous Title 1] and [Previous Title 2]".

If a journal must be divided into 2 or more new journals and its title is continued for 1 of the journals, that journal should continue its ISSN and volume numbering; the new journals should apply the recommendations for a new journal, but the cover, table of contents, and masthead should carry for 1 year the statement "Continues in part [Original Title, ISSN]".

Any changes in title should be called to the attention of all subscribers with a notice mailed separately from issues of the journal or journals making the change.

CHANGE IN TRIM SIZE

A change in trim size should be made only with the 1st issue of a volume. The former trim size (and other aspects of format) should be applied, however, to any parts of a volume supplied after the 1st issue of the volume with the new size, for example, a volume index or a supplement that is part of the preceding volume.

CHANGE IN FREQUENCY OF ISSUES

A new frequency of issues (for example, weekly instead of the previous monthly) should begin with a new volume, with the change prominently announced (on the cover and the table of contents) in the last issue in the previous frequency and the 1st issue with the change.

IRREGULARITIES

Any changes other than the kinds considered in this section should be called to the attention of readers in the 1st affected issue, on the cover, at the head of the table of contents, and on the masthead page, with a description of the change and its expected duration.

When issues are combined or separated into parts, the issue number should carry in its various locations (cover, table of contents, masthead) clear identification representing the change. For example, 2 combined issues could be represented by "Volume 17, Issues 1 and 2", an issue split into 2 parts by "Volume 17, Part 1 of Issue 5" and "Volume 17, Part 2 of Issue 5".

JOURNALS IN TRANSLATION

28·28 A translated journal should, as far as possible, have the same characteristics as the original version (trim size, design, format, volume and issue numbers, frequency), particularly when the scientific contents are identical except for language. If a translated version represents only selected parts of the original, the title should be followed by a brief note indicating the issue or issues represented, for example "Contents selected from Volume 19, Issues 1 and 2".

The translated title should accurately represent the meaning of the original title and should be accompanied in its various locations by the original title in a subordinate size.

REPRINTS AND OFFPRINTS

28·29 Reprints and offprints should retain the trim size, format, and pagination of the original. Two or more articles combined in a single reprint may carry new, additional pagination that begins on the 1st right-hand text page with arabic numeral 1, but the original page numbers, within parentheses (round brackets), should accompany the new page numbers.

The usual bibliographic identifiers on the title pages of articles should also appear on the reprints. If a reprint of 1 or more articles is issued under a new cover, the cover should also carry full titles, author designations, and bibliographic identifiers.

ADVERTISING

28·30 The pagination for advertising pages should be in the same sequence as that of other pages also likely to be discarded when the journal is bound for library use. A system of pagination distinguishing these pages from the text pages can help librarians and binderies identify pages to be excluded from bound volumes; see section 28·10.

Advertising that might be confused with the scientific text of the journal should be refused, especially if it is formatted to resemble text pages.

POSTAL REQUIREMENTS

28·31 National postal services may have requirements for the periodic publication of notices concerning ownership, circulation, and other characteris-

Table 28·8 Title page of a scientific article: necessary elements (see sections 28·32–28·37)

Elements	Notes
Title	Terms describing the subject.
Author statement (byline)	All author names, in the sequence preferred by the authors or stipulated by the journal; surnames accompanied by given names (preferably) or all of their initials. See section 28·33.
Author affiliation	The institution (or institutions) that was the site of the reported research or study. Current affiliations and addresses can be provided in an appendix.
Abstract of the article	Abstract in the format and within the length limits specified by the journal.
Bibliographic reference	The reference by which the article should be identified in bibliographies published in the subject field of the journal.
Beginning text	Depending on availability of space.
Footnotes	May be needed for extensive group authorship or affiliation lists or to call attention to appended current affiliations and addresses.
Footline elements	Journal title, volume and issue numbers, copyright notice.
Initial page number	

tics of journals; the frequency and required location in the journal of such notices may also be specified. The postal service in the country of publication should be consulted for detailed information on such requirements.

JOURNAL ARTICLES

28·32 This section applies to aspects of format and style that are common to the principal kind of articles in scientific journals. Special additional considerations are needed specifically for research reports (see section 28·43) and for other kinds of articles (see section 28·44) such as reviews, editorials, letters to the editor, and book reviews.

TITLE PAGE

The 1st page of an article should carry all the information needed to enable a reader to rapidly identify the contents, author(s), and origin of the article and its location in the journal. These elements are summarized in Table 28·8.

Article Title

The title should be as informative as possible within the limit of length, which should be stipulated by the journal in its information-for-authors page. A title should be straightforwardly descriptive and eschew hyperbolic rhetoric. A title may be a declaratory statement if it does not imply a generalizability of the reported findings beyond what is supported by the described evidence. An interrogatory title can indicate the question considered, such as the subject of the research reported or the matter considered in an article like an editorial.

> Large-scale Eradication of Rabies in Southern Belgium with Recombinant Vaccinia–Rabies Vaccine
>
> [not "Large-scale Eradication of Rabies with Recombinant Vaccine"; readers may want to know that the campaign was regional, not national or international, and the specific kind of vaccine]
>
> The VASE Exon Downregulates the Neurite Growth-promoting Activity of NCAM 140
>
> Does Astronomy Really Need the Hubble Space Telescope?

The title should start with a word or term representing the most important aspect of the article, with the following terms in descending order of importance, if possible. Terms subordinate to the subject of the research, for example, a description of the research design, can be carried in a subtitle. Hyphenated terms should as much as possible be avoided or their hyphens removed (if the meaning is not thus distorted) because elements of hyphenated terms may not be detected by some computer search programs. Likewise, terms with possessive forms ("Crohn's disease") should be changed to nonpossessive forms ("Crohn disease").

> Rabies Eradicated in Southern Belgium with Recombinant Vaccinia–Rabies Vaccine
>
> [or "Rabies Eradicated with Recombinant Vaccinia–Rabies Vaccine in Southern Belgium"]
>
> Hypertension in Young-Adult Black Males Treated with Hydrochlorothiazide or Weight Reduction
>
> A Randomized, Controlled Trial [subtitle]
>
> [not "A Randomized, Controlled Trial of Treatment of Hypertension in Young-Adult Black Males with Hydrochlorothiazide or Weight Reduction"; move the study-design descriptor to a subtitle]

Standard terms from formal scientific nomenclatures should generally be preferred to common or nonstandard terms. If alternative or synonymous terms might be used by bibliographic searchers, authors should consider using both terms in the title; for example, titles with common names of plants or animals might include the proper taxonomic names, and titles with generic drug names might include the trade names if the research was concerned with specific products.

The North American Distribution of the House Fly (*Musca domestica*)
The Incidence of Adverse Effects from Phenytoin (Dilantin)

The conventions for typefaces appropriate to scientific nomenclature should also be applied in titles, for example, the species name *Musca domestica* in the example above.

Use of abbreviations in titles should be avoided, but when they are needed (as in reports of trials of multidrug treatments), they should be explained parenthetically in the abstract (if it appears on the title page. Widely accepted standard abbreviations and symbols (for example, "DNA", "pH") may be acceptable.

Sequential papers derived from the same study can be titled to indicate their commonality.

Predicted Atmospheric Warming: North America
Predicted Atmospheric Warming: South America

Also see section 28·25. Journals should discourage the serial titling of papers with no commonality except for their origin in a laboratory concerned with a general area of research.

Punctuation in titles should be kept to a minimum. Commas are acceptable for separating the elements of a series of related terms (see section 4·15); colons are acceptable for coupling main and subordinate elements in a title (see section 4·11). Semicolons and dashes should generally not be used.

For capitalization in titles, see sections 8·6–8·12.

Author Statement (Byline)

28·33 Only those persons qualifying for authorship by standards stated in the scientific community (ACS 1986; APA 1983; Huth 1986, 1988; ICMJE 1993) should be listed as authors. In general, these standards ensure that persons credited as authors can take public responsibility for the report through having participated in designing the research, carrying it out, and writing or revising drafts of the article. Journals can help ensure that author statements in their articles represent responsible authorship by requiring that putative authors sign statements that they can vouch for the authenticity and validity of their articles through adequate participation in the research and in reporting it.

For large-scale research involving many investigators, authorship can be indicated by a collective (corporate) title.

The Stanford–CERN Collaboration
The French Fullerene Development Group
The Chinese Ecologic Research Coalition
The Scottish Oncologic Study Network

For such articles, further information on participants is usually desirable; the minimum is the name and address of the investigator who will respond to inquiries about, or criticism of, the reported research. An appendix can list persons (and their affiliations) who were responsible for writing the paper and who signed the voucher described above (perhaps as a "Writing Committee" or "Principal Investigators"); additional listings can identify participants by function, such as "Detector Design", "Statistical Analysis", and so on. Alternatively, the journal can maintain in its editorial office the names and addresses of participants in the reported research and be prepared to require them to respond publicly in the journal to inquiries or criticisms.

Authors should be identified by at least 1 name (not just initials) in addition to the surname (family name) so as to reduce ambiguity in author names in indexing services. If space for names must be limited, 2 or more initials for names preceding the surname should be used if available.

The basis of the sequence of names should be stipulated by the journal, for example, alphabetic by surname, or the order of decreasing degrees of responsibility for the research and reporting it. The basis should be consistently applied within the journal and should be stated in its information-for-authors sheet. Note that surnames are the 1st element of a name in some cultures; in parts of the world where the surname is the last element of a personal name, the sequence in name elements should be consistent for all names. For alphabetization of names, see section 30·15; for the use of small-capital letters to indicate the surname in Chinese, Hungarian, and other names conventionally placing the family name 1st, see section 9·7.

A journal must establish its own policy on inclusion or exclusion of academic degrees and honorific designations with author names. The scientific standard that the validity and importance of research should stand on its evidence and not on the "authority" of authors suggests that academic degrees and honorific designations should not appear on scientific articles.

Author Affiliation and Site of Research

28·34 The institution or institutions in which the author or authors carried out the reported research or, for other kinds of articles, the intellectual work forming the basis for the article was completed should be identified under the author statement. For a large group of authors and multiple institutions, this information may have to be given in a footnote on the article's title page. Affiliations not directly related to reported research should not be included. In addition, the responsible institution(s) is (are) often identified in a line beginning with "From the . . ." when the authors' affiliations do not make that clear.

To enable interested persons in the scientific community to communicate with the authors, their affiliations, addresses, and other information (such as telephone numbers, E-mail addresses) should be provided, but this information could consume too much space on the title page and is more conveniently provided in an appendix at the end of the article. If it is thus placed, a footnote on the title page should indicate the location. A journal may choose to limit such information to that for a single author who agrees to be responsible for responding to all communications.

Abstract

28·35 The abstract should come after the author and affiliation statements and before the text. It should be no longer than the length limit stipulated in the journal's information-for-authors sheet. In some disciplines the major indexing services provide a maximum length by stipulating the length it will allow in its database or the length at which a longer abstract will be truncated.

In general, abstracts should be single paragraphs and carry no subheadings. An exception is the structured abstract developed by Haynes and associates (AHWGCAML 1987) and required by many journals reporting clinical trials and publishing review articles.

The content and sequence should accurately and objectively represent the text and include the major elements of the methods, findings, and conclusions. Abstracts of research reports should be informative (ANSI 1979), giving specific summaries of all elements of content. For reviews and other similarly long and wide-scope articles, abstracts may have to be indicative (ANSI 1979), simply sketching out the topics of the article and not summarizing evidence and conclusions. Abbreviations should not be used unless they are understood when standing alone (like "DNA", "pH", "USA"). Abstracts should not include bibliographic references or tabulated data.

Bibliographic Reference

28·36 The title page of an article should include the bibliographic reference by which the article should be represented in bibliographies. The format should be that specified by the journal for references in its articles; see Chapter 30 for recommended formats. The reference is conveniently placed at the end of the abstract. In an electronic journal it may be placed more conveniently for most readers at the top of the 1st display screen, above the title of the article. Square-bracketing the reference will indicate that it was added by the journal and was not provided by the author.

Text on the Title Page

28·37 The text will usually begin on the title page unless the title or author and affiliation statements are very long. The page design, layout, and typefaces should enable readers to distinguish readily between the abstract and the beginning of the text. Generally, figures and tables should not be placed on the title page lest they occupy too much space for clear presentation of the other elements.

The Footline

The title page should carry in its running foot or footline(s) the title of the journal, the volume and issue numbers, the issue date, and the copyright notice for the article.

TEXT

Text Subheadings

28·38 Readers scannng articles are helped by subheadings for parts of text that indicate their specific content. Subheadings can indicate what structural element of the article the section represents (for example, "Methods") or inform on the subject of the section (for example, "Antisepsis Antibodies"). Additionally, subheadings break up what would otherwise be long, uninterrupted blocks of type that give pages a "gray look", which is intimidating to readers.

Subheadings can represent different hierarchical levels of sections. First-level subheadings indicate the main divisions of text within the article. Second-level subheadings indicate divisions of subject matter within those main divisions, and so on. In general, a heading should be followed by at least 2 or more subheadings at the next level down, not by a single subheading.

Subheadings representing different hierarchical levels must be readily distinguished by readers. This need must govern the choices of typefaces and their styles, sizes, and weights. A designer or typographer should be consulted for choices. In general, the subheadings for all levels should be in, or closely related to, the typeface of the text or in a contrasting, but consistent, typeface (sans serif with serif-type text, for example). The hierarchy of subheadings can also be suggested by their placement: centered in the type column above the following text, flush with the left edge of a column and spaced above the following text, or flush with the left edge but running into the paragraph. A typical choice is illustrated below.

1st level: small capitals	METHODS Xxx xxxxx xxx xxxxxxx xxxx xxxxx xxx x xxxx xxxx.
2nd: capitals and lowercase	Chemical Analyses
3rd: italic small capitals	*CALCIUM*
4th: italic capitals and lowercase	*Ionized Calcium* Xx xxxx xxxx xxxxxxxx xx xxxx x. . .

Decisions on how many levels are needed should hinge on the length of sections. Relatively short sections of text may be structured clearly enough with paragraphing that represents topical divisions almost as well as low-level subheadings. In general, most journals will find 2 levels adequate.

Citations of References

For recommendations on the location and spacing of citations in text sentences and tables, see sections 30·11 and 30·12.

Footnotes and Endnotes

28·39 Although the placement of footnotes at the bottom of the pages carrying their citations may be convenient for readers, this format raises composition costs and is increasingly not used in scientific journals. It is still used in some books; for recommendations on footnotes and footnote signs, see sections 31·13 and 31·14.

The functions of footnotes can often be served in journal text with parenthetic statements when extended comment is not needed.

> The patients in Group 1 were treated with a sodium salt of panmalignomycin (supplied by Northeast Pharmaceuticals; Boston,

Massachusetts) and with prednisone; Group 2 patients were treated with a microsomal form of the potassium salt (supplied by Occidental Drugs, Incorporated; Seattle, Washington) and the same dosage form of prednisone.

Endnotes serve the same function as footnotes and are useful when extended parenthetic comment is needed but is too long or too frequent and would irritatingly interrupt the text. They can be placed at the end of the text under a heading such as "Notes". References in the text to the notes can be superscript lowercase letters, a device that avoids confusion with superscript numbers for text citations of references (see sections 31·13 and 31·14). The corresponding notes are identified in the "Notes" section with preceding on-the-line letters.

The patients were treated with a sodium salt of panmalignomycin[a] and with prednisone; Group 2 patients were treated with a microsomal form of the potassium salt[b] and the same dosage form of prednisone.

Notes

a. Northeast Pharmaceuticals; Boston, Massachusetts. Note that . . .

b. Occidental Drugs, Incorporated; Seattle, Washington. This form is also . . .

This style is likely to be applied more frequently with the development of electronic journals having display software that provides rapid hypertext jumps between text citations and the cited text (references, notes).

Some journals combine references and notes in a single section and cite both in the text with a single sequence of numbers. The journal *Science,* published by the American Association for the Advancement of Science, is a widely available example of this style.

Tables and Figures

28·40 Tables and figures in journal articles can generally be governed by the same considerations of style and format applicable to those in books; see Chapter 29. Note, however, that the larger dimensions of journal pages may tempt persons responsible for page layout to put too many tables or figures on individual pages, thus producing pages that may visually confuse the reader as to the sequence of text. A useful general principle is to place tables or figures at the top of pages and the text at the bottom. When an article has many figures, some may be combined with appropriate cropping into multipart figures that can be treated in layout as single figures.

NOTES, APPENDIXES, ADDENDA, AND ACKNOWLEDGMENTS

28·41 Various sections for functions subordinate to the text proper of an article are usually placed between the text and the reference section. A "notes" section is described above in section 28·39.

Appendixes

Appendix sections can be used for aspects of an article's subject that may be needed by some readers but are too long and detailed to be put in the text; they would interrupt the flow of information for many readers. Content of this kind includes long descriptions of unusual methods, long quotations from cited documents, questionnaires, and current addresses of authors. An appendix should be headed with a specific title, for example, "Appendix 1: Ethical Standards of Some Major Scientific Societies", not simply "Appendix 1".

Addenda

An addendum section is usually used for information an author needs to add late in the publication sequence, such as evidence from a just-published paper by another author, additional validating evidence from the author's own work, an overlooked reference, and so on.

Acknowledgments

An acknowledgment section can carry notices of permission to cite unpublished work, identification of grants and other kinds of financial support, and credits for contributions to the reported work that did not justify authorship.

REFERENCES AND BIBLIOGRAPHY

28·42 In scientific publication "references" usually means the bibliographic descriptors of documents cited in the text; "citations" is reserved for the indicators in the text pointing to the references (see section 30·1). The term "bibliography" can be reserved for references to works relevant to the subject of the article and possibly useful to the reader wishing to pursue its topic further; these are references not cited in the article. Another option is that used in this manual: a "References" section divided into 2 subsections, "Cited References" and "Additional References".

For the listing, sequencing, and formatting of references see Chapter 30.

SPECIAL FORMATS

Research Reports

28·43 A wide variety of formats may be needed for formal reports of research, and no single format can serve all possible needs. A journal covering a clearly defined field of science should, if possible, stipulate 1 or more specific formats most useful for that field. These formats should be summarized on the journal's information-for-authors page; see section 28·23 and Table 28·6. Nevertheless, some general principles can govern the structuring and formatting of reports in many fields.

Most scientific articles must have the structure of critical argument (Huth 1990): a question or hypothesis is posed, the evidence, pro and con, bearing on the answer is presented and assessed, and an answer is reached. Reports of scientific research generally follow this structure, but include additional elements. First, the means by which the investigator gathered his or her own evidence bearing on the question or hypothesis is described in detail sufficient to enable another investigator to replicate the research. Second, the evidence gathered by the investigator is presented separate from that available in the scientific literature.

These considerations have led to a format widely used in many scientific fields: division and sequencing of a report so that it explicitly appears with sections that can be headed sequentially Introduction, Methods, Results, and Discussion. These subheadings need not appear, or appear but not in this form; for example, a report can open with a paragraph serving the functions of an introduction but not headed with "Introduction". If the materials used in the research must be described in detail, the methods section can be titled "Methods and Materials". Further, a research report is more likely to be clearly understood if the research is described in the sequence in which it was conceived, designed, and carried out, its results analyzed, and its conclusions reached; this chronologic sequence corresponds to the sequence of critical argument. Therefore, whatever headings are assigned sequentially to sections within a report, they should have been selected with these principles applied. Table 28.9 indicates a typical sequence of headings and the functions served by the sections they head.

Table 28·9 Sections of a research report: typical headings and functions

Heading for section	Function of section; comments
Introduction	Describes the state of knowledge that gave rise to the question examined by, or the hypothesis posed for, the research. States the question (not necessarily as an explicit question) or hypothesis.
Methods and Materials	Describes the research design, the methods and materials used in the research (subjects, their selection, equipment, laboratory or field procedures), and how the findings were analyzed. Various disciplines have highly specific needs for such descriptions, and journals should specify what they expect to find in a methods section.
Results	Findings in the described research. Tables and figures supporting the text.
Discussion	Brief summary of the decisive findings and tentative conclusions. Examination of other evidence supporting or contradicting the tentative conclusions. Final answer. Consideration of generalizability of the answer. Implications for further research.
References	Sources of documents relevant to elements of the argument and describing methods and materials used.

Review Articles

28·44 Properly conceived review articles have the same basic structure of critical argument as a research report. Their subheadings should make clear the subtopics considered sequentially. Explicit description of the methods and standards applied in selecting the references cited helps review articles meet the same intellectual standards (Mulrow and others 1988) as research reports; such a description can be carried in a methods section near the beginning of the review or in an appendix.

Editorials

Because of their brevity, editorials usually do not need text subheadings even though their intellectual structure should have the elements of critical argument. Many journals do not indicate authorship of editorials. Those that do can choose whether to place the authorship statement (see section 28·33) under the title, to emphasize the authority of the author, or, for deemphasis, at the end of the text.

Letters to the Editor

Letters are usually published largely as submitted, hence with the typical structure of a letter. Each letter or group of letters on a single topic should be headed with a short title to clarify for scanners of a letters section what topics are covered. Many journals edit the salutations of letters to a standard form such as "To the Editor:". An editor should decide how much identification should be given the authors of letters: name only, name and academic degrees, name and postal address, or some other combination.

Book Reviews

The structure of a book-review section and its contents should be designed for the convenience of the reader. A section with many reviews might have them grouped by topic, with appropriate headings for groups. The bibliographic, price, and source information about the book should head a review. More useful for this bibliographic heading than the format for references is the sequence of title and then author; see section 30·16.

REFERENCES

Cited References

[ACS] American Chemical Society. 1986. ACS ethical guidelines to publication of chemical research. In: Dodd JS, editor. 1986. The ACS style guide: a manual for authors and editors. Washington: ACS. p 217–22.

[AHWGCAML] Ad Hoc Working Group for Critical Appraisal of the Medical Literature. 1987. A proposal for more informative abstracts of clinical articles. Ann Intern Med 106:598–604.

[ANSI] American National Standards Institute, Subcommittee 10 on Periodicals: Format and Arrangement. 1977. American national standard for periodicals: format and arrangement: ANSI Z39.1-1977. New York: ANSI.

[ANSI] American National Standards Institute. 1979. American national standard for writing abstracts: ANSI Z39.14-1979. New York: ANSI. Available from NISO Press Fulfillment, PO Box 338, Oxon Hill MD 20750-0338.

[ANSI–NISO] American National Standards Institute, National Information Standards Organization. 1992. Permanence of paper for publications and documents in libraries and archives: American national standard Z39.48-1992. New Brunswick (NJ): Transaction Publishers. Available from NISO Press Fulfillment, PO Box 338, Oxon Hill MD 20750-0338.

[APA] American Psychological Association. 1983. Publication manual of the American Psychological Association. 3rd ed. Washington: APA. Chapter 1, Content and organization of a manuscript, Characteristics of authorship and articles; p 20.

Huth EJ. 1986. Guidelines on authorship of medical papers. Ann Intern Med 104:269–74.

Huth EJ. 1988. Scientific authorship and publication: process, standards, problems, suggestions. Washington: Inst of Medicine.

Huth EJ. 1990. How to write and publish papers in the medical sciences. 2nd ed. Baltimore: Williams & Wilkins. Chapter 4, Critical argument and the structure of scientific papers; p 55–8.

[ICMJE] International Committee of Medical Journal Editors. 1993. Uniform requirements for manuscripts submitted to biomedical journals. JAMA 269:2282–6.

[ISI] Institute for Scientific Information. 1983. ISI checklist for journal editors and publishers. Philadelphia: ISI Pr.

[ISO] International Organization for Standardization. 1986. Documentation — international standard serial numbering (ISSN): ISO 3297-1986 (E). Geneva: ISO.

[ISO] International Organization for Standardization. 1988. Documentation and information: ISO standards handbook 1. 3rd ed. Geneva: ISO.

Mulrow C, Thacker S, Pugh J. 1988. A proposal for more informative abstracts of review articles. Ann Intern Med 108:613–5.

[NISO] National Information Standards Organization. 1985. Preparation of scientific papers for written or oral presentation: Z39.16(R1985). New Brunswick (NJ): Transaction Publishers. Available from NISO Press Fulfillment, PO Box 338, Oxon Hill MD 20750-0338.

[NISO] National Information Standards Organization. 1990. Printed information on spines: Z39.41-1990. New Brunswick (NJ): Transaction Publishers. Available from NISO Press Fulfillment, PO Box 338, Oxon Hill MD 20750-0338.

[NISO] National Information Standards Organization. 1992. International standard serial numbering (ISSN): Z39.9-1992. New Brunswick (NJ): Transaction Publishers. Available from NISO Press Fulfillment, PO Box 338, Oxon Hill MD 20750-0338.

Additional Reference

O'Connor M. 1986. How to copyedit scientific books and journals. Baltimore: Williams & Wilkins.

29 Books, Technical Reports, and Monographs

> Books must follow sciences, and not sciences books.
>
> —Francis Bacon, *A Proposal for Amending the Laws of England,* 1616

> A well-turned-out book has a serenity which hides every vestige of the disarray which has so often marked its gestation. The number of things that can go wrong in the making of a book is so great, it is small wonder that none has ever been produced entirely free of fault.
>
> —Brooke Crutchley, *To Be a Printer,* 1980

In scientific publishing, books, technical reports, and monographs generally share their main characteristics. Typically they have 3 sections: the pages of the opening section, usually called "front matter" or "preliminary pages", the main body of the publication or "text" (also "subject matter"), and accessory closing pages, usually called "back matter" (Tables 29·1–29·3).

A technical report differs from a book in that it is designed to convey results of basic or applied research and may combine book and article formats. Technical reports are often of book length and complete unto themselves; they usually include, however, an abstract; methods, results, and discussion sections; conclusions; and references. The term "monograph" is often used for publications on particular and relatively narrow subjects; such publications usually have more the character of a book than of a long scientific article.

Table 29·1 Components of books: the front matter

Component	Page location and number	Notes
Page for half-title of the book	recto, i	See section 29·2.
Page facing half-title page		See section 29·2
Verso of half-title page	verso, ii	Various uses; see section 29·2
Title page	recto, iii	Required; may be pages ii and iii, see section 29·3
Copyright page	verso, iv	See sections 29·4–29·8.
Copyright notice		Required; see section 29·4.
History of publication		See section 29·5
Cataloguing-in-Publication		See section 29·6
International Standard Book Number		Required; see section 29·7
Acknowledgments and Permissions		See section 29·8
Abstract		See section 29·8
Printer		See section 29·8
Dedication	recto, v	See section 29·9
Contents list	recto (initial page), v or vii	Required; see section 29·10
Errata		See section 29·11
Illustrations list; tables list		See section 29·12
Abbreviations list		See section 29·12
Foreword	recto	See section 29·13
Preface	recto	See section 29·13
Miscellaneous front matter		See section 29·14
Limited editions		
Contributors		
Chronology		

Table 29·2 Components of books: the text

Component	Initial page and number	Notes
Introduction [optional]	[1]	See section 29·15
Epigraph [optional]		See sections 29·9, 29·15
Second half-title [optional]		See section 29·15
Parts [optional]		See section 29·16
Chapter 1	recto, 1 or 3	See section 29·17
Chapters		
Title pages	recto [preferably]	See section 29·17
Text pages		See sections 29·17–29·19
Subheadings		See section 29·17
Section numbering		See section 29·18
Running heads		See section 29·19

Table 29·3 Components of books: the back matter

Component	Preferable initial page[a]	Notes
Addendum	recto	See section 29·20
Appendixes	recto	See section 29·20
Notes	recto	See section 29·20
Glossary	recto	See section 29·20
Bibliography or references	recto	See section 29·21
Index or indexes	recto	See section 29·22
Colophon page	verso	See section 29·22

[a]For economy of paper, these sections (except for the colophon page) may start on verso, rather than recto, pages.

BOOKS

29·1 There is no single formal standard for the structure and format of books. In the United States, *The Chicago Manual of Style* (UCP 1993) has become, because of its detailed description of conventional practice in book layout and formats, the *de facto* standard. The legalities and customs of publishing in Canada are described in *The Publisher's Path* (NLC 1989). British practices are well reflected in Chapters 7–9 of *Copy-editing: The Cambridge Handbook for Editors, Authors and Publishers* (Butcher 1992).

Many individual standards published by the International Organization for Standardization (ISO standards) are relevant to aspects of book formatting and have been gathered into *Documentation and Information: ISO Standards Handbook 1* (ISO 1988). Some of the standards issued by the National Information Standards Organization are also relevant, particularly for the United States, and are cited in this chapter.

TERMINOLOGY

Book publishers and printers use many technical terms, some of which should be known by authors as well as by editors. Among the most widely used are "recto", meaning a right-hand page, and "verso", meaning a left-hand page. A printed page number is called a "folio" (which has, however, additional meanings); a page number at the bottom of the page is a "drop folio". Definitions of additional technical terms can be found in glossaries in the Cambridge and Chicago manuals (Butcher 1992, UCP 1993).

BOOK SECTIONS: FRONT MATTER

29·2 The usual components of the front matter (the "preliminary" pages of the book) and their usual sequence are listed in Table 29·1. Other sequences may be suitable. The components highly likely to be used by most readers should start on rectos (right-hand pages), but this rule can be ignored for some components if economical use of pages is a necessity. The pages of the front matter are usually given lowercase roman-numeral folios (page numbers). If the front matter begins with a half-title page, the folio for that page is "i"; if the 1st page is the title page, it instead is designated "i" (but these numerals are usually not printed on the half-title page, the verso of the half-title page, and the title page).

The Half-title and the Facing Page

The right-hand page (recto) preceding the title page is usually used for the half-title (also "bastard title" or "false title"). The half-title page, an artifact from when books were printed without covers, may be eliminated for economy of pages. Usually the page carries only the main title with no subtitle and no author's name. It sometimes carries an epigraph or series name. The half-title should be the same as that on a half-title page placed to immediately precede the text. If the book is divided into parts or units, each may be preceded by a half-title page bearing the title of the part.

The verso (left-hand page) facing the half-title page is usually blank but sometimes is used to carry an "ad card". An "ad card" (also known as "the book card", "card page", "face title") lists other books by the author, usually with phrasing such as "John Jones is also the author of . . ." or "Other books by John Jones are . . .". Alternatively, this page can carry the title and list of the series in which the book is published.

Frontispiece

A frontispiece, an illustration setting the tone of the book by representing an important aspect of the book (for example, a portrait of the subject of a biography), faces the title page.

A book without such an illustration may use this verso page facing the title page for information of the kind otherwise carried on the verso page facing the half-title page (other books by the author; other books in the series represented by this book).

Title Page

29·3 The title page presents the full title of the book. This page should be designed before the rest of the front matter, and it and all other preliminary pages should harmonize with the rest of the book. The full title should follow conventions of title capitalization and scientific conventions such as those for symbolization and italicization ; see sections 8·6–8·12. The title and the subtitle (if there is one) should be clearly delineated by consistent type style and size, any needed punctuation, and the layout of the page.

> CBE
>
> Style Manual for Authors and Editors
>
> [Is the correct title *CBE Style Manual for Authors and Editors* or just *Style Manual for Authors and Editors*?]

Using capital and lowercase letters will make proper capitalization of the title clear to readers not familiar with the conventions of capitalization. A subtitle should be below the main title and be in a smaller type size; a superior position can suggest that the subtitle is the main title or part of it.

The page should include not only the full title of the book but also the names of all authors or editors, the publisher's name and location (city or cities), and the year of publication. The title page may also carry additional elements: the name of a translator, editor, illustrator, author of the foreword or introduction, and a colophon (publisher's trademark or graphic device used as an embellishment). If the book has been revised or represents a new edition, the title page should carry the number of the edition.

> CBE Style Manual
>
> 3rd edition, revised and enlarged

The arrangement of elements depends on the book. Usually the title comes 1st, but if the author is famous, his or her name may be placed above the title. In general, the proper sequence of elements from the top down is as follows: title, subtitle, the author or editor statement, the edition identifier, the publisher's name, the publisher's location, the year of publication. In scholarly monographs, the author (or editor) statement may include a short statement of the author's academic or research affiliation.

Occasionally a book designer may want to add graphic interest and specificity to a book's title page by designing an expansive title "spread" that spans both pages ii and iii. If so, the design should, preferably, carry the "imprint" information (publisher, place of publication, and date of publication) on page iii.

Copyright Page

29·4 The verso of the title page is often called "the copyright page" because it is usually the location of the copyright notice. This page is also used to carry additional detailed information that identifies the book more fully than its title page.

COPYRIGHT NOTICE, UNITED STATES

Although the copyright notice is placed on the copyright page (the verso following the title page), it may be on the title page itself. The pre-1976 US copyright law specified that the notice appear on the title page or the page immediately following. This is no longer required, but custom prevails. The copyright notice must consist of the word "Copyright" or the abbreviation "Copr" or the symbol ©, accompanied by the name of the copyright holder and the year of publication. The order of the 3 elements is not important. The copyright symbol is not appropriate for books published before 1955 or for copyright renewals.

Each subsequent edition (but not reprintings of a particular edition) must be copyrighted. If the copyright owner is the same, only the new date is added. A new holder of the copyright must be represented.

> Copyright Joan Author 1980. All rights reserved [or] © 1980 by Joan Author
>
> © 1980 1985 1990 Joan Author
>
> Copyright Estate of Joan Author 1985. All rights reserved

If the actual year of publication is more than 1 year before the stated date, the correct date must be stated when the copyright is registered.

If the book is published under a new and different title, the previously published title must appear on the copyright page as well as on the front flap of the jacket and must be mentioned in catalogues and circulars (in compliance with Federal Trade Commission regulations). Adding "All rights reserved" gives copyright protection in most South American countries, under the Buenos Aires Convention. Adding "published simultaneously in Canada by . . ." followed by the publisher's name gives copyright protection, under the Berne Convention.

If the book was published after 1 January 1978, the term of the copyright is for the author's life plus 50 years. If it was published before 1978, the 1st term covers 28 years but is renewable for an additional 47 years. Information on securing copyright on a book or manuscript in the United

States can be obtained from the Register of Copyright, Copyright Office, Library of Congress, Washington DC 20559.

For additional details on copyright, consult Chapter 4, "Rights and Permissions", *The Chicago Manual of Style* (UCP 1993), and section 3.7, "Copyright permissions and acknowledgements", *Copy-editing* . . . (Butcher 1992).

COPYRIGHT NOTICE, CANADA

Copyright in Canada is automatically acquired at the creation of an original work. Under the provisions of the Canadian Copyright Act, it is not necessary to indicate on the work that the author holds copyright. To retain copyright protection from the time of 1st publication, under the provisions of the Universal Copyright Convention in countries that require registration (the United States, for example) all copies of the author's work must be marked (see above) (CCAC 1988).

HISTORY OF PUBLICATION

29·5 The copyright notice is often followed by statements representing the year of 1st publication, the years of subsequent editions, and year of each impression. For further details consult Chapter 1 of *The Chicago Manual of Style* (UCP 1993).

CATALOGUING-IN-PUBLICATION (CIP)

29·6 Cataloguing-in-Publication (CIP) is a prepublication cataloguing program established in 1971 by the Library of Congress to enable libraries to catalogue books immediately on their arrival. CIP data are also distributed to the Library of Congress' MARC (**Ma**chine **R**eadable **C**ataloguing) tape subscribers for use in selecting titles for purchase.

As soon as the final title has been decided on, the contents page completed, and the ISBN assigned (see section 29·7), the CIP form provided by the Library of Congress should be filled out and sent with a copy of the galley proof or enough material from the manuscript for the Library to catalogue the book properly to the following address: Descriptive Cataloguing Division, Library of Congress, Washington DC 20540. The Library needs about a month to catalogue the book. The CIP data should appear on the copyright page line-for-line as it appears on the cataloguing card.

The National Library of Canada does not have requirements for what copyright and other publication data must appear in Canadian publications. The Library does strongly recommend that an ISBN (see section 29·7) and a CIP statement appear. Canada's sole legal requirement of publishers (under the National Library Act) is that 2 copies of any book published in Canada, or manufactured wholly or partly in Canada, be sent to the National Library within 1 week of publication. This action is called legal deposit (NLC 1988).

INTERNATIONAL STANDARD BOOK NUMBER (ISBN)

29·7 The International Standard Book Number (ISBN) uniquely identifies a book and thus facilitates handling orders and keeping track of inventory by computer. The identifying series of 10 numbers should appear on the jacket and at the foot of the outside back cover as well as on the copyright page. The 1st number indicates the country; the next group, the publisher; the next, the title; and the last number is a computer-assigned "check digit" that ensures the accuracy of the preceding numerals.

ISBN 0-87044-763-7

[a book published in the United States, "0", by the National Geographic Society, "87044"; the "763" identifies the title; the "7" is the check digit]

An ISBN identifies 1 title, or edition of a title, from 1 publisher and is unique to that title or edition. If the book is published in paperback and in clothbound editions, a separate ISBN is assigned to each. When a work is published in 2 or more volumes, a separate ISBN is assigned to each volume unless the work is sold only as a set; nevertheless, the ISBN should be printed on the copyright page of each volume.

In the United States, the ISBN Agency is administered by the RR Bowker Company. Bowker gives each publisher a block of numbers, and the publisher assigns each book a number from that block. For more information, contact Standard Book Number Agency, RR Bowker Co, 1180 Avenue of the Americas, New York NY 10036.

COUNTRY IN WHICH PRINTED

A required statement of the country in which the book was printed is conventionally placed on the copyright page.

Printed in the United States of America   Printed in Canada

PERMANENCE OF PAPER STATEMENT

When the paper used in a book meets the standards for the permanence of paper set forth in *Permanence of Paper for Publications and Documents in Libraries and Archives* (NISO 1992), this fact should be represented on the copyright page by the statement, "This paper meets the requirements of ANSI/NISO Z39.48-1992 (Permanence of Paper)". See Table 28·5 for the compliance symbol that should appear with this statement.

ACKNOWLEDGMENTS AND PERMISSIONS

29·8 An author's recognition of persons, groups, and contributions required to produce the book is stated in the acknowledgment. An acknowledgment that involves copyright may appear on the copyright page, as is necessary for anthologies; if the acknowledgment is too long for 1 page, it should begin on the copyright page and run to the next page.

Permission is clearance from a copyright owner for the author to quote passages or reproduce illustrations from another publication. When permissions are extensive, it is sometimes desirable to include them on the copyright page, but they should not duplicate acknowledgments given elsewhere. Photo credits may also appear here instead of with the illustrations.

ABSTRACT

Some nonfiction books on narrow topics not accurately reflected by the book's title can usefully carry an abstract on the verso of the title page or on the right-hand page following it. If separate abstracts are supplied for chapters, they should appear on the page preceding the start of a chapter or on its opening page.

PRINTER

The name of the compositor, printer, or binder, or any combination thereof, may be included on the verso of the title page.

ADDITIONAL INFORMATION ABOUT THE COPYRIGHT PAGE

Additional information on appropriate contents of the copyright page can be found in *The Chicago Manual of Style* (UCP 1993). Information on contents for the United Kingdom can be found in Butcher (1992).

Dedication

29·9 Whether a book will carry a dedication, an inscription to honor or compliment a patron, relative, or friend, and what it will say are the author's choices; brevity should be encouraged. The dedication usually faces the copyright page, that is, on the 1st right-hand page following the title page, but sometimes it is on the copyright page or the 1st recto page following the contents list. If it is on a recto page, it should be followed by a blank verso page.

Contents List (Table of Contents)

29·10 The list of contents sets forth the titles (usually only the chapter titles) representing the content of the book. It simplifies finding particular materials in the text and back matter and clarifies the organization of the book. If the number of subheadings for chapters is not too great and their listing would assist readers in finding needed content rapidly (as in a reference book), they can be given under the chapter titles; the design of the contents list should, then, be such as to enable readers to distinguish readily chapter titles from subheadings.

The numbers of beginning pages of chapters (and sections of chapters if these are also listed) are conveniently indicated immediately after the chapter (and section) titles. Titles of parts of the book (chapter groups) are not, however, followed by beginning-page numbers. Formerly titles were always placed flush left and the page numbers flush right, but now that leaders (dots leading from the title to the page number) have gone out of vogue, a closer placement of titles and page numbers allows the reader's eye to quickly pick out the appropriate page number. Traditionally the contents list has followed all front matter except an introduction, but if the book contains a large amount of front matter, the list should appear as early as possible and list everything that follows. The contents list contains no reference to items that precede it. This section can be simply titled "Contents". Its ease of use will depend greatly on the qualities of typography and layout established by the designer of the book.

Note that European practice often places the contents list at the end of the book.

Errata

29·11 If a list of errors in the book is needed, placing it immediately after the contents list will increase the probability that it will be seen by readers. If the book has already been bound when the errors are discovered, a separate leaf may be tipped in there. Each erratum notice should give the page number, line number, error, and correction.

> page 85, line 8: For "Einstein's theory" read "Einstein's wife's theory".

This section reports only substantive, never simply typographic, errors.

Lists of Illustrations and Tables

29·12 Decisions on whether to list illustrations and tables and their page numbers will depend on the importance of directing readers to them from the front matter rather than simply from the text for which they are relevant. The front matter of a heavily illustrated book might be burdened by long lists. But a reference book in which tables are major sources of information could helpfully list its tables. If the illustrations are on unfolioed (unnumbered) inserts, their location should be identified with phrasing such as "facing page . . ." or "following page . . .". If tables are listed, the list comes after the list of illustrations; table titles may be shortened for the listing.

Both kinds of lists should either be a continuation of the contents list and in harmony with its design or start on recto pages immediately following the contents list.

Foreword

29·13 The foreword is often of a scope similar to that of an author's preface but is by someone other than the author. It often serves to place the book in the context of the related literature and to point out merits of the book as seen by an authority other than the author (or editor). It typically runs to no more than 2 pages and ends with its author's name. Note that "Foreword" is not "Foreward" or "Forward".

Preface

The preface is the author's comment about the book as a whole and describes its purpose, sources, and extent. It could be a background note on why the author wrote the book or on the process of writing. It differs from the introduction in that the introduction discusses the text and prepares the reader for the contents or explains it. Customarily the author ends by acknowledging indebtedness to other authorities or to persons who helped write or proof the book or who contributed information. The preface is not part of the text and is numbered with the appropriate sequence of lowercase roman numerals. An editor's preface, if used, usually precedes the author's preface. The current preface is placed before prefaces to any previous editions.

Miscellaneous Front Matter

LIMITED EDITIONS

29·14 A limited edition is a book of which a stated number of copies are printed. The certificate of limited edition (also known as the limit page or limit notice) is an announcement of the number of copies printed and the number of the particular copy.

> This edition is limited to 500 copies of which this is number . . .

The number is written in by hand and may be accompanied by the signature of the author, printer, or publisher. The location of this notice is selected by the publisher. It often appears on the copyright page, the colophon page (the very last printed page), or on the recto of a special page before the title page.

CONTRIBUTORS

For multi-authored books in which only the editor's name appears on the title page, it is appropriate to list the contributors with statements of their affiliations and other relevant information following the contents list. If each author is given a biographical note, the notes are placed at the end of the book. An alternative is to list the contributors' names only in the contents and give a note about each author at the beginning of the appropriate chapter.

CHRONOLOGY

For some books, a list of events important to the text is helpful. Such information might alternatively appear on an endpaper.

BOOK SECTIONS: TEXT

29·15 The 1st page of the text begins the sequence of pagination with arabic numerals. If this 1st page does not carry substantive text, the folio (page number) is omitted and the 1st folio appears on the 1st substantive text page.

[for example]	Part 1	[is on page 1 but the folio is omitted]
	Chapter 1	[begins with the 3rd page and carries the folio "3"]

In general, components of the text (parts and chapters) should begin on recto pages.

Although the text section of many books simply starts with the 1st chapter, which should begin on a recto page, some books may carry several types of content preceding the actual text but on text-numbered pages.

Introduction

The introduction, part of the text, is usually set like the text. It defines the limits and organization of the work and states the purpose, scope, and author's approach. Sometimes it is signed by its author, but if the author of the book also wrote the introduction, the signature is not necessary. This manual carries 2 introductory chapters, and they make up "Part 1 Introduction"; the 1st defines the purpose, scope, and design of the manual, and the 2nd gives a historical background. The manual proper begins with Chapter 3.

Epigraph

The epigraph, a quotation intended to convey a theme or tone developed in the text that follows, could appear on the title page or on the back of the dedication. It could replace the 2nd half-title or be on its verso and face the 1st page of text. Its source need not be documented beyond the name of the quoted author and, perhaps, the title of the quoted work.

> When a man knows he's to be hanged in a fortnight it concentrates his mind wonderfully.
> —Ben Jonson

Epigraphs may be used for the same purpose for individual chapters; these should appear on chapter title pages between the title and the chapter's text.

Second Half-Title

The 2nd half-title (if repeated) marks the end of the front matter and the beginning of the text. If a 2nd half-title page is used, it is usually the 1st page designated with arabic numerals, the preceding pages being numbered in lowercase roman numerals. Note, however, that the folio for this page may not be used and only implied by placing a folio only on the 1st page of the text proper.

The 2nd half-title may be a repetition of the 1st half-title. If the text of the book is divided into parts, a 2nd half-title page is not needed; its function has been performed by the 1st part-title page.

Parts

29·16 Chapters may be logically grouped into numbered parts (or sections). The part number and title appear on a right-hand page preceding the part; its verso is usually blank. Chapters within parts are numbered consecutively throughout the book and do not begin with number 1 in each part. Each part may have an introduction. Part numbers should be in arabic numerals.

Chapters

29·17 Most prose books are divided into chapters, not necessarily of similar length. Each usually starts on a new page, preferably recto (right-hand page), but the verso (left-hand) location can be used for economical use of pages. Note that beginning a chapter on a recto page reduces costs for offprints because the printer need not reimpose pages to print the offprints of each chapter.

A chapter should begin with a chapter number and title (chapter head) placed above the text at its beginning. The opening page carries a drop folio (page number at the bottom of the page) and no running head (short title printed at the top of the page). In a multiauthored book, each author's affiliation or other identification is usually given in a footnote, if not in a list of contributors in the front matter. If a source reference is also used, the author's identification precedes it.

Abstracts of chapters are not commonly used. If they are needed, they should appear at the head of the chapter beneath its title (and author statement, if one is needed). Alternatively an abstract could be placed on the verso page facing the recto page representing the chapter's 1st page.

SUBHEADINGS

As an aid to the reader, chapters may be subdivided with subheadings, secondary subheadings, and tertiary subheadings. They should be succinct and meaningful, accurately representing the sequence and, with secondary and lower subheadings, hierarchies of content. Subheadings are usually set on a line separate from the 1st line of the section. A designer may have to place the lowest level subheading on the 1st line of a section's text.

The levels of subheadings should be made clear by appropriate choices of type styles; the design should make them clearly distinguishable from the text proper. Subheadings should not contain citations of references or notes.

The 1st sentence after a subheading should be complete unto itself and not assume that the reader has read the subheading.

Subheadings	[not] Subheadings
Subheadings should not contain references . . .	They should not contain references . . .

SECTION NUMBERING

29·18 Sections and subsections of chapters may be numbered to help guide readers quickly to parts of the text. This device is especially useful in a reference book, like this manual, unlikely to be read sequentially but used mainly to get answers to specific questions. A common method is double numeration: the number of a section consists of the number of the chapter, a demarcation point (in this manual a raised period applied to avoid using a pseudodecimal point; see section 4·4, item 6, and section 4·6, item 7), and the number of the section within the chapter. For example, section 4·8 is the 8th section of chapter 4; 4·12 is the 12th section of chapter 4.

Another method ignores chapter numbers: the primary subheadings are numbered consecutively thoughout the book, the secondary subheadings are numbered consecutively under each primary subheading, and the tertiary subheadings are numbered consecutively under each secondary subheading. Thus, 25·10·7 might appear in chapter 6.

For books such as instruction manuals, a 3rd method is to use the chapter number followed by a demarcation point and a paragraph number.

RUNNING HEADS (OR RUNNING FEET)

29·19 Running heads are headings across the tops of the pages (and running feet, across pages below the text) intended to indicate content. Often the title of the book appears on the left-hand pages and a short title for the chapter on the right-hand pages. They are discretionary and may be omitted where they serve no purpose. Running heads are omitted on display pages (that is, chapter title pages, pages of illustrations).

An alternative scheme is a running head for the chapter at the top of verso pages and running heads for sections of chapters at the top of recto pages.

FOOTNOTES

See "Notes", section 29·20.

BOOK SECTIONS: BACK MATTER

29·20 The back matter (also "reference matter") constitutes the pages following the main text of the book. It continues with arabic numerals and includes any of the components described below; the sequence is generally that shown here.

Addendum

Brief supplemental data that became available too late for inclusion in text may be added at the end of the book in a section placed immediately after the text and headed "Addendum". Such a section should be listed in the book's "Contents". In general, use of such a section should be avoided; its contents are likely to be overlooked by readers as they go through the text.

Appendixes

Appendixes (sometimes called "annexes") can serve to carry supplemental material that illustrates, enlarges on, or otherwise supports the text without distracting readers from the main line of the text's exposition:

letters, lists, tables, documents, forms, speeches, detailed protocols, questionnaires. The 1st appendix begins on a recto page and is preceded by a half-title ("Appendix" or "Appendixes" as appropriate) if space permits.

An appendix may be placed at the end of its respective chapter, especially in a multiauthored book, if it is needed for that chapter and not for the others. This also applies where offprinting is anticipated and each author's appendix will appear with the chapter to which it relates. When the appendix follows its chapter, it may begin on either a left- or a right-hand page, or it may run on at the end of the chapter's text.

A supplement is more extensive and may be issued at a later date than the text. If 2 or more appendixes are carried, they are numbered or lettered consecutively.

Notes

Notes consist of references and other information of interest to scholars and researchers. They usually follow the chapter to which they refer, but if they are placed at the end of the book they should be grouped by chapter. Endnotes are usually set 1 or 2 points smaller than the text. The numbering of notes usually begins anew with "1" with each chapter.

Footnotes on text pages should be avoided. They increase the difficulty and cost of page make-up, reduce text areas on facing pages to uneven depths, disarrange the orderly appearance of the page, and may interrupt the flow of reading.

Glossary

A glossary is a dictionary of terms and concepts used in the book. It usually precedes the bibliography. Sometimes each entry is capitalized, but capitalizing all entries eliminates the distinction between proper and other nouns.

Bibliography or References

29·21 The bibliography lists works on the subject of the book and may comprise mainly works used to write the book. If they comprise only the works the author cites in the text, they are designated "references". See Chapter 30 for forms of citations and arrangement of the elements of references.

If individual chapters are to be made available as offprints, the references should preferably appear at the end of the chapter in which cited. The same choice may be desirable when large numbers of references are needed and the reader will find them more readily at the end of the relevant chapter than in a long compilation in the back matter.

Index

29·22 An index is an alphabetic list following the text that cites names, places, and topics discussed in the book and gives the page number on which the discussion occurs. When section numbers are used in chapters, they may be used rather than page numbers so as to direct the reader more quickly to the contents being sought.

Long books citing large numbers of authors may preferably carry separate author and subject indexes. With this division, the author index precedes the subject index.

The index can be set in smaller type than the text, in 2 or more columns. Main headings should be clearly distinguished from subheadings. Note the importance of capitalizing only those entry terms that are properly capitalized as proper nouns; see section 31·36. See sections 31·28–31·37 for additional details of format and style for indexes.

Colophon Page

The publisher's name and location is the imprint and should appear on the title page; see section 29·3. The publisher's trademark is the colophon ("finishing touch"). Sometimes a paragraph of information regarding the book's design and manufacture is included here. The imprint and trademark are on the title page; the descriptive colophon (if there is one) is on the last page (a recto or verso page following the last page of the book, in scholarly books the last page of the index). Lately, the title page and the copyright page have taken the place of the colophon in recording this information. Now the colophon is being revived as a place to record typographic details of finely made books, with notes on the designer, the typeface, and the paper. Also frequently, but incorrectly, it is used for the trade emblem or device of a printer or publisher (required by British law, but not American, to appear in the book).

If the book is a limited edition, the limit notice with number of copies produced may appear on the colophon page.

PAGINATION

29·23 All text pages are numbered consecutively whether the folio (page number) is expressed or not. The most common location for the page number is at the top of the page and flush with the outside margin. If it is at the bottom of the page, it is called a drop folio. The opening page of each

Table 29·4 Components of books: the externals

Component	Notes
Cover	See section 29·24
Spine	See section 29·24
Endpapers	See section 29·25
Jacket	See section 29·26

chapter carries a drop folio. The outer location (top or the "drop folio" location) is to be preferred to centered locations because of its being more readily seen by scanners of a book.

The front matter (everything preceding the 2nd half-title page) is numbered with lowercase roman numerals. The drop folio location is preferred.

Folios are not expressed on pages with only illustrations or tables.

In multi-volume works, pagination may be either consecutive throughout the volumes, or may begin anew with each volume.

COVER

Spine

29·24 The top 3-quarters of the spine, as the book sits on a shelf, comprises the bibliographic identification area; and the remaining quarter is shared by the library identification area (location for the library classification number) above the publisher identification area. If the book is too thin for horizontal type, the title should read down the spine (top to bottom), that is, from left to right when the cover lies face up. (See Table 29·4.)

Bibliographic identification may include the book's title; author(s) or editor(s); series title, volume, or part number. The publisher's identification area includes the publisher's name, logo, or other information.

The extent to which this information can be carried will depend on the size of the book, which determines the height and width of the spine.

Endpapers

29·25 The endpapers (the folded paper half of which is pasted to the inside cover of a hardcover book and whose other half is adjacent to the 1st or last leaf of the book) are convenient places for maps, chronologies, pronounciation tables, and other information that will be consulted frequently. Bear in mind that a paperback book will have no endpapers.

Jacket

29·26 Information that must appear on the back cover and jacket is the 4-part, 10-character ISBN (International Standard Book Number), which also appears on the copyright page; see section 29·7.

The jacket (or dust-wrapper) fits around the cover. Its original purpose was to protect the book before sale, but it has become an important advertisement for the book, and designers now invest considerable effort on it. Editors should be concerned with the accuracy of all the text on the jacket. The left inside flap of the jacket (the inside front of the jacket) may contain a summary of the book; the right flap (the inside back of the jacket) usually has a brief biography of the author, perhaps with a photograph.

TECHNICAL REPORTS AND MONOGRAPHS

Technical reports and monographs are scientific or scholarly articles in book form. The following recommendations are based on ANSI (1987) recommendations, with particular attention to the elements that differ from those in a book. Most of the components of a technical report are akin to those of books (see Table 29·5), but there are some special needs, which are detailed below. The preceding sections on books are relevant to the components that reports and books have in common.

FRONT MATTER

Cover

29·27 The cover identifies the subject of the report and indicates whether it contains classified or proprietary information. The relevant information includes the following elements.

1 Report number, a unique number placed in the upper left-hand corner.
2 Report title and subtitle.
3 Author(s), principal investigator(s), editor(s), or compiler(s).
4 Performing organization (author affiliation): the organization responsible for the research.
5 Publication date.
6 Type of report and period covered, if applicable.
7 Contract or grant number, if applicable.
8 Sponsoring or issuing organization (if different from the performing

Table 29.5 Components of technical reports; see section 29·27

Component	Initial page and number
Cover and front matter	recto, i [see relevant sections for books: 29·2–29·14]
Notices	
Abstract	recto, iii
Contents list	recto, v
Figures list	recto
Tables list	recto
Foreword	recto
Preface and acknowledgments	recto
Lists of symbols, abbreviations	recto
Text	
Summary	recto, 1
Introduction	recto
Methods, assumptions, and procedures	
Results and discussion	
Recommendations	
References	
Back matter	[see relevant sections for books: 29·20–29·23]
Appendixes	recto
Bibliography	recto
Glossary	recto
Index	recto
Distribution list	
Externals: the cover	

organization): the organization that funded the research and usually controls report publication and distribution.

Depending on the quality of paper and the cover color and design, the information specified above may be readily carried on the cover. Some technical reports are bound with a more durable cover with a design and color that preclude its carrying such detailed information; in such cases the information designated as items 1–6 should be carried on the 1st recto page, which would then correspond to the title page of a book.

Notices

Special notices on the cover or its verso call attention to particular conditions for the report: its security classification, restricted distribution, or proprietary status; that findings are preliminary; that the report is a draft or working paper intended to elicit comments and ideas; that the report is a preprint of a paper to be presented at a professional meeting; or legal conditions, for instance, use of brand or trade names. If the report is copyrighted (an unusual event), the notice appears here. Disclaimers should be avoided, but if necessary, they go here.

Abstract

The abstract states in approximately 200 words the purpose, scope, and major findings of the report. It contains no references or illustrations and can be understood when seen freestanding. The abstract should be on the next recto page following the front cover-page.

TEXT

Summary

The actual technical report begins with a brief (500 to 1000 words) summary, which restates the principal results, conclusions, and recommendations the report makes, but introduces no new material.

Report Text

The rest of the report follows the format and style of a journal article.

BACK MATTER

Distribution List

The elements of a report's back matter are the same as for a book, except that here the back matter ends with a list of names and affiliations of persons who will be sent a copy of the report.

REFERENCES

Cited References

[ANSI] American National Standards Institute. 1987. American national standard for information sciences–scientific and technical reports–organization, preparation, and production: ANSI Z39.18-1987. New York: ANSI. For availability, see "Standards for Editing and Publishing" in Appendix 3.

Butcher J. 1992. Copy-editing: the Cambridge handbook for editors, authors and publishers. 3rd ed. Cambridge (UK): Cambridge Univ Pr.

[CCAC] Consumer and Corporate Affairs Canada. 1988. Copyright: questions and answers. Ottawa: Minister of Supply and Services Canada.

[ISO] International Organization for Standardization. 1988. Documentation and information: ISO Standards Handbook 1. Geneva: ISO.

[NISO] National Information Standards Organization. 1992. Permanence of paper for publications and documents in libraries and archives: American national standard Z39.48-1992. New Brunswick (NJ): Transaction Publishers. For availability, see "Standards for Editing and Publishing" in Appendix 3.

[NLC] National Library of Canada. 1988. Legal deposit: a guide. Ottawa: Minister of Supply and Services Canada.

[NLC] National Library of Canada. 1989. The publisher's path. Ottawa: Minister of Supply and Services Canada.

[UCP] University of Chicago Press. 1993. The Chicago manual of style. 14th ed. Chicago: UCP.

Additional References

[ANSI] American National Standards Institute. 1980. American national standard for title leaves of a book: ANSI Z39.15-1980. New York: ANSI.

[ANSI–NISO] American National Standards Institute, National Information Standards Organization. 1990. American national standard for printed information on spines: ANSI/NISO Z39.41-1990. New Brunswick (NJ): Transaction Publishers. For availability, see "Standards for Editing and Publishing" in Appendix 3.

Lee M. 1979. Bookmaking: the illustrated guide to design/ production/ editing. 2nd ed. New York: RR Bowker.

[NISO] National Information Standards Organization. 1984. American national standard for describing books in advertising, catalogs, promotional materials, and book jackets: ANSI Z39.13-1979(R1984). New Brunswick (NJ): Transaction Publishers. For availability, see "Standards for Editing and Publishing" in Appendix 3.

[NISO] National Information Standards Organization. 1988. Book numbering: Z39.21-1988. New Brunswick (NJ): Transaction Publishers. For availability, see "Standards for Editing and Publishing" in Appendix 3.

[NISO] National Information Standards Organization. 1990. Standard technical report number (STRN): format and creation: Z39.23-1990. New Brunswick (NJ): Transaction Publishers. For availability, see "Standards for Editing and Publishing" in Appendix 3.

O'Connor M. 1986. How to copyedit scientific books and journals. Baltimore (MD): Williams & Wilkins.

Peters J, editor. 1983. Bookman's glossary. 6th ed. New York: RR Bowker.

Strong WS. 1992. The copyright book: a practical guide. 3rd ed. Cambridge (MA): MIT Pr.

[UNESCO] United Nations Educational, Scientific, and Cultural Organization. 1981. The ABC of copyright. New York: Unipub.

White JV. 1988. Graphic design for the electronic age. New York: Watson-Guptill (Xerox Pr).

Williamson H. 1983. Methods of book design: the practice of an industrial craft. 3rd ed. New Haven: Yale Univ Pr.

Xerox Corporation. 1988. Xerox publishing standards: a manual of style and design. New York: Watson-Guptill Publications.

30 Citations and References

> . . . one foot rests when one walks, leaving prints another may trace as he follows.
>
> —Augustine of Hippo, *De Trinitate* [*On The Trinity*], Book 11: Chapter 6, 399–422 CE

> I mention the source, because I do not care to strut in borrowed plumes.
>
> —Alessandro Manzoni, *I Promessi Sposi* [*The Betrothed*], Chapter 11, 1827

> This is the rationale for . . . seemingly arbitrary rules: in prescribing the exact form and sequence in which the required data are to be communicated, they make it more likely that the data will be fully and accurately communicated and that lapses will be readily discerned.
>
> —Gertrude Himmelfarb, *NY Times Book Review,* 1991

30·1 In this and other chapters the bibliographic descriptors of documents cited in the text, tables, or legends of figures of a document are called "references". The term "citations" is reserved for the brief formal indications in text, tables, or figure legends of the documents cited and represented in the references listed at the end of the article, chapter, or book. Note that this usage differs from that in *National Library of Medicine Recommended Formats for Bibliographic Citation* (NLM 1991), the main source for the recommendations in this chapter. The Library is hereafter referred to in this chapter as the "NLM".

Two general systems for formatting references and citing the documents they represent are widely used in the scientific literature: the citation-sequence system and the name-year system. Both have been applied with many variations. Which system is selected by an editor or publisher should be determined by judgments on its advantages and disadvantages for the expected readership. The recommendations here represent both systems; they have been drawn up to guide authors and editors to reduce the number of variations in the formats of references, regardless of which system is used.

Not all of the possible forms of references for all the various kinds of documents are represented in the recommendations here. See Appendix Table 30·1 for the additional coverage in the NLM source.

THE CITATION-SEQUENCE SYSTEM

30·2 In the citation-sequence (C-S) system, the numeric citations in text, tables, and figure legends are the numbers that identify the references (to articles, chapters, books) listed at the end of the document. References are numbered and listed in the sequence in which they are 1st cited.

> Modern scientific nomenclature really began with Linnaeus in botany[1], but other disciplines[2,3] were not many years behind in developing various systems[4–7] for nomenclature and symbolization.

Subsequent citations of the same document use the same number as that of its initial citation.

ADVANTAGES

The main advantage of the C-S system is that citation numbers interrupt only minimally the reading of text. This advantage is conspicuous when continuous and long sequences of citations (as in review articles) are represented by the 1st and last numbers of a sequence separated by an en-dash rather than by explicit citation of all reference numbers (or by names and years in the other system). Using numbers as citations also saves space, paper, and cost.

DISADVANTAGES

There are at least 3 disadvantages to using the C-S system. Readers usually will have to turn to the reference list to find out exactly whose work is being cited. When authors must add or delete references at a late stage in revising an unsubmitted paper or revising a paper to meet criticisms of the editor or peer reviewers, they or their typists may be burdened with reordering and renumbering references and with the necessary retyping; this disadvantage may disappear when manuscripts are prepared with bibliographic computer programs. The visibility of author names is lower when they appear only in the references and not in citations in the text.

THE NAME-YEAR SYSTEM

30·3 In the name-year (N-Y) system, citations in text, tables, and figure legends of the references listed at the end of the document (article, chapter, book) consist of the surname of the author, by which the reference is alphabetized in the list, and the year of publication of the document. The name and date are enclosed within parentheses (round brackets).

> By contrast, the several antisera that have been raised against Sp1, a defined RNA polymerase II transcription factor (Kadonaga 1986), stain exclusively the nucleus . . .

Variations in formats needed for multiple authors and special cases to identify the author and publication year are given in the detailed recommendations below.

This system is widely known as "the Harvard system", a term that does not reflect its structure and hence is not to be preferred to "the name-year system". The origin of the term "the Harvard system" has been described by a former chairman of the Council of Biology Editors (Chernin 1988).

ADVANTAGES

There are several advantages to the N-Y system over the C-S system. For authors preparing manuscripts, adding or removing references from the reference list will lead only to adding or deleting citations and will not force a renumbering of citations as in the citation-sequence system. In some disciplines, readers may be able to identify documents readily from the citations and will thus not have to turn to the references. Even for readers not familiar with the cited literature, the years in the citations will convey some historical perspective on the development of concepts and methods being discussed. Cited authors are likely to be more pleased with this system.

DISADVANTAGES

But there are also disadvantages to the N-Y system. When large numbers of documents must be cited at particular points in the text, the long string of citations within parentheses may be highly irritating to readers as a gross interruption in the text. This occurs most often in review articles and in other kinds of articles with statements summarizing many documents, for example, individual case reports. A similar interruption can occur with very long author names, such as those of organizations and committees indicated as authors without abbreviation.

The rules for sequence of citations, punctuation within citations, and alphabetization of reference lists are more complex than the rules for citing by number and ordering references by sequence of 1st citation.

THE BASIS FOR RECOMMENDATIONS

30·4 The many variations of both systems found in the scientific literature can each be justified rationally to some degree. The decision to recommend the reference formats described below was based on 5 considerations.

1 The formats should be built as much as possible on rational and logical principles and easily applied to new documents.
2 Decisions on details in the formats should be facilitated by reference to comprehensive, authoritative, and readily available sources.
3 The formats should be readily applicable to many kinds of documents.
4 The formats and the information they carry should be economical of space, time, labor, and cost in preparing manuscripts and typesetting for publication. Punctuation not needed for functional reasons should be eliminated.
5 Differences between formats for the 2 systems should be minimal.

The formats for references are based on the formats in *National Library of Medicine Recommended Formats for Bibliographic Citation* (NLM 1991). This exhaustive document is valuable for additional guidance. Note that some details of the recommended formats in this manual differ at a few points from those developed by the NLM; the reasons for the differences are explained as they are described.

The NLM's recommendations are directly related to standards of the American National Standards Institute and the National Information Standards Organization (ANSI–NISO). The recommendations of the NLM are also the basis for the recommendations of the International Committee of Medical Journal Editors published in their "Uniform Requirements" documents (ICMJE 1993).

Other citation-reference systems are used, notably one listing references alphabetically according to the same principles as the name-year system but numbered in alphabetic order, and cited in the text with the reference numbers. Publishers and editors preferring that system could readily use the formats for references given here.

CITATIONS

CITATIONS IN THE CITATION-SEQUENCE SYSTEM

30·5 The citation of a reference in the C-S system is the number assigned to it in the reference list; the references are listed in the sequence in which they are initially cited, 1st throughout the text and then in tables and figure legends. Citation numbers not in a continuous numeric sequence are separated by commas with no spaces. For more than 2 numbers in a continuous sequence, the 1st and last numbers are connected by an en-dash (or a hyphen if the en-dash is not available); if there are only 2 consecutive numbers, they are separated by a comma. The numbers are preferably placed superscript at each point of citation; the superscript position eliminates confusion between citation numbers and parenthetic numbers that are not citations. The superscript numbers should be 1 or 2 points smaller than the size of the text type.

> has been shown[1] to replace IL-3 for the transient growth of factor-dependent cells . . .
>
> have been shown[1,2,5,7,11–15] to abrogate the requirements of T cells . . .

For documents that have not been seen by the author and are being cited from a subsequent document, the number for the reference to the original document is followed without spacing by the parenthetic statement "cited in" with the reference number for the source.

> The original description[12(cited in 13)] apparently was a classic of taxonomic detail.

Full references should be given for both documents, and the reference for the original document should include a closing note indicating the reference to the document in which it was cited. Note, however, that such citation of original documents not seen by the author should, in general, be discouraged.

In publications whose style requires placing citation numbers on the line and within parentheses instead of in the superscript position, numerals other than citation numbers that appear within parentheses should be accompanied by a term, unit, or symbol that unambiguously distinguishes them from citation numbers.

> and in our case series (12 patients) we found (37) that . . .
>
> and the high value (12 g/L) previously reported (17) was later shown to have been affected by . . .

On-the-line, parenthetic citations for references reduce time at the keyboard compared with time needed to put in superscript numbers, but computer bibliographic programs may carry out superscripting readily.

CITATIONS IN THE NAME-YEAR SYSTEM

General Format: Personal and Organizational Authors

30·6 A citation in the N-Y system includes the name of the author or authors of the cited document and the year of publication.

> By contrast, the several antisera that have been raised against Sp1 (Kadonaga 1986), a defined RNA polymerase II transcription factor, stain the nucleus . . .

If the date of publication in the reference includes, in addition to the year, the month and the day, generally the year will suffice in the citation.

The references are listed alphabetically by authors' surnames (last names) in the reference list; see section 30·15. Organizational names cited as authors are preferably given in an abbreviated form (see section 30·33), but in a text with relatively few citations, full names may be acceptable.

> the recently published document on requirements for manuscripts (ICMJE 1991) should be consulted for . . .
>
> [or]
>
> the recently published document on requirements for manuscripts (International Committee of Medical Journal Editors 1991) should be consulted for . . .

The abbreviation must, of course, appear as the initial element in the author field of the reference; see section 30·19. Note that if the abbreviation of the organizational name is put within square brackets in the reference, the brackets are omitted in the citation. The same rule applies to the added element "[Anonymous]". The bracketed information is, however, used to alphabetize the reference list.

[citation]	The most vigorous defense of the policy (Anonymous 1990) recently appeared in . . .
[reference]	[Anonymous]. Our need for a national energy policy [editorial]. Technol Sci 1990;72:13–4.

Note that the recommendation for using "Anonymous" is not drawn from the NLM's recommendations and is based solely on the need for an author designation in citations in the name-year system; see section 30·19.

Document Cited Near Author Name in Text

30·7 If documents by the same author (or author group) are cited close to the mention of the author's name in the text and no uncertainty as to author identification could arise, the citation can be limited to the publication year.

> When Smith's sequence of studies (1958, 1963a, 1963b, 1967) is examined closely . . .

Same-Author Documents Published in Different Years

References by the same author published in different years are distinguished in their citations by the years in chronologic sequence.

> Smith's studies of *Culex* species (Smith 1967, 1978) have shown that . . .
>
> When his subsequent studies were published (Smith 1979, 1980, 1982), it became clear that . . .

Note that the years are separated by commas and spaces; see section 30·9 for punctuation rules.

Same-Author Citations: 2 or More Documents in the Same Year

For 2 or more references by the same author published in the same year, an alphabetic designator is added to the year in the reference and in the citation; such designators begin with the letter "a".

> and the most recent work (Dawson 1986a, 1986b) on this problem has . . .

In this circumstance, the sequence of designations of the references by "a", "b", and so on should be determined by the sequence of publication dates, earliest to latest, which usually can be identified for articles in journals from volume and issue numbers.

Authors with Identical Surnames

When the authors of 2 documents published in the same year have identical surnames (used in alphabetization in the reference list), their initials are given in the citation.

> and the most recent work (Dawson GL 1986; Dawson WM 1986) has shown a . . .

Two-Author Citations

30·8 If a cited document has 2 authors, both names are given, separated by "and".

and the most recent work (Dawson and Briggs 1986) on this problem has . . .

Note that this rule is followed despite the identification of the authors in the reference as "Dawson GT, Briggs MF"; the rationale is the avoidance of a comma between the names in the citation, the comma being reserved to indicate 2 or more references by the same author in a citation.

If both authors have the same surname, their initials are added.

(Smith TL and Smith UV 1990)

Use of the ampersand ("&"; see section 4·41) instead of "and" in such citations is discouraged. The "&" is less likely than "and" to be known by nonanglophone readers.

Three-or-More-Author Citations

Documents with 3 or more authors are cited by the 1st author's name followed by "and others" and the publication year.

but later studies (Dawson and others 1987) established that . . .

If the 1st author names and the years are identical in several references, enough author names are added to eliminate ambiguity.

(Smith, Jones, and others 1990)

(Smith, Jones, Thomas, and others 1990)

The preference for "and others" over "et al." or "et alii" is in accordance with the preference stated elsewhere in this manual for English terms and abbreviations over abbreviations based on Latin; see section 10·5.

Multiple Citations at 1 Point

When 2 or more documents with different authors are cited within a parenthetic citation, they are cited in chronologic sequence from earliest to latest. Those published in the same year should be sequenced alphabetically by author name(s).

and the main contributors (Dawson and Briggs 1974; Dawson and Jones 1974; Smith AL 1978; Smith GT 1978; Smith and others 1978; Tyndall and others 1978; Zymgomoski 1978; Brown 1980) established beyond a doubt that . . .

References Cited from Another Source

For documents that have not been seen and are cited from a subsequent document, the original author's name and year are given with "cited in" and the name and date for the source.

> The original description (Powell 1858, cited in Forbes 1872) was a classic of taxonomic detail.

Full references should be given for both documents, and the reference for the original document should include a closing note indicating the reference that was the source; see "Notes" in section 30·37. As noted in section 30·5, such citation through secondary sources should, in general, be discouraged.

Punctuation in Citations

30·9 A comma followed by a space separates citations of different references by the same author, or authors, represented by the same-year references ("1986a, 1986b"), or of different-year references ("1985, 1986").

> and additional work (Briggs 1986, 1987, 1988) has shown that . . .
>
> and additional work (Dawson and Briggs 1986a, 1986b) has shown that . . .
>
> and additional work (Dawson and Briggs 1984, 1987) has shown that . . .

A comma also separates author names accompanied by initials in citations with more than 2 author names.

> (Smith GL, Jones GC, and others 1990)

A semicolon followed by a space separates citations of references by different authors.

> and additional work (Dawson 1988; Briggs 1989) has shown that . . .
>
> and additional work (Smith 1981; Briggs 1982, 1983; Dawson and Briggs 1984; Briggs and others 1990) has shown that in most cases studied by this method . . .
>
> and additional work (Smith 1981, 1987; Dawson and Briggs 1989) has shown that . . .

If a page reference is given in a citation, it is separated from the year by a comma and a space.

> Our preference for this method was confirmed (Smith 1982, p 73) by the finding that . . .

Unreferenced Citations

30·10 Occasionally the source of information that is not available in published or archival form, such as personal communications and oral presentations at a meeting, may have to be indicated. The nature and source of the cited information should be identified in the running text by an appropriate statement. If the source of the information is placed within parentheses (round brackets, as for a citation), a term, or terms, must indicate clearly that the citation is not represented in the reference list.

> and most meningiomas proved to be inoperable (a 1943 letter from RS Grant to me; unreferenced, see "Notes"), but a few were not.

Note that the author must seek to provide the publisher with written permission from the cited person (if living) or from the organization to cite the information if it is carried in a document that is not accessible to scholars. The permission should be acknowledged in an "Acknowledgments" or a "Notes" section that follows the text of an article or is placed at the end of a book's main text; such statements may include helpful additional details, such as the reason for the communication, the location of a manuscript if it is not generally available, the title, location, and exact date of a meeting, and so on.

Such citations are not represented in the reference list unless the document is available to scholars in a depository (library, archives) or from a depository on order in a microfiche, microfilm, xerographic, or electronic format. Unpublished documents that are thus available can be cited and referenced like published documents; see section 30·67.

Note that the NLM does authorize citations of, and references for, papers presented at meetings. It recommends that personal communications be identified in the text (and this is suggested at the head of this section) or in footnotes (discouraged by this manual and most publishers).

LOCATION AND SPACING OF CITATIONS

In Text

30·11 A citation in the text should immediately follow the title, word, or phrase to which it is directly relevant. The relevant word or phrase may represent the essence of the relevant concept or finding; it may represent the referenced document. Ambiguity about what concept is being referenced can be avoided by not placing citations at the end of long clauses or sentences. Citations should be separated from adjacent text elements by single spaces, but they are not spaced from a following punctuation mark.

> The observation that *abl* can also abrogate IL-2 dependence (Cook 1987), at least under some conditions, suggests that tyrosine phosphorylation may be involved in IL-2 signal transduction.
>
> [or]
>
> Cook's report (1987) that *abl* can also abrogate IL-2 dependence, at least under some conditions, has led to the suggestion . . .
>
> [not "The observation that *abl* can also abrogate IL-2 dependence . . . suggests that tyrosine phosphorylation may be involved in IL-2 signal transduction (Cook 1987)."]

The same principles for location apply to citations in the citation-sequence system.

In Tables

30·12 Citations in tables are usually most appropriately put in footnotes (see section 31·14). If citations must appear within the field of a table as a column or row, an appropriate column or stub heading should identify them as indicating references. If they are in the citation-sequence system, the superscript convention is used.

	[N-Y system]	[C-S system]
[Source]	[Citation]	[Citation]
The ICSU Study	Schmidt and Schwartz 1978	The ICSU Study[12]
Darlington	Darlington 1980	Darlington[13]

Note that superscript citation numbers should not be attached to numbers in the field of a table because they could be misinterpreted as exponents (powers). Superscript alphabetic symbols can represent footnotes to the table where the references can be cited with superscript numbers. See section 31·14 for additional explanation.

In Figures

Citation numbers should not be used on figures (charts, graphs, illustrations). If citations are needed to support data or methods relevant to a figure, they should be put in the figure's legend; see section 31·14.

REFERENCES: PRINCIPLES FOR LISTING

THE REFERENCE LIST

30·13 The references to documents cited in text, tables, or figures should be listed under the heading "References" or "Cited References". References to documents that are not specifically cited in the text but that served as sources should be listed alphabetically by author(s) under the heading "Additional References". References for additional reading or other purposes that are not related to a citation should be listed under a heading such as "Additional Reading" or "Bibliography" that distinguishes these documents from source documents.

SEQUENCE OF REFERENCES

Citation-Sequence System

30·14 In the citation-sequence (C-S) system, references are listed and numbered in the sequence in which each referenced document is 1st cited in the text, tables, and figures (illustrations). References are numbered in tables and figures in order after their 1st mention as though they were part of the text. For example, if the last citation number in text before the 1st mention of a table is 22, the numbering of 5 new references in the table begins with 23 and ends with 27; the next citation in the text (if this is the 1st mention of the reference) will be 28, and so on. Citation numbers for figures and tables are usually in their legends and footnotes, so also see section 31·14. Since, however, citations in text usually far outnumber citations in tables or figure legends, and since tables and figures in a published document may—for reasons of design and format—not be positioned close to the places in the text where they are 1st mentioned, citations in tables and figures may be more conveniently treated as a 2nd numeric sequence following the last citation in the text.

Name-Year System

30·15 The sequence (order) of references in the reference list is determined by 2 basic rules.

1 Alphabetic sequencing (ordering) is determined by the 1st author's surname (family name) and then, if necessary, by letter-by-letter alphabetic sequencing determined by the initials of the 1st author and beginning letters of any following surnames. This is the sequence also applied in indexes; see section 7·1 and Table 7·1 for guidance on determining the proper sequence for surnames.

2 When the author designation (name or names) is completely identical in 2 or more references, these references are sequenced by publication date (earliest to latest).

Note the following additional considerations for special cases.

1 Author surname (family name). Particles such as "de", "la", "van", "von" are generally treated in English-language scientific publications as part of the surname and govern the alphabetization; this rule assumes that the prefix is part of the author name as it appears on the cited document; this principle applies whether the prefix is capitalized or not. This principle is not invariable. Names better known without the particle (such as "Beethoven" for "van Beethoven") can be alphabetized by the main element of the name. Some European publications omit all such particles for alphabetization. For additional guidance, see Table 7·1.

2 Diacritical marks, accents, and special characters based on the roman alphabet are ignored; a marked letter is treated as if it were not marked, and ligated letters are treated as if not ligated. Note that to simplify rules for English-language publications this rule ignores some conventions in some non-English languages.

Å [treated as] A   Ø [treated as] O   æ [treated as] ae

Ç [treated as] C

Note, however, that alphabetic sorting by computer programs may not follow this convention; in that case, alphabetization will have to be carried out manually. Anglicized forms of names will be alphabetized by the letters representing the original form, for example, an original umlaut (such as "ue" for "ü").

3 Initial elements of surnames that are abbreviations are alphabetized as written, not by the full term or name the abbreviation represents. This rule assumes that the abbreviation was used in the author name on the cited document. A space in the family name is ignored whether it is punctuated or not.

St Louis
[follows "Stickley" and is not alphabetized as if "St" were "Saint"]

4 Names of Celtic origin beginning with "M'", "Mac", or "Mc" are alphabetized in the order of the initial letters in such names, not as if "M'" and "Mc" represented "Mac". The apostrophe in "M'" is ignored for alphabetic sequencing.

5 The square brackets enclosing abbreviations for organizational author-names are ignored for alphabetic sequencing.

An example of a sequence of references determined by these rules on names and years is shown below.

Jones GK. 1973.

Jones HP. 1974.

Jones HP. 1976.
Jones QT. 1987a.
Jones QT. 1987b.
MacVeagh GT, Rosen HR, Smith MM. 1945.
McDowell RI, Rosen HR, Smith MM. 1945.
[MGS] Massachusetts Geological Survey. 1948.
M'Veagh ET, Rosen HR, Smith MM. 1931.
Neighbors J. 1876.
[NLM] National Library of Medicine. 1991.
Smith AL, Briggs GT. 1965.
Smith AL, Dawson MT. 1955.
Smith GT. 1981.
Stickley BT. 1977.
St Louis CH. 1972.
Vandelow MB. 1965.
von Grautschwitz ETA. 1990.

For decisions on what elements are defined as the surname (family name) in various languages, see section 7·1 and Table 7·1.

REFERENCES: GENERAL PRINCIPLES FOR FORMATTING

IN REFERENCE LISTS IN THE CITATION-SEQUENCE SYSTEM

30·16 In the citation-sequence system (C-S), the general sequence of information in a reference is author name, title, and additional items (including year of publication) as specified below in sections 30·19–30·68.

IN REFERENCE LISTS IN THE NAME-YEAR SYSTEM

The sequence of information in references in the name-year system (N-Y) differs from that in references in the citation-sequence system only in the placement of the year: the author names are followed by the publication year, which is then followed by the other items in the same sequence except for year. This is the sole difference in formats, other than the adjustment needed because of the shift of the year. Note that this sequence is not authorized in the recommendations of the NLM but is needed in the name-year system to facilitate finding references from their citations.

FORMATS IN BOOK REVIEWS AND OTHER SIMILAR PRESENTATIONS

In some kinds of text, the references may have to display content and sequences that differ from those in reference lists and many bibliographies. For example, for the convenience of readers interested foremost in the subject of a book being reviewed and then in the identity of the book's author, reviews may be headed with a description that gives the title and then the author's name, the name of the publisher, and date of publication. Normal capitalization rules should be preferred to the simplifications used in references.

[the heading above a book review]

An Atlas of Rhinoplasty. 2nd ed. Thomas Jones and Smith Shockley. New York: Scott & Williams; 1989. 447 pages. $107.95.

[in a reference list]

Jones T, Shockley S. An atlas of rhinoplasty. 2nd ed. New York: Scott & Williams; 1989.

The same principle can be applied to entries in a "Books Received" list. The exact choice of bibliographic data to be included for each book is likely to depend on the needs of the publication's audience; librarians and booksellers, for example, would probably want to know the International Standard Book Number (ISBN; see section 29·7).

The annotated bibliography at the end of this manual (Appendix 3) applies this principle because the subject of a listed book as indicated by its title will probably be of greater interest to most readers of this appendix than the name of the author.

TYPOGRAPHY

30·17 The recommendations in this chapter do not specify type styles for references but simply indicate what information they should include, in what sequence, and with what punctuation. Some editors may choose to use italic or boldface type for some bibliographic elements to help the reader distinguish among them in a reference.

You CH, Lee KY, Chey RY, Menguy R. Electrogastrographic study of patients with unexplained nausea, bloating and vomiting. *Gastroenterology* 1980 Aug;79(2):311–4.

Voet D, Voet JG. **1990.** Biochemistry. New York: J Wiley.

Too many variations in type styles may confuse more than help. Intelligent choices may depend on ascertaining reader preferences for such devices.

For economy of space, references are usually set in a smaller type size than that of the main text. For example, references for a text in 12-point type could be in 10-point type.

REFERENCES: CONTENT, FORM, AND SEQUENCE, WITH EXAMPLES

GENERAL CONSIDERATIONS, INCLUDING PUNCTUATION

30·18 The sections below specify the information (bibliographic elements) to be included in references to various kinds of documents, the form of the elements, and the sequence. Some of the optional recommendations of the NLM representing relatively infrequent needs or minor variations in kinds of documents are noted but not fully described; the full recommendations are available in *National Library of Medicine Recommended Formats for Bibliographic Citation* (NLM 1991), which should be consulted for details. Deviations here from the NLM recommendations are noted and the rationales are explained.

Punctuation is determined by the following general principles.

1 Each group of bibliographic elements (such as the author names) closes with a period; such a group is designated a "bibliographic field". Note, however, that to save space the NLM omits the period at the end of a journal title; see section 30·19; this convention is applied here. Also note that in this manual's recommendations periods are not used with the abbreviations "p" for "page" and "nr" for "number"; this is a deviation applied so as to confine the use of periods in bibliographic references to delineating bibliographic fields.

2 A comma separates items of equal importance within a group such as author names in a reference to a journal article.

3 A semicolon separates items not directly related. In the field of elements of publication, that is, the date (year and month) and the volume-and-page data, the semicolon precedes the volume number; it is not regarded as following the publication date. This point is especially relevant in adapting citation-sequence formats for name-year formats.

4 A colon indicates that the data that follow it are subordinate to the data that precede the colon, such as a subtitle following the main title and a colon, and page numbers following a volume number and a colon.

5 If names are spelled out in an inverted-name format, as in an optional format for book references, surnames (family names) are followed

by commas, and each author's full name is followed by a semicolon if another author's name follows it.

Examples of references for the citation-sequence system are indicated by C-S and for the name-year system by N-Y. It is understood that in a list of references, each C-S reference begins with a number that indicates the order in which it is 1st cited in the text or its position in an alphabetic list of references.

JOURNAL ARTICLES

General Formats

30·19 The bibliographic elements, their sequence, punctuation, and spacing for most references to journal articles are as follows.

C-S Author(s). Article title. Journal title year month;volume number(issue number):inclusive pages.

N-Y Author(s). Year. Article title. Journal title volume number(issue number):inclusive pages.

For weekly journals the "month" datum in the C-S system includes the day of the month.

[C-S]

Steiner U, Klein J, Eiser E, Budkowski A, Fetters LJ. Complete wetting from polymer mixtures. Science 1992 Nov 13;258(5085):1122–9.

[N-Y]

Thomison JB. 1988. Uniform requirements for manuscripts [editorial]. South Med J 81(8):947.

Note that "Year" is not followed by a period in the C-S system (it is in the N-Y system). Also note that in the name-year system the semicolon is omitted before the volume number because it is not needed to separate that number from the date data; this is a deviation for this system from the NLM format.

Examples of Formats

STANDARD JOURNAL ARTICLE

30·20 C-S You CH, Lee KY, Chey RY, Menguy R. Electrogastrographic study of patients with unexplained nausea, bloating and vomiting. Gastroenterology 1980 Aug;79(2):311–4.

C-S Steiner U, Klein J, Eiser E, Budkowski A, Fetters LJ. Complete wetting from polymer mixtures. Science 1992 Nov 13;258(5085):1122–9.

Inclusion of the month (and day for weeklies) and issue number is recommended by the NLM. Omitting the month (and day for weeklies) and issue number is an option for journals with continuous pagination in volumes. This option is used in this manual.

C-S You CH, Lee KY, Chey RY, Menguy R. Electrogastrographic study of patients with unexplained nausea, bloating and vomiting. Gastroenterology 1980;79:311–4.

C-S Steiner U, Klein J, Eiser E, Budkowski A, Fetters LJ. Complete wetting from polymer mixtures. Science 1992;258:1122–9.

Note that in the C-S system, the form of the citation is always the superscripted number of the reference, which is the number representing the sequence in which the reference was 1st cited (see sections 30·2 and 30·18; therefore, the form of citation of C-S references will not be specified with every example of a C-S reference.

N-Y You CH, Lee KY, Chey RY, Menguy R. 1980. Electrogastrographic study of patients with unexplained nausea, bloating and vomiting. Gastroenterology 79:311–4.
Form of Citation: (You and others 1980) [or] (You et al. 1980)

N-Y Steiner U, Klein J, Eiser E, Budkowski A, Fetters LJ. 1992 Nov 13. Complete wetting from polymer mixtures. Science 258(5085):1122–9.
Form of Citation: (Steiner and others 1992)

[or]

Steiner U, Klein J, Eiser E, Budkowski A, Fetters LJ. 1992. Complete wetting from polymer mixtures. Science 258:1122–9.

ORGANIZATION AS AUTHOR

C-S Scandinavian Society for Clinical Chemistry and Clinical Physiology, Committee on Enzymes. Recommended method for the determination of γ-glutamyltransferase in blood. Scand J Clin Lab Invest 1976;36:119–25.

N-Y [SSCCCP] Scandinavian Society for Clinical Chemistry and Clinical Physiology, Committee on Enzymes. 1976. Recommended method for the determination of γ-glutamyltransferase in blood. Scand J Clin Lab Invest 36:119–25.

Form of Citation: (SSCCCP 1976)

ANONYMOUS AUTHOR

C-S [Anonymous]. Epidemiology for primary health care. Int J Epidemiol 1976;5:224–5.

N-Y [Anonymous]. 1976. Epidemiology for primary health care. Int J Epidemiol 5:224–5.

Form of Citation: (Anonymous 1976)

TYPE OF ARTICLE

C-S Smith KL. New dangers in our field [editorial]. Am J Nucl Eng 1991;13:15–6.

N-Y Smith KL. 1991. New dangers in our field [editorial]. Am J Nucl Eng 13:15–6.

Form of Citation: (Smith 1991)

ARTICLE ON DISCONTINUOUS PAGES

C-S Crews D, Gartska WR. The ecological physiology of the garter snake. Sci Am 1981;245:158–64, 166–8.

N-Y Crews D, Gartska WR. 1981. The ecological physiology of the garter snake. Sci Am 245:158–64, 166–8.

Form of Citation: (Crews and Gartska 1981)

ARTICLE IN A JOURNAL PAGINATED BY ISSUE

30·21

C-S Eliel EL. Stereochemistry since LeBel and van't Hoff: part II. Chemistry 1976;49(3):8–13.

N-Y Eliel EL. 1976. Stereochemistry since LeBel and van't Hoff: part II. Chemistry 49(3):8–13.

Form of Citation: (Eliel 1976)

ARTICLE IN A SUPPLEMENT TO AN ISSUE

C-S Gardos G, Cole JO, Haskell D, Marby D, Paine SS, Moore P. The natural history of tardive dyskinesia. J Clin Psychopharmacol 1988;8(4 Suppl):31S–37S.

N-Y Gardos G, Cole JO, Haskell D, Marby D, Paine SS, Moore P. 1988. The natural history of tardive dyskinesia. J Clin Psychopharmacol 8(4 Suppl):31S–37S.

Form of Citation: (Gardos and others 1988)

ARTICLE IN A SUPPLEMENT TO A VOLUME

C-S Magni F, Rossoni G, Berti F. BN–52021 protects guinea-pig from heart anaphylaxis. Pharm Res Commun 1988;20 Suppl 5:75–8.

N-Y Magni F, Rossoni G, Berti F. 1988. BN–52021 protects guinea-pig from heart anaphylaxis. Pharm Res Commun 20 Suppl 5:75–8.

Form of Citation: (Magni and others 1988)

Format Details

AUTHOR(S)

30·22 The surname (family name) comes 1st, followed by initials. When there are 2 to 10 authors, all should be named, including the 10th author; if there are more than 10 authors, the 1st 10 are listed, followed by "and others". The surname and initials of an author are separated only by a space; successive names are separated from each other by a comma and a space. Authors should note that some journals depart from this NLM recommendation and allow fewer author names.

If the author is anonymous, the author name is represented by "[Anonymous]".

In the name-year system, if the author is an organization the full name is preceded by an abbreviation for it within square brackets to simplify the in-text citation. Names in nonroman alphabets are romanized. Note that organizational names should be given in descending hierarchical order; this requirement may call for breaking and reordering the elements of an organizational author name as it appears on the referenced document. For example, "Committee on Enzymes of the Scandinavian Society for Clinical Chemistry and Clinical Physiology" becomes "Scandinavian Society for Clinical Chemistry and Clinical Physiology, Committee on Enzymes". Note that the hierarchy need not include all possible elements; the highest should be the one most likely to be known by the readership as having a major public responsibility for the content of the document. For example, although the National Library of Medicine is hierarchically under the US Department of Health and Human Services, a journal article that is on bibliographic services and identified as coming from the Reference Section of the National Library of Medicine would be given the authorship designation "National Library of Medicine, Reference Section", not "US Department of Health and Human Services, National Library of Medicine, Reference Section". If the organizational name is identical with that of 1 or more other organizations, a geographic modifier should be added for clarification: "Beth Israel Hospital (Boston)"; there are many Beth Israel hospitals in the United States.

Some citation-sequence systems, notably the one widely known as "the Vancouver system" (ICMJE 1993), specify fewer authors to be named before "and others" is substituted for the rest.

Note that the recommendations for square-bracketed abbreviations of organizational names and for "[Anonymous]" as an author name are not NLM recommendations. These deviations were introduced here to facilitate the formation of citations in the name-year system and useful alphabetization in reference lists.

ARTICLE TITLE AND TYPE

30·23 In article titles, only the 1st word and proper nouns and proper adjectives are capitalized. Obvious exceptions are capitalized abbreviations and symbols in the original title (such as HIV–1, DDAVP, pH). Note also that some chemical prefixes must be kept in lowercase type with only the prefixed compound name capitalized, for example, "*o*-Methylphenol" representing an ortho-substituted compound; if the lowercase italicized "*o*" were capitalized because it is the 1st letter of the term, the term would become "*O*-methylphenol", an oxygen-substituted compound (see section 16·13).

For articles published in non-English languages, the original language is used, but titles in nonroman alphabets, for example, Cyrillic or Greek, are romanized. Translated titles (for example, ones originally in Chinese or Japanese) are placed within square brackets. Internal punctuation and scientific conventions such as italics for a species name in the original title are retained.

The article type can be indicated within square brackets after the title but within the title field, for example, "[editorial]", "[letter]". A translated title already within brackets can be followed by an em-dash and the term for the article type.

> The case for a single-payer system [editorial].
>
> [Minamata disease: recent cases—editorial].

JOURNAL TITLE

30·24 Single-word titles are given in unabbreviated form. Multiple-word titles are abbreviated in the forms given in one of the standard sources based on the relevant standard of the National Information Standards Organization (ANSI–NISO 1985): *Serials Sources for the BIOSIS Data Base* (BIOSIS [annual]); *List of Journals Indexed in Index Medicus* (NLM [annual]). Many British journals prefer to use the *World List of Scientific Periodicals* (Brown and Stratton 1963–5). Title words, unabbreviated and abbreviated, are capitalized. Abbreviations are not followed by periods. Titles in nonroman alphabets are romanized.

For rules for abbreviating journal titles see Appendix 1.

DATE OF PUBLICATION

30·25 For the date of publication, the year is given in full; for a 2-year span, only the last 2 digits of the 2nd year are given. The NLM recommendations specify additional data for date; this manual departs from this requirement by offering the option of including only the year except for referenced journals without continuous pagination throughout a volume; see sections 30·20 and 30·21. This option is offered in part to simplify adapting the NLM formats for the name-year system. When a full publication date is required, it includes the month (or months-span or season) and day of the month (if identified on the journal). Month names longer than 3 letters are abbreviated to their 1st 3 letters; see section 12·7 and, for the abbreviations, Table 12·2. Season names are spelled out and capitalized. Month and season spans are represented by an en-dash, as in "Jan–Feb" and "Fall–Winter"; when the en-dash is not available, the hyphen can be used.

With the growing use of bibliographic software programs, which carry fields for dates, more authors may begin to record exact-date information and be able to include month and day in references. If the day and month are included, the date should be in the order "year month day".

VOLUME NUMBER

30·26 Numeric terms and roman numerals must be converted to arabic numerals: "Sixtieth" to 60th; "XC" to 90.

SUPPLEMENT, PART, OR SPECIAL NUMBER

Additions (supplements, parts, or special numbers) to regular issues of a volume or to a volume are indicated with the abbreviations Suppl, Pt, or Spec Nr and any numeric (arabic) or alphabetic designations, such as "Suppl 1", "Pt 2".

ISSUE NUMBER

Only arabic numerals should be used for issue numbers.

PAGINATION

30·27 Inclusive pages are given; duplicate digits are not repeated, for example, "137–9" (for "137 through 139"), "1196–201" (for "1196 through 1201"). The numbers for the beginning and ending pages are separated by an en-dash (hyphen if the en-dash is not available). Alphabetic designations that are a part of page numbers are retained; duplicate digits are repeated when the letter follows the numbers ("123A–124A"), but the initial rule applies when the letter precedes the numbers ("H34–5").

OTHER POSSIBLE ELEMENTS

30·28 The bibliographic elements described above will adequately represent most journal articles. Some additional elements will have to be carried within the reference to serve generally infrequent needs. The NLM recommends that the type-of-medium and edition elements be included; microfilming may occasionally accidentally omit some pages, and different editions of a journal may have differences in pagination and other elements. Below are some of the other possible elements to be included in references.

Author affiliation: Within parentheses after each author's name.

Retraction statement: A reference within square brackets after the article title.

Type of medium: Type of microform.

Edition: Indication after journal title, within parentheses, for example, "(British ed)".

Numeration, title, and pagination of part: Reference to a part, such as particular pages within an article, a figure, a table, an appendix.

Physical description: Designation of the kind of microform version.

Language: Specification of a non-English language.

Notes: Statement at the end of the reference.

For details on how to add these elements to references, consult the NLM's *Recommended Formats for Bibliographic Citations* (NLM 1991); see Appendix Table 30·1 of this chapter.

JOURNALS: SPECIAL ISSUES AND JOURNAL TITLES

30·29 Occasionally references are needed for entire issues on a specific subject or to a journal in its entirety (as in subject bibliographies); see Appendix Table 30·1.

BOOKS

30·30 Note that "books" here can apply to smaller publications frequently known as "booklets", "pamphlets", "brochures", or "reports", but see sections 30·45–30·49 for special requirements for scientific and technical reports.

General Formats

The bibliographic elements, their sequence, punctuation, and spacing for most references to a book are as follows.

C-S: Author(s) [or editor(s)]. Title. Place of publication: publisher name; Year. Number of pages.

N-Y: Author(s) [or editor(s)]. Year. Title. Place of publication: publisher name. Number of pages.

Examples of Formats

BOOK WITH AUTHORS

30·31 C-S Voet D, Voet JG. Biochemistry. New York: J Wiley; 1990. 1223 p.

[Note: In the C-S system, the form of citation is always the superscript number of the reference; therefore, the form is not specified for every example of C-S references.]

N-Y Voet D, Voet JG. 1990. Biochemistry. New York: J Wiley. 1223 p.
Form of Citation: (Voet and Voet 1990)

BOOK WITH EDITORS

C-S Gilman AG, Rall TW, Nies AS, Taylor P, editors. The pharmacological basis of therapeutics. 8th ed. New York: Pergamon; 1990. 1811 p.

N-Y Gilman AG, Rall TW, Nies AS, Taylor P, editors. 1990. The pharmacological basis of therapeutics. 8th ed. New York: Pergamon. 1811 p.

Form of Citation: (Gilman and others 1990)

BOOK WITH FULL NAMES FOR AUTHORS OR EDITORS

C-S Voet, Donald; Voet, Judith G. Biochemistry. New York: J Wiley; 1990. 1223 p.

C-S Gilman, Alfred G; Rall, Theodore W; Nies, Alan S; Taylor, Palmer, editors. The pharmacological basis of therapeutics. 8th ed. New York: Pergamon; 1990. 1811 p.

N-Y Voet, Donald; Voet, Judith G. 1990. Biochemistry. New York: J Wiley; 1223 p.

Form of Citation: (Voet and Voet 1990)

N-Y Gilman, Alfred G; Rall, Theodore W; Nies, Alan S; Taylor, Palmer, editors. 1990. The pharmacological basis of therapeutics. 8th ed. New York: Pergamon. 1811 p.

Form of Citation: (Gilman and others 1990)

Author names with initials are recommended for reference lists that also include references to journal articles, for which names with initials are preferred. Consistency in style within a reference list reduces work for typists and editors and reduces chances for errors in punctuation.

BOOK WITH AUTHOR, EDITOR, TRANSLATOR, AND NOTE

30·32 C-S Luzikov VN. Mitochondrial biogenesis and breakdown. Galkin AV, translator; Roodyn DB, editor. New York: Consultants Bureau; 1985. 362 p. Translation of: Reguliatsiia formirovaniia mitokhondrii.

N-Y Luzikov VN. Mitochondrial biogenesis and breakdown. 1985. Galkin AV, translator; Roodyn DB, editor. New York: Consultants

Bureau. 362 p. Translation of: Reguliatsiia formirovaniia mitokhondrii.

Form of Citation: (Luzikov 1985)

ORGANIZATION AS AUTHOR; NOTE AS SERIES STATEMENT

C-S ASTM Committee E–8 on Nomenclature and Definitions. Compilation of ASTM standard definitions. 3rd ed. Philadelphia: American Soc for Testing and Materials; 1976.

C-S International Organization for Standardization. Statistical methods. Geneva: International Standards Organization; 1979. (ISO standards handbooks: 3).

N-Y [ASTM] ASTM Committee E–8 on Nomenclature and Definitions. 1976. Compilation of ASTM standard definitions. 3rd ed. Philadelphia: American Soc for Testing and Materials.

Form of Citation: (ASTM 1976)

N-Y [ISO] International Organization for Standardization. 1979. Statistical methods. Geneva: ISO; (ISO standards handbooks: 3).

Form of Citation: (ISO 1979)

PLACE OF PUBLICATION CLARIFIED

C-S Mazria E. The passive solar energy book: a complete guide to passive solar home, greenhouse and building design. Emmaus (PA): Rodale; 1979. 435 p.

N-Y Mazria E. 1979. The passive solar energy book: a complete guide to passive solar home, greenhouse and building design. Emmaus (PA): Rodale. 435 p.

Form of Citation: (Mazria 1979)

MICROFORM

C-S Heath DF. Organophosphorus poisons: anticholinesterases and related compounds [microfilm]. Elmsford (NY): Microforms International; 1961. 1 reel: 16 mm.

N-Y Heath DF. 1961. Organophosphorus poisons: anticholinesterases and related compounds [microfilm]. Elmsford (NY): Microforms International. 1 reel: 16 mm.

Form of Citation: (Heath 1961)

Format Details

AUTHOR(S), EDITOR(S)

30·33 The surname is 1st, followed by a space and the initials. Successive names are separated by a comma and a space. Optionally, 1st names can be used instead of initials. The name of an editor is followed by a comma, a space, and the designation "editor"; the rule for author names applies to a sequence of names representing 2 or more editors. Names in nonroman alphabets are romanized. Unnamed authors are represented by "[Anonymous]". The term "[Anonymous]", a deviation from the NLM recommendations, is an element needed for the name-year system of citation and alphabetization of reference lists. In the name-year system, organization-author names are preceded by an abbreviation within square brackets to simplify the in-text citation.

The name of the organization to be represented as author is usually taken from the title page of the book; if more than 1 organization is represented there, the 1st should be the one most likely to be identified publicly as directly responsible for the book; an organization hierarchically above the responsible subdivision need not be named if it is not likely to be identified as having the primary responsibilities represented by the book. For example, the National Library of Medicine is directly responsible for its annually published *List of Journals Indexed in Index Medicus*; even though the Library is part of the Public Health Service, which is part of the US Department of Health and Human Services, the author of the *List* should be shown as "National Library of Medicine". If a unit functionally within the responsible organization is also designated on the title page it should be named next, for example, "National Library of Medicine, Public Services Division".

Note that the NLM recommendations specify using 1st names of authors rather than initials. The preference for initials in this manual is to reduce differences between formats of references to journal articles and to books or parts of books and to save space. Also note that 1st names followed by surnames (family names) may be preferred for formats in such documents as book reviews and books-received lists; see section 30·16.

TITLE

30·34 A title should be in its original language with the original punctuation; a translation can follow within square brackets. A title in a nonroman alphabet should be romanized. Only the 1st word and any proper nouns and proper adjectives are capitalized.

The title is taken from the book's title page. A subtitle can follow the title in the reference after a colon and a space; if the main title contains a colon introducing a subsidiary part, that colon should be changed to a comma, and the subtitle should follow the main title after a colon and a space. This modification is recommended so as to keep all title elements within a single title-field, closed with a period.

EDITION

An edition field may be needed for books published in 2nd or later editions; it should follow the title with the form, for example, "2nd ed" ("edition" abbreviated). A designation such as "Second Edition" should be converted to the numeric form, "2nd ed". Descriptive edition terms should be abbreviated, for example, "New revised edition" as "New rev ed". First editions not known to be followed by new editions usually need not be identified as "1st ed", but if later editions have been published, the 1st edition should be identified as such.

PLACE OF PUBLICATION

30·35 The place of publication is the city or town indicated on the title page as the location of the publisher. If more than 1 place is named, use the 1st named or the place likely to be most accessible to readers wanting to get in touch with the publisher. The name of the publisher's state, province, or country can be added within parentheses to clarify the location; the 2-letter postal service abbreviation can be used for a state or province or the ISO 3166-recommended abbreviation (ISO 1988) for a country. Such an addition is not needed for widely known cities such as Chicago, Paris, Edinburgh. Use English versions of place names: "Turin" for "Turino"; "Vienna" for "Wien".

If the place of publication is not identified on the title page or the verso of the title page or elsewhere in the book but can be inferred from the location of the publisher, that place name can be given within square brackets. If no place can be identified, it can be given as "[place unknown]".

PUBLISHER

30·36 The publisher is the company or organization responsible for issuing the book and is usually identified on the title page or its verso. When an organization and 1 of its units are given, the higher level should be listed 1st, for example, "University of Pennsylvania, Institute of Health Economics". If no publisher can be identified from the title page, its verso, or elsewhere, "[publisher unknown]" serves as the publisher element. Names should be given in their original language, but nonroman-alphabet names should be transliterated or translated into English. The original elements of the name, including punctuation or no punctuation, except for the abbreviations suggested below, should be retained.

Publisher names can be abbreviated according to rules in Appendix 2, which also includes examples to show how they are applied. An editorial office should compile its own list of preferred abbreviated forms to maintain consistency in publisher names. In the name-year system, the abbreviation within square brackets preceding the organizational author name can be used if that organization was the publisher.

YEAR (DATE OF PUBLICATION)

The year of publication is usually found on the title page; if not there, it will probably be stated on the verso of the title page as either the publication date or the copyright date; the publication date is preferred. The NLM-recommended reference style also includes month of publication if given, as in "1988 Jan"; this datum is rarely available in books but is likely to be found in technical reports (see section 30·45).

PAGINATION

30·37 Pagination is given as the total number of pages, including back-matter such as the index, in arabic numerals followed by "p"; when a "p" is followed by a period, that period is the marker for the close of the bibliographic field. Front-matter pages, such as contents list, preface, and so on, that are paginated with roman numerals need not be represented for most bibliographic purposes in publication. Note that "p" followed by a number or numbers (for a page sequence) represents a specific page or pages, not a total number of pages; see section 30·39.

1237 p [read as "one thousand two hundred and thirty-seven pages"]
p 1237 [read as "page one thousand two hundred thirty-seven"]
p 17–28 [read as "pages seventeen through twenty-eight"]

NOTES

As the last element, the notes field can be used optionally for additional information likely to clarify the reference with regard to accompanying materials, original source of a reprint or new version, the citing by another reference of a original document not seen by the author, sponsorship, availability, ordering information, and so on. If the statement identifies the series in which a book is published, the statement is enclosed within parentheses.

OTHER POSSIBLE ELEMENTS

30·38 The elements described above will adequately represent most books. Some additional elements, as shown below, can be carried within the reference to serve generally infrequent needs.

Type of medium: A square-bracketed statement can be placed as an element after the title to indicate the type (such as microfiche, microfilm, microcard).

Secondary author(s): For books with a primary author (or authors), an element can be added just before the place of publication to indicate editors, translators, or other contributors not serving as primary authors. The name (or names) is in the format for primary authors and is followed by a term for the function, such as "editor" or "editors", "translator". If more than 1 kind of secondary author is named (for example, both editor and translator), their names are given in the order in which they appear in the book.

Physical description: For example, details of the microform.

Series: Description of the series in which the book has been published.

Language: Notation of the language of the book if it is not English.

For details on how to add these elements to references, consult the NLM *Recommended Formats* . . . (NLM 1991); see Appendix Table 30·1 of this chapter.

BOOKS: PARTS AND CONTRIBUTIONS

30·39 Specific parts of books (such as chapters, page sequences, individual pages, tables, and illustrations) may need to be referred to. This section gives examples of such references and also of references to parts contributed by authors other than the editors or authors of most of the rest of the book. Some authors refer to parts of books as "Contributions to Books".

General Formats

When the author of a part is the author of the book, the sequence, punctuation, and spacing for most references to a part (other than a volume) are as follows.

C-S Author(s) [or editor(s)]. Title. Place of publication: publisher name; year. Kind of part and its numeration, title; pages.

N-Y Author(s) [or editor(s)]. Year. Title. Place of publication: publisher name. Kind of part and its numeration, title; pages.

A reference to the volume of a book takes the following general form. Note that numeration must be in arabic numerals.

C-S Author(s) [or editor(s)]. Title. "Volume" with numeration, title. Place of publication: publisher name; year.

N-Y Author(s) [or editor(s)]. Year. Title. "Volume" with numeration, title. Place of publication: publisher name.

When the author of the relevant part of a book is not an author of the book, the sequence, punctuation, and spacing for references to that part are as follows.

C-S Author(s) of the part. Title of the part. In: author(s) [or editor(s)]. Title of the book. Place of publication: publisher; year of publication. Pages of the part.

N-Y Author(s) of the part. Year. Title of the part. In: author(s) or editor(s). Title of the book. Place of publication: publisher. Pages of the part.

The large number of possible variations in these kinds of reference precludes giving more than a few examples.

The styling of reference elements is in general the same as for references to entire books (see sections 30.30–30·38), but additional elements with their own styling are needed. For an entire volume, the designation of volume and the title are placed between the book's title and the place of publication. For a part or contribution to a book, the numeration and title are as found in the book; the title for a volume, chapter, table, illustration, or other relevant part follows the general term for the part and its number; needed specific pages follow, with numerals as prescribed for journal articles (see section 30·22).

Examples of Formats

VOLUME WITH SEPARATE TITLE

30·40 C-S Cajori F. A history of mathematical notations. Volume 2, Notation mainly in higher mathematics. Chicago: Open Court; 1929.

N-Y Cajori F. 1929. A history of mathematical notations. Volume 2, Notation mainly in higher mathematics. Chicago: Open Court.

Form of Citation: (Cajori 1929)

CHAPTER OR OTHER PART WITH SEPARATE TITLE BUT SAME AUTHOR(S)

C-S Hebel R, Stromberg MW. Anatomy of the laboratory rat. Baltimore: Williams & Wilkins; 1976. Part C, Digestive system; p 43–54.

N-Y Hebel R, Stromberg MW. 1976. Anatomy of the laboratory rat. Baltimore: Williams & Wilkins. Part C, Digestive system; p 43–54.

C-S Singleton P, Sainsbury D. Dictionary of microbiology and molecular biology. 2nd ed. New York: J Wiley; 1987. Plasmid; p 682–3.

N-Y Singleton P, Sainsbury D. 1987. Dictionary of microbiology and molecular biology. 2nd ed. New York: J Wiley. Plasmid; p 682–3.

CHAPTER OR OTHER PART WITH DIFFERENT AUTHORS

C-S Kuret JA, Murad F. Adenohypophyseal hormones and related substances. In: Gilman AG, Rall TW, Nies AS, Taylor P, editors. The pharmacological basis of therapeutics. 8th ed. New York: Pergamon; 1990. p 1334–60.

N-Y Kuret JA, Murad F. 1990. Adenohypophyseal hormones and related substances. In: Gilman AG, Rall TW, Nies AS, Taylor P, editors. The pharmacological basis of therapeutics. 8th ed. New York: Pergamon. p 1334–60.

CONFERENCE PUBLICATIONS: PROCEEDINGS

General Formats

30·41 References to entire proceedings should include the following information with this sequence, punctuation, and spacing.

C-S Editor(s). Title of publication or conference. Name of conference (if not the 2nd element); inclusive dates of conference; place of conference. Place of publication: publisher; date of publication. Total number of pages.

N-Y Editor(s). Date of publication. Title of publication or conference. Name of conference (if not the 2nd element); inclusive dates of conference; place of conference. Place of publication: publisher. Total number of pages.

The elements are styled, in general, as in references to books; see sections 30·30–30·38.

Examples of Formats

PROCEEDINGS WITH TITLE DIFFERENT FROM CONFERENCE NAME

30·42 C-S Vivian VL, editor. Child abuse and neglect: a medical community response. 1st AMA National Conference on Child Abuse and Neglect; 1984 Mar 30–31; Chicago. Chicago: American Medical Assoc; 1985. 256 p.

N-Y Vivian VL, editor. 1985. Child abuse and neglect: a medical community response. 1st AMA National Conference on Child Abuse and Neglect; 1984 Mar 30–31; Chicago. Chicago: American Medical Assoc. 256 p.

Form of Citation: (Vivian 1985)

PROCEEDINGS WITHOUT SEPARATE TITLE

C-S Ferguson RM, Sommer BG, editors. Proceedings of a Conference on the Clinical Management of the Renal Transplant Recipient with Cyclosporine; 1985 Nov 3–5; Palm Springs, FL. Orlando (FL): Grune & Stratton; 1986. 216 p.

N-Y Ferguson RM, Sommer BG, editors. 1986. Proceedings of a Conference on the Clinical Management of the Renal Transplant

Recipient with Cyclosporine; 1985 Nov 3–5; Palm Springs, FL. Orlando (FL): Grune & Stratton. 216 p.

Form of Citation: (Ferguson and Sommer 1986)

PROCEEDINGS WITHOUT EDITOR

C-S [Anonymous]. Symposium on Nonhuman Primate Models for AIDS; 1989 Oct 11–13; Portland, OR. [Portland]: Oregon Regional Primate Center; 1989. 115 p.

N-Y [Anonymous]. 1989. Symposium on Nonhuman Primate Models for AIDS; 1989 Oct 11–13; Portland, OR. [Portland]: Oregon Regional Primate Center. 115 p.

Form of Citation: (Anonymous 1989)

Format Details

EDITORS

30·43 The same format should be used for names of editors as for authors of journal articles and for book authors or editors; see sections 30·22 and 30·33. If no personal editors are identified, an organization may be named as editor; if no personal editor or organization is identified, the editor is "[Anonymous]". Note that the NLM recommendations call for 1st names rather than initials. The deviation here is recommended to reduce differences between formats of references to journal articles and to books or parts of books and to supply the editor element in text citations in the name-year system.

Title and Type of Medium

The title may be a specific subject title or a general title for the conference. If the 2 titles differ, the conference title is given as the 3rd element. Conference titles are formal names and are capitalized accordingly. Subject titles have only initial-word capitalization. Ordinal designations are converted to numeric ordinals, for example, "Third" to "3rd". Microforms are indicated within square brackets at the end of the 1st title element, for example, "[microfiche]". If the only title is the conference name, the medium designator follows the place of the meeting.

DATES OF CONFERENCE

Dates should be in the sequence "year month day"; months are 3-letter abbreviations without periods; examples are "1988 Jan 24–26" and "1987 Jan 31–Feb 2".

PLACES OF CONFERENCE AND PUBLICATION

The style for books is recommended; see section 30·35. An institutional name such as "University of Arizona" may precede the city or town name, as in "University of Arizona, Tempe (AZ)", for the place of conference.

PUBLISHER, DATE OF PUBLICATION, TOTAL NUMBER OF PAGES, NOTES

The rules for books should be applied; see sections 30·36 and 30·37.

CONFERENCE PRESENTATIONS (PAPERS) AND ABSTRACTS

General Format

30·44 The structure of references to conference papers and abstracts is parallel to the structure of references to book chapters (sections 30·39 and 30·40), the title of the paper being followed by the general format for references to conference proceedings.

For oral presentations published as "papers".

C-S Author(s) of paper. Title of paper. Connective phrase [In]: editors of proceedings. Title of publication or name of conference, or both; inclusive dates; place of conference. Place of publication: publisher; date of publication. Inclusive pages of paper. (Notes [if needed, as for series title]).

N-Y Author(s) of paper. Date of publication. Title of paper. Connective phrase [In]: editors of proceedings. Title of publication, or name of conference, or both; inclusive dates; place of conference. Place of publication: publisher. Inclusive pages of paper. (Notes [if needed, as for series title]).

For abstracts.

C-S Author(s) of abstract. Title of abstract [abstract]. Connective phrase [In]: editors of proceedings. Title of publication or name of conference; inclusive dates; place of conference. Place of publication: publisher; date of publication. Page of program or proceedings. Abstract number.

N-Y Author(s) of abstract. Date of publication. Title of abstract [abstract]. Connective phrase [In]: editors of proceedings. Title of publication or name of conference; inclusive dates; place of conference. Place of publication: publisher. Page of program or proceedings. Abstract number.

Most of the elements are styled as they are in references to book chapters and to conference proceedings; see sections 30·39–30·42. The connective phrase is "In:". The inclusive page numbers follow "p" and the rule for pages of journal articles. Appended to the title element is the statement "[abstract]"; the format for "Abstract number" is "Abstract nr xxxx.".

Examples of Formats

PAPER FROM A PROCEEDINGS WITH SEPARATE TITLE

C-S Meyer B, Hermanns K. Formaldehyde release from pressed wood products. In: Turoski V, editor. Formaldehyde: analytical chemistry and toxicology. Proceedings of the symposium at the 187th meeting of the American Chemical Society; 1984 Apr 8–13; St Louis, MO. Washington: American Chemical Soc; 1985. p 101–16. (Advances in chemistry series; 210).

N-Y Meyer B, Hermanns K. 1985. Formaldehyde release from pressed wood products. In: Turoski V, editor. Formaldehyde: analytical chemistry and toxicology. Proceedings of the symposium at the 187th meeting of the American Chemical Society; 1984 Apr 8–13; St Louis, MO. Washington: American Chemical Soc. p 101–16. (Advances in chemistry series; 210).

Form of Citation: (Meyer and Hermanns 1985)

PAPER FROM A PROCEEDINGS WITHOUT SEPARATE TITLE

C-S Kalter RJ. Macro and micro economic implications of bovine somatotropin on the dairy industry. In: BIO EXPO 86: proceedings; 1986 Apr 29–May 1; Boston. Stoneham (MA): Butterworth; 1986. p 203–15.

N-Y Kalter RJ. 1986. Macro and micro economic implications of bovine somatotropin on the dairy industry. In: BIO EXPO 86: proceedings; 1986 Apr 29–May 1; Boston. Stoneham (MA): Butterworth. p 203–15.

Form of Citation: (Kalter 1986)

ABSTRACT

C-S Mendez MF, Manon-Espaillat R, Lanska DJ, Burstine TH. Epilepsy and suicide attempts [abstract]. In: American Academy of Neurology 41st annual meeting program; 1989 Apr 13–19; Chicago. Cleveland (OH): Edgell Communications; 1989. p 295. Abstract nr PP369.

N-Y Mendez MF, Manon-Espaillat R, Lanska DJ, Burstine TH. 1989. Epilepsy and suicide attempts [abstract]. In: American Academy of Neurology 41st annual meeting program; 1989 Apr 13–19; Chicago.

Cleveland (OH): Edgell Communications. p 295. Abstract nr PP369.

Form of Citation: (Mendez and others 1989)

SCIENTIFIC AND TECHNICAL REPORTS AND THEIR PARTS

General Format

30·45 Most of the references for reports are similar in content and style to those for books, but additional information is likely to be needed for adequate identification.

C-S Author(s) (performing organization). Title. Place of publication: publisher or sponsoring organization; date of publication. Report number. Contract number. Total number of pages. Availability statement.

N-Y Author(s) (performing organization). Date of publication. Title. Place of publication: publisher or sponsoring organization. Report number. Contract number. Total number of pages. Availability statement.

Examples of Formats

30·46 The formats illustrated below can be applied to documents issued by any government agency, university unit, or private research institute.

US GOVERNMENT REPORT WITH SEPARATE PERFORMING ORGANIZATION

C-S Cooper LN (Department of Physics, Brown University, Providence RI). Theoretical and experimental research into biological mechanisms underlying learning and memory. Final progress report 1 Aug 88–31 Jul 89. Washington: Air Force Office of Scientific Research; 1990 Apr 24. Report nr AFOSR-TR–90–0672. Contract nr AFOSR–88–0228;2305;B4. 19 p. Available from: NTIS, Springfield, VA; AD-A223615.

N-Y Cooper LN (Department of Physics, Brown University, Providence RI). 1990. Theoretical and experimental research into biological mechanisms underlying learning and memory. Final progress report 1 Aug 88–31 Jul 89. Washington: Air Force Office of Scientific Research. Report nr AFOSR-TR–90–0672. Contract nr AFOSR–88–0228;2305;B4. 19 p. Available from: NTIS, Springfield, VA; AD-A223615.

Form of Citation: (Cooper 1990)

US GOVERNMENT REPORT WITHOUT SEPARATE PERFORMING ORGANIZATION

C-S National Institutes of Health (US) [NIH]. Report of the Human Fetal Tissue Transplantation Panel, consultants to the Advisory Committee to the Director, NIH. Final report. Bethesda (MD): NIH; 1988 Dec. 2 volumes. Available from: NTIS, Springfield, VA; PB90–155268, PB90–155276.

N-Y [NIH] National Institutes of Health (US). 1988 Dec. Report of the Human Fetal Tissue Transplantation Panel, consultants to the Advisory Committee to the Director, NIH. Final report. Bethesda (MD): NIH. 2 volumes. Available from: NTIS, Springfield, VA; PB90–155268, PB90–155276.

Form of Citation: (NIH 1988)

US GOVERNMENT-SPONSORED REPORT ISSUED BY PERFORMING ORGANIZATION

C-S Moray NP, Huey M. Human factors research and nuclear safety. Washington: National Academy Pr; 1988. Contract nr NRC–04–86–301. 122 p. Available from: NTIS, Springfield, VA; PB89–175517. Sponsored by the Nuclear Regulatory Commission.

N-Y Moray NP, Huey M. 1988. Human factors research and nuclear safety. Washington: National Academy Pr. Contract nr NRC–04–86–301. 122 p. Available from: NTIS, Springfield, VA; PB89–175517. Sponsored by the Nuclear Regulatory Commission.

Form of Citation: (Moray and Huey 1988)

NONGOVERNMENT REPORT

C-S Johns Hopkins University [JHU], Applied Physics Laboratory. Biomedical research, development, and engineering at the Johns Hopkins University Applied Physics Laboratory. Annual report 1 Oct 78–10 Sep 79. Laurel (MD): JHU; 1979 Oct. Report nr JHU/APL/MQR–79. 74 p.

N-Y [JHU] Johns Hopkins University, Applied Physics Laboratory. 1979 Oct. Biomedical research, development, and engineering at the Johns Hopkins University Applied Physics Laboratory. Annual report 1 Oct 78–10 Sep 79. Laurel (MD): JHU. Report nr JHU/APL/MQR–79. 74 p.

Form of Citation: (JHU 1979)

Format Details

AUTHOR(S) OR EDITOR(S)

30·47 See section 30·33 for the styles for personal names and organizational names. When specific authors or editors are not named on the title page, the sponsoring or publishing organization can serve as author for citation purposes; when 2 or more units representing a unitary hierarchy are named as author, the higher unit comes first.

The author's affiliation or the site of the reported work can be identified within parentheses at the end of the author name, but a performing organization is not named here if it is the same as the publisher or sponsoring organization.

PUBLISHER OR SPONSORING ORGANIZATION

Note that "publisher" here often means the organization responsible for issuing the report rather than the publisher in the usual sense. For example, in a technical sense the Superintendent of Documents is the publisher for many documents issued by departments, agencies, or other units of the US government, but the issuing unit is named as the publisher. The publisher name can be represented by an abbreviation that has been inserted as a square-bracketed addition in a preceding element; note, however, that this form is not sanctioned by the relevant ANSI–NISO and ISO standards.

REPORT NUMBER

30·48 A report's number should open with the connective phrase "Report nr" followed by the alphanumeric identifier assigned by the sponsoring organization. The identifier often includes an alphabetic abbreviation identifying the organization followed by numeric indicators of year of issuance and the place of the report in the sequence of other reports of the same category for the organization.

Note that the NLM recommends "No." as the abbreviation for "number" but that this manual recommends "Nr" and "nr". If, however, the document uses the abbreviation "No.", that form should be used. The preference for "Nr" and "nr" comes from its being an accurate abbreviation of the English word "number" and thus is to be preferred over an abbreviation from another language ("No." from the French "numéro" or the Italian "numero"); see section 10·3, item 3. Also note that "nr" is a standard abbreviation for "number" in Germanic languages. No period is used after "Nr" and "nr" so as to reserve the period in bibliographic references as a marker for the end of a bibliographic field.

CONTRACT (GRANT, ORDER) NUMBER

30·49 An identifier for financial support (contract, grant, order) should open with a connective phrase such as "Contract nr" followed by the usually alphanumeric identifier.

AVAILABILITY STATEMENT

Many reports are not directly obtainable from the "publisher" or sponsoring organization but must be ordered from a separate agency. The relevant information should be introduced by the connective phrase "Available from:".

PARTS OF REPORTS

The format and style for references to parts of reports are, in general, the same as for parts of books; see sections 30·39, 30·40.

DISSERTATIONS, THESES, AND THEIR PARTS

General Format

30·50 References to dissertations, theses, and similar academic documents follow the general format as references to books but with some other specific elements.

C-S Author. Title [type of publication]. Place of institution: institution granting the degree; date of degree. Total number of pages. Availability statement.

N-Y Author. Date of degree. Title [type of publication]. Place of institution: institution granting the degree. Total number of pages. Availability statement.

Examples of Formats

C-S Ritzmann RE. The snapping mechanism of *Alpheid* shrimp [dissertation]. Charlottesville (VA): University of Virginia; 1974. 59 p. Available from: University Microfilms, Ann Arbor, MI; AAD74–23.

N-Y Ritzmann RE. 1974. The snapping mechanism of *Alpheid* shrimp [dissertation]. Charlottesville (VA): University of Virginia. 59 p. Available from: University Microfilms, Ann Arbor, MI; AAD74–23.

Form of Citation: (Ritzmann 1974)

Format Details

TYPE OF PUBLICATION

The type of publication is specified within square brackets at the end of the title. In some instances an indication of the degree level may be needed.

[dissertation] [DPhil thesis] [MSc thesis]

PUBLICATION DATA

The elements of place of institution, institution, and date of degree substitute for place of publication, publisher name, and date of publication.

AVAILABILITY STATEMENT

An availability statement includes information as to where the document can be found or borrowed if the source is other than the institution's own library; it may specify a vendor or another library.

BIBLIOGRAPHIES

General Format

30·51 References to bibliographies have the general format of references to books but may contain some other specific information.

C-S Compiler(s). Title [type of publication]. Place of publication: publisher; date of publication. Total number of pages. Availability. Notes.

N-Y Compiler(s). Date of publication. Title [type of publication]. Place of publication: publisher. Total number of pages. Availability. Notes.

Examples of Formats

C-S Gluckstein FP, Glock MH, Hill JG, compilers. Bovine somatotropin [bibliography]. Bethesda (MD): National Library of Medicine, Reference Section; 1990. 53 p. Available from: US Government Printing Office, Washington; Stock nr 817–006–00013–2. 1097 citations; 1986 Jan through 1990 Oct.

N-Y Gluckstein FP, Glock MH, Hill JG, compilers. 1990. Bovine somatotropin [bibliography]. Bethesda (MD): National Library of Medicine, Reference Section. 53 p. Available from: US Government Printing Office, Washington; Stock nr 817–006–00013–2. 1097 citations; 1986 Jan through 1990 Oct.

Form of Citation: (Gluckstein and others 1990)

Format Details

COMPILER(S)

The author element is the name of the designated compiler(s), usually a personal name, but an organization may be designated as compiler. An editor may be indicated in the publication rather than a compiler and should be so designated.

TITLE

A square-bracketed "[bibliography]" is added to the title element if the title contains no term describing the type.

AVAILABILITY AND NOTES

If the bibliography is available only from a source other than the publisher, the information relevant to securing copies should be added. The notes field can carry additional description, such as number of citations, annotations, indexes, but such data may not be needed in reference lists.

PATENTS

General Format

30·52 References are composed mainly of information specific to patents.

C-S Name of the inventor of the patented device or process, the word "inventor"; assignee. Title. Patent descriptor. Date issued [year month day].

N-Y Name of the inventor of the patented device or process, the word "inventor"; assignee. Date issued [year month day]. Title. Patent descriptor.

Examples of Formats

C-S Harred JF, Knight AR, McIntyre JS, inventors; Dow Chemical Company, assignee. Epoxidation process. US patent 3,654,317. 1972 Apr 4.

N-Y Harred JF, Knight AR, McIntyre JS, inventors; Dow Chemical Company, assignee. 1972 Apr 4. Epoxidation process. US patent 3,654,317.

Form of Citation: (Harred and others 1972)

In some contexts a citation by patent number may be suitable.

C-S US patent 3,654,317. 1972 Apr 4.

N-Y US patent 3,654,317. 1972 Apr 4.

Form of Citation: (US patent 3,654,317 1972)

Format Details

INVENTOR(S)

This equivalent of the author element gives the inventor name(s) followed by a comma, a space, and the role designator, "inventor".

ASSIGNEE(S)

This element gives the name(s) of the person(s) or legal entity (such as a company) to whom the rights accruing from the patent are assigned; it closes with the role designator, "assignee".

PATENT DESCRIPTOR

The descriptor must include the name of the country issuing the patent; the name can be a standard abreviation. The patent may be designated as such, or a specific document name may be appropriate. The unique patent number must be given.

NEWSPAPER AND MAGAZINE ARTICLES

General Format

30·53 References to newspapers and magazines (nonscholarly periodicals) include the same general information as journal-article references but refer to specific details of newspaper or magazine style and format.

C-S Author(s). Article title. Newspaper title and date of publication;section designator:page number(column number).
C-S Author(s). Article title. Magazine title and date of publication:page numbers.
N-Y Author(s). Date of publication. Article title. Newspaper title;section designator:page number(column number).
N-Y Author(s). Date of publication. Article title. Magazine title:page numbers.

In the name-year system, the citation in most instances need include only the year of publication.

Examples of Formats

SIGNED NEWSPAPER ARTICLE

C-S Rensberger B, Specter B. CFCs may be destroyed by natural process. Washington Post 1989 Aug 7;Sect A:2(col 5).

N-Y Rensberger B, Specter B. 1989 Aug 7. CFCs may be destroyed by natural process. Washington Post;Sect A:2(col 5).

Form of Citation: (Rensberger and Specter 1989)

UNSIGNED NEWSPAPER ARTICLE

C-S [Anonymous]. Gene data may help fight colon cancer. Los Angeles Times 1990 Aug 24;Sect A:4.

N-Y [Anonymous]. 1990 Aug 24. Gene data may help fight colon cancer. Los Angeles Times;Sect A:4.

Form of Citation: (Anonymous 1990)

Note that periods are omitted after abbreviations for standard terms such as "Section" and "Column".

MAGAZINE ARTICLE

C-S Lu C. A small revelation: Newton has arrived at long last. Macworld 1993 Sep:102–6.

N-Y Lu C. 1993 Sep. A small revelation: Newton has arrived at long last. Macworld:102–6.

Form of Citation: (Lu 1993)

Format Details

AUTHOR(S)

For an article without an identified author, use "[Anonymous]". Note that this usage is not recommended by the NLM but is recommended here to facilitate citations in the name-year system.

ARTICLE TITLE

A secondary or subsidiary title can follow the main title after a colon and a space. Specific kinds of articles can be identified as such by an appropriate term within square brackets at the end of title, for example "[editorial]".

NEWSPAPER OR MAGAZINE TITLE

If the place of publication of the newspaper or magazine needs to be identified because of the publication's obscurity or ambiguity for a readership, it can be given within parentheses at the end of the newspaper or magazine title. The same style can be used for statement of a specific edition, which may be needed because of differing pagination in 2 or more editions of the same issue. Newspaper titles that include place names may include standard abbreviated forms. The titles of popular magazines should not be abbreviated.

New York Times [or] NY Times

DATE OF PUBLICATION

The date is given as year month day; the month is given as the full English name or a 3-letter abbreviation. If the magazine carries volume and issue numbers (as on scholarly journals), these may be used instead of the date and in a format like references to scholarly journals.

PAGINATION

The style of pagination is the same as for journal articles; discontinuous pages are separated by a comma and a space.

MAPS

General Format

30·54 The format for references to maps substitutes area information for the usual author name; this exception also applies to the citation in the name-year system.

C-S Area represented. Title [type of map]. Place of publication: publisher; date of publication. Physical description.

N-Y Area represented. Date of publication. Title [type of map]. Place of publication: publisher. Physical description.

Examples of Formats

SHEET MAP

C-S Indonesia. Malaria DDT spraying programs [demographic map]. Washington: US Army Map Service; 1953. 3 sheets.

N-Y Indonesia. 1953. Malaria DDT spraying programs [demographic map]. Washington: US Army Map Service. 3 sheets.

Form of Citation: (Indonesia 1953)

MAP IN ATLAS

C-S China. Stomach (male) cancer mortality, 1973–1975, by county [demographic map]. In: Atlas of cancer mortality in the People's Republic of China. Shanghai: China Map Pr; 1979. p 53–4. Color, scale 1:12,000,000.

N-Y China. 1979. Stomach (male) cancer mortality, 1973–1975, by county [demographic map]. In: Atlas of cancer mortality in the People's Republic of China. Shanghai: China Map Pr. p 53–4. Color, scale 1:12,000,000.

Form of Citation: (China 1979)

Format Details

AREA

The geographic or political area represented is, for US usage, usually named at the country level followed by a comma and space and then the smaller area unit, for example, "France, Paris"; for units within the

United States the state is the initial part of the element and is followed by the smaller unit, for example, "Maryland, Montgomery County". If the smaller unit is not named on the map, a designator may have to be devised; the devised name should be preceded by "Section", for example, "United States, Section Mid-Atlantic". English equivalents are used for non-English names, for example, "Italy, Rome" for "Italia, Roma".

TITLE, TYPE OF MAP, AND DESCRIPTION

The type of map should be designated within square brackets at the end of the title. Physical descriptions (such as numbers of sheets, sheet size, scale, and so on) follow the date of publication in the C-S system and the publisher name in the N-Y system.

OTHER ELEMENTS

Additional kinds of identification may be needed.

Cartographer: Placed after title.

Edition: Placed after title (or after cartographer if named).

Series statement: Within parentheses at end of reference.

Availability: Placed after series statement.

LEGAL, GOVERNMENT, AND AGENCY DOCUMENTS

Legal Documents

30·55 The NLM *Recommended Formats* . . . (NLM 1991) does not provide its own recommendations but illustrates the conventions used in law in the United States. References in the legal literature to legal documents differ greatly in style from those generally used in science and the humanities. The style is described in detail in *A Uniform System of Citation* (HLRA 1992). Even though these references may be barely comprehensible to most readers in science, they identify legal documents accurately for retrieval from law and general libraries. The structure of a legal reference gives the name of a case or title of a statute and then the publication in which the case is reported or the statute is published.

Note that periods not marking the ends of bibliographic fields are allowed in these formats as idiosyncrasies of the established formats in the legal literature.

Note that these formats represent the standard system in the United States and may not be applicable to legal documents in other countries.

Examples

CASES

Jackson v. Metropolitan Edison Co., 348 F. Suppl. 954, 956–58 (M. D. Pa. 1972)

[The abbreviation "F. Suppl." stands for *Federal Supplement.*]

Willis v. Thomas, 600 P.2d 1079, 1083 (Alaska 1979)

[The abbreviation "P.2d" stands for *Pacific Reporter.*]

STATUTES

National Environmental Policy Act of 1969, § 102, 42 U.S.C. § 4332 (1982)

[The abbreviation "U.S.C." stands for *United States Code.*]

Citation of Legal References

The citation-sequence system cites such references with its convention of superscript numbers representing the references.

In the name-year system the citation should include the 1st reference element (title of case; name of statute) followed by year of publication.

Forms of Citation: (*Willis v. Thomas* 1979)
(National Environmental Policy Act of 1969, 1982)

Note that these recommendations depart from the general rule for the name-year system of citations that use the author name.

Legislative Documents

30·56 Similar conventions are applied for references to unenacted and enacted bills (that is, bills and acts) and resolutions, and to hearings and reports. The format is an abbreviated designator for the bill, resolution, or report, followed by a descriptor for the legislative session. Hearings are represented by a title rather than an abbreviated designator. Citations can use the initial designation, or an abbreviated version of it, and the date. Again, these formats represent styles in the United States and may not be relevant for references to legislative documents in other countries.

A US SENATE BILL, UNENACTED

S. 2830, 96th Cong., 2d Sess. § 8 (1980)

Form of Citation: (S. 2830, 1980)

AN ENACTED STATE (UTAH) JOINT RESOLUTION

H.R.J. Res 1, 40th Leg., 2d Spec. Sess., 1974 Utah Laws 7

Form of Citation: (H.R.J. Res. 1, 1974)

A US SENATE HEARING

Toxic Substances Control Act: Hearing on S. 776 Before the Subcomm. on the Environment of the Senate Comm. on Commerce, 94th Cong., 1st Sess. 343 (1975)

Form of Citation: (Toxic Substances Control Act 1975)

Other Government and Agency Documents

30·57 The great number of types of other documents in the United States precludes stating recommendations here for references for all possible needs. Many reports of scientific importance can be referenced with the formats recommended in sections 30·45–30·49 for government and government-sponsored reports. The NLM *Recommended Formats . . .* (NLM 1991) gives examples of references to US Senate and House of Representative reports; see Appendix Table 30·1.

Guidance on formats for governmental and agency documents in other countries and for documents issued by international organizations such as the United Nations can be found in *The Complete Guide to Citing Government Information Resources: A Manual for Writers and Librarians* (Garner and Smith 1993), the Canadian *Bibliographic Style Manual* (Thibault 1990), the Australian *Style Manual for Authors, Editors and Printers* (AGPS 1988), and *A Guide to Writing for the United Nations* (UN 1984).

AUDIOVISUAL PUBLICATIONS AND MATERIALS

General Format

30·58 For the citation-sequence system, the format differs from formats for journal articles and books by having the title as the 1st field.

C-S Title [type of medium]. Author(s) or editor(s) or any combination of these. Producer (if corporate and different from publisher). Place of publication: publisher; date of publication. Physical description. (Series statement if available). Accompanying material. Availability statement.

N-Y Author(s) or editor(s) or any combination of these. Date of publication. Title [type of medium]. Producer (if corporate and different from publisher). Place of publication: publisher. Physical description. Series statement (if available). Accompanying material. Availability statement.

This departure for the name-year system from the NLM's recommendation to place the title field in the 1st position is adopted to maintain the general rule for citations: (Author name Year).

Not all of this information is needed in some cases. For example, a reference in a research report to a commercially available videocassette applied in the research to instruct subjects in the study may be entirely adequate without the addition of physical description, accompanying material, and notes.

Examples of Formats

AUDIOCASSETTE

30·59 C-S Topics in clinical microbiology [audiocassette]. Clark R and others, editors; American Society for Microbiology, producers. Baltimore: Williams & Wilkins; 1976. 24 audiocassettes: 2-track, 480 min. Accompanied by: 120 color slides, 2 × 2 in; 1 guide.

N-Y Clark R and others, editors. 1976. Topics in clinical microbiology [audiocassette]. American Society for Microbiology, producers. Baltimore: Williams & Wilkins. 24 audiocassettes: 2-track, 480 min. Accompanied by: 120 color slides, 2 × 2 in; 1 guide.

Form of Citation: (Clark and others 1976)

VIDEOCASSETTE

30·60 C-S New horizons in esthetic dentistry [videocassette]. Wood RM, editor. Visualeyes Productions, producer. [Chicago]: Chicago Dental Society; 1989. 2 videocassettes: 170 min, sound, color with black and white, 1/2 in. (Clinical topics in dentistry; Nr 46). Accompanied by: 1 guide. Available from: Great Plains National Instructional Television Library, Lincoln, NE.

N-Y Wood RM, editor. 1989. New horizons in esthetic dentistry [videocassette]. Visualeyes Productions, producer. [Chicago]: Chicago Dental Society. 2 videocassettes: 170 min, sound, color with black and white, 1/2 in. (Clinical topics in dentistry; Nr 46). Accompanied by: 1 guide. Available from: Great Plains National Instructional Television Library, Lincoln, NE.

Form of Citation: (Wood 1989)

Format Details

AUTHOR(S), EDITOR(S), INDIVIDUAL PRODUCER(S), CORPORATE PRODUCER

30·61 The general format for the author field is surname and initials as for journal articles and books. If the field must carry the names of author(s) and editor(s), place the author(s) name(s) in the 1st position followed by a semicolon and a space and the editor name(s) in the same format but followed as needed by an indicator of function, such as "editor". If no persons are named for these functions, the publisher name is given as the author name; see section 30·28 for formats.

Capitalization for the names of corporate producers is the same as for proper names.

TITLE AND TYPE OF MEDIUM, EDITION, PLACE OF PUBLICATION, PUBLISHER, DATE OF PUBLICATION, SERIES STATEMENT, AVAILABILITY, LANGUAGE

The relevant formats are those for books; see sections 30·27–30·34. Examples of appropriate terms for type of medium are "audiocassette", "filmstrip", "motion picture", "videocassette", "videodisk", "videotape".

PHYSICAL DESCRIPTION

The physical description field can include the number of cassettes, reels, or other kinds of package units, followed after a colon and a space by the running time, designators for sound and color character ("silent", "sound" ; "color", "black and white"); and the dimension(s) of the units (tape or film width, speed, and so on).

ACCOMPANYING MATERIALS

This element opens with "Accompanied by:" followed by a space and the relevant statement, such as "1 instructor's guide, 6 p.".

ELECTRONIC PUBLICATIONS

30·62 Many kinds of publications and source materials are available in electronic formats; some are transportable (such as computer disks) and others are available through telecommunication ("online"). Detailed specifications of appropriate references and example references are given in the NLM *Recommended Formats* . . . (NLM 1991); also see Appendix Table 30·1.

Most references in the scientific literature to publications in electronic formats are likely to be to journal (serial) articles, to books or monographs and their parts, and to computer programs. Note that the titles of electronic publications that have been trademarked or copyrighted and are fully capitalized on the screen or on the title page of the documentation are included that way in the reference.

Electronic-Journal Articles

30·63 The format recommended by the NLM for references to electronic-journal articles is like that for print-journal articles (see sections 30·19–30·21).

C-S Author(s). Title of article. Abbreviated journal title [type of medium] date of publication; volume number(issue number):pagination. Availability statement. [Date of accession if needed].

N-Y Author(s). Date of publication. Title of article. Abbreviated journal title [type of medium]; volume number(issue number):pagination. Availability statement. [Date of accession if needed].

Some journals in electronic formats may allow changes within documents; if so, the date the journal was accessed should be given within square brackets as the last field, with the format "Accessed year month day". This location differs from that recommended by the NLM and is recommended here to minimize the risk of confusion with the publication date.

Note that some electronic journals may not use the conventions of "volume", "issue", and "pages"; for these, the critical identifiers for an article may be the "date of publication" (year month day) and a designator such an accession or document number. Because electronic-journal publication is still in its infancy, new varieties of formats may have to be developed; the following examples illustrate the NLM recommendations.

C-S Cisler S. MediaTracks. Public Access Comput Syst Rev [serial online] 1990;1(3):109–15. Available from: Public Access Computer Systems Forum PACS-L via the INTERNET. Accessed 1990 Nov 29.

N-Y Cisler S. 1990. MediaTracks. Public Access Comput Syst Rev [serial online];1(3):109–15. Available from: Public Access Computer Systems Forum PACS-L via the INTERNET. Accessed 1990 Nov 29.

Form of Citation: (Cisler 1990)]

C-S Harrison CL, Schmidt PQ, Jones JD. Aspirin compared with acetoaminophen for relief of headache. Online J Therap [serial online] 1992 Jan 2;Doc nr 1:[4320 words; 10 paragraphs]. 5 figures; 10 tables.

N-Y Harrison CL, Schmidt PQ, Jones JD. 1992 Jan 2. Aspirin compared with acetoaminophen for relief of headache. Online J Therap [serial online];Doc nr 1:[4320 words; 10 paragraphs]. 5 figures; 10 tables.

Form of Citation: (Harrison and others 1992)

Electronic Books (Monographs) and Their Parts

30·64 The general formats are like those for books and parts of books; see sections 30.30–30·40. The type-of-medium designator and a statement on availability will usually be needed. The date "accessed" (consulted) is needed to maintain an adequate "paper trail" (precise identification of a document as it existed at a particular time; some electronic documents may be modified from time to time.

Note that the name-year examples deviate from the NLM recommendations to establish an author name at the beginning of the reference for its use in the citation.

ELECTRONIC BOOKS (MONOGRAPHS)

C-S MARTINDALE ONLINE [monograph online]. London: Pharmaceutical Society of Great Britain; 1989 [updated 1989 Dec]. Available from: Dialog Information Services, Palo Alto, CA. Related to the publication Martindale, The Extra Pharmacopoeia. Accessed 1990 Jan 10.

N-Y [PSGB] Pharmaceutical Society of Great Britain. 1989. MARTINDALE ONLINE [monograph online]. London: PSGB; [updated 1989 Dec]. Available from: Dialog Information Services, Palo Alto, CA. Related to the publication Martindale, The Extra Pharmacopoeia. Accessed 1990 Jan 10.

Form of Citation: (PSGB 1989)

PARTS OF ELECTRONIC BOOKS (MONOGRAPHS)

C-S THE MERCK INDEX ONLINE [monograph online]. 10th ed. Rahway (NJ): Merck;1984 [updated 1989 Jan]. Acyclovir; monograph nr 140 [44 lines]. Available from: BRS Information Technologies, McLean, VA. Accessed 1990 Dec 7.

N-Y Merck. 1984. THE MERCK INDEX ONLINE [monograph online]. 10th ed. Rahway (NJ): Merck;1984 [updated 1989 Jan]. Acyclovir; monograph nr 140. Available from: BRS Information Technologies, McLean, VA. Accessed 1990 Dec 7.

Form of Citation: (Merck 1984)

Computer Programs

30·65 The general format for citation-sequence references to computer programs has the title field in the 1st position if no author is identified; for the name-year system, the publisher can be named as "author". Note that "edition" is frequently designated as "version".

C-S Title [type of medium]. Edition. Place of publication: publisher; date of publication. Physical description. Accompanying material. Notes.

N-Y Title [type of medium]. Date of publication. Edition. Place of publication: publisher. Physical description. Accompanying material. Notes.

Not all of this information is needed in some cases. For example, a reference in a research report to a standard computer program applied in the research for statistical analysis may be entirely adequate without additional physical description, accompanying material, and notes.

Note in the examples below that the names of personal authors of computer programs are placed 1st to bring these formats into consistency with most formats for print materials with personal authors; this is a deviation from the NLM recommendations.

COMPUTER PROGRAM WITHOUT NAMED AUTHOR AND WITH TRADEMARKED TITLE

C-S GRATEFUL MED [computer program]. Version 5.0. Bethesda (MD): National Library of Medicine; 1990. 5 computer disks: 5 1/4 in.; or 2 computer disks: 3 1/2 in. Accompanied by: 1 user's guide; 1 troubleshooting guide. System requirements: IBM PC family or fully compatible computer; DOS 2.0 or higher; Hayes Smartmodem or fully compatible modem; 384K RAM required, 512K RAM recommended; 1 or more floppy drives; hard disk with a minimum of 2 MB of free space strongly recommended.

N-Y [NLM] National Library of Medicine. 1990. GRATEFUL MED [computer program]. Version 5.0. Bethesda (MD): NLM. 5 computer disks: 5 1/4 in.; or 2 computer disks: 3 1/2 in. Accompanied by: 1 user's guide; 1 troubleshooting guide. System requirements: IBM PC family or fully compatible computer; DOS 2.0 or higher; Hayes Smartmodem or fully compatible modem; 384K RAM required, 512K RAM recommended; 1 or more floppy drives; hard disk with a minimum of 2 MB of free space strongly recommended.

Form of Citation: (NLM 1990)

COMPUTER PROGRAM WITH PERSONAL AUTHORS

C-S Rosenberg V, Ghalambor C, Rycus P, Thomas R. PRO-CITE [computer program]. Version 1.4. Ann Arbor (MI): Personal Bibliographic Software; 1988. 3 computer disks: color, 5 1/4 in. Accompanied by: 1 manual. System requirements: IBM PC, XT, AT, PS/2, or any 100% compatible computer; 320K RAM; DOS 2.0 or higher.

N-Y Rosenberg V, Ghalambor C, Rycus P, Thomas R. 1988. PRO-CITE [computer program]. Version 1.4. Ann Arbor (MI): Personal Bibliographic Software. 3 computer disks: color, 5 1/4 in. Accompanied by: 1 manual. System requirements: IBM PC, XT, AT, PS/2, or any 100% compatible computer; 320K RAM; DOS 2.0 or higher.

Form of Citation: (Rosenberg and others 1988)

CLASSICAL RELIGIOUS AND SECULAR LITERATURE

30·66 References to well-known works published in countless editions and to their parts usually need not include any more information than the author (as with well-known plays and poems), the work, and, for parts of works, the specific location in the work (chapter and verse[s]; book and line[s]). References may have to specify a particular edition (representing a particular translation or edited version, for example); such references should be formatted with the same structure as references to books (monographs) or parts of books; see section 30·40.

Religious Works

GENERAL REFERENCE

C-S The Bible. I Corinthians 12:1–11.

N-Y The Bible. I Corinthians 12:1–11.

Form of Citation: (The Bible, I Corinthians 12:1–11)

SPECIFIC REFERENCE

C-S The New English Bible: New Testament. Oxford and Cambridge: Oxford Univ Pr, Cambridge Univ Pr; 1961. I Corinthians 12:1–11. p 294–5.

N-Y [Anonymous translator]. 1961. The New English Bible: New Testament. Oxford and Cambridge: Oxford Univ Pr, Cambridge Univ Pr. I Corinthians 12:1–11. p 94–5.

Form of Citation: (Anonymous translator 1961)

Classical Literature

GENERAL REFERENCE

C-S The Iliad. 18:601–14.

N-Y The Iliad. 18:601–14.

Form of Citation: (The Iliad 18:601–14)

SPECIFIC REFERENCE

C-S Homer. The Iliad; Book 18:601–14. In: Bryant WC, translator; Simons SE, editor. The Iliad of Homer translated into English blank verse. Boston: Houghton Mifflin; 1916. p 185.

N-Y Homer. The Iliad; Book 18:601–14. In: Bryant WC, translator; Simons SE, editor. 1916. The Iliad of Homer translated into English blank verse. Boston: Houghton Mifflin. p 185.

Form of Citation: (Homer, Bryant translator 1916)

UNPUBLISHED DOCUMENTS

30·67 Documents generally available to scholars in an archives or from a depository can be cited and referenced. The general format is similar to that of references for other documents, but the availability information must be given; a document identifier should be given if the author is not known. If the author is known, the author name comes 1st. The recommendations here draw in part on *The Chicago Manual of Style* (UCP 1993), which can be consulted for further guidance. Two formats are possible.

1 C-S Author. Title and date. Physical description. Availability.
N-Y Author. Date. Title. Physical description. Availability.

2 C-S and N-Y Title or document identifier if available. (Title if document identifier is given). Date. Physical description. Availability.

Unpublished Documents: Examples of References

LETTER

C-S Darwin C. [Letters to Sir William Jackson Hooker, 1863]. Located at: Archives, Royal Botanic Gardens, Kew, London, England.

N-Y Darwin C. 1863. [Letters to Sir William Jackson Hooker]. Located at: Archives, Royal Botanic Gardens, Kew, London, England.

Form of Citation: (Darwin 1863)

MANUSCRIPT

C-S Marcianus document 299. [Greek symbols for chemical substances and technical works. 10th century]. Located at: San Marco Library, Venice, Italy.

N-Y Marcianus document 299. [10th century]. [Greek symbols for chemical substances and technical works.] Located at: San Marco Library, Venice, Italy.

Form of Citation: (Marcianus document 299)

FORTHCOMING ("IN PRESS") DOCUMENTS

30·68 Some journals are willing to let authors cite documents not yet published but reliably identified as being in the process of publication. Because "in press" refers only to paper publication, the term "forthcoming" is preferable for electronic and other newer media as well as for journals and newspapers. If the publication date is known, it can be included in the reference. The general format should be the same as for references to published documents of the same kind. Whatever data are available, such as author, date of publication, title, place of publication, publisher, should be given. The closing piece of information is "Forthcoming", not "In press".

Date Known

C-S Cohen M. Zidovudine interaction with probenecid. AIDS Res Hum Retroviruses 1995. Forthcoming.

N-Y Cohen M. 1995. Zidovudine interaction with probenecid. AIDS Res Hum Retroviruses. Forthcoming.

Form of Citation: (Cohen 1995)

Date Not Known

C-S Cohen M. Zidovudine interaction with probenecid. AIDS Res Hum Retroviruses. Forthcoming.

N-Y Cohen M. Zidovudine interaction with probenecid. AIDS Res Hum Retroviruses. Forthcoming.

Form of Citation: (Cohen forthcoming)

Editors willing to accept references of this kind should have written assurance from the publisher that the document is indeed about to be published.

Appendix Table 30·1 Additional types of references illustrated in *National Library of Medicine Recommended Formats for Bibliographic Citation* (NLM 1991) on the pages shown below.

Type of document	Page
Journal article	
Article containing retraction	6
Article retracted	6
Author affiliation included	5
Author name: full names	8
Authors: 3 or more [ICMJE style] (ICMJE 1993)	9
Date at end of reference	9
Edition added to journal title	6
Issue with no volume	7
Issues, multiple numbers	7
Language of article other than English: roman and nonroman alphabets	5
Microform	6
Multiple years, months in date	7
New series: volume number	7
No issue or volume	7
Page numbers in full	9
Page numbers in roman numerals	8
Page numbers not given	8
Page numbers with letters	8
Pagination, discontinuous	8
Part in an issue	7
Part in a volume	7
Part of an article: specific page or pages; figure; table	8
Punctuation: periods with journal title abbreviations	9
Retracted article, see "Article containing retraction", "Article retracted"	6
Season with year	7
Spacing within all elements	9
Special number: year or volume	6
Journal: special issue	
Proceedings as an article	11
Proceedings as a monograph	11
Special subject as an article	11
Special subject as a monograph	11
Journal as a publication entry ("journal title")	
Ceased publication	16
Current publication: reference with optional information	16
Current publication: standard reference	16
Language of title other than English	16
Languages, multiple	17
Microform	17
Publisher with a subsidiary division	16
Supplement to a journal	16
Title with edition	16
Title with subtitle	16

Appendix Table 30·1 (*cont.*)

Type of document	Page
Book	
Author, organizational: place clarified	23
Author, organizational: subsidiary unit named	23
Authors: more than 3	23
Book also published as journal article	25
Book in a series with an editor	25
Language of title other than English	24
Microform	25
Physical description	25
Place of publication, publisher, and date of publication unknown	25
Publication date unknown	25
Publisher unknown	25
Volumes: more than 1	25
Book component	
Chapter or other component with no title	31
Chapter or other component in a volume with a separate title	32
Chapter or other component in a book in microform	32
Volume with editor or translator	31
Volume with no separate title	31
Conference proceedings	
Institutional name with conference location	43
Journal, proceedings as part of a	44
Language: other than English, with translation	43
Microform	44
Notes, supplementary	44
Place of publication and publisher unknown	44
Translator given	43
Conference paper	
Author address	50
Authors, full names	50
Language other than English, with translation	51
Paper also published as a journal article	51
Scientific and technical report and parts	
Language other than English	59
Microform	59
Part of a report	59
Report in a series	59
Report, state government	58
Dissertation, thesis, and parts	
Language other than English	64
Microform	64
Part of a dissertation or thesis	64
Bibliography	
Compiler: organization	69
Edition statement	69
Editor	69
Language other than English	69

Appendix Table 30·1 (*cont.*)

Type of document	Page
Patent	
Inventor also an assignee	74
Non-US patent with specific document type	74
Non-US patent with translation	74
Newspaper article	
Dateline	78
Edition named	78
Language other than English	78
Pagination with section letter	78
Volume and issue numbers included	78
Map	
Cartographer named	82
Location where map is available	82
Series named	82
Legal or government document	
US Senate or House of Representatives report	84
Audiovisual publication and material	
Chart	90
Filmstrip	90
Journal	94
Motion picture	90
Slide	90
Videodisk	89
Electronic publication and sources	
Bulletin board, contributions with authorship	155
Bulletin board, entire	150
Computer program, accompanied by other than a manual	106
Computer program, additional medium	105
Computer program, available from other than the publisher	106
Computer program, no version specified	105
Computer program, published by organization, subsidiary unit named	105
Computer program, update given	105
Database, contributions with authorship	126
Database, entire	112–3
Database, part	120
Journal (serial), entire	131
Mail	161
Unpublished document: personal communication, talk	
Personal communication	190
Talk (unpublished meeting paper)	186
Forthcoming document	
Book (monograph), edition indicated	182
Book (monograph), organization as author	182
Book (monograph), volume indicated	182
Journal article, article type indicated	178

Appendix Table 30·1 (*cont.*)

Type of document	Page
Journal article, author address	178
Journal article, medium type indicated	178
Journal article, organization as author	178
Proceedings of a conference	182

CITED REFERENCES

[AGPS] Australian Government Publishing Service. 1988. Style manual for authors, editors and printers. Canberra: AGPS.

[ANSI–NISO 1985] American National Standards Institute, National Information Standards Organization. 1985. Abbreviations of titles of publications: American national standard ANSI/NISO Z39.5–1985. Bethesda (MD): NISO Pr. Available from NISO Press Fulfillment, PO Box 338, Oxon Hill MD 20750–0330, USA; fax 301 567-9553.

BIOSIS. [annual]. Serials sources for the BIOSIS database. Philadelphia: BIOSIS.

Brown P, Stratton GB, editors. 1963–5. World list of scientific periodicals published in the years 1900–1960. London: Butterworths.

Chernin E. 1988. The "Harvard system": a mystery dispelled. Brit Med J 297:1062–3.

Garner DL, Smith DH. 1993. The complete guide to citing government information resources: a manual for writers and librarians. Rev ed. Bethesda (MD): Congressional Information Service.

[HLRA] Harvard Law Review Association. 1992. A uniform system of citation. 15th ed. Cambridge (MA): HLRA.

[ICMJE] International Committee of Medical Journal Editors. 1993. Uniform requirements for manuscripts submitted to biomedical journals. JAMA 269:2282–6.

[ISO] International Organization for Standardization. 1988. International standard codes for the representation of names of countries: ISO 3166–1988 (E/F). Geneva: ISO. Also available in: Documentation and information: ISO standards handbook. 3rd ed. Geneva: ISO;1988.

[NLM] National Library of Medicine. 1991. National Library of Medicine recommended formats for bibliographic citation. Bethesda (MD): NLM.

[NLM] National Library of Medicine. [annual]. List of journals indexed in Index Medicus. Bethesda (MD): NLM.

Thibault D. 1990. Bibliographic style manual. Ottawa: National Library of Canada.

[UCP] University of Chicago Press. 1993. The Chicago manual of style. 14th ed. Chicago: UCP.

[UN] United Nations. 1984. A guide to writing for the United Nations. New York: UN.

31 Accessories to Text: Tables, Figures, and Indexes

A good table is worth hundreds of words. A bad table may confuse more than it communicates.

—Ian Montagnes, *Editing and Publication: A Training Manual,* 1991

One picture is worth more than ten thousand words.

—Chinese proverb, [not dated; in *Familiar Quotations,* 13th ed., John Bartlett, 1955]

Index-learning turns no student pale,
Yet holds the eel of science by the tail.

—Alexander Pope, *The Dunciad,* 1728

The needs of readers of both journals and books are often served by content that supports the text proper: tabular data, illustrations (figures), and indexes. Careful attention to the formatting and details of style for these accessories can help support their value.

TABLES

31·1 Because of the frequent need in scientific publications to present large amounts of numeric and other descriptive data, tables prepared to carry data should be built up and presented in accord with principles that can make them readily and accurately understood.

Tables can be used to present various kinds of information.

1 Data for which precise numeric values are important (as opposed to conveying trends or proportions, for which a figure is often more effective).
2 A large number of numeric values in a compact form.
3 A summary of information.
4 Information too complex to be easily or concisely explained in text or shown in a figure.

31·2 Clear, easy-to-read tables are seldom easy to design, and tables can be the most difficult and time-consuming part of a manuscript to edit. Large, complex tables can take 2 to 3 times longer to edit than an equivalent number of pages of scientific text. Although the basic components of a table can be described and explained (as they are below), the design and editing of a table often call for creativity rather than the rigid application of rules. But some guidelines can be helpful.

1 A table should be complete enough to be understood without continual reference to the text, but it should contain only the data needed for the reader's understanding.
2 A table should be as simple as possible. When a simple structure is not possible with the information to be presented, the goal should be to provide a table that, although complex, is orderly and logically organized.
3 There should be some logical basis for the sequences of columns and rows.
4 The units, symbols, and data of the table must be consistent with those in the text.
5 Tables containing similar types of information should have parallel formats.
6 The same data should not be presented in both tables and figures.
7 Data should not be put in a table if they can be adequately presented in a few sentences of text. This may be the correct decision when the proposed table would have only 1 or 2 columns and only 2 or 3 rows.
8 When the design of a table is not satisfactory, redesigning the entire

table is sometimes easier than struggling to edit the existing table into a better form.

PARTS OF A TABLE

31·3 A table has 5 major parts; see Figure 31·1.

1 Number and title.
2 Column headings.
3 The stub (containing the row headings).
4 The field (containing data).
5 Footnotes.

Three full-width horizontal rules separate major sections of a table: the 1st rule between the table number and title and the column headings, the 2nd between the headings and the field, and the 3rd between the field and the footnotes. Vertical lines to separate the stub from the field or the columns from each other are not recommended; they would add to the cost of composition and are not needed if due care is taken with formatting columns and their headings; also see section 31·18.

Number and Title

31·4 Each table in a document must have a unique number and title. Table numbers should be in arabic numerals and be assigned in the order in which the tables are referred to in the text. The numbers are normally consecutive through the document, but in large documents they may reflect the number of the section or chapter of which they are part ("Table 1·1", "Table 1·2", and so on). The word "Table" and its number are followed by 2 spaces and the title. No period follows the title unless explanatory text is part of the title block.

The title of each table must be unique, succinct, and informative; it should be a phrase, not a sentence, but the phrase should be capitalized in sentence-style. The title should not be a list of the column headings of the table; it preferably specifies a category or class that encompasses the variables in the table. In large documents containing a series of similar tables, special care is needed to ensure that the titles allow the reader to distinguish among individual tables. The comprehensibility of closely related tables when considered without direct reference to the text may be helped by the use of common initial phrasing in titles, with a phrase specific for the particular table placed after a colon.

Table nr Typical table title is short without ending punctuation

[rule]

	Spanner head [rule]			Spanner head [rule]		
Stub head	Col head	Col head	Col head	Col head	Col head	Col head

[rule]

[field]

[stub]	▼	▼	▼	▼	▼	▼
Row head	[column]	[column]	[column]	[column]	[column]	[column]
Row subhead	xxx	xxx	xxx	xxx	xxx	xxx◀[row]
Row subhead	xxx	xxx	xxx	xxx	xxx	xxx◀[row]
Row subhead	xxx	xxx	xxx	xxx	xxx	xxx◀[row]
Row subhead	xxx	xxx	xxx	xxx	xxx	xxx◀[row]
Row head						
Row subhead	xxx	xxx	xxx	xxx	xxx	xxx◀[row]
Row subhead	xxx	xxx	xxx	xxx	xxx	xxx◀[row]
Row subhead	xxx	xxx	xxx	xxx	xxx	xxx◀[row]
Total	xxx	xxx	xxx	xxx	xxx	xxx◀[row]

[field]

[rule]

[a] [footnote] Source: xxxxxxxxxxxxxxxxxxxxxxxxxxxxx. [period]
[b] [footnote] xxxxxxxxxxxxxxxxxxxxxxxxxxxxxxxxx. [period]
[c] [footnote] xxxxxxxxxxxxxxxxxxxxxxxxxxxxxx. [period]

Table 31·1 Typical fatty acid composition of selected dietary fats and oils (wt %)

	Saturated			Unsaturated		
Fat, oil	Palmitic	Stearic	Other	Oleic	Linoleic	Other[c]
Animal fat						
Lard	29.8	12.7	1.0	47.8	3.1	5.6
Chicken	25.6	7.0	0.3	39.4	21.8	5.9
Butter	25.2	9.2	25.6	29.5	3.6	7.2
Beef	29.2	21.0	3.4	41.1	1.8	3.5
Vegetable oil						
Corn	8.1	2.5	0.1	30.1	56.3	2.9
Peanut	6.3	4.9	5.9	61.1	21.8	—[a]
Olive	10.0	3.3	0.6	77.5	8.6	—
Soybean	9.8	2.4	1.2	28.9	50.7	7.0[b]
Coconut	10.5	2.3	78.4	7.5	trace	1.3

[a]Source: Adapted from Ref yy.
[b]Not reported.
[c]Mostly linolenic acid.

Figure 31·1 The major parts of a table

Table 1 Infectious diseases in China: incidence by socioeconomic class
Table 2 Infectious diseases in China: incidence by region
Table 3 Infectious diseases in Japan: incidence by socioeconomic class
Table 4 Infectious diseases in Japan: incidence by region

Terms in the title should correspond to those used in the text, and if abbreviations are used, they should have been introduced in the text before being used in a table title.

Depending on house or document style, the table number and title may be in a bold face or not, and may be placed flush with the table's left edge or centered on the table. About 0.6 cm (0.25 inch; 1.5 picas) should separate the last line of the table title from the 1st horizontal rule of the table.

31·5 Some journals require that each table title or its accompanying headnote include a summary of the experimental conditions applying to the data. This style can result in a "paragraph" of information as large as the table itself, and providing such a summary for each table takes up valuable space. Routine restatement of experimental conditions for each table should not be needed if this information has been clearly provided in the methods section of the document and the text that refers to the table makes clear from which experiments the data come. When information specific to a table is needed, it should be provided in a footnote keyed to the title rather than in an extended title or headnote. If the extended title or headnote is required by house style, the table number and a concise title should be presented in bold type, with the remaining information (whether experimental conditions or table legend) distinguished from the title by use of a normal typeface.

Column Headings

31·6 Column headings (also called "heads" or "boxheads") identify the entries in the columns of the table; each column of a table, including the stub, must have a heading. Columns and their headings often (but not always) are used to display the dependent variable being presented in the table, so that like data are compared down the columns.

A heading consists of a word or short phrase descriptive of the entries in the column, followed (if needed) by the appropriate units set within parentheses. Sentence-style capitalization is used for the headings. Occasionally a column will not need a descriptive word or phrase and the heading will consist solely of a unit designation; in such cases, the unit is not enclosed in parentheses.

Table xx Summary of vitamin contents

Vit. D (IU)	Vit. A activity (IU/kg)	Vit. E activity (IU/kg)

[becomes]

Vit. D (IU)	Activity (IU/kg)	
	Vit. A	Vit. E

Figure 31·2 Consolidation of column headings into spanners

Symbols for variables should be used alone as headings with caution; the meaning of the symbol must be immediately clear to the reader from the table title and from the accompanying text. In particular, the use of "*n*" or "*N*" alone should be avoided; use instead an explicit heading: "Number of subjects", "Nr of subjects", "Nr of samples", "Patients (*n*)", or some similar phrase.

31·7 Because horizontal space on a page is often at a premium, headings should make extensive use of abbreviations, symbols, and other short forms. Abbreviations and symbols used only in the table must be identified in footnotes. If such an abbreviation recurs only in other tables on the same or the facing page, it need be identified only in the 1st table of the series. The definitions (in footnotes) may have to be repeated for tables not adjacent to the table with the initial definitions.

To save more space, common elements of adjacent column heads can be gathered into a "spanner" (also called a "straddle heading" or a "decked heading"), with the heading specific to each column becoming a subheading under the spanner. Units are taken into the spanner if they apply to every column the spanner encompasses. If, however, the units are the only common element among 2 or more columns, the unit should not be set alone as a spanner above the column headings. A horizontal rule ("spanner rule") below the spanner runs the width of the columns to which the spanner applies (Figure 31·2).

Whenever a spanner is used, every column under the spanner must have its own subheading. The practical limit to the use of spanners is 2 levels of subheadings. If a spanner would encompass all the data columns in a table, the spanner is unnecessary; it can be eliminated and the information incorporated into the title of the table (although it is often already there and the spanner is simply duplicating part of the title). A spanner never covers the stub column.

Table xx Measured and calculated values of KA series samples

Sample	Temperature (°C)	L	FC index (dyn/cm^2)	W^a
KA-100	20	2.17	3.472	0.86
KA-100	40	3.53	4.774	0.86
KA-102	20	2.04	5.962	0.86
KA-102	40	3.46	4.627	0.86
KA-104	20	1.86	8.388	0.86
KA-104	40	3.29	5.981	0.86

[a]Calculated value.

Table xx Measured and calculated values of KA series samples

Sample	L	FC index (dyn/cm^2)	W^a
KA-100			
20 °C	2.17	3.472	0.86
40 °C	3.53	4.774	0.86
KA-102			
20 °C	2.04	5.962	0.86
40 °C	3.46	4.627	0.86
KA-104			
20 °C	1.86	8.388	0.86
40 °C	3.29	5.981	0.86

[a]Calculated value.

Figure 31·3 Consolidation of a table by subordination of headings within the stub

The Stub

31·8 The left-most column of a table is called the “stub”, and like other columns it carries a column heading above it. The stub contains the headings (“row heads”) that are words or phrases describing the entries in a row, are units applicable to the row, or are information on experimental conditions. The row headings follow sentence-style capitalization, and units following heading words or phrases are within parentheses; units that alone form row headings are presented without parentheses. The rows and row headings often represent the independent variables being presented.

The headings in the stub should contain all the constant information applying to each row of the table. When the information in the first 2 or 3 columns of a table consists of constants or experimental conditions rather than data, this information should be incorporated into the stub by using subordination, shown by indenting (Figure 31·3). Footnotes may also carry such information, alone or in combination with subordination of the row headings.

Note in Figure 31·3 that although the column headed W^a contains the same value in each row, this column is not brought into the stub because, as a calculated value, it is data rather than a constant. However, in such cases, consider whether it would be simpler and just as clear to state in the text that "All calculated values of W equaled 0.86."

The Field

31·9 The field, or body, of a table contains the information the author wants to present. This information may be numbers, text, or symbols; exactly what information is being presented should be clear from the table title. Each entry is at the intersection of a column and a row, the intersection being termed a "cell".

ALIGNMENT OF ENTRIES

31·10 Each column of entries should be aligned with its respective heading, either flush left or centered. On a table that continues on more than 1 page, the widest column entry is used as the basis for positioning the heading on every page of the table, even if that entry occurs on only 1 page.

If a column contains only numeric entries and all are in the same units, the entries are aligned on the decimal point. If the entries are all numbers of fewer than 4 digits (9999 or less), the digits may be unspaced. If at least 1 entry is greater than 10 000, all 4-digit entries should also be spaced with 3-digit groupings (see section 11·3). If the entries in a column do not carry the same units (which is sometimes unavoidable), the entries may be aligned flush left, or aligned flush right in the column; if there is a convenient, common element such as a ± symbol, those elements are aligned (Figure 31·4). The major concern should be that the format of presentation not lead the reader to infer relationships or comparisons that are not valid.

Numbers that are summed should be presented in columns rather than rows. Totals that are not the exact sum of the numbers in a column or row should be identified with a explanatory footnote (for example, "Values in the columns may not add up to the totals because of rounding.")

Table xx Abundances of elements in meteorite samples A and B

Element	A	B	Relative abundance, A:B
Chromium (ppm)	96.6 ± 1.2	2250	0.043
Iridium (ppb)	13.1 ± 4.7	514	0.025
Selenium (ppm)	17.3 ± 8.0	19.5	0.89
Zinc (wt %)	3.46 ± 0.01	0.030	114

Figure 31·4 Alignment of columns containing values in different units

TEXT ENTRIES

31·11 When words make up the field entries in a table, they should be as concise as possible. Entries requiring several lines should be left-justified and single-spaced. A blank line is virtually always needed between rows in a table containing only text, and generally each entry should use sentence-style capitalization (but see section 31·16). Entries consisting of complete sentences are seldom effective for concisely conveying information, but when they are needed, they should be capitalized and punctuated in sentence style.

EMPTY CELLS

31·12 Tables may have empty cells, that is, cells for which the author has no information to present. If a value can logically be expected in a cell and a datum is not available, an em-dash (—) should be placed in the cell to indicate that the lack of an entry is deliberate. In a table of quantitative data in which plus and minus symbols appear, the em-dash should be replaced with another symbol or abbreviation. Because the symbol used to fill an empty cell must be unambiguous, even the recommended em-dash should be defined in a footnote to the table, for example, "The em-dash (—) represents no data collected."

A cell should be left blank only if an entry would not make logical sense. For example, in a table of chemical compositions of pure compounds, the intersection of a column labeled "Sodium (wt %)" with a row labeled "$Ca(OH)_2$" should remain blank because there is no sodium in calcium hydroxide. If, however, the table presents trace impurities in chemical compounds, this cell should not remain blank, because a preparation of calcium hydroxide could conceivably contain some sodium.

Leaving cells blank (particularly if there is only 1 empty cell) can cause the reader uncertainty about whether a cell was meant to be empty or is empty because of a typographic error. In a large table with many empty cells, blanks are less likely to be taken as errors. The reader may, however, have difficulty in following rows from the stub to entries several columns across the page, particularly if the table must appear on a "landscape page" (table turned counterclockwise 90°). Further, an entry may have been omitted by mistake, but there is no way for the reader to recognize the omission. When a table has many blank cells, the table may need to be redesigned.

When an author supplies an abbreviation that conveys the reason some data are not available, the abbreviation(s) must be unambiguous. "ND" can mean "not determined" or "not detectable" as well as "no data" or "not done". If the phrases "not detectable" and "not determined" are both needed, they should not be represented by "ND" and "nd"; the reader is too likely to assume that the change from upper case to lower or vice versa was inadvertent and too unlikely to read the 2nd footnote explaining the difference, particularly if these abbreviations occur in different tables. In any case involving confusing possibilities, the entry in the table cell should be a footnote letter in brackets, [x], or if the brackets would be confusing, "Note *x*" should be placed in the cell and the explanation in the "*x*" footnote.

If several abbreviations or symbols are used in the table field, all can be defined in a general footnote keyed to the table title.

Footnotes

31·13 Footnotes are used when their information will not fit into the logical structure of the table and is not readily available in the accompanying text. Superscript lowercase letters should be used as signs directing the reader to the footnotes of a table, for 3 reasons.

1 More letters are available to serve as footnote signs than devices such as asterisks and daggers and avoid the frequent need to double these devices to produce an adequate number of signs for a table.
2 The alphabetic sequence of letters provides a readily recognizable order of use.
3 Because most tables contain numeric rather than word entries, superscript letters provide less possibility for misreading or for the appearance of typographic errors than do superscript numerals serving as footnote signs. Also, superscript numerals are not usable as footnote signs in a document using superscript numbers as citations.

Footnote letters are assigned starting with the table title, which may be amplified with a footnote that applies to all the data in the table. Assignment of letters then moves through the column headings (beginning with the heading for the stub) from left to right. If a table contains spanners,

Table xx Sample table headings

	Spanner 1[b]					Spanner 2[d]	
	Subhead 1[e]		Subhead 2				
Stub-head[a]	Sub-sub-head 1[g]	Sub-sub-head 2	Sub-sub-head 3[h]	Sub-sub-head 4[i]	Head[c]	Sub-head 3	Sub-head 4[f]

Figure 31·5 Assignment of footnote letters to column headings (including stub) is hierarchical

the uppermost spanners and any columns with unspanned headings are on the same hierarchical level as the stub heading, and footnote letters are assigned to this level from left to right. Next come 1st-level subheadings from left to right, and then (if needed) 2nd-level subheadings, left to right (Figure 31·5). Finally, footnote signs are assigned through the table stub and field, beginning with the heading of the 1st row and moving to the right, then to the heading of the 2nd row and through that row, and so on through the rest of the table to the end.

Footnotes are placed in alphabetic order below the bottom rule of the table, each starting on a separate line. The indented or superscript letter is placed flush with the left side of the table. The footnote may be flush against or spaced from the footnote-sign letter, depending on house style. If flush, carryover lines should be aligned with the 1st letter of the 1st line of the note, so the footnote letters are clearly visible. If the footnotes are numerous but short and the table is wide, several columns of footnotes may be positioned across the bottom of the table to avoid a wasteful vertical series of short lines below the table.

31·14 If a source line is needed to acknowledge the source of the information in a table, that line usually appears before any other footnotes to the table. The form of the credit line depends on the style of reference citations used in the document and the specificity needed.

Source: Reference 10. Source: Adapted from Table 3 in Reference 2.
Source: Smith 1989. Source: Recalculated from Table 14 in Smith (1989).

When only a few cells contain data from a different source, an overall source line can be used, and the data from the minor source can be credited by footnotes to the individual data. When the data in a table come equally from several sources, no overall source line is normally used, and the sources are credited in the column or row headings. Sometimes footnotes to the headings are needed (also see section 30·14).

When a table is reproduced from a published source, including the source in the reference list is not sufficient. Permission to reproduce the table must be obtained from the copyright holder, and an appropriate statement must be made in a credit line accompanying the table.

[table title] Table 3 Chemical composition of widely used pesticides[a]
[footnote] [a] Reproduced from Smith (1989) by permission of the copyright holder, the author.

The statement should have the form and content requested by the grantor of the permission, and a copy of the permission letter should be submitted with the manuscript.

PREPARING TABLES

Design

31·15 The question of which data should be in rows and which in columns has no simple answer. One guideline is that like data are best presented in the columns because comparisons are made more easily by looking down a column. This guideline implies that the row headings represent the independent variables, and the column headings represent the dependent variables. The final choice of the format that will be easiest for the reader may have to take into account the constraints of the page, the width of the cell entries, the number of items, and the number of properties being compared. In considering the possibilities, note that the row headings can be arranged in more levels of subordination than column headings and spanners.

For example, when a table contains many variables with different units, the question arises as to whether those variables should be placed in the columns (with units in the column headings) or in the rows (with units in the stub). For a table comparing 15 physical and chemical properties of methanol, gasoline, and diesel fuel, the page size would probably dictate 3 columns, 1 for each fuel, and 15 rows, 1 for each property, with the appropriate units in the row headings, even though comparisons would then have to be made across the rows.

31·16 All of the headings and cell entries in tables consisting solely of text need to begin with capital letters for visual ease, and conventional table design calls for initial capitals to begin each heading or each cell entry that is a word. Yet technical tables often contain words, abbreviations, or symbols in which capitalization (or the lack thereof) is an essential element of recognition: for example, pH, cDNA, Fe, c-*jun,* AU, sin, log, newton (the unit, not Newton the person). When the reader will not be confused or misled, table headings and cell entries may be capitalized in sentence style. Such a table generally contains only common words or technical words having no special scientific style associated with them (see, for example, Figure 31·1). If the reader might be misled by the use of capitals for words that should not be capitalized in text, initial capitals should not be routinely used (for example, see Table 11·4). Furthermore, technical tables in which entries are fragments of sentences often are aesthetically more pleasing set in lowercase type (see Table 11·9).

Within a scientific document, a simple policy of sentence-style capitalization for alphabetic elements in tables is difficult to apply consistently because of the frequency with which ambiguous capitalization can arise. The primary guideline must be to avoid ambiguity in how terms would be capitalized in text. The secondary guideline should be to make the table visually easy to read. The recommendations above for sentence-style capitalization of column and row headings are based on the latter guideline. The reader will note various capitalization styles for the tables in this manual, the choice having depended on the types of terms to be displayed.

Designation of Units

31·17 If all the values in the table field are given in the same units, that unit is placed within parentheses after the title. If any complete column or row of the table carries units that differ from the rest of the entries, no unit appears after the title; each column (or row) has its unit shown in the column (or row) heading. If a small number of scattered entries (as a guideline, fewer than 10%) have different units, the units for the majority of entries may be placed after the title, and those units that differ can be so designated in a footnote for each individual entry. When, however, such entries become too numerous, the many footnote signs render worthless any overall unit designation after the title; they also make the table visually unappealing.

Column headings normally provide the reader with a single unit applying to all of the entries of a column. When the numeric values are extremely large or small, the 1st choice of a multiplier to present these values is the appropriate SI prefix to the unit. The SI prefix is preferred even when the range of values will create entries that would not normally be used (for example, "0.008 km"), if the numbers can then all be entered in terms of 1 unit.

When the information in a column is not expressed in SI units (for example, the number of cells in a culture or the number of occurrences of an event), very large or very small numbers are expressed with a multiplier in the column heading. Such multipliers can easily cause confusion and uncertainty; therefore, they must be used correctly. The multiplier is placed in the heading with a times symbol in front of the name of the quantity (not in front of the unit for the quantity) in the form "10^n × 'name of quantity' ('name of unit')".

10^{-9} × Distance (parsecs) $10^5 \times W$

Thus, for these examples, the number 23 in a table cell under the column heading "10^{-9} × Distance (parsecs)" would represent "23×10^9 parsecs" and 15.4 in a table cell under the column heading "$10^5 \times W$" would represent "0.000154" (the calculated value of the dimensionless quantity W).

The form "Cells (× 10^3)" should not be used, because one can never be certain whether the table entries have already been multiplied by the author or are to be multiplied by the reader to produce the correct value. Although the correct value can sometimes be recognized by the absurdity of the alternative if the reader is knowledgeable in the subject, there is no reason to use a fundamentally ambiguous notation. When a table presents only nonscientific information such as budget data, the traditional form of heading, "$ (millions)" may be used.

When the range of values in a column is extremely wide, it may be necessary to use a heading that requires the entries in each cell to be in full scientific notation (for example, 1.596×10^{-4}). Because of the space problems presented by the width of such entries, every effort should be made to avoid this option by redesigning the table or by judiciously selecting data to be presented.

Spacing

31·18 Tables in draft manuscripts should be double-spaced throughout. Tables in final copy are normally set with single-spacing of lines to provide the most compact format. Routine double-spacing of tables in published form is unnecessary except for tables of text entries, in which blank lines between rows help the reader discern different text elements.

In large tables of numeric data, particularly those containing closely spaced columns, it can be helpful to the reader to break the rows with a blank line for visual ease. Such breaks are usually inserted about every 5 lines, but in a particular case the optimal break may come after as few as 3 or as many as 6 lines. For full-page tables that would carry over to the following page if such blank lines were inserted, a light screen (shading) can be used over alternating sets of rows, providing a visual marker without the loss of vertical space.

Tables were once set with vertical rules separating the columns and horizontal rules separating the rows, but the extraordinary expense of setting vertical rules in traditional typesetting brought about the elimination of such rules. Although the capabilities of today's word-processing and desktop-publishing systems allow rules to be placed with little effort, the use of more than 3 full-width horizontal rules is seldom necessary.

Judicious spacing between columns allows the table to be easily read and presents a visually simpler appearance than vertical rules in virtually every case. A space between adjacent columns of 0.5 cm (3/16 inch; 1.25 picas) provides adequate visual separation; a space of 0.6 cm (1/4 inch; 1.5 picas) is preferable for ease of reading. These spaces are based on measurements between the widest entries (whether in the head or the field) in adjacent columns, so many tables will have more than minimum spacing in many rows. Observing minimum spacing is most important when the column headings are short and the field values are the elements closest to each other.

Alignment

31·19 Column headings should be flush left or centered over their respective columns, depending on house style. They can be broken and stacked vertically to save horizontal space as necessary (see Figure 31·6). Spanner rules extend over the widest elements of the outside columns they span, whether those elements are in the subheadings or in the field entries.

When a row heading in a single-spaced table carries over to a 2nd line, that line should be indented; entries needing only a single line opposite a multiple-line row heading should be placed opposite the 1st (unindented) line of the heading. Occasionally, circumstances (for example, text tables with entries of several lines each) or aesthetics will dictate that a table be set with blank lines between rows. In such cases, carryover lines in the stub need not be indented, and single-line entries opposite a multiple-line row heading should again be set opposite the 1st line of the heading.

When stub headings need to be indented, make sure that the indents are large enough to be easily visible (at least 0.5 cm [3/16 inch; 1.25 picas; 15 points]). Although cut-in headings may occasionally be needed in complex tables to help visually organize a large expanse of the table field, most tables will be made easier to read by moving such headings to flush left as major headings of the stub and indenting the subordinate stub headings (see Figure 31·6)

POSITIONING TABLES

31·20 Tables are placed in a document as close as possible to their 1st mention in the text. They should normally be positioned at the top or the bottom

Table xx Mineral survey samples meeting minimum specifications

Location	Mineral-containing samples (%)	Mineral content (wt % ± SE)
	Survey 1	
Idaho	56.4	35.1 ± 7.9
Montana	42.8	38.8 ± 7.6
Wyoming	44.3	25.4 ± 6.7
	Survey 2	
Idaho	51.4	32.6 ± 6.6
Montana	52.6	29.0 ± 9.2
Wyoming	46.3	27.1 ± 8.5
	Survey 3	
Idaho	58.1	34.8 ± 5.9
Wyoming	52.5	33.1 ± 6.9

Table xx Mineral survey samples meeting minimum specifications

Location	Mineral-containing samples (%)	Mineral content (wt % ± SE)
Survey 1		
Idaho	56.4	35.1 ± 7.9
Montana	42.8	38.8 ± 7.6
Wyoming	44.3	25.4 ± 6.7
Survey 2		
Idaho	51.4	32.6 ± 6.6
Montana	52.6	29.0 ± 9.2
Wyoming	46.3	27.1 ± 8.5
Survey 3		
Idaho	58.1	34.8 ± 5.9
Wyoming	52.5	33.1 ± 6.9

Figure 31.6 Consolidation of cut-in field headings into the stub

of a page, but preferably at the top to let the text serve as a visual anchor for the bottom of the page.

About 1 cm of space (3/8 inch; 2 picas) should separate either the last line of text from the title of a table below it or the last footnote of a table from the 1st line of text following it. Tables should be centered horizontally in the available space. Tables that occupy most of the width of a page should be centered between the margins; tables occupying a single column of type should be centered within the boundaries of that column. The horizontal rules of a table should not be extended to fill a column of type if the table requires only a part of the column width; the visually disconcerting result will be that the table will "float" within the overextended rules.

Tables are more often limited in placement by width than by length. A table that is too wide for the standard page will often fit in landscape orientation; it should be rotated 90° counterclockwise so that the title is at the left of the page. The need to turn the document to read the table, however, means that this option should be used sparingly. Often a table slightly too wide for a page can be photographically reduced to a size that remains legible and yet fits horizontally on the vertical page in normal orientation; such a solution is preferable to a landscape page. If photoreduction is used and 2 tables appear on the same page, both should appear at the same percentage reduction.

Horizontal extension of a table across facing pages can present problems in alignment, but such tables can be successfully used. The row headings should be repeated on the continuation page. Further extension that requires the reader to follow rows by turning pages becomes tedious and should be avoided. Extremely wide tables should be presented on foldout pages. Tables can be extended vertically by continuing onto succeeding pages with less difficulty. The table number (but not the title) plus "Continued" and the column headings are repeated at the top of each continuation page.

Some problems in fitting a wide table into a page in normal (horizontal, rather than landscape) orientation can be solved by reorganizing the table so that the column headings become stub headings and the stub headings become column headings (though the change violates the general rule of using the stub headings for independent variables and columns for dependent variables.

FIGURES

31·21 The term "figures" here includes graphs, charts, photographs, maps, and other types of illustrations. The following sections on figures are devoted mainly to editorial considerations in the use of figures in documents. More detail on the characteristics of various kinds of illustrations and their proper preparation can be found in *Illustrating Science* (CBE 1988).

Figures are used to present main kinds of information.

1 Data for which trends or proportions are the important characteristic.
2 Visual aids to the understanding of complex concepts.
3 Drawings or photographs of items, places, or procedures under discussion.

References in text to figures should be parenthetic. If the reference to a figure begins a sentence, "Figure" should be spelled out; in other positions an abbreviation may be used, but note that there is no logic to abbreviating "figure" and writing out "table" in text references.

> The distillation apparatus (Fig. 2) was custom-made.
>
> [or]
>
> The distillation apparatus (Figure 2) was used to collect the data shown in Table 3.
>
> Figure 2 shows the distillation apparatus.

The phrasing of the 3rd example should be used only when the figure literally shows a drawing or photo of something physical. Avoid rhetorical forms stating that a figure is itself evidence; it contains evidence. For example, avoid statements such as "Figure 5 shows that separation was complete after 20 h." or, worse, "Figure 13 proves that additive X was the most effective." Change these to a form such as "Additive X was the most effective (Figure 13)." After a figure has been referred to once, succeeding references to the same figure should be "see" references.

> In this section we present the reaction times from Experiments 2 and 3 (Figures 8 and 9, respectively); these times were much faster than those in Experiment 1 (see Figure 4).

31·22 The terms, symbols, and abbreviations used in figures should be the same as those used in the text. Standard symbols (such as for SI units) and symbols previously defined in the text may be used freely. Symbols or abbreviations specific to a figure should be defined in a key or in a note that is part of the caption. In a series of figures within a document, symbols, shadings, and line patterns should be used consistently. For example, if open circles and open squares represent control and treated groups in initial figures, other symbols (such as open and filled triangles) should not be used for the same groups in later figures. If the early and later figures represent 2 parallel sets of experiments, however, the later set might logically be represented by filled circles and squares. If it is necessary to provide visually distinct symbols for different variables over a series of figures, keeping constant the meanings of open and filled symbols greatly aids the reader's understanding.

NUMBERING AND CAPTIONS

31·23 The figure caption includes the word "Figure" followed by a space and the figure number followed by 2 spaces and the title or legend of the figure. These are placed below the figure, separated from the lowest element of the figure by about 0.6 cm (0.25 inch; 1.5 picas). About 1 cm of space (3/8 inch; 2 picas) should separate the last line of the caption from text below it. The caption may be bold or not and may be in the text font or a complementary font, depending on house style. The caption may be centered on the figure or aligned flush with the figure's left edge.

Figures are numbered with arabic numerals assigned in consecutive order as the figures are referred to in the text. The numbering generally is consecutive through short documents, but for long manuscripts, reports, or books, numbering by section or chapter (as in this manual) may be more flexible and helpful to readers. The figure number is followed by 2 spaces and the figure title.

A figure title should informatively describe in a short form the content of the figure; it is usually a sentence fragment. The title is capitalized in sentence style, and no period follows the title unless additional information is presented in the title block. The figure title may contain abbreviations and symbols that have been defined in the text. Parenthetic phrases within the title should be avoided. If the title is long enough to take more than 1 line, the 1st line of the title should be the longest. The 2nd and succeeding lines may align with the letter F in "Figure" if the caption is flush with the figure's left edge or may be indented. If the caption is centered on the figure, each line should be centered. Consider both textual sense and the appearance of the caption when deciding where to break the lines of the title; no line should be wider than the figure. If a source must be identified, it is set within parentheses as the last element of the caption.

Figure 6·1 The parts of a table (adapted from *Journal of Food Science*
1981; 46(2):661)

For graphs, titles should not simply repeat the axis labels in the form "ordinate and abscissa" or "ordinate as a function of abscissa"; rather, the title should be a phrase describing what the data show. For example, a graph of pressure and flow rate might be titled "Permeability of . . . ". Titles of figures should not begin with a phrase describing the type of figure: for example, "Photograph of . . ."

Some journals require that each figure caption be accompanied by a summary of the experimental conditions applying to the figure. This can result in a paragraph of information that may be as large as the figure itself, and providing such a summary for each figure takes up valuable space. Routine restatements of experimental conditions for each figure should not be necessary if this information has been clearly provided in the methods section of the document and if the text that refers to the figure makes clear from which experiments the data come. When information specific to a figure is needed, it should be a brief note; otherwise, it should be provided in the text. If the extended summary is required by house style, the figure number and a title should be presented in bold type, with the remaining information (whether experimental conditions or figure legend) distinguished from the title by use of a normal typeface.

A single caption should be provided for multipart (composite) figures, with necessary information about the separate parts provided by their individual labels. If the separate parts of the figure call for so much information that individual captions are needed, the composite should be divided into separately numbered figures.

PLACEMENT

31·24 Figures are placed in a document as close as possible to the 1st point in the text at which they are referred to. If the figure will not fit on the page with the 1st reference to it, it should go on the next available page.

Figures should be at the top or the bottom of a page or column. The top position should be preferred; it allows the text to visually anchor the bottom of the page. Text may appear on any page carrying 1 or more figures, but the minimum amount of space available for text should be no less than 5.1 cm (2 inches, 12 picas); fewer lines on a page of figures can be overlooked too easily. When a figure appears on a page with text, about 1 cm of space (3/8 inch; 2 picas) should separate the text from the top of the figure or the bottom of the caption.

Figures should be placed on a page to be read in the normal orientation of the page. Only if a figure is too wide for the page and cannot be reduced should it be rotated 90° counterclockwise for reading by the same rotation of the journal or book.

Figures should be centered horizontally in the space they occupy; that is, a figure occupying most of the width of a page is centered between the margins, and a narrow figure occupying 1 column is centered within the boundaries of that column.

Most figures are reduced from their original size for printing. A primary concern in reduction is that symbols in the figures be easily distinguishable and lettering remain legible. As a guideline for figures being presented on an 8.5- by 11-inch page, capital letters should be between 0.15 and 0.3 cm (1/16 and 1/8 inch; 0.4 and 0.8 picas) high when the figure is at its final size for printing.

PRODUCTION AND REPRODUCTION

31·25 Drawings and graphs should be prepared to reflect the same quality as the text they accompany. Although software packages make graphics capabilities widely available, their use does not guarantee high-quality artwork. For important documents, professional artists should prepare the figures. Figures prepared for oral presentations as slides and those prepared for use in printed documents are seldom interchangeable. Figures for use as slides carry titles as part of the figure and must be visually simple; figures for publication have titles within the caption, and they may have to be more complex and carry more information than presentation slides.

When many original figures will appear in the same document, all the figures should be prepared in accordance with a single set of specifications. The caption style, axis labels of graphs, lettering size and style, and other details should be consistent throughout the document. The acceptability of a variety of figure styles should be limited to review documents in which the figures are reproductions from separate, previously published works. When figures from previously published documents are used, permission to reproduce them must be obtained from the copyright holder; copies of the permissions should be submitted with the manuscript.

GRAPHS

31·26 Graphs should be simple and contain no more information than is needed to make the author's point. Several curves can be displayed on a single graph as long as they are not so close as to create a problem in following any of the individual curves; the norm is 3 to 4, the practical maximum, about 8. When data points are plotted as well as curves, fewer curves can be plotted without causing difficulties for the reader.

Graphs are normally designed so that the vertical axis (ordinate, *y* axis) represents the dependent variable and the horizontal axis (abscissa, *x* axis) represents the independent variable. The range of values on the axes should be slightly larger than the range of values being plotted, so that the entire set of data points falls within the field of the graph and yet most of the range of values is used. If an extremely large range must be covered and cannot be practically shown with a continuous scale, a discontinuity in an axis can be shown with paired diagonal lines (—//—) indicating a missing extent of the range.

The numbers used to mark axes should be chosen to be simple, appropriate multiples of the quantity graphed; use multiples of 2, 5, or 10 whenever possible. Divisions of the graph axes should be shown with interval (tick) marks rather than with a grid over the entire graph. The numbers represented on the axes appear outside the field of the graph, immediately opposite and centered on their respective tick marks (left of the left vertical axis and below the horizontal axis). The numbers on both axes should be the same size and should all read horizontally. Unlabeled tick marks should not be used to show subdivisions between the numbered marks. When the axis scale includes values less than 1, a zero is required before each decimal point.

Complex graphs may need a right-hand vertical axis to carry a 2nd vertical scale. The curves of such a graph must be clearly identified with the right or left axis, either in the figure caption, in a key, or by the use of arrows pointing from the curve to the appropriate axis.

Axis labels identify the variable plotted on the axis in a word or phrase; the labels should be in sentence-style capitalization. Each label is centered on the length of its axis. The axis label for the vertical axis is rotated 90° counterclockwise and placed so that the lettering reads parallel to the axis; vertical axis labels should never be placed to read vertically downward. Units of measurement for the variable are placed in parentheses after the wording. When a multiplier is needed for the axis values, the quantity is multiplied by the appropriate factor, not the unit of measurement.

10^{-3} × Radioactivity (counts/min)

[not "Radioactivity (counts/min × 10^{-3})"]

Axis labels should be accurate statements of the variable displayed, for example, "Number of cells" rather than "Cell number". Occasionally an axis shows a complex quantity for which a suitable label consists only of a unit; in such a case, the unit abbreviation is not placed within parentheses: "ng/cell·d" (equals "nanograms per cell per day"). If an axis is labeled with only a simple unit, the appropriate quantity should be added: for example, a label of "km" for "kilometers" should usually be "Distance (km)".

The combination of figure caption and axis labels should provide enough information to make the figure comprehensible without immediate reference to the text. For example, in a figure titled "Employed persons in Chicago as a percentage of total population, 1986–1991", a suitable axis label might be "Employed (%)". In the case of a composite figure of 6 parts entitled "Employed persons in major cities as a percentage of total population, 1986–1991", each graph might need an axis label such as "Employed persons in Chicago (%)". As alternatives, the simpler label "Employed (%)" might be suitable if each of the 6 graphs included the city name as a key within the boundaries of the graph or if each graph carried a sublabel (A, B, and so on) with the corresponding city given in the overall figure caption: "A. Chicago", "B. New York", and so on.

Graphs may be fully boxed or may show only the 2 (or 3) axes needed for the variables, depending on the document style; whatever style is used should be consistent throughout the document. Data curves should be labeled directly; if this is impractical, a key may be included within the limits of the axes in a blank area of the graph. Only as a last resort should the key be put in a note following the figure title.

PHOTOGRAPHS

31·27 Photographs should show the item of interest without distraction; in most cases, photographs need to be cropped to ensure that the reader's attention is directed correctly. Indicators of scale are often needed to make photos meaningful. Such indicators may be included in the photograph itself (for example, a geology hammer showing the scale of a rock formation) or added later. Scales are most useful when they are visual, such as a bar denoting a given length in a micrograph, rather than mathematical (a statement of magnification such as "1000×"). Visual indicators also eliminate a need to recalculate the magnification if the original photo must be reduced for printing. Marks such as scale bars, arrows, or labels should always be shown on transparent overlays to the photograph; these marks should never be added directly onto the face of the photo to be reproduced.

Lettering in labels identifying parts of an illustration should be consistent with the lettering in other document figures. Reduction of a photograph or drawing before printing should be taken into account so that the final size of the letters is suitable for a printed page (final height of capital letters, 0.15 to 0.3 cm).

INDEXES

31·28 An index should not be compiled until all page numbers (folios) or section numbers, or both, have been assigned. A number of decisions must be made as the indexing is started. The major decisions include conceptual bases for the structure of the index and the probable needs of the users of the index. An index, whether for a volume of a journal or for a book, is compiled to enable readers to find information easily; therefore, the indexer must have a clear view of who these readers are likely to be as a guide to what terms to select for the index, what cross-references should be inserted, and what depth of indexing will be most appropriate. This manual does not discuss how to deal with these questions; guidance can be found in texts (Lancaster 1991, Wellisch 1991) on the entire process of indexing; another helpful source is Chapter 17 of *The Chicago Manual of Style* (UCP 1993). After these initial decisions come some decisions on the structure of the index (see Table 31·1). These must be made at the outset.

Table 31·1 Checklist for decisions on content and formatting of indexes

Peculiarities of topic and particular audiences: Is a political document being indexed, with a need to index each person's name at any mention in text; a geographic document, for which each place name should be indexed at mention; a historical document, in which dates should be incorporated?
Is only a subject index needed? Or should an index of cited authors also be prepared?
Number of index entry-terms selected per page of text (depth of indexing)?
Levels of index entries: main headings only, main headings with subheadings, main headings with subheadings with sub-subheadings?
Alphabetization of entries word-by-word or letter-by-letter?
Inclusive or initial page numbers? Or section numbers? Or both?
Separate inclusive page numbers with an en-dash or a hyphen?
For inclusive page numbers, repeat digits (for example: p 110–118, 110–18, 110–8; Sections 31·10–31·18, 31·10–18, 31·10–8)?
Subheadings stacked or run-in?
Number of spaces to indent subheadings?
Number of spaces to indent the 2nd line of entries that extend longer than 1 line ("turnovers")?
Punctuation between the entry and the page or section number?
Number of spaces between entry and page or section number?
Need to observe scientific conventions, such as italicization of species names, small capitals for some chemical-name prefixes, and other possibilities?
Italicize "see" and "see also"?
Commas between page or section numbers?

ENTRY AND SUBENTRY TERMS

31·29 Indexes can have different numbers of levels for entry terms: 1 level (only single-level entries), 2 levels (top-level terms and subterms), or, rarely, more levels.

The next possible variations are subentry terms that are stacked (indented under the main entry terms, each subentry term flush left) or run-in (continued on the same line as their main entry and in paragraph form, a space-saving device). In both formats, the subheadings are alphabetized.

eponymic terms
  capitalization 90
  clinical genetic syndromes 237
  rules for use 132–133

[or]

eponymic terms: capitalization 90; clinical genetic syndromes 237; rules for use 132–3

For main entries and subentries, only nouns with no preceding articles and modifiers should serve as index terms.

Albright syndrome 237 [not "the Albright syndrome 237"]

Occasionally, for a particular audience, the term will correctly begin with an adjective. In such cases, often the noun the adjective modifies is so general that without the adjective(s) it would be meaningless.

red blood cell herpes virus ground truth

North Pole polar bear

31·30 Because the reader will infer that a subentry means that particular aspect of the entry at the next higher level (usually a main entry at the top level), a preposition indicating the relation of the subentry to the entry above is seldom needed. If a preposition is needed for clarity, it should be at the end of the entry, not the beginning. If a term is repeated with different pronouns, they are taken into account in alphabetization.

metastases	[not] metastases
liver, from 45–92	from liver 45–92
liver, to 1–5	to liver 1–5

31·31 Whether to use common abbreviations and acronyms as main headings also depends on whether the prospective audience will understand them.

AIDS	[or]	immunodeficiency syndrome, acquired
	[or]	acquired immunodeficiency syndrome
DADDS	[or]	*N,N′*-(sulfonyldi-4,1-phenylene)bisacetamide
DNA	[or]	deoxyribonucleic acid

Alphabetization

31·32 Entries in an index are alphabetized by 1 of 2 systems. In the letter-by-letter system, all words in a term are considered as a whole (as if run together). In the word-by-word system, terms beginning with the same word or abbreviation are grouped and then alphabetized by the sequence of letters in the 2nd word.

LETTER-BY-LETTER OR WORD-BY-WORD?

Alphabetizing terms letter by letter is generally preferred.

[letter-by-letter]	[word-by-word]
Saint Augustine	Saint Augustine
Sanhedrin	San Jacinto
sanitarium	San Xavier
San Jacinto	Sanhedrin
Sanskrit	sanitarium
Santa Claus	Sanskrit
Santa Isabela	Santa Claus
San Xavier	Santa Isabela

Note that some computer programs for producing indexes may not automatically produce an accurate letter-by-letter sequence and the sequence may have to be corrected.

PREFIXES

31·33 For chemical names beginning with a numeral, the numeral is ignored in alphabetizing, but the entries are kept in numeric order. Other descriptive prefixes (for example, chloro- , α-, β-, D-, L-) are ignored. Note that alphabetization carried out by computer sorting may place terms with such prefixes at the end of the alphabetic sequence along with other non-standard spellings and hyphenated words. This kind of sequencing will call for rearranging those terms in the proper alphabetic sequence.

butanol
1-butanol
1-butanol , 1-cyclohexyl-
1-butanol, 2-cyclohexyl-
2-butanol
tert-butanolysis
16,17-butanomorphinan
1-butanone
2-butanone
2-butanone, 3,3-dimethyl-
2-butanone, 3-hydroxy-
8,9-butano–9-nonanolide
Butanox
butanoyl chloride
butanoyl disulfide
butanoyl nitrite

"MAC", "MC", AND "M'"

31·34 In British practice, names beginning with "Mac" or "Mc" or "M'" may be alphabetized as though they all began with "Mac". This manual recommends a letter-by-letter alphabetic sequence for such names; this sequence assumes that a reader searching for a name in an index knows how the name is spelled.

[recommended sequence]	[British sequence]
Mably, G	Mably, G
Macalister, D	McAdam, J
MacArthur, A	McAdook W
Macbeth, R	Macalister, D
MacBride, E	McAllister, A
MacWhirter, J	McAlpine, W
Macy, J	McAneny, G
Madach, I	MacArthur, A
McAdam, J	Macbeth, R
McAdook W	McBey, J
McAllister, W	MacBride, E
McAlpine, A	McClintock, F
McAneny, G	M'Clintock, J
McBey, J	MacWhirter, J
McClintock, F	Macy, J
M'Clintock, J	Madach, I

ABBREVIATIONS

31·35 Abbreviations are sometimes alphabetized separately at the beginning of an index of their respective alphabetic sections. This manual recommends placing them in a normal letter-by-letter sequence.

NON-ENGLISH ROMAN CHARACTERS

In some European languages, words beginning with a letter or letter with a diacritical mark may be placed in a sequence not expected by anglophones. For example, in a Danish dictionary words beginning with the ligature "æ" follow words beginning with "z"; words beginning with "ø" follow those beginning with "æ"; words beginning with "å" follow those beginning with "ø". As in the recommendations for alphabetization of author names in reference lists (see item 2 in section 30·15), diacritical marks should be ignored in alphabetizing non-English words for English-language indexes; the terms should be treated as beginning with a letter (or letters) not diacritically marked or ligated.

Capitalization

31·36 Some indexers always capitalize main-entry terms and set subentries in lower case. In the sciences, capitalization or noncapitalization of terms may be needed to differentiate unequivocally among classes of what the terms represent. For example, capitalizing proprietary names of drugs distinguishes them from the noncapitalized nonproprietary names (see sections 19·2–19·8). Therefore, indexes of scientific publications should not arbitrarily capitalize all entry terms, but should apply the rules for use of capital or lowercase initial letters specified elsewhere in this manual, such as those, for example, for drug names and for trade names of commercial products; see sections 8·15, 8·25, and 19·3.

PUNCTUATION

31·37 Punctuation is generally not needed with 1-level or stacked multilevel indexes and single initial-page numbers or section numbers following entry terms; spacing suffices.

anatomy 57
bacteriology 62

Commas are needed to separate multiple initial-page or inclusive-page numbers.

anatomy 57, 69, 101   bacteriology 62, 64–7, 89

If the index will contain entries with numbers in them, it may be visually helpful to separate the entry term from the location number by a comma. Note, however, that in indexes that contain entries where punctuation is meaningful to the concept, as in an index with chemical names, extra space may be preferred to the extra comma.

history
  14th century, 12–13
  1799–1825, 40, 44–47
  1850–1890, 52–57, 66, 70
  Freemasons, 99, 102

$C_2H_4O_2$ 199–200, 204
$C_5H_{10}O_2$ 272–274, 410
dextrin–1,6-glucosidase 5, 7, 10
vitamin A 11
vitamin B_{12} 7–9, 11

PAGE NUMBERS

31·38 The indexer decides at the outset whether to give only the initial page number (as from a computer-compiled index or the simpler procedure for a human indexer) or inclusive numbers (the initial through the final-page numbers on which the information appears). Separating ranges of numbers with an en-dash (–) is visually helpful, but if the en-dash is not available, the hyphen (-) can be substituted. With regard to the use of inclusive page numbers, the indexer will have already decided on whether to give the complete number, or to give only the nonrepeated digits (for example: "112–117 or 112–7; 555–9; 559–62").

SECTION NUMBERS

31·39 Some texts are organized by section number. They can be indexed with or without page numbers. The same decisions about repeating digits will have to be made.

age 203·1 (4)	[or]	age 203·1
AIDS		AIDS
certification 203·5 (17–20)		certification 203·5
confidentiality 203·6 (21)		confidentiality 203·6
precautions 203·7 (27)		precautions 203·7
anemia 400·3 (23)		anemia 400·3

CROSS-REFERENCING

31·40 A helpful index includes references to related entries or to the term under which the information has been indexed, indicated by "see" and "see also". "See" and "see also" are often italicized to make clear that they are not entry terms, and entries with these prefatory words usually appear at the end of the subheadings. Placement and typeface, as well as spacing, should be decided at the outset. It is inconsiderate to refer the reader to a 2nd entry and then give only 1 page number; it would be better to cite the page number in both places. Some information should be indexed as a subentry under a main entry and again as a main heading.

Pennsylvania
  Amish 25–29
  barns 25, 31, 33–35
  Lake Erie 20–22
  rivers 11–15

barns
  Amish 4–5
  New England 5–10
  Pennsylvania 25, 31, 33–35
  tertiary use 40–45
  see also houses

BIA *see* Bureau of Indian Affairs

CLARIFYING HOMONYMS

Words or names with the same spelling but with different referents, for example, the names of 2 persons spelled exactly the same, may have to be indexed. When 2 or more such homonyms are found, some means should be found to differentiate them so as to not misdirect the user of the index to an undesired part of the text. Such means might include identification of position, birthplace, or birth date, or of type of object even if the indexed text does not carry such information.

Jones, John [American chemist]	condenser [electrical]
Jones, John [English explorer]	condenser [chemical]
Jones, John [New Zealand geologist]	

REFERENCES

Cited References

[CBE] Council of Biology Editors, Scientific Illustration Committee. 1988. Illustrating science: standards for publication. Bethesda (MD): CBE. Available from Council of Biology Editors; 11 South LaSalle Street, Suite 1400; Chicago IL 60603, USA. Telephone, 312 201-0101; fax, 312 201-0214.

Lancaster FW. 1991. Indexing and abstracting in theory and practice. Champaign (IL): Univ Illinois, Graduate School of Library and Information Science.

[UCP] University of Chicago Press. The Chicago manual of style. 14th ed. 1993. Chicago: UCP.

Wellisch HH. 1991. Indexing from A to Z. New York: HW Wilson.

Additional References

[ASM] American Society for Microbiology. 1991. ASM style manual for journals and books. Washington: ASM. Tables; p 107–28.

[ASI] American Society of Indexers. 1992. Indexer evaluation checklist. Port Aransas (TX): ASI.

Bertin J. 1983. Semiology of graphics: diagrams, networks, maps. Madison (WI): Univ Wisconsin Pr.

Bonura LS. 1994. The art of indexing. New York: J Wiley.

Brock MH. 1990. A researcher's guide to scientific and medical illustrations. New York: Springer-Verlag.

Butcher J. 1992. Copy-editing: the Cambridge handbook for editors, authors and publishers. 3rd ed. Cambridge (UK): Cambridge Univ Pr.

Cleveland WS. 1985. The elements of graphing data. Monterey (CA): Wadsworth Advanced Books and Software.

Cleveland WS. 1993. Visualizing data. Murray Hill (NJ): AT&T Laboratories. Available from Hobart Pr, Summit, NJ.

Hansen WR. 1991. Suggestions to authors of the reports of the United States Geological Survey. Washington: US Government Printing Office. Tables; p 216–22.

Hodges ERS. 1989. The Guild handbook of scientific illustration. New York: Van Nostrand Reinhold.

[NISO] National Information Standards Organization. 1984. Basic criteria for indexes: Z39.4-1984. New Brunswick (NJ): Transaction Publishers. For availability see "Standards for editing and publishing" in Appendix 3.

Ore O. 1990. Graphs and their uses. Rev, updated. Washington: Mathematical Assoc of America.

Spilker B, Schoenfelder J. 1989. Presentation of clinical data. New York: Raven.

Tufte ER. 1983. The visual display of quantitative information. Cheshire (CT): Graphics Pr.

Tufte ER. 1990. Envisioning information. Cheshire (CT): Graphics Pr.

PART 5 The Publishing Process

32 Typography and Manuscript Preparation, Manual and Electronic

Various kinds of typefaces (such as roman, italic, and boldface), kinds of characters (such as capital letters, small capitals), and arrangements of characters (such as superscript and subscript letters and numerals) convey special meanings in scientific publication. These conventions are summarized in Chapter 9 and presented in detail in chapters of Part 3. Authors should know how manuscripts are marked by copy editors to indicate these conventions to the printer but also should be able themselves to indicate to the copy editor, if necessary, the need for a typographic convention that will convey the intended meaning of the text.

Copy editors may have to mark up manuscripts additionally for other details in publication style and format such as the size of type to be used in titles, headings, text, and tables, the leading (spacing) between type lines, and other particulars that specify the format of a journal, book, report, or article. Some of these specifications and how they are to be indicated are likely to have been established by the publisher and printer, especially for journals. Some specifications may have to be indicated, however, for particular manuscripts.

This manual does not attempt to describe all of the decisions, steps, and means that must be applied in preparing a book or even shorter documents for publication. Detailed guidance on these aspects of manuscript preparation can be found in such books as *The Chicago Manual of Style* (UCP 1993), *Copy-Editing: The Cambridge Handbook for Editors, Authors and Publishers* (Butcher 1992), and *Webster's Standard American Style Manual* (Morse 1985).

CHARACTERISTICS OF TYPE

32·1 The chief characteristics of type with which all editors should be familiar are typeface styles and weights, type sizes, and the spacing of characters and lines. These elements must be specified to define the appearance of printed matter and often have to be indicated in marking up manuscripts for the printer. Generally, authors need not be concerned with this information, but those with some knowledge of typographic characteristics may better understand how their manuscripts have to be prepared for the printer.

The Chicago Manual of Style (UCP 1993) is an excellent, convenient source of detail on elements of typography. *The Elements of Typographic Style* (Bringhurst 1992) has clear, concise descriptions of practical and aesthetic aspects of typographic design and of typefaces.

TYPEFACE STYLES AND WEIGHTS

32·2 A typeface (sometimes called simply "face") is a particular coherently designed group of sets of letters, numerals, and, often, additional characters such as punctuation marks and other symbols used together as needed in printing text. A font is a complete set of the characters of a typeface in a given size. The meanings of "typeface" and "font" have been blurred in recent years in the microcomputer and desktop-publishing communities, where "font" is often applied where "typeface" has generally been used in the printing industry and among typographers.

The classifications of typefaces are complex. Typefaces used in printing English-language texts are generally based on the style of capital letters used in the classical Roman culture and related lowercase letters developed later; hence the faces are generally called "roman typefaces". Typefaces based on the style of lowercase letters developed in Italy early in the 16th century are generally known as "italic typefaces". A second classification is that of serif and sans serif faces. Serif typefaces are those in which letters of the alphabet have cross-elements terminating the main strokes of letters; serifs were a characteristic of classical Roman letters and define present roman typefaces; sans serif typefaces have basic characteristics of roman faces but lack serifs terminating letter strokes.

Council of Biology Editors
[roman (serif) typeface]

European Association of Science Editors
[sans serif typeface]

A 3rd classification is that of upright and slanted (oblique) faces. Typical roman typefaces are upright (not slanted). Letters in italic faces have slanted vertical axes. These characteristics have led to the term "italic" also being applied to slanted (oblique) letters of a basically roman typeface, such as those produced in many word-processing programs. Sections 9·2 and 9·3 summarize general and scientific uses of italic type. This manual follows the convention of using "roman" to mean "upright" and "italic" to mean slanted.

Scientific Style and Format
["italic" (slanted) version of a roman (serif) typeface]

Scientific Style and Format
["italic" (slanted) version of a sans serif typeface]

Each particular typeface is given an identifying name, which may be that of the designer, or may refer to a geographic or historical origin, or may have been coined to be evocative of the design's aesthetic character.

This statement is set in Times.
[roman typeface named Times; it is a serif face related in its design to that of a face designed for *The Times* of London]

This statement is set in Helvetica.
[a sans serif face named Helvetica]

There are many other varieties of typefaces for the roman alphabet, some with scientific applications, such as Fraktur faces (sometimes called "black-letter script") based on Germanic origins (see sections 3·2 and 11·14), faces for phonetic symbols, and other faces not used in scientific publication, such as script typefaces designed to simulate handwritten letters.

Additional variations available within many typefaces are useful in scientific publishing; an example is that of "small capitals", letters with the basic structure of capital letters but of the same height as lowercase letters such as the letter "x" ("ex") or "m"; see section 9·7 for uses of small capitals. A particular typeface may give rise to closely related faces differing in their "weight", a characteristic determined mainly by the width of letter strokes; such a derived face, for example, with heavy or wide strokes is known as a "boldface" version.

Council of Biology Editors
[capital and small-capital letters]

This statement represents a bold italic font of Helvetica.
[boldface]

Section 9·4 summarizes uses of boldface type in scientific publication.

TYPE SIZES AND SPACING

32·3 Linear dimensions in the printing industry have for many years been based on a system of measurement that has "the point" as its basic unit; the larger unit in the system is "the pica". The point and the pica are defined as fractions of an inch.

1 point = 1/12 pica = 1/72 inch [approximately 0.35 mm]

1 pica = 12 points = 1/6 inch [approximately 4.2 mm]

6 picas = 72 points = 1 inch [approximately 25.4 mm]

Points are used to describe the size of characters in a font; to specify spacing between characters, between words, and between lines of type; and to specify the thickness of ruled lines.

Picas are used to describe the dimensions of a type page (the block of text on a page), the spacing between columns of type (sometimes given in ems; see the next paragraph), and other elements in the design of pages. Note, however, that the outside dimensions of an entire page (type page and its margins) are stated in inches (in those parts of the publishing and printing industries still working with nonmetric units).

The size of a font is defined in points by the maximal vertical dimension of its characters measured from the lowest point of "descenders" of characters (such as the part of the letter "j" extending below the bottom of lowercase characters such as the letters x ("ex") and m ("em") in a line of type) to the highest point of "ascenders" (such as the upward stroke of the letter "h" that goes above the upper line of lowercase characters such as x and m). The height of the lowercase characters without ascenders or descenders such as x and m is called the "x-height".

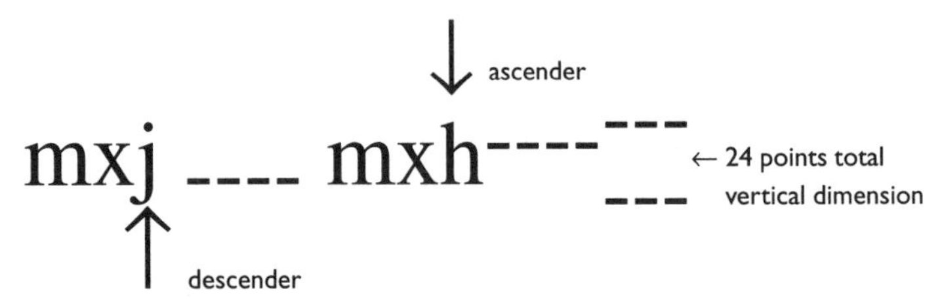

This is 14-point Times. This is 12-point Times.
This is 10-point Times. This is 18-point Times.

The basic unit of horizontal linear dimensions is the "em", a name derived from the width of the capital letter "M". The "en" has traditionally been defined as a length half that of the em. These 2 units are not so rigidly defined in today's photocomposition systems. Ems and ens are used to specify horizontal distances for such design elements as indentions, word- and letter-spacing, and dashes (see sections 4·20 and 4·22 for the em-dash and the en-dash).The term "em quad" is used for the symbol designating a horizontal space in a type line equal to a whole-number multiple of an em: "1-em quad" for a space in a line equal to 1 em; "2-em quad" for a space equal to 2 ems.

MANUSCRIPT MARKING FOR THE PRINTER

32·4 In the older systems of publication, manuscripts are, in general, sent to the printer with 2 kinds of marks. One is that kind of mark needed to correct typing errors, misspellings, grammatical errors, and elements of publication style (such as units of measurement) not in accordance with that of the publisher ; these are copyediting marks. The marking needed to specify type styles and sizes, spacing, and other typographic details is called "markup". The evolution of computer applications in the preparation of manuscripts and typesetting of pages has modified these procedures in a number of ways.

MANUAL MARKING: COPYEDITING AND MARKUP

32·5 The specifications for a journal's paper, design, and detailed typographic characteristics are usually established by the journal's publisher or editor in consultation with the printer selected to produce the journal or with a professional designer not on the printer's staff. The specifications may be developed for a new journal about to begin publication or for a change in design of an existing journal.

The specifications are developed in detail and can then help the journal's manuscript editors decide what needs to be marked on manuscripts to instruct the printer's compositors and what compositors will understand from the agreed-on specifications without manuscript markup. Table 32·1 summarizes the details most often established for a journal's specifications.

Authors should be expected to mark a manuscripts for italics, boldface, small capitals, special characters, and other details before submitting it; a requirement of this kind should be made clear in the journal's information-for-authors page with a section on manuscript style and format; see section 28·23 and Table 28·6. Table 32·2 illustrates the marking needed to specify type styles. Manuscripts prepared with word-processing programs that readily produce italic characters, boldface, and other type variations may not need such marking by authors.

Before a manuscript is sent to the printer for typesetting, the manuscript editor (also called "copy editor", or "subeditor" in British par-

Table 32·1 Specifications for a journal: dimensions, paper, design, typographic details

Characteristic	Notes
Trim size	Overall dimensions of the cover and pages: width and height.[a]
Paper: kind and weight	Selection determined by kind and quality of paper needed and cost; the kind influences the quality of halftone illustrations.
Layout of pages	
Type-page margins	Space around the type page (area for text).
Columns	1-column, 2-column, or 3-column layout of the type page.
Typefaces	Determined in part by aesthetic considerations and judgments on readibility, but needs for fonts of special characters, as in chemistry, mathematics, and physics, may limit the possible choices.
Sizes	For text[b], references, tables, titles, and legends for figures.
Leading	Space between lines, for each size of type.
Title page for articles	Layout of necessary elements[c].
Text format	Indentions; spacing between paragraphs and around figures and tables.
Text heads and subheads	Typefaces and sizes; decision on number and treatment of different levels[d].
Table characteristics	Rules, spacing, footnote signs[e].
Design of special pages	Contents page, masthead page, information-for-authors page[f].
Cover design	Decisions on layout, title elements, and other details[g].

[a]See section 28·16.
[b]See section 28·22.
[c]See sections 28·32–28·37.
[d]See section 28·38.
[e]See section 31·13.
[f]See sections 28·20, 28·21, 28·23.
[g]See sections 28·18, 28·19.

Table 32·2 Typographic conventions, markings for them in manuscript, and cross-references to relevant sections in this manual[a]

Convention	Mark	Example	Printed appearance	Footnc
Italic type	Single underline	Escherichia coli	*Escherichia coli*	b
Boldface type	Wavy underline	A·B	**A·B**	c
Capital letter	Triple underline	professor of geology	Professor of Geology	d
Small capital	Double underline	d-cystathionine	D-cystathionine	e

[a]Additional marks frequently needed by manuscript editors for giving instructions to printers are represen Table 33·1.
[b]For additional discussion of the use of italic type, see sections 9·2, 9·3.
[c]For additional discussion of the use of boldface type, see section 9·4.
[d]For additional discussion of the use of capital letters, see sections 9·5, 9·6.
[e]For additional discussion of the use of small-capital letters, see section 9·7.

lance) checks the manuscript for adherence to the journal's style specifications (such as the style details set forth in this manual), for correctness and proper formatting of references, for correctness of spelling and punctuation, for adherence to scientific conventions, and for other needs in copyediting. If the copyeditor cannot be sure of the author's intent with any detail of the manuscript, questions should be resolved with the author, preferably before the manuscript is sent to the printer. Alternatively, the copyeditor can decide on the apparently correct interpretation and put a query for the author in a marginal note on the manuscript (to be sent to the author with proof). If substantive changes may change the meaning of the text, the manuscript must be returned to the author for approval of changes (speedily done with facsimile [fax] transmission) or the changes discussed with the author by telephone.

32·6 In the markup, elements in the manuscript for which specifications have been defined and which are readily identified by the typesetter need not be marked: for example, running text in roman-alphabet letters and with normal capitalization, single-level headings, article titles, and abstracts. Elements about which there may be ambiguity (such as superscripts, subscripts, special characters, and different levels of headings) or for which there are no specifications must be marked. Instructions to the printer should be immediately adjacent to the element in question and circled to indicate that they are not editing revisions to the manuscript.

Queries to the author that arise during copyediting should be clearly distinguishable from directions to the printer (markup) and from copyediting marks. They can, for example, be placed within a square enclosure. They must be marked specifically for the author's attention: for example, "?A:" (for "query to the author") or "Au: . . .?". With manuscripts submitted on diskettes, the copyeditor may be able to return a copyedited manuscript with queries to the author placed on separate sheets at the end of the manuscript or in an accompanying letter to the author.

Corrections should be marked in the line or immediately above it. Authors are expected to submit double-spaced manuscript with ample margins because copyeditors need space for editing and for markup notations. Figure 32·1 is an example of a manuscript showing both copyediting and markup.

ELECTRONIC MARKUP

32·7 With the development of computerized typesetting and page formatting came the possibilities of having type styles and formats applied automatically to components of books and journals. These possibilities have be-

agents transmembrane potentials and concentrations of the text species in the bathing solution on E, r_a and r_p may be elevated.

ital c+lc

B. APPLICATION TO LEVEL FLOW

Level-flow of a test species in the stionary state occurs when X = 0. Accordingly, in this case Eq. (19) reduces to the following:

$$\frac{s\Omega}{1 - EXP(-s\Omega)} = 1 - A\Omega s\Omega \qquad (22)$$

8pt #238

The graphical solution of the above equation is shown in Fig. 3. Note that $s\Omega = 0$ is always a solution. However, it is a physical tolution of the equation only when $A\Omega = -0.5$ since the slipe of the exponential curve is -0.5 at s = 0b.

Au: figure 2 meant?

Au: delete?

The slope of the exponential curve approaches 1.0 as a approaches − infinity. Thus −1.0 is the minimum permissible value of $A\Omega$. The active trans-port process must be properly oriented to correspond to the direction of the virual level-flow index. Thus, if the unifirectional fluxes are chosen so that J_{v1} is negative then E must also be negative.

Au: s, not a?

---

b L'hospital's rule has been used to obtain this slope.

Au: spelling OK?

8pt #238

come realities through systems in which tags that designate the functional elements of text, equations, and tables are applied to manuscripts as a final step in their being prepared in a digital format for printing. The computer is programmed to identify the tags in the electronic (digitized) manuscript and to automatically apply stipulated specifications (such as for typeface, type weight, type size, and so on) to the tagged components in its typesetting function. One additional value of the tags is in their converting manuscripts in this format into a kind of database so that the text, tables, and equations can be manipulated to facilitate additional uses of the material so tagged. An economically more important value is in the options for using such tagged texts for publication in media other than paper books and journals, for example, in the CD-ROM format and in electronic online systems.

Some of these systems have been developed by printing firms and are "proprietary"; manuscripts tagged with their conventions may not be "readable" in other proprietary systems. Fortunately, standards that may lead to uniformity and interchangeability in electronic (digitized) manuscripts have been developed. The important American document setting forth the standards for preparing and tagging electronic manuscripts is *Electronic Manuscript Preparation and Markup* (ANSI–NISO 1991). This standard draws on the international standard, ISO 8879 (cited in ANSI–NISO 1991), that defines the Standard Generalized Markup Language, widely known as "SGML".

This manual does not go beyond briefly describing and illustrating electronic markup (tagging). Four notably useful, practical sources of detailed information on the tagging of document components, including special characters not represented in the ASCII character set (see item 3 in section 32·8), are the publications prepared by the Association of American Publishers: *Reference Manual on Electronic Manuscript Preparation and Markup* (AAP 1989d), *Author's Guide to Electronic Manuscript Preparation and Markup* (AAP 1989a), *Markup of Mathematical Formulas* (AAP 1989b), and *Markup of Tabular Material* (AAP 1989c).

◀ **Figure 32·1** A page of edited manuscript with queries from the copyeditor to the author (marked with "Au:" and placed within square enclosures ["boxed"]) and directions to the printer (circled). Superscripts are marked with ∨ and subscripts with ∧, as are apostrophes and commas, respectively, when handwritten. Characters to be removed are crossed out and marked with a delete mark by hand and ones to be inserted are indicated with a caret (∧) at the point of insertion. Potentially confusing zeroes and the letter "oh" are identified. Any letter to be changed from capital to lowercase is marked with a diagonal stroke. Capitals are indicated by triple underlining. Italics are indicated by single underlining. Indentions are marked with em-quad marks (see Table 33·1).

Additional information can be found in manuals published by The University of Chicago Press (UCP 1986, 1993).

Categories and Structures of Tags

32·8 Three types of tags serve different functions.

1 Generic identifiers indicate the structural components of a document such as the title, the abstract, a paragraph, a footnote, a bibliographic reference.
2 Emphasis-font tags identify elements within designated structural components for which special typographic treatment is desired, such as italics for a text phrase, boldface for page numbers in a reference, and so on.
3 Entity references indicate specific characters included in the ASCII character set (and the ISO 646 set) and other characters (such as greek letters and mathematical symbols) that are not available through some computer keyboards. ASCII (**A**merican **S**tandard **C**ode for **I**nformation **I**nterchange) characters are those letters, numerals, punctuation marks, and other symbols defined by *International Standard ISO 646, 7-Bit Coded Character Set for Information Processing Interchange* [ISO 646-1983 (E)], except for a few minor departures needed to adapt the ISO standard for American use.

32·9 Generic identifiers and emphasis-font tags are formed, in general, of abbreviations for the names of elements being designated enclosed within angle brackets ("less than" and "more than" symbols; see section 4·29), the angle brackets serving as delimiters for the tag. Thus "<fn>" identifies a footnote; "<scp>" identifies small capitals. The tags with angle brackets (< >) enclosing only the abbreviation are "start-tags" that indicate the beginning of the structural element. Some structural elements must be closed with an "end-tag"; these tags use the same abbreviation as the start-tag for the element but the abbreviation is preceded by a slash (slant line; see section 4·34).

<p> [a start-tag; start of a paragraph]

</p> [an end-tag; end of the paragraph]

Entity references have an ampersand (&; see section 4·41) as the opening delimiter, an abbreviation for the character or symbol being designated, and a semicolon (;) as the closing delimiter; an example is "& THgr;" for the capital greek letter theta (Θ).

Table 32·3 gives additional examples of tags for these 3 types. Figure

Table 32·3 Examples of tags for electronic markup of manuscripts[a]

Tag	Element represented by the tag
Manuscript component	
<abs>	abstract [start tag]
</abs>	end of abstract [end-tag]
<aff>	affiliation statement (organization name, location, and so on) [start tag]
</aff>	end of affiliation statement [end-tag]
<bm>	back matter [start tag]
<fn>	footnote [start tag]
<h>	heading [start tag]
<p>	paragraph [start tag]
</p>	end of paragraph [end-tag]
<phi>	acid-free paper indicator [start tag]
<toc>	table of contents [start tag]
<ti>	title [start tag]
Elements needing special typographic treatment	
<b>	bold
<gk>	greek text
<it>	italic
<pp>	page number of reference
<vid>	volume identification data
Special characters	
&	& (ampersand)
†	† (dagger)
‡	‡ (double dagger)
÷	÷ (division symbol)
£	£ (pound sterling)
§	§ (section symbol)
×	× (multiplication symbol)
™	™ (trademark)
¥	¥ (yen)

[a]See sections 32·7–32·9.

32·2 illustrates how they can be applied in markup of manuscripts for electronic typesetting and formatting or other treatments.

32·10 A document to be published is categorized by a "document type declaration" (DTD). The 3 standard document types are books, articles, and serials (generally known to authors and readers as "journals"). A publisher designates the elements that can be used in each document type, and this designation then can be checked by the computer program used to typeset and format for adherence of a tagged manuscript to the styles and formats specified for the document type it represents. The 3 standard document types can be used to establish other document types such as conference proceedings, dissertations, and technical reports.

Manuscript as Tagged

```
<article>
<at1>The Objectives of the New CBE Style
Manual
<au><fnm>Eduoard<snm>Diderot
<aff><onm>Council of Biology Editors
<cty>Chicago, <sbd>Illinois <pc>60601-
4298, <cny>USA</aff>
<sec>
<h1>Introduction</h1>
<p>Through the past 2 or 3 decades the
conceptual boundaries between the
traditional scientific disciplines have
become blurred. Nomenclatures and
symbolizations once confined to a
particular discipline came to be needed
in other fields. Plant physiology, for
example, now needs the symbols of
thermodynamics.</p>
```

Figure 32·2 Example of a manuscript to be typeset and formatted by a computer-driven program able to interpret the instructions of a tagging system based on the principles of the Standard Generalized Markup Language (SGML) as interpreted by ANSI–NISO (1991).

Manuscript as Typeset and Formatted

> The Objectives of the New CBE Manual
>
> Eduoard Diderot
> *Council of Biology Editors, Chicago, Illinois 60601-4298, USA*
>
> INTRODUCTION
>
> Through the past 2 or 3 decades the conceptual boundaries between the traditional scientific disciplines have become blurred. Nomenclatures and symbolizations once confined to a particular discipline came to be needed in other fields. Plant physiology, for example, now needs the symbols of thermodynamics.

Figure 32·3 The typeset copy resulting from the SGML-tagged manuscript in Figure 32·2.

Preparation of Tagged Manuscripts

32·11 Manuscripts submitted as a file on a computer diskette can be converted to tagged copy by inserting the needed tags using the characters available on standard keyboards. Note that the copyediting needed for correction of errors in text and imposition of the publisher's requirements for publication style should be carried out before this step; the electronic markup should be carried out after the changes required in copyediting have been completed.

The process can be simplified greatly and its accuracy effectively controlled by using computer programs specifically designed for tagging and verifying the tagging for the document type. One such program, "SoftQuad Author/Editor" (SoftQuad Inc., Toronto, Canada M8X 2W4) is used for all the steps in manuscript preparation, tagging, and verification. Another program, "WordPerfect® Intellitag" (WordPerfect Corporation, Orem UT 84057-2399, USA), operates on texts prepared with some standard word-processing programs and imported into "Intellitag". Such programs will undoubtedly be joined by similar programs as the demand for SGML tagging increases.

The ease with which a manuscript can be tagged is enhanced by its being submitted both as a properly prepared file on a computer diskette and as a paper manuscript ("hard copy"). Authors should be told by publishers what characteristics of diskette files are needed. For journals this information should be given in the information-for-authors page (see section 28·23 and Table 28·6). The exact requirements may depend on the tagging and computer-composition systems used by the publisher and the printer. The information that should be given to authors is likely to include these considerations.

1 The size and capacity of acceptable diskettes.
2 The types and models of computers to be used by authors or other persons preparing the file on a diskette.
3 The word-processing programs that are acceptable; whether only ASCII files or formatted files; and the preferred version of the word-processing program.

The information for authors may have to include specific, more detailed instructions. These might include a requirement for submission of paper manuscripts along with diskettes, a requirement for absolute identity between the contents of the paper manuscript and the diskette file, injunctions against using a lowercase "ell" (l) to represent the numeral "one" (1) or a capital "oh" (O) to represent the numeral "zero" (0), and requirements against allowing special bibliographic citation codes to remain in texts and using automatic hyphenation. The information needed to fully identify a diskette and its file should be specified.

REFERENCES

Cited References

[AAP] Association of American Publishers. 1989a. Author's guide to electronic manuscript preparation and markup. Version 2.0. Rev ed. Dublin (OH): AAP. Available from EPSIG [Electronic Publishing Special Interest Group], c/o OCLC, 6565 Frantz Road, Dublin OH 43017-0702; telephone 614 764-6000.

[AAP] Association of American Publishers. 1989b. Markup of mathematical formulas. Version 2.0. Rev ed. Dublin (OH): AAP . [See the availability note in AAP 1989a.]

[AAP] Association of American Publishers. 1989c. Markup of tabular material. Version 2.0. Rev ed. Dublin (OH): AAP. [See the availability note in AAP 1989a.]

[AAP] Association of American Publishers. 1989d. Reference manual on electronic manuscript preparation and markup. Version 2.0. Rev ed. Dublin (OH): AAP. [See the availability note in AAP 1989a.]

[ANSI–NISO] American National Standards Institute, National Information Standards Organization. 1991. Electronic manuscript preparation and markup: American national standard for electronic manuscript preparation and markup, ANSI/NISO Z39.59-1988. Bethesda (MD): NISO.

Bringhurst R. 1992. The elements of typographic style. Vancouver (BC): Hartley & Marks.

Butcher J. 1992. Copy-editing: the Cambridge handbook for editors, authors and publishers. 3rd ed. Cambridge (UK): Cambridge Univ Pr.

Morse JM, editor. 1985. Webster's standard American style manual. Springfield (MA): Merriam-Webster.

[UCP] University of Chicago Press. 1986. Chicago guide to preparing electronic manuscripts. Chicago: UCP.

[UCP] University of Chicago Press. 1993. The Chicago manual of style. 14th ed. Chicago: UCP.

Additional References

Judd K. 1990. Copyediting: a practical guide. 2nd ed. San Francisco (CA): William Kaufmann.

Montagnes I. 1991. Editing and publication: a training manual. Manila: International Rice Research Inst; Ottawa: International Development Research Centre.

O'Connor M. 1986. How to copyedit scientific books and journals. Baltimore (MD): Williams & Wilkins.

White JV. 1988. Graphic design for the electronic age: the manual for traditional and desktop publishing. New York: Watson-Guptill.

33 Proof Correction

33·1 Most publishers provide authors with the printer's proofs of the text (article, chapter, book) being published, including proofs of any illustrations. A proof is an impression on paper of text in typeset form and illustrations as prepared for publication.

Proofs are usually accompanied by the original manuscript that shows the changes made by the editor or copyeditor. The author is responsible for checking the proof of the typeset text and illustrations against the originals and for making any corrections on proof. Usually this is the last time the author sees the text and illustrations before publication, so thoroughness is critical. Proofs must be returned to the printer, editor, or publisher by the specified deadline; responses by telephone or fax are sometimes acceptable.

Editors should inform authors when to expect the arrival of proofs so that if an author will not be available at that time he or she can ask a colleague to check them. Manuscripts prepared in "camera-ready" form for offset printing must be proofread carefully before being submitted to the publisher, because authors will not see proofs of papers prepared for this type of publication.

Some publishers do not supply proofs and simply send the author the copyedited manuscript before typesetting; in this system the author is expected to check on the correctness of any changes made by the copyeditor and respond to any queries raised.

KINDS OF PROOF

TEXT AND COMPOSED TABLES

33·2 The text of the manuscript and any accompanying tabular material may be set in type by various systems. In hot-metal composition, the standard system for many years, type is cast from molten metal with a Linotype or Monotype machine. The metal lines or pieces of type that are produced are placed in long trays called galleys, and proof is produced by direct impression of the inked type onto paper, on which changes are marked. After corrections have been made, the type is assembled in the format of pages. In recent years, composition by computer systems has come to replace hot-metal composition, and proof in these systems is generally prepared in the format of pages and supplied to authors as page proof. A clear and concise description of several of these systems can be found in *The Chicago Manual of Style* (UCP 1993).

ILLUSTRATIONS

33·3 The kinds of proof provided for continuous-tone (halftone) illustrations and line drawings depend on the printing process used.

For illustrations that are to be printed by letterpress, images are photographically transferred and engraved onto metal plates. Proofs are made by direct transfer onto paper from the inked engravings. An author may be sent a set of engraver's proofs separate from the text or galley proofs that show the engravings assembled, with the legends below the illustrations to which they apply. Galley-stage proofs of halftone illustrations show less contrast and detail than the illustrations will show on the printed page; an engraver's proof, on coated or glossy paper of high quality, is better evidence of the quality of a halftone. The impression of an illustration on the journal or book page is not likely to have the brilliance of the engraver's proof unless the paper used in the journal or book is of a quality equal to that used for the proof.

For illustrations that are to be printed by the offset process, prints of negatives that will be used to prepare the press plates serve as proof. The quality of such proofs varies since they may represent carefully prepared prints or merely photocopies. With halftones, the author is usually sent a print of a quality close to that of the illustration as it will appear in the journal or book.

CHECKING PROOFS

TEXT AND COMPOSED TABLES

33·4 Authors should be instructed not to mark the original manuscript returned with proofs. It is the record of what the printer was asked to typeset and the basis for the billing to the publisher for the author's and editor's alterations. Authors should also be instructed not to erase or change any of the queries, suggestions for changes, and corrections placed in the margins of a proof by the editor or a proofreader. Likewise, authors should be instructed not to cross out or cut off the letters and numerals on proofs that include the number of the proof and other information needed by the printer; the instructions to authors can indicate that they will be deleted when the pages are made up.

Authors should be expected to read proofs at least twice. In the 1st reading the author should follow the text of the proof while another person reads the manuscript aloud. In the 2nd reading the author should read the proof alone.

The instructions to authors on the handling of proofs should point out that at this stage only the correction of errors is expected and that the proof stage is not the time for making trivial changes, improving prose style, adding new material, or making deletions. They can be reminded that corrections cost more than initial composition, and errors may be introduced; correcting a photocomposed line usually necessitates recomposing at least the line with the error. With some photocomposition systems, even small changes require that entire paragraphs be rehyphenated and a new set of prints prepared.

33·5 Printer's errors (PEs) should be corrected at no cost to the journal or author. Author's alterations (AAs) are usually charged to the publisher and ultimately may have to be charged to the author. Authors and editors should indicate printer's errors by marking PE near the correction. If text must be revised, the number of characters added (each space equals a character) should be balanced, if possible, by deletion of an equal number of characters (and spaces). In computer-assisted systems, character-for-character replacement is less critical. The text is corrected on a text-editing terminal, and paragraphs are rerun.

33·6 Authors may want to add at the proof stage some text on observations made since submitting the manuscript that they believe to be important. The decision to allow such additions must be that of the editor; a copyeditor usually should not allow such additions without the editor's permission. Adding new content to a peer-reviewed article under an old "received" date is generally considered to be unethical if that content has not been judged for its acceptability by the peer reviewers or the scientific editor. The editor may suggest including a dated addendum containing the new material, which will obviate the need for changes in the text.

Accuracy

33·7 Proofs must be carefully checked against the original text for accuracy of equations and numeric data (especially in tables) and proper spelling and punctuation, separation of paragraphs, order of headings, and citation of references, figures, and tables. Attention should be paid to the appropriate location of tables and figures in relation to their 1st mention in the text.

Typography

Letters and paragraphs must be checked closely for possible repetition or transposition. Defective characters occur less frequently with photocomposition than with hot-metal composition. Copyeditors at the publisher's office and at the printing plant are responsible for correcting poor typography and alignment, but authors should verify the proper alignment of statistical data, chemical formulas and equations, and mathematical expressions. Symbols should be checked for agreement with recognized conventions.

Word Divison

33·8 End-of-line word breaks on proofs should be checked carefully; they may not be the same as in the typescript. Computer systems for composition are likely to have their own rules for hyphenating words at line ends. The extent to which computers are programmed to handle exceptions to general rules differs, so special care should be taken with checking proofs of text prepared in this way. In the American system, word division is based on pronunciation, not on the etymologic derivation of words as in the British system. In checking divided words, particular attention should be paid to Latin words and scientific terms.

Every Latin word is divided into as many syllables as it has separate vowels and digraphs. The consonant "h" between 2 vowels always joins the 2nd vowel (*mi-hi, co-hors*); an "x" between 2 vowels always joins the preceding vowel (*sax-um, ax-il·la*); "ch", "ph", and "th" are never separated (*Te-thys*); "gl", "tl", and "thl" are always separated (*Ag-la-o- ne-ma, At-las, ath-let-i-cus*). A single consonant between the last 2 vowels of the word or between any 2 unaccented vowels joins the final or 2nd vowel (*pa-ter*); but a single consonant before or after an accented vowel joins the accented syllable (*i-tin'·e·ra, dom'-i·nus*). Two consonants between vowels must be separated (*cor-pus, for-ma*); for 3 consonants between vowels, the last joins the vowel that follows (*emp-tor*); for 4 consonants between vowels, 2 are joined to each vowel (*trans-trum*). When a compound word is divided into syllables, the component parts are separated when the 1st part ends with a consonant (*su·per-est, sub-i·tus, trans-i·tur*); otherwise, it is divided as if it were a simple word (*dil·i-gens*). As in other aspects of Latin grammar, there are exceptions to these rules, but they are too technical or unusual to merit treatment here.

Special care must be taken with scientific words that need hyphens as part of their proper structure; chemical names are the most frequently used terms of this kind. When such words must be broken at the end of a line, the break should, if possible, use the word's proper hyphen for the end-of-the-line hyphen and not repeat a hyphen at the beginning of the following line.

> l-glutamine amido- ligase [not "l-glutamine amido- -ligase"]

If an equation or other mathematical expression is in a line of running text and must be broken at the end of a line, the break should be made, if possible, so that an operator symbol (such as the "equals" symbol) is at the end of the line to function as the equivalent of a hyphen and is not at the beginning of the following line.

> . . . and if $x =$ $3.51y + 75.89$, then . . .
> [not ". . . and if x- $= 3.51y + 75.89$"]

A break that produces part of a word that can be read as a word in its own right and with a meaning that might be confusing, amusing, or offensive in the context should be avoided if possible. If a break cannot be avoided, the break should avoid the possibly disturbing effect even if a standard rule for breaks must be violated; often an inappropriate break can be avoided by rearranging the sentence.

Detailed guidance on word breaks can be found in *The Chicago Manual of Style* (UCP 1993) and *Hart's Rules for Compositors and Readers at the University Press* (OUP 1983).

ILLUSTRATIONS

33·9 The proof of each line drawing should be examined for the absence of numerals, letters, lines, or parts of lines that may unintentionally have been routed out by the engraver; some letters or numerals may have been "filled in" with ink. Any correction needed in a halftone or a line drawing must be indicated clearly. Some kinds of flaws in a line drawing can be easily corrected by the engraver; superfluous dots and lines can be tooled or routed out, and slightly broken lines and characters can be repaired. The amount of contrast in some halftone engravings can be improved without remaking the engraving. Adding to or otherwise revising an illustration necessitates making a new cut. Corrections on illustrations for offset printing usually involve rephotographing the original rather than correcting the master print. Care must be taken to have the identifying figure number appear near each illustration. The proper orientation of a figure may have to be identified by writing "Top" in the upper margin of the figure's proof. Some journals require that magnifications for figures stated in legends be revised to accommodate any reduction or enlargement that occurred during preparation of halftone engravings.

MARKING PROOFS

33·10 Corrections should be made with standard proofreaders' marks. The American system (ANSI–NISO 1991) differs slightly from the British (BSI 1976, Butcher 1992) system. There is no standardized system of European proofreaders' marks, many of which differ from those of the American and British systems; most European printers recognize American and British marks or provide authors with lists of marks they prefer. See Table 33·1 for a comparison of American and British marks.

Proof corrections must be made in the margins of the galley or page proofs (not above lines of type as in correcting a manuscript; see section 32·6 and Figure 32·1). In examining proofs for needed changes, the typesetter or compositor scans the margins of proofs and makes only the alterations that are indicated there.

Marks for corrections should be made in colors different from any already on the proof or in the colors specified in instructions that accompany the proofs. For example, if the printer's proofreader has marked corrections in green, the printer's errors can be indicated in red and needed author's alterations in blue.

CORRECTION MARKS NEEDED

33·11 For each correction, at least 2 marks are needed on the proof: 1 or more in the text (in-line) and 1 or more in the margin (marginal) nearest the in-line mark; see Figure 33·1. One of the in-line marks may be a caret (‸) to show where an addition is to be made; another may be a line drawn through a character or a word to be deleted. A marginal mark indicates what the change is to be. It may be 1 or more characters or words to be inserted, or a proofreader's symbol, such as a space or delete sign. Each mark should be made on an imaginary extension (preferably right) of the

Table 33·1 American and British proofreaders' marks and symbols[a]

Instruction	American marginal mark	American in-line mark
Delete	℘	the ~~red~~ book
Close up	⁀‿	the bo‿ok
Delete and close up	℘ (closed up)	the b~~o~~ook
Restore deletion	stet	the ~~red~~ book (dotted underline)
Insert in line	red	the ^ book
Substitute in line	red	the ~~black~~ book
	e	th~~a~~ book
Insert space in line	#	the|book
Equalize spacing	eq #	the ✓ yellow ✓ book
Lead (space between lines)	# [or] ld	The red book / was lost.
Remove leads (space) between lines	℘# [or] ℘ld	The red book / was found.
Insert hair space or thin space	hr# [or] thin #	100/000
Begin new paragraph	¶ [or] ⌊	The red book was lost. ¶The black book was found.
Run paragraphs together	no ¶	The black book was lost. ⤸ The red book was found.
Insert 1-em quad (indent)	□	^The red book
Insert 2-em quad (indent)	□□ [or] [2]	^was found
Insert 3-em quad (indent)	□□□ [or] [3]	^at night.
Move to left	⊏	⊏ the book
Move to right	⊐	the⊐book
Center	ctr	⊐ the book ⊏
Move up	⌐¬	⌐the book¬
Move down	⌊⌋	⌊the book⌋

Table 33·1 (*cont.*)

British marginal mark	British in-line mark	Corrected text
	the red book	the book
	the bo ok	the book
	the boook	the book
	the red book	the red book
red	the book	the red book
red	the black book	the red book
e	tha book	the book
	thebook	the book
	the yellow book	the yellow book
extend text mark	The red book was lost.	The red book was lost
extend text mark	The red book was found.	The red book was found.
thin	100000	100 000
	The red book was lost. The black book was found.	The red book was lost. The black book was found.
	The black book was lost. The red book was found.	The black book was lost. The red book was found.
1	The red book	The red book
2	was found	was found
3	at night.	at night.
	the book	the book
	the book	the book
	the book	the book
	the book	the book
	the book	the book

Table 33·1 (*cont.*)

Instruction	American marginal mark	American in-line mark
Align vertically	‖ [or] align	The book was lost in the fog.
Align horizontally	= [or] straighten	The book was found.
Transpose	tr.	Teh book was found. The found book was.
Spell out	(sp)	He arrived 1st.
Push down quad (spacing material)	⊥	the ■ book
Reset broken letter	×	the book
Turn right side up	9	the book
Lowercase letter	lc	the Green book
Capitalize as marked	cap	The good Book
Set in small capital	sc	D-glucose
Set in italic type	ital	The Good Book
Set in roman type (upright type)	rom	the *book*
Set in boldface type	bf	The Good Book
Set in lightface type	lf	The **book**
Set in capitals and small capitals	C+SC	Dong geng
Set in boldface italics, capitals and lowercase	bf ital c+lc	a style manual
Wrong font; reset	wf	body type
Reset as superscript (superior)	2	$e = mc2$
Reset as subscript (inferior)	2	H2S
Insert as superscript (superior)	b	1203
Insert as subscript (inferior)	2	HO
Period (full stop)	⊙	Read the book
Comma	,	leaves, buds and branches
Semicolon	;	Think then decide.

Table 33·1 (*cont.*)

British marginal mark	British in-line mark	Corrected text
||	The book was lost in the fog.	The book was lost in the fog.
=	the book was found.	The book was found.
	Teh book was found. The found book was	The book was found. The book was found
first	He arrived 1st.	He arrived first.
	the book	the book
x	the book	the book
	the book	the book
	the Green book	the green book
	The good Book	The Good Book
=	D-glucose	D-glucose
	The Good Book	*The Good Book*
	the *book*	the book
	The Good Book	**The Good Book**
	The **book**	The book
	Dong geng	DONG Geng
	a style manual	***A Style Manual***
⊗	body type	body type
	$e = mc$	$e = mc^2$
	HS	H_2S
	1203	1203^b
	HO	H_2O
	Read the book	Read the book.
,	leaves, buds and branches	leaves, buds, and branches
;	Think then decide.	Think; then decide.

Table 33·1 (*cont.*)

Instruction	American marginal mark	American in-line mark
Colon	:	Read these books‸
Hyphen	=/=	graft‸versus‸host disease
Apostrophe	’	*Donovans Demise*
Double quotation marks	“/”	He said‸book‸.
Single quotation marks	‘/’	“Don’t cry‸Fire!‸”
Question mark	?	Is this your book
En-dash (rule)	1/N	pages 10‸15
Em-dash	1/M	His book‸find it!
3-em dash	3/M	His reply,“nuts”.
Parenthesis marks	(/)	report‸Smith 1992‸was
Brackets (square brackets)	[/]	the book‸a manual‸
Slash (slant line, oblique)	/	5 ms

[a]Sources: ANSI–NISO (1991) and BSI (1976).

line to which it applies. To indicate several corrections for a single line, arrange them in sequence from left to right in the nearer margin, and separate adjacent corrections by a slash (slant line).

EUROPEAN SYMBOLS

33·12 European proofreaders use a series of symbols to indicate several corrections in a single line of type: [symbols]. The symbol is placed through the type to be corrected in the proof, and the same symbol is placed before the correction that is in the margin.

LONG CORRECTIONS

33·13 If the correction or insertion consists of more than 1 or 2 lines, it should be typed on a separate sheet that is then attached to the proof at the appropriate place with tape, never with a paper clip, pin, or staple. Where the new copy is to be inserted must be indicated clearly. Each insert should be marked with a letter and the number of the proof sheet to which it belongs, for example, “Insert A for proof sheet 2”. The correc-

Table 33·1 (*cont.*)

British marginal mark	British in-line mark	Corrected text
:	Read these books	Read these books:
|=| |=|	graft versus host disease	graft-versus-host disease
'	*Donovans Demise*	*Donovan's Demise*
" "	He said book.	He said "book".
' '	"Don't cry Fire!"	"Don't cry 'Fire'!"
?	Is this your book	Is this your book?
1en	pages 10 15	pages 10–15
1em	His book find it!	His book—find it!
3em	His reply, 'nuts'!	His reply, ———!
( )	report Smith 1992 was	report (Smith 1992) was
[ ]	the book a manual	the book [a manual]
⊘	5 ms	5 m/s

tions must be neatly made so as to be clearly legible. If a blanket order such as "set *Rosa* in italics throughout" is called for, each occurrence of the change must be marked on the proof; the printer cannot be expected to make corrections that are indicated only in an accompanying letter.

There may be queries to an author from the editor or the printer in the margins of the proof. An author who wants to have the line remain as set should draw a line through the query, but not erase it. If the editor or a proofreader has indicated a change that is not acceptable, a line can be drawn through the suggested change and accompanied by marginal note "OK as set" or "stet"; dots should be placed under the material that must be retained. Notations that are not to be set in type, such as a note to the printer, must be circled. All queries should be answered.

RETURNING PROOFS

33·14 Authors must return proofs to whomever the publisher stipulates in the instructions usually sent by the printer or publisher with the proofs. If illustrations are to be remade the printer must have the original artwork.

Proof Marked for Corrections

ctr: bf] Fawns Versus Food [

It is basic in animal biology that far more young are produced than necesary to carry on the species. This is true of ants elephants, people and deer. The better nourished a doe is, the more fawns she produces, and the betler chances her fawns have for survival after birth. One of the principles of deer herd management, or raising livestock can be briefly stated if on a givon amount of food, we carry a smaller number of bred females overwinter, each one will be better fed. 10 well fed does will produce at least as many fawns as 15 half-starved ones. This has been proven beyond question. Michigan is no exception to this rule. In the upper peninsula the a■rage rate of fawn production■is 14 or 15 fawns per year from every 10 breeding does . . . and in Southern Michigan fawn production jumps to 20 per 10 does.

—Michigan Whitetails, 1959

Proof after Corrections

Fawns Versus Food

It is basic in animal biology that more young are produced than are necessary to carry on the species. This is true of ants, elephants, people, and deer. The better nourished a doe is, the more fawns she produces, and the better chances her fawns have for survival after birth. One of the principles of deer herd management, or livestock raising, can be stated briefly: If, on a given amount of food, we carry a smaller number of bred females over winter, each one will be better fed. Ten well-fed does will produce at least as many fawns as 15 half-starved ones. This has been proved beyond question.

Michigan is no exception to this rule. In the Upper Peninsula the average rate of fawn production is 14 or 15 fawns per year from every 10 breeding does . . . and in southern Michigan fawn production jumps up to 20 per 10 does.

—MICHIGAN WHITETAILS, 1959.

Figure 33·1 Marked and corrected proof. The upper example shows a proof block of text marked for correction with American proof-correction marks (ANSI–NISO 1991). If more than 1 instruction must be written in the margin next to a line of type, the instructions are separated by slashes (slant lines). The lower example shows the resulting corrected copy.

REFERENCES

Cited References

[ANSI–NISO] American National Standards Institute, National Information Standards Organization. 1991. Proof corrections: American National Standard proof corrections, ANSI/NISO Z39.22-1989. Bethesda (MD): NISO Pr. Available from NISO Press Fulfillment, PO Box 338, Oxon Hill MD 20750–0338, USA; fax 301 567-9553.

[BSI] British Standards Institution. 1976. British Standard copy preparation and proof correction, BS 3261. Part 2. Specifications for typographic requirements, marks for copy preparation and proof correction, proofing procedure. London: British Standards Inst. Available from: American National Standards Institute, 1430 Broadway, New York NY 10018. For additional sources, consult the section "Standards for Editing and Publishing" in Appendix 3, "Annotated Bibliography".

Butcher J. 1992. Copy-editing: the Cambridge handbook for editors, authors and publishers. Cambridge (UK): Cambridge Univ Pr.

[OUP] Oxford University Press. 1983. Hart's rules for compositors and readers at the University Press. 39th ed. New York: OUP.

[UCP] University of Chicago Press. 1993. The Chicago manual of style. 14th ed. Chicago: UCP.

Additional References

Judd K. 1990. Copyediting: a practical guide. 2nd ed. San Francisco (CA): William Kaufmann.

Morse JM, editor. 1985. Webster's standard American style manual. Springfield (MA): Merriam-Webster.

O'Connor M. 1986. How to copyedit scientific books and journals. Baltimore (MD): Williams & Wilkins.

Appendixes

APPENDIX

1 Abbreviated Forms of Journal Titles

AI·1 In section 30·24, Chapter 30, “Citations and References”, the recommendations for formatting references to journal articles specify that journal titles be abbreviated. Standard forms of abbreviations for a large number of journal titles in the biological and medical scientific fields are readily found in 2 major compilations: *Serial Sources for the BIOSIS Previews® Database* (published annually by BIOSIS, 2100 Arch Street, Philadelphia PA 19103–1399, USA) and *List of Journals Indexed in Index Medicus* (published annually by the National Library of Medicine, 8600 Rockville Pike, Bethesda MD 20894, USA). The titles of journals not represented in these 2 compilations can be abbreviated in accordance with the rules applied by these organizations; these rules are based on the *American National Standard Z39.5–1985: Abbreviation of Titles of Publications.*

RULES FOR ABBREVIATING TITLES

AI·2 The title-abbreviation rules of the National Library of Medicine based on the ANSI standard have been summarized by Arenales and Sinn (1989); they are relevant to the journal-title abbreviations called for in section 30·24 and are digested below.

1 The title to be abbreviated is the title proper: the name of the journal in its fullest form. The names of sponsoring organizations are excluded unless the name is syntactically connected to a generic term for a serial publication such as “Annals”, “Journal”, “Bulletin”, “Archives”.

   Bull Univ Nebr State Mus
   [“Bulletin of the University of Nebraska State Museum”]

   If the full title is preceded by an abbreviated form serving as an acronym, only the full title is selected for abbreviation unless the acronym is the only title appearing elsewhere throughout the journal.

2 Articles, conjunctions, and prepositions are dropped unless they are part of a personal or place name, are in a scientific or technical term, or are a part of a standard phrase.

   In Vitro Cell Dev Biol [“In Vitro Cellular and Developmental Biology”]

Table A1·1 Standard abbreviations of frequently used journal title words[a]

Word	Abbreviation	Word	Abbreviation
Abstracts	Abstr	Entomology	Entomol
Academy	Acad	Environment	Environ
Acta	Acta [no change]	Environmental	Environ
		Ergebnisse	Ergeb
Advances	Adv	European	Eur
Agricultural	Agric	Fortschritte	Fortschr
Agriculture	Agric	Genetic	Genet
Agronomy	Agron	Genetics	Genet
Akademie	Akad	Geologica	Geol
American	Am	Geological	Geol
Anales	An	Giornale	G
Annales	Ann	Immunology	Immunol
Annals	Ann	Indian	Indian [no change]
Annual	Annu		
Archives	Arch	International	Int
Asociacion	Asoc	Italian	Ital
Association	Assoc	Italiano	Ital
Australia	Aust	Japanese	Jpn
Beiträge	Beitr	Journal	J
Biochemie	Biochem	Laboratoire	Lab
Biochemistry	Biochem	Laboratorio	Lab
Boletin	Bol	Laboratory	Lab
Bollettino	Boll	Marine	Mar
Botanical	Bot	Mathematical	Math
Botany	Bot	Memoires	Mem
British	Br	Memoirs	Mem
Bulletin	Bull	Memorias	Mem
Bureau	Bur	Memories	Mem
Canadian	Can	Meteorology	Meteorol
Canadienne	Can	Microbiology	Microbiol
Chemistry	Chem	Molecular	Mol
Chimie	Chim	Monatsschrift	Monatsschr
Chinese	Chin	Monographs	Monogr
College	Coll	Museum	Mus
Communications	Commun	Nacional	Nac
Comptes Rendus	C R	National	Nat
		Natur	Nat
Comunicaciones	Comun	Natural	Nat
Contribuciones	Contrib	Naturale	Nat
Contributions	Contrib	Naturelle	Nat
Current	Curr	New Zealand	NZ
Danish	Dan	Newsletter	Newsl
Dansk	Dan	Norsk	Nor
Deutsche	Dtsch	Norwegian	Norw
Developmental	Dev	Notas	Notas [no change]
Entomological	Entomol		

Table A1·1 (*cont.*)

Word	Abbreviation	Word	Abbreviation
Nuclear	Nucl	Scientific	Sci
Organic	Org	Social Science	Soc Sci
Organisation	Organ	Societa	Soc
Organization	Organ	Societe	Soc
Palaeonotology	Palaeontol	Society	Soc
Pharmaceutical	Pharm	Spectrometry	Spectrom
Pharmacological	Pharmacol	Studies	Stud
Pharmacology	Pharmacol	Swedish	Swed
Physics	Phy	Sveriges	Sver
Physiology	Physiol	Symposium	Symp
Proceedings	proc	Transactions	Trans
Progress	Prog	Travaux	Trav
Psychology	Psychol	Trudy	Tr
Publications	Publ	Universidad	Univ
Review	Rev	University	Univ
Revista	Rev	Voprosy	Vopr
Rivista	Riv	Wissenschaftliche	Wiss
Sbornik	Sb	Zoological	Zool
Schweizer	Schweiz	Zoologische	Zool
Schweizerische	Schweiz	Zoology	Zool
Science	Sci	Zeitschrift	Z

[a]The examples of terms for disciplines (words ending in "ology") illustrate the general process of truncation for such terms. Cognate terms in most of the European languages are abbreviated similarly, as can be seen in this table. Both of these points indicate how abbreviations of terms not in this list can be confidently abbreviated by analogy.

3 Single-word titles are not abbreviated.

Nature [for "Nature"]   Science [for "Science"]

4 Abbreviation is preferably by truncation: at least the 2 final letters of a word are dropped.

Geol Surv Finl Bull [from "**Geol**ogical **Surv**ey of **Finl**and **Bull**etin"]

J Heterocycl Chem [from "**J**ournal of **Heterocycl**ic **Chem**istry"]

5 Cognates and variant forms with the same stem in the same, or a related language, are represented by the same abbreviation.

Entomol Abh [for "**Entomol**ogische **Abh**andlungen"]

Entomol Am [for "**Entomol**ogica **Am**ericana"]

Entomol Gaz [for "**Entomol**ogist's **Gaz**ette"]

Entomol Phytopathol App
[for "**Entomol**ogie et **Phytopathol**ogie **Appl**iquees"]

6 Words may also be abbreviated by contraction (omission of internal letters).

Ztg [for "**Zeit**u**ng**"]   Ctry [for "**C**oun**try**"]

7 Single-syllable words and words of 5 or fewer letters (in singular form) are not usually abbreviated unless they appear in the *List of Serial Title Word Abbreviations.* The list does have some exceptions to the 5-letter criterion, for example, "Hum" for "Human".

Blood Cells [for "Blood Cells"]

8 The initial letter of each abbreviated word is capitalized.
9 All punctuation, including apostrophes, within titles is omitted; abbreviations are followed by a space and not by a period (full stop).

Biomater Artif Cells Artif Organs
[for "**Biomater**ials, **Artif**icial **Cells,** and **Artif**icial **Organs**"]
Childs Brain [for "**Child's Brain**"]

10 All diacritical marks are omitted.
11 Ampersands and dashes are omitted; hyphenated terms are treated as 2 words if each can stand alone.

The source for standard abbreviated forms is the *List of Serial Title Word Abbreviations* published by ISDS International Centre, 20 rue Bachaumont, 75002 Paris, France. See Table A1·1 for the standard abbreviations of many frequently used terms. For additional details consult the article by Arenales and Sinn (1989) and the *American National Standard Z39.5–1985: Abbreviation of Titles of Publications* (available from NISO Press Fulfillment, PO Box 338, Oxon Hill MD 20750–0338, USA; fax 301 567–9553).

CITED REFERENCE

Arenales D, Sinn S. 1989. How to amputate: rules for journal title abbreviations. CBE Views 12:106–8.

APPENDIX

2 Abbreviations of Publisher Names

RULES FOR ABBREVIATING NAMES

A2·1 The names of publishers can be abbreviated in references (see section 30·36) and other formats such as tables in which brevity is needed. The following rules are suggested; they are followed by examples of how the rules are applied and by a list of additional examples.

1 Omit prefatory articles such as "The" and "Les"; omit connectives such as "of" and "de" except in the names of professional organizations designated as publisher.
2 For names of publishers beginning or ending with conventional terms or abbreviations for commercial publishing organizations such as "Cie", "Co.", "Éditions", "Inc.", "Ltd.", "Press", "S. A.", "Verlag", omit such terms, abbreviations for them, or preformed abbreviations.
3 If omitting an element of a name might cause ambiguity or uncertainty about the publisher's identity, retain that element. For example, retain "Press", abbreviated "Pr", and "University", abbreviated "Univ", with university press names because a university itself can be a publisher and not necessarily publish through a press associated with it. Retain "Éditions" in "Les Editions INSERM" because "INSERM" represents a different entity. The name of a press that is not widely known as a press should retain an identifier such as "Press" in the abbreviated form "Pr" (or "Prs" for "Presses").
4 The names of professional and scientific societies not known primarily as publishers can be retained in full, with abbreviations for generic terms such as "Assoc" (for "Association"), "Coll" (for "College"), "Inst" (for "Institute"), and "Soc" (for "Society"). See Appendix 1 for recommended forms.
5 If omitting 1 or more terms would leave only an adjectival form for the name, retain the other term or terms in full or abbreviated form.
6 If the publisher is also shown as the author and the author element in the reference opens with a square-bracketed abbreviation preceding the full author name, as recommended for the name-year system (see section 30·33), that abbreviation can be used as the publisher element.
7 Initials in publisher names can be treated as they are in author names.
8 Terms with widely known abbreviations can be replaced by them.

See Appendix 1 for standard abbreviations of frequently used generic terms such as "Society" and "University".

The following examples illustrate these rules.

[Rule(s)]	[Full name]	[Recommended form]
2	Éditions Flammarion	Flammarion
	Little, Brown and Company	Little, Brown
	McGraw-Hill Book Company	McGraw-Hill
	Merriam-Webster Inc., Publishers	Merriam-Webster
2 and 3	Springer Publishing Company	Springer Publishing
	Springer-Verlag	Springer-Verlag
1, 2, and 3	Les Éditions INSERM	Éditions INSERM
	The MIT Press (the press of the Massachusetts Institute of Technology]	MIT Pr
1 and 3	The University of Chicago Press	Univ Chicago Pr
	APS Press	APS Pr
	Jackdaw Press	Jackdaw Pr
1 and 4	American Chemical Society	American Chemical Soc
	American College of Physicians	American Coll of Physicians
5	Academic Press	Academic Pr
6	International Organization for Standardization	ISO (when the author element is "[ISO] International Organization for Standardization")
2 and 7	John Wiley & Sons	J Wiley
	W. B. Saunders Company	WB Saunders
8	United States Pharmacopeial Convention	US Pharmacopeial Convention

These rules differ slightly from the recommendations of the National Library of Medicine so as to reduce the use of unclear amputated names and to simplify punctuation.

To reduce the need to consult these rules, an editorial office should compile its own list of names and abbreviations of names of publishers appearing frequently in the references in its publications.

ADDITIONAL EXAMPLES

A2·2 This list is illustrative and is not intended to be comprehensive.

[Publisher name]	[Recommended form]
A. A. Balkama	AA Balkama
Addison-Wesley Publishing Company	Addison-Wesley

[Publisher name]	[Recommended form]
Alan R. Liss	AR Liss
American Institute of Physics	American Inst[a] of Physics
American Mathematical Society	American Mathematical Soc
The Analytic Press, Inc.	Analytic Pr
Atheneum Publishers	Atheneum
Blackwell Scientific Publications, Inc.	Blackwell Scientific
Butterworth-Heinemann	Butterworth-Heinemann
Cambridge University Press	Cambridge Univ Pr
Churchill Livingstone, Inc.	Churchill Livingstone
Cornell University Press	Cornell Univ Pr
CRC Press, Inc.	CRC Pr
David R. Godine, Publisher	DR Godine
Douglas & McIntyre	Douglas & McIntyre
Dover Publications, Inc.	Dover
Éditions Cepadues	Cepadues
Éditions Elf-Aquitaine	Editions Elf-Aquitaine
Elsevier Science Publishing Co., Inc.	Elsevier Science
Futura Publishing Co., Inc.	Futura
The Galileo Press	Galileo
Geological Society of America	Geological Soc of America
Graphics Press	Graphics Pr
Harper & Row, Publishers, Inc.	Harper & Row
Harvard University Press	Harvard Univ Pr
Henry Holt & Co., Inc.	Henry Holt
Inkata Press Pty Ltd	Inkata
John Wiley & Sons	J Wiley
The Johns Hopkins University Press	Johns Hopkins Univ Pr
Jones & Bartlett Publishers, Inc.	Jones & Bartlett
The Keynes Press	Keynes
Longman Group	Longman
Macmillan Publishing Co., Inc.	Macmillan
McGraw-Hill, Inc.	McGraw-Hill
Merck & Co., Inc.	Merck

[a]Inst stands for "Institute" and "Institution".

[Publisher name]	[Recommended form]
Modern Language Association of America	Modern Language Assoc of America
National Academy Press	National Acad Pr
New York Academy of Sciences	NY Acad of Sciences
Oxford University Press	Oxford Univ Pr
Pergamon Press	Pergamon
Plenum Publishing Corp.	Plenum
Presses Universitaires de France	Prs Univ France
Raven Press	Raven
The Reader's Digest Association, Inc.	Reader's Digest Assoc
Routledge, Chapman & Hall	Routledge, Chapman & Hall
Sage Publications, Inc.	Sage
The Shoe String Press, Inc.	Shoe String
Sinauer Associates	Sinauer
Smithsonian Institution Press	Smithsonian Inst[a] Pr
SPB Academic Publishing BV	SPB Academic Publishing
State University of New York Press	State Univ New York Pr
St. Martin's Press	St Martin's
Van Nostrand Reinhold Company	Van Nostrand Reinhold
VCH	VCH
W. H. Freeman & Company	WH Freeman
Williams & Wilkins	Williams & Wilkins

[a]Inst stands for "Institute" and "Institution".

APPENDIX

3 Annotated Bibliography

The works listed here are books and other sources widely useful in scientific writing and publishing. Additional sources with narrower scope are listed under "References" at the ends of chapters. Not all entries have annotations, but the absence of an annotation should not be taken as implying that the work has less value than annotated entries.

Note that the sequence of elements in the references here differs from that used in references appended to chapters. The title is placed 1st because in this context it is a more informative element than the author name. Author initials and surnames are in the normal sequence, not the inverted sequence generally used in references. The year of publication is placed last as in the format for references in the citation-sequence style for books (see section 30·36).

STYLE MANUALS

GENERAL

Bibliographic style manual. D Thibault. Ottawa: National Library of Canada; 1990. Available from Canadian Government Publishing Centre, Supply and Services Canada, Ottawa, Canada K1A 0S9.

A comprehensive, detailed manual with recommendations based on *ISO Standard 690, Documentation, bibliographic references: content, form, and structure.* Although it recommends formats with parallels to those in the US *National Library of Medicine recommended formats . . .* (see below under "Scientific"), it cannot substitute for the NLM's manual. May be useful to editors seeking alternative formats.

The Canadian style: a guide to writing and editing. Department of the Secretary of State of Canada. Toronto: Dundern Pr; 1985.

In addition to the usual content on general style matters such as abbreviation, capitalization, punctuation, and other details of publication style, this manual gives special attention to Canadian preferences in usage, to bilingual matters, and to sexual and ethnic stereotyping. Has an extensive, selective bibliography.

Chicago guide to preparing electronic manuscripts. Chicago: Univ Chicago Pr; 1987.

Among the earliest of the manuals specifically concerned with computer-assisted writing and publishing.

The Chicago manual of style. 14th ed. Chicago: Univ Chicago Pr; 1993.

For many years, this manual has set standards for style and format in scholarly publishing. This edition has the same general scope and arrangement as earlier editions, but carries much more detail on many aspects of publication style. The opening part covers in detail editorial procedure in book publishing; the closing part gives an extensive account of book design and production. Extensive glossary; detailed index.

Editing Canadian English. L Burton, C Cragg, B Czarnecki, SK Paine, S Pedwell, IH Phillips, K Vanderlinden [authors for the Freelance Editors' Association of Canada]. Vancouver: Douglas & McIntyre; 1987.

Has much the same scope and detail as *The Canadian Style,* described above. An appendix discusses in detail peculiarly Canadian questions of style.

The Gregg reference manual. WA Sabin. Lake Forest (IL): Glencoe; 1992.

A thorough, comprehensive style manual that covers alphabetic filing in detail and many grammatical topics in addition to the usual scope of a general style manual. Includes formats for reports, letters, and other kinds of business documents. Convenient as a desk reference because of its spiral-wire binding, and attractive in part for its relatively low price.

Hart's rules for compositors and readers at the University Press Oxford. 39th ed. Oxford (UK): Oxford Univ Pr; 1983.

A short but detailed style manual for scholarly publishing. Especially useful for guidance in British style.

A manual for writers of term papers, theses, and dissertations. 5th ed. KL Turabian. Chicago: Univ Chicago Pr; 1987.

Special attention to the scholarly apparatus characteristic of academic documents. Especially useful for its sections on citations of, and references to, documents of the Canadian, British, and US governments and of the United Nations. Has an extensive set of sample pages illustrating appropriate formats and specifications.

The MLA style manual. WS Achtert, J Gibaldi. New York: Modern Language Assoc of America; 1985.

Covers scholarly procedure in writing and publishing, details of publication style, and citation and reference formats. More useful for the literature of the humanities than that of science. Much of the same content is presented in *The MLA Handbook for Writers of Research Papers,* 3rd ed, J Gibaldi and WS Achtert, Modern Language Assoc of America; 1988.

Style manual for authors, editors and printers. 4th ed. Australian Government Publishing Service. Canberra: AGPS Press; 1988.

A detailed manual on publication style (spelling, capitalization, typographic conventions, bias-free usage, references, numbers, measurement), preparing copy for printing, parts of a publication, processes of printing and bookmaking. A good source for usage characteristic of Australia and other Commonwealth countries; has a glossary and a detailed appendix on honorifics and their proper sequences.

United States Government Printing Office style manual 1984. Washington: US Government Printing Office; 1984.

Helpful with many details of style relevant mainly to public documents, but has sections useful in scientific writing, such as the extensive list of insect and plant names and the section on formation of compound words. The extensive section on foreign languages includes alphabets (with diacritical marks), rules for syllabication, capitalization, punctuation, hyphenation, cardinal and ordinal numbers, and commonly used terms for dates and times.

Webster's standard American style manual. Springfield (MA): Merriam-Webster; 1985.

Has a scope similar to that of *The Chicago Manual of Style;* offers less detail on style relevant in the humanities but more on scientific style. Almost half of the manual offers guidance in indexing, copyediting, proof correction, and book production. Has a glossary, bibliography, and index.

SCIENTIFIC

Most of these manuals specify style and format for the journals published by the issuing organization, but their styles may be useful for guidance on style in other publications in the same discipline and in other scientific fields. Their scopes are indicated by the title or the name of the issuing organization.

The ACS style guide: a manual for authors and editors. JS Dodd. Washington: American Chemical Soc; 1985.

AIP style manual. 4th ed. New York: American Inst of Physics; 1990.

American Medical Association manual of style. 8th ed. Baltimore: Williams & Wilkins; 1989.

ASM style manual for journals and books. Washington: American Soc for Microbiology; 1991.

In addition to its detailed sections on style and nomenclature in bacteriology and virology, this manual is notably useful for its extensive and detailed sections on illustrations and tables.

Author's guide to the journals of the American Meteorological Society 1983. Boston: American Meteorol Soc; 1983.

CBE style manual: a guide for authors, editors, and publishers in the biological sciences. 5th ed. CBE Style Manual Committee. Bethesda (MD): Council of Biology Editors; 1983.

The predecessor to this manual. The opening chapters provide guidance in the preparation of manuscripts of journal articles.

Guidelines for preparing electronic manuscripts: A_{M}S-L^{A}T$_{E}$X. Providence (RI): American Mathematical Soc; 1991.

The IAU style manual (1989): the preparation of astronomical papers and reports. GA Wilkins. Paris: International Astronomical Union; 1989.

Manual de estilo. O Vilarroya, editor. Barcelona, Spain: Doyma; 1993.

A detailed manual of medical style for Spanish-language papers and books. Includes chapters on writing journal papers, editorial processes, and ethics. The special problems of scientific prose in Spanish get close attention. The conventions for scientific notation and symbolization are much the same as in English. Heavily referenced.

A manual for authors of mathematical papers. Providence (RI): American Mathematical Soc; 1990.

Mathematics into type: copy editing and proofreading of mathematics for editorial assistants and authors. Providence (RI): American Mathematical Soc; 1979.

Medical style & format: an international manual for authors, editors, and publishers. EJ Huth. Philadelphia (PA): ISI Pr; 1987. Available from Williams & Wilkins, Baltimore (MD).

Metric editorial handbook: CSA special publication Z372–1980. Rexdale, Toronto: Canadian Standards Assoc; 1980.

A detailed guide to correct and uniform editorial style in the presentation and formatting of SI units.

National Library of Medicine recommended formats for bibliographic citation. Bethesda (MD): National Library of Medicine; 1991.

The basis for most of the recommendations in Chapter 30, "Citations and References".

Physical Review style and notation guide. A Waldron, P Judd. Woodbury (NY): American Physical Soc; 1983.

Publication manual of the American Psychological Association, 4th ed. Washington: American Psychological Assoc; 1994.

Heavily revised from the 3rd edition. Offers guidance on writing papers for psychology, reporting statistics, refining prose style, avoiding biased writing, and following ethical principles. In addition to the expected content on style matters (references, abbreviations, and other details), describes the editorial and production procedures of the association's journals.

Publications handbook and style manual. Madison (WI): American Soc of Agronomy, Crop Science Soc of America, Soil Science Soc of America; 1988.

Science and technical writing: a manual of style. P Rubens, editor. New York: Henry Holt; 1992.

Nomenclature, symbolization, and style in scientific fields. Chapter 2 offers guidance in many problems of prose style. The chapters on mathematical style and on illustrations, tables, charts, and diagrams are notably helpful. Has an extensive appended bibliography, but recommendations on details of scientific style are not supported by documentation in the text.

Science editors' handbook. PH Enckell, editor. London: European Assoc of Science Editors; 1993 [and following years].

Individual chapters are issued to members of the European Association of Science Editors. As of mid-1993, 2 chapters had been published: "Symbols for physical quantities" by GI Ågren, and "Zoological nomenclature" by PC Barnard.

Suggestions to authors of the reports of the United States Geological Survey. WR Hansen. Washington: US Government Printing Office; 1990.

The authoritative US manual for geology and related earth sciences. Especially valuable for chronostratigraphic nomenclature and style.

Uniform requirements for manuscripts submitted to biomedical journals. International Committee of Medical Journal Editors. JAMA 1993 May 5;269:2282–6.

The recommendations for formats of references (the "Vancouver style") are similar to those in Chapter 30, "Citations and References".

WHO editorial style manual. Geneva: World Health Organization; 1993.

A short but comprehensive manual. Recommendations on formats for references are essentially those of the International Committee of Medical Journal Editors (the "Vancouver style"; see the publication listed immediately above) and those recommended in Chapter 30 of this manual. Scientific notation and symbolization are covered briefly. Has useful sections on nondiscriminatory language and politically and legally sensitive topics. Helpful information on proper formal names of countries and international agencies. Prefers British spellings. References to source documents.

DICTIONARIES

GENERAL

English-language dictionaries differ widely in their particular values. The best compromises among the values of scope, up-to-dateness, compactness, and price may be *The American Heritage Dictionary* and *The Oxford Encyclopedic English Dictionary,* but a number of other dictionaries have their own distinctive values.

The American Heritage dictionary of the English language. 3rd ed. Boston: Houghton Mifflin; 1992.

Notable for its helpful usage notes, notes on synonyms, and biographic and geographic entries.

Chambers English dictionary. Edinburgh: W&R Chambers; 1992.

Longman dictionary of the English language. 2nd ed. Harlow, Essex CM20 2JE, United Kingdom: Longman; 1991.

Merriam-Webster's collegiate dictionary. 10th ed. Springfield (MA): Merriam-Webster; 1993.

Perhaps the best compact desk dictionary for US usage. The appendixes include abbreviations and symbols for the chemical elements; compilations of foreign words and phrases, biographic and geographic names, signs and symbols, and a short guide on punctuation, capitalization, and forms of address.

The new shorter Oxford English dictionary on historical principles. Oxford (UK): Oxford Univ Pr, Clarendon Pr; 1993.

Has some of the characteristics of its big 20-volume father, *The Oxford English Dictionary on Historical Principles,* for example, a generous use of quotations illustrating meaning and usage, but it is up-to-date in its representation of scientific vocabulary. This would be the best choice for an editorial office wishing to have at least 1 "unabridged" dictionary.

Oxford dictionary of current idiomatic English: Volume 1: Verbs with prepositions and particles. AP Cowie and R Mackin. Oxford (UK): Oxford Univ Pr; 1975.

Detailed presentation of phrasal verb forms, some of which may not be represented in standard English dictionaries.

Oxford dictionary of current idiomatic English. Volume 2: Phrase, clause and sentence idioms. AP Cowie, R Mackin, IR McCaig. Oxford (UK): Oxford Univ Pr; 1983.

Idioms that are not standard but nevertheless may be used in formal writing for their figurative value.

The Oxford encyclopedic English dictionary. Oxford (UK): Oxford Univ Pr, Clarendon Pr; 1991.

In addition to the expected coverage of English-language vocabulary, notable for the geographic and biographic entries, and for extensive treatment of compound terms and phrasal verbs . Many of the appendixes present fundamentals of scientific nomenclature and symbolization.

Random House unabridged dictionary. 2nd ed. New York: Random House; 1993.

A major US dictionary, notably useful for its illustrative examples of usage, definitions of phrasal verbs, and suggestions of synonyms.

SCIENTIFIC AND SPECIALIZED

Abbreviations dictionary. 8th ed. R De Sola. Boca Raton (FL): CRC Pr; 1992.

Academic Press dictionary of science and technology. C Morris, editor. San Diego (CA): Academic Pr; 1992.

Extensive coverage of both single and compound scientific and technical terms (including those of biological nomenclature) , but the entry terms often do not show the relevant style conventions (such as italicization of genus names). The appendixes provide symbols and nomenclature relevant to astronomy, biology, chemistry, geology, and physics, as well as a chronology of science since 1403.

Cambridge dictionary of science and technology. PMB Walker. New York: Cambridge Univ Pr; 1990.

Cambridge world gazetteer; a geographical dictionary. D Munro. New York: Cambridge Univ Pr; 1990.

Dictionary of biochemistry and molecular biology. 2nd ed. J Stenesh. New York: J Wiley; 1989.

Dictionary of microbiology and molecular biology. 2nd ed. P Singleton and D Sainsbury. New York: J Wiley; 1993.

A dictionary of scientific units: including dimensionless numbers and scales. HG Jerrard and DB McNeil. London: Chapman & Hall; 1992.

The Facts on File dictionary of science. EB Uvarov, A Isaacs. New York: Facts on File; 1986.

Henderson's dictionary of biological terms. 10th ed. E Lawrence, editor. New York: Halsted Pr.

International dictionary of medicine and biology. SI Landau, editor. New York: J Wiley; 1986.

This encyclopedic work is undoubtedly the most comprehensive dictionary in the English language for medicine and related aspects of biology and chemistry. It observes the convention of italicization for species names, but gives possessive forms of single-name eponymic terms. It is beginning to show its age; "AZT" is explained only as "Ascheim-Zondek test" and "zidovudine" is not an entry. Despite such minor evidences of aging, it belongs in every editorial office concerned with medical and basic medical-science texts.

The language of biotechnology: a dictionary of terms. JM Walker, ME Cox. Washington: American Chemical Soc; 1988.

McGraw-Hill dictionary of scientific and technical terms. 5th ed. New York: McGraw-Hill; 1993.

Similar to the *Academic Press Dictionary of Science and Technology* (see annotation above) but with more illustrations.

The Macmillan dictionary of measurement. M Darton, J Clark. New York: Macmillan; 1994.

Described on its dust jacket as covering "values, volumes, frequencies, temperatures, and speeds", this new comprehensive dictionary covers in detail archaic, nonscientific, and scientific units of measurement, units of currency, and many related topics difficult to find in other sources.

The Oxford dictionary for scientific writers and editors. Oxford (UK): Oxford Univ Pr, Clarendon Pr; 1991.

A much smaller dictionary than the Academic Press and McGraw-Hill dictionaries, but it could probably answer most editorial queries, notably for explanations of symbols and abbreviations.

The Oxford dictionary of abbreviations. Oxford (UK): Oxford Univ Pr; 1992.

A compact but comprehensive dictionary with obvious emphasis on abbreviations of British origin but amply covering abbreviations from the Commonwealth countries and the United States. Scientific abbreviations and symbols are well represented; conventions such as italicization and boldfacing are stipulated where appropriate, but the conventions are not applied to the entry letters. Notably useful for the abbreviations representing academic diplomas and degrees, honorific titles, and military ranks.

Stedman's abbrev.: abbreviations, acronyms & symbols. Baltimore: Williams & Wilkins; 1992.

Stedman's medical dictionary. 25th ed. Baltimore: Williams & Wilkins; 1990.

Carries definitions of terms relevant to bacteriology, biochemistry, and virology in addition to strictly medical terms. Does not apply some scientific style conventions, such as italicization of genus and species names.

Webster's new geographical dictionary. Springfield (MA): Merriam-Webster; 1988.

GUIDES TO USAGE AND PROSE STYLE

Many books offer guidance on accurate usage and the writing of clear prose. This list is far from exhaustive and includes only guides with a broad scope.

The Columbia guide to standard American English. KG Wilson. New York: Columbia Univ Pr; 1993.

Defines and illustrates grammatical and syntactical terms and punctuation. Many entries illustrate the nuances in meanings of closely related words. Has a dictionary format.

The complete plain words. E Gower; S Greenbaum, J Whitcut, revisers. DR Godine; 1990. A dictionary of modern English usage. 2nd ed. HW Fowler; E Gower, reviser. Oxford: Oxford Univ Pr; 1965.

Two classic guides to clear and precise statement and the relevant principles. British in orientation but valuable throughout the anglophone world.

Lauther's complete punctuation thesaurus of the English language. H Lauther. Boston: Branden; 1991.

A comprehensive and exhaustive guide to punctuation, structured by elements needing punctuation rather than by punctuation marks: words, phrases and adverbial clauses, sentences, quotations, time statements, questions, lists, numbers, titles, and names. One section considers the marks directly; other sections deal with hyphenation, capitalization, and abbreviation.

Longman guide to English usage. S Greenbaum, J Whitcut. Burnt Mill, Harlow, Essex, England: Longman; 1989.

Merriam-Webster's dictionary of English usage. Springfield (MA): Merriam-Webster; 1989.

A comprehensive guide to punctuation, grammatical principles, and word usage.

Notes on the composition of scientific papers. TC Allbutt. London: British Medical Assoc, Keynes Pr; 1984.

A classic originally published in 1904. Fluent, graceful, and often gently humorous essays on what makes good prose good and bad prose bad: vocabulary, usage, syntax, and other elements that determine the quality of prose style in English.

Revising prose. 3rd ed. RA Lanham. New York: Macmillan; 1992.

A short, pithy, witty guide to revising sentences with attention to their shape, length, rhythm, and sound. Lanham shows how to apply his "Paramedic Method" to cutting down "The Lard Factor" and turning "The Official Style" into plain English.

The scientist's handbook for writing papers and dissertations. AM Wilkinson. Englewood Cliffs (NJ): Prentice-Hall; 1991.

A comprehensive treatise on the structure of scientific prose, the writing of scientific documents, the preparation of manuscripts, and the subsequent steps in publication. Includes content on accurate usage.

Style: toward clarity and grace. JM Williams. Chicago: Univ Chicago Pr; 1990.

A contemporary classic on how to write clear, accurate, and efficient prose and on how to clean out turgidity, obscurity, verbiage, and confusing sequence.

The use and abuse of the English language. [Formerly: The reader over your shoulder]. R Graves, A Hodge. New York: Paragon House; 1990.

A thorough guide to analyzing and correcting unclear, verbose, and confusingly sequenced prose. The principles set forth are as applicable to scientific prose as to prose in other fields. Generously illustrated with examples of flawed style.

EDITING: PROFESSIONAL AND REDACTORY

Copy-editing: the Cambridge handbook for editors, authors and publishers. 3rd ed. J Butcher. Cambridge (UK): Cambridge Univ Pr; 1992.

A rationally organized, clearly written, and thorough guide to preparing a book for publication. Many of its principles and procedures could be adapted for preparation of manuscripts of articles for journal publication. The valuable appendixes cover non-English alphabets and abbreviations, mathematical and phonetic symbols, electronic typescript information, and checklists for steps in copyediting. The recommended formats for bibliographic references and some recommendations for style differ from those in this manual, but this is a valuable book for all editorial offices.

Copyediting: a practical guide. 2nd ed. K Judd. San Francisco: William Kaufmann; 1990.

Not as comprehensive as Butcher's book, described immediately above, but useful for most kinds of copyediting.

Editing: an annotated bibliography. BW Speck. Westport (CT): Greenwood Pr; 1991.

Editing and publication: a training manual. I Montagnes. Manila: International Rice Research Inst and Ottawa: International Development Research Center; 1991.

The subtitle, "A Training Manual", obscures the wide scope of this detailed guide to planning for publication, revising prose for clarity and precision, copyediting manuscripts, selecting formats and designs, and seeing publications through production. Although it was prepared specifically for developing countries, it would be valuable for instruction and reference even in countries with well-established publishing operations. A companion guide for instructors is available: *Editing and Publication: A Handbook for Trainers.*

Editorial forms: a guide to journal management. Journal Procedures and Practice Committee, Council of Biology Editors. Bethesda (MD): Council of Biology Editors; 1987.

A compilation of 30 forms designed to expedite the editorial process from the original submission of a manuscript, through peer review, to final disposition of the work.

How to copyedit scientific books and journals. M O'Connor. Baltimore: Williams & Wilkins; 1986.

A compact but thorough guide to principles and procedures for copyeditors. Checklists for the various steps in copyediting. Describes differing American and British conventions.

How to edit a scientific journal. CT Bishop. Baltimore: Williams & Wilkins; 1984.

A thorough look by an experienced editor at all aspects, both policy and management, of running scholarly journals in science. Chapters 3, 4, and 5 cover editorial boards and both internal and external reviewing.

EDITORS' NEWSLETTERS

Copy editor: the national newsletter for professional copy editors. New York: Copy Editor (PO Box 604, Ansonia Station, New York NY 10023-0604, USA).

A bimonthly. Short articles on usage, new words, jargon, vocational help. Short book reviews; questions–answers and letters sections. Wide coverage of many details in usage and style.

The editorial eye. Alexandria (VA): EEI [Editorial Experts, Incorporated] (Suite 200, 66 Canal Center Plaza, Alexandria VA 22314-5507).

A monthly. Similar to *Copy Editor* but with generally longer articles.

MEASUREMENT

Canadian metric practice guide. Canadian Standards Association. Rexdale, Toronto: Canadian Standards Association; 1989. National Standard of Canada CAN/CSA-Z234.1-89.

A guide to international recommendations on names and symbols for quantities and units of measurement. DA Lowe. Geneva: World Health Organization; 1975.

Units of measurement: ISO standards handbook 2. 2nd ed. Geneva: International Organization for Standardization; 1982.

Fifteen ISO standards relevant to reporting scientific measurements: metric usage and the reporting of numbers for all sciences; quantities, their definitions, and their symbolic representation, considered by subjects of measurement or by scientific discipline. These standards have been superseded by new editions issued in 1992 (see "Quantities and units" standards listed below under "Standards for Editing and Publishing"), but this compilation remains a useful and compact source.

NOMENCLATURE AND TAXONOMY

Biological nomenclature. 3rd ed. C Jeffrey. New York: Cambridge Univ Pr; 1992.

A broad treatment of nomenclature in all biological fields, with particular attention to taxonomic implications.

HANDBOOKS, ENCYCLOPEDIAS, AND OTHER REFERENCE WORKS

CRC handbook of chemistry and physics. 74th ed. Boca Raton (FL): CRC Pr; 1993.

A massive, classic handbook of quantitative and descriptive data. Also covers geophysics, astronomy, acoustics, and mathematics. One section covers nomenclature and symbolization.

Geigy Scientific Tables. 8th ed. 6 volumes. West Caldwell (NJ): CIBA-Geigy; 1991–1993.

An exhaustive and authoritative compilation of tabular and descriptive information relevant to clinical bacteriology, clinical medicine, laboratory medicine, medical science, nutrition, physics, physical chemistry, mathematics, and statistics. Individual volumes are referenced in the appropriate chapters of this manual. A remarkably low-priced reference work.

McGraw-Hill concise encyclopedia of science & technology. 2nd ed. New York: McGraw-Hill; 1989.

The same scope as the encyclopedia listed below but with much shorter articles. The appendix covers units of measurement, conversion factors, scientific notation and symbolization, and classification of living organisms.

McGraw-Hill encyclopedia of science & technology. 7th ed. 20 volumes. New York: McGraw-Hill; 1992.

Detailed and comprehensive coverage of pure and applied science. Most articles have short bibliographies of additional sources. The index volume includes an index of topics by discipline and technical field and a section on scientific notation with conversion factors and symbols.

The Oxford companion to the English language. T McArthur, editor. New York: Oxford Univ Pr; 1992.

A 1-volume encyclopedic source on the history of English, on definitions of linguistic and lexicographic terms, and on grammar, punctuation, usage, alphabets, regional differences, and many other aspects of published and spoken English. Thoroughly cross-referenced. A valuable work for all anglophones who want to know their language better.

Publishers directory. 15th ed. New York: Gale Research; 1994.

Synopsis and classification of living organisms. New York: McGraw-Hill; 1982.

Encyclopedic compilation of descriptions of taxa down to, and including, families. Taxa described for characteristics, biology, ecology, distribution. Purely fossil taxa excluded. Index of all described taxa.

Van Nostrand's scientific encyclopedia. 7th ed. New York: Van Nostrand Reinhold; 1988.

A comprehensive work akin to the *McGraw-Hill Concise Encyclopedia of Science & Technology,* described above.

STANDARDS FOR EDITING AND PUBLISHING

In the United States, standards issued by the National Information Standards Organization (NISO) can be purchased from NISO Press Fulfillment, PO Box 338, Oxon Hill MD 20750-0338, USA (telephone, 800 282-6476, 301 567-9522; fax, 301 567-9523). A NISO-developed and approved standard becomes an American National Standard after the American National Standards Institute (ANSI) verifies that the process for approval has met the ANSI criteria.

Standards of the International Organization for Standardization (ISO) can be purchased from American National Standards Institute, 11

West 42nd Street, New York NY 10036-8002, USA (telephone 212 642-4900).

In Canada, the standards issued by the Canadian Standards Association and by ANSI, ISO, and other standards organizations can be purchased from the Standards Council of Canada, 45 O'Connor Street, Ottawa ON K1P 6N7, Canada (telephone 613 238-3222).

British standards can be purchased from the British Standards Institute, Linford Wood, Milton Keynes MK14 6LE, England, UK (telephone 0908-220 022); in the United States, they can be purchased from the American National Standards Institute.

Bibliography of publications designed to raise the standard of scientific literature. Paris: UNESCO; 1963.

359 annotated references to books, standards, and pamphlets on language, composition, technical writing, editing, printing, publishing, information retrieval, indexing, and publication conventions and style. Indexes to subjects, authors, and languages of publication.

Citing and referencing published material: BS 5605:1990. Linford Wood, Milton Keynes, England (UK): British Standards Inst; 1990.

Copy preparation and proof correction: recommendations for preparation of typescript copy for printing: BS 5261:Part 1:1975. Linford Wood, Milton Keynes, England (UK): British Standards Inst; 1975.

Copy preparation and proof correction: specification for typographic requirements, marks for copy preparation and proof correction, proofing procedure: BS 5261:Part 2:1976. Linford Wood, Milton Keynes, England (UK): British Standards Inst; 1976.

Copy preparation and proof correction: specification for marks for mathematical copy preparation and mathematical proof correction and their uses: BS 5261:Part 3:1989. Linford Wood, Milton Keynes, England (UK): British Standards Inst; 1989.

Documentation and information: ISO standards handbook 1, 1988. 3rd ed. Geneva: International Organization for Standardization; 1988.

Standards for vocabulary, terminology, character sets, transliteration, library procedure, and documentation. Some are relevant to scientific style and format.

Electronic manuscript preparation and markup: ANSI/NISO Z39.59-1988. New Brunswick (NJ): Transaction Publishers; 1991. Availability: see the note at the head of this section on NISO Press.

Permanence of paper for publication and documents in libraries and archives: Z39.48-1992. Bethesda (MD): NISO Pr; 1992.

Quantities and units — part 0: general principles: ISO 31-0:1992(E). Geneva: International Organization for Standardization; 1992.

Quantities and units — part 1: space and time: ISO 31-1:1992(E). Geneva: International Organization for Standardization; 1992.

Quantities and units — part 2: periodic and relation phenomena: ISO 31-2:1992(E). Geneva: International Organization for Standardization; 1992.

Quantities and units — part 3: mechanics: ISO 31-3:1992(E). Geneva: International Organization for Standardization; 1992.

Quantities and units — part 4: heat: ISO 31-4:1992(E). Geneva: International Organization for Standardization; 1992.

Quantities and units — part 5: electricity and magnetism: ISO 31-5:1992(E). Geneva: International Organization for Standardization; 1992.

Quantities and units — part 6: light and related electromagnetic radiations: ISO 31-6:1992(E). Geneva: International Organization for Standardization; 1992.

Quantities and units — part 7: acoustics: ISO 31-7:1992(E). Geneva: International Organization for Standardization; 1992.

Quantities and units — part 8: physical chemistry and molecular physics: ISO 31-8:1992(E). Geneva: International Organization for Standardization; 1992.

Quantities and units — part 9: atomic and nuclear physics: ISO 31-9:1992(E). Geneva: International Organization for Standardization; 1992.

Quantities and units — part 10: nuclear reactions and ionizing radiation: ISO 31-10:1992(E). Geneva: International Organization for Standardization; 1992.

Quantities and units — part 11: mathematical signs and symbols for use in the physical sciences and technology: ISO 31-11:1992(E). Geneva: International Organization for Standardization; 1992.

Quantities and units — part 12: characteristic numbers: ISO 31-12:1992(E). Geneva: International Organization for Standardization; 1992.

Quantities and units — part 13: solid state physics: ISO 31-13:1992(E). Geneva: International Organization for Standardization; 1992.

Recommendation for references to published materials: BS 1629:1989. Linford Wood, Milton Keynes, England (UK): British Standards Inst; 1989.

SI units and recommendations for the use of their multiples and of certain other units: ISO 1000:1992(E). Geneva: International Organization for Standardization; 1992.

Specification for abbreviation of title words and titles of publications: BS 4148:1985. Linford Wood, Milton Keynes, England (UK): British Standards Inst; 1985.

Technical manuals: specification for presentation of essential information: BS 4884:Part 1:1992. Linford Wood, Milton Keynes, England (UK): British Standards Inst; 1992.

GRAPHICS AND DESIGN

Envisioning information. E Tufte. Cheshire (CT): Graphics Pr; 1990.

Extends the thorough consideration of effective graphic presentation and its principles that Tufte developed in his *The Visual Display of Quantitative Information* (see annotation below). This book offers the wider view of representing spatial and structural information that does not directly represent quantitative data.

Graphic design for the electronic age. JV White. New York: Watson-Guptill Publications, Xerox Pr; 1988.

In addition to well-illustrated sections on design and layout, includes extensive descriptions of the characteristics of type, compilations of frequently used symbols, and summaries of the uses of punctuation.

The Guild handbook of scientific illustration. ERS Hodges, editor. New York: Van Nostrand Reinhold; 1989.

Chapters on media for illustration and on considerations in illustrating specific categories of biological and medical subjects. Separate chapters cover charts, diagrams, and maps. "Guild" is the Guild of Natural Science Illustrators.

Illustrating science: standards for publication. Scientific Illustration Committee, Council of Biology Editors. Bethesda (MD): Council of Biology Editors; 1988.

Covers artwork, photography, and graph and map construction applied to illustrating scientific subjects and information. Includes chapters on procedure, standards, and legal and ethical considerations.

Presentation of clinical data. B Spilker, J Schoenfelder. New York: Raven; 1989.

Despite its orientation to medical research, the principles and devices described for structuring tables of numeric and descriptive data and graphically representing experimental designs and quantitative information (data and statistics) are applicable in other fields.

Semiology of graphics: diagrams, networks, maps. J Bertin; WJ Berg, translator. Madison (WI): Univ Wisconsin Pr; 1983.

An exhaustive, scholarly, heavily illustrated treatise on theory and concrete principles for displaying quantitative and structural information in graphic form.

Symbol sourcebook: an authoritative guide to international graphic symbols. H Dreyfuss. New York: Van Nostrand Reinhold; 1984.

In addition to extensive illustrations of symbols widely used in nonscientific fields, has sections on graphic symbols in astronomy, biology, chemistry, geology, mathematics, medicine, meterology, and physics. The section on symbols grouped by graphic forms enables one to ascertain a symbol's meaning from its characteristics of form (such as circle, triangle, and so on).

The visual display of quantitative information. ER Tufte. Cheshire (CT): Graphics Pr; 1983.

An elegantly and profusely illustrated treatise on faulty and effective graphic displays of data in scientific, political, sociologic, and popular literature; the text develops principles in large part from the illustrations.

Visualizing data. WS Cleveland. Murray Hill (NJ): AT&T Laboratories; 1993. Distributed by Hobart Pr, Summit (NJ).

A thorough consideration of how to present quantitative data, with particular attention to multivariate data with statistical complexities. An extension of his earlier monograph, *The Elements of Graphing Data* (Hobart Pr).

Xerox publishing standards: a manual of style and design. New York: Watson-Guptil Publications; 1988.

Mainly a text on design, layout, and typography, but has a comprehensive section on the organization of books, technical reports, and other types of comprehensive publications

SERIAL-TITLE ABBREVIATIONS

Abbreviations of names of serials reviewed in Mathematical Reviews. Providence (RI): American Mathematical Soc; 1991.

List of journals indexed in Index Medicus. Bethesda (MD): National Library of Medicine; [annual].

Serials sources for the BIOSIS data base. Philadelphia: BIOSIS; [annual].

GUIDES TO INFORMATION SOURCES

The author's guide to biomedical journals: complete manuscript submission instructions for 185 leading biomedical periodicals. New York: MA Liebert; 1993.

Composite index for CRC handbooks. 3rd ed. Boca Raton (FL): CRC Pr; 1991.

A collective index covering the numerous handbooks published by CRC Press. A CD-ROM enables rapid access to the contents of the index.

Directory of publications resources, 1993–94. Alexandria (VA): EEI [Edit Experts]; 1993.

A guide to reference books and other sources on writing, usage, editing, production; to training programs; to professional associations; and to software programs.

Guide to information sources in the botanical sciences. EB Davis. Littleton (CO): Libraries Unlimited; 1987.

Handbooks and tables in science and technology. 3rd ed. RH Powell, editor. Phoenix (AZ): Oryx Pr; 1994.

An annotated bibliography covering over 3000 sources. In addition to covering sources of quantitative data, describes sources like *Bergey's Manual* and *The Merck Index.*

Information sources in chemistry. 4th ed. RT Bottle, JFB Rowland. London: Bowker-Saur UK; 1992.

Information sources in the earth sciences. 2nd ed. JE Hardy and others. London: Bowker-Saur UK; 1990.

Information sources in the life sciences. 4th ed. HV Wyatt, editor. London: Bowker-Saur UK; 1994.

Information sources in the medical sciences. 4th ed. LT Morton, S Godbolt, editors. London: Bowker-Saur UK; 1992.

Information sources in physics. DF Shaw. London: KG Saur; 1985.

Information sources in science and technology. CD Hunt. Littleton (CO): Libraries Unlimited; 1988.

Science and technology annual reference review. HR Malinowsky, editor. Phoenix (AZ): Oryx Pr; [annual].

Index

Note: entries in quotation marks refer to words as such; page numbers followed by f refer to figures; page numbers followed by t refer to tables; page numbers followed by n refer to footnotes; entries preceded by a hyphen refer to suffixes; entries followed by a hyphen refer to prefixes; entries preceded and followed by hyphens refer to infixes.